Advanced Air and Noise Pollution Control

Advanced Air and Noise Pollution Control

Edited by

Lawrence K. Wang, PhD, PE, DEE

Zorex Corporation, Newtonville, NY
Lenox Institute of Water Technology, Lenox, MA
Krofta Engineering Corp., Lenox, MA

Norman C. Pereira, PhD

Monsanto Corporation (Retired), St. Louis, MO

Yung-Tse Hung, PhD, PE, DEE

Department of Civil and Environmental Engineering
Cleveland State University, Cleveland, OH

Consulting Editor

Kathleen Hung Li, MS

HUMANA PRESS ✳ TOTOWA, NEW JERSEY

Library of Congress Cataloging-in-Publication Data

Library of Congress Cataloging-in-Publication Data
Advanced air and noise pollution control / edited by Lawrence K. Wang, Norman C. Pereira, Yung-Tse Hung ; consulting editor Kathleen Hung Li.
    p. cm. — (Handbook of environmental engineering ; v. 2)
  Includes bibliographical references and index.
  ISBN 1-58829-359-9 (alk. paper) eISBN 1-59259-779-3
  1. Air—Pollution. 2. Air quality management. 3. Noise pollution. 4. Noise control. I. Wang, Lawrence K. II. Pereira, Norman C. III. Hung, Yung-Tse. IV. Handbook of environmental engineering (2004) ; v. 2.
  TD170 .H37 2004 vol. 2
  [TD883]
  628 s—dc22
  [628.5   2003023705

Preface

The past 30 years have seen the emergence worldwide of a growing desire to take positive actions to restore and protect the environment from the degrading effects of all forms of pollution: air, noise, solid waste, and water. Since pollution is a direct or indirect consequence of waste, the seemingly idealistic demand for "zero discharge" can be construed as an unrealistic demand for zero waste. However, as long as waste exists, we can only attempt to abate the subsequent pollution by converting it to a less noxious form. Three major questions usually arise when a particular type of pollution has been identified: (1) How serious is the pollution? (2) Is the technology to abate it available? and (3) Do the costs of abatement justify the degree of abatement achieved? The principal intention of the *Handbook of Environmental Engineering* series is to help readers to formulate answers to the last two questions.

The traditional approach of applying tried-and-true solutions to specific pollution problems has been a major contributing factor to the success of environmental engineering, and has accounted in large measure for the establishment of a "methodology of pollution control." However, realization of the ever-increasing complexity and interrelated nature of current environmental problems renders it imperative that intelligent planning of pollution abatement systems be undertaken. Prerequisite to such planning is an understanding of the performance, potential, and limitations of the various methods of pollution abatement available for environmental engineering. In this series of handbooks, we will review at a tutorial level a broad spectrum of engineering systems (processes, operations, and methods) currently being utilized, or of potential utility, for pollution abatement. We believe that the unified interdisciplinary approach in these handbooks is a logical step in the evolution of environmental engineering.

The treatment of the various engineering systems presented in *Advanced Air and Noise Pollution Control* will show how an engineering formulation of the subject flows naturally from the fundamental principles and theory of chemistry, physics, and mathematics. This emphasis on fundamental science recognizes that engineering practice has in recent years become more firmly based on scientific principles rather than its earlier dependency on the empirical accumulation of facts. It is not intended, though, to neglect empiricism when such data lead quickly to the most economic design; certain engineering systems are not readily amenable to fundamental scientific analysis, and in these instances we have resorted to less science in favor of more art and empiricism.

Since an environmental engineer must understand science within the context of application, we first present the development of the scientific basis of a particular subject, followed by exposition of the pertinent design concepts and operations, and detailed explanations of their applications to environmental quality control or improvement. Throughout the series, methods of practical design calculation are illustrated by numerical examples. These examples clearly demonstrate how organized, analytical reasoning leads to the most direct and clear solutions. Wherever possible, pertinent cost data have been provided.

Our treatment of pollution-abatement engineering is offered in the belief that the trained engineer should more firmly understand fundamental principles, be more aware of the similarities and/or differences among many of the engineering systems, and exhibit greater flexibility and originality in the definition and innovative solution of environmental pollution problems. In short, the environmental engineer should by conviction and practice be more readily adaptable to change and progress.

Coverage of the unusually broad field of environmental engineering has demanded an expertise that could only be provided through multiple authorships. Each author (or group of authors) was permitted to employ, within reasonable limits, the customary personal style in organizing and presenting a particular subject area, and consequently it has been difficult to treat all subject material in a homogeneous manner. Moreover, owing to limitations of space, some of the authors' favored topics could not be treated in great detail, and many less important topics had to be merely mentioned or commented on briefly. All of the authors have provided an excellent list of references at the end of each chapter for the benefit of the interested reader. Since each of the chapters is meant to be self-contained, some mild repetition among the various texts is unavoidable. In each case, all errors of omission or repetition are the responsibility of the editors and not the individual authors. With the current trend toward metrication, the question of using a consistent system of units has been a problem. Wherever possible the authors have used the British system (fps) along with the metric equivalent (mks, cgs, or SIU) or vice versa. The authors sincerely hope that this doubled system of unit notation will prove helpful rather than disruptive to the readers.

The goals of the *Handbook of Environmental Engineering* series are: (1) to cover the entire range of environmental fields, including air and noise pollution control, solid waste processing and resource recovery, biological treatment processes, water resources, natural control processes, radioactive waste disposal, thermal pollution control, and physicochemical treatment processes; and (2) to employ a multithematic approach to environmental pollution control since air, water, land, and energy are all interrelated. Consideration is also given to the abatement of specific pollutants, although the organization of the series is mainly based on the three basic forms in which pollutants and waste are manifested: gas, solid, and liquid. In addition, noise pollution control is included in this volume of the handbook.

This volume of *Advanced Air and Noise Pollution Control*, a companion to the volume, *Air Pollution Control Engineering*, has been designed to serve as a basic air pollution control design textbook as well as a comprehensive reference book. We hope and expect it will prove of equally high value to advanced undergraduate or graduate students, to designers of air pollution abatement systems, and to scientists and researchers. The editors welcome comments from readers in the field. It is our hope that this book will not only provide information on the air and noise pollution abatement technologies, but will also serve as a basis for advanced study or specialized investigation of the theory and practice of the unit operations and unit processes covered.

The editors are pleased to acknowledge the encouragement and support received from their colleagues and the publisher during the conceptual stages of this endeavor. We wish to thank the contributing authors for their time and effort, and for having patiently borne our reviews and numerous queries and comments. We are very grateful to our respective families for their patience and understanding during some rather trying times.

The editors are especially indebted to Dr. Howard E. Hesketh at Southern Illinois University, Carbondale, Illinois, and Ms. Kathleen Hung Li at NEC Business Network Solutions, Irving, Texas, for their services as Consulting Editors of the first and second editions, respectively.

Lawrence K. Wang
Norman C. Pereira
Yung-Tse Hung

Contents

3 Carbon Sequestration

Contributors

JAMES P. CHAMBERS, PhD • *National Center for Physical Acoustics and Department of Mechanical Engineering, University of Mississippi, University, MS*

CHEIN-CHI CHANG, PhD, PE • *District of Columbia Water and Sewer Authority, Washington, DC*

JIANN-LONG CHEN, PhD, PE • *Department of Civil, Architectural, Agricultural, and Environmental Engineering, North Carolina A&T State University, Greensboro, NC*

WEI-YIN CHEN, PhD • *Department of Chemical Engineering, University of Mississippi, University, MS*

JAMES E. ELDRIDGE, MS, ME • *Lantec Product, Agoura Hills, CA*

ALI GÖKMEN, PhD • *Department of Chemistry, Middle East Technical University, Ankara, Turkey*

INCI G. GÖKMEN, PhD • *Department of Chemistry, Middle East Technical University, Ankara, Turkey*

THOMAS C. HO, PhD • *Department of Chemical Engineering, Lamar University, Beaumont, TX*

YUNG-TSE HUNG, PhD, PE, DEE • *Department of Civil and Environmental Engineering, Cleveland State University, Cleveland, OH*

PAUL JENSEN • *BBN Technologies, Cambridge, MA*

ROBERT L. KANE, MS • *Office of Fossil Energy, U.S. Department of Energy, Washington, DC*

DANIEL E. KLEIN, MBA • *Twenty-First Strategies, LLC, McLean, VA*

KATHLEEN HUNG LI, MS • *NEC Business Network Solutions, Inc., Irving, TX*

L. YU LIN, PhD • *Department of Civil and Environmental Engineering, Christian Brothers University, Memphis, TN*

NGUYEN THI KIM OANH, DrEng • *Environmental Engineering and Management, School of Environment, Resources and Development, Asian Institute of Technology, Pathumthani, Thailand*

MASAAKI OKUBO, PhD • *Department of Energy Systems Engineering, Osaka Prefecture University, Sakai, Osaka, Japan*

NORMAN C. PEREIRA, PhD (RETIRED) • *Monsanto Company, St. Louis, MO*

JERRY R. TARICSKA, PhD, PE • *Environmental Engineering Department, Hole Montes, Inc., Naples, FL*

LAWRENCE K. WANG, PhD, PE, DEE • *Zorex Corporation, Newtonville, NY, Lenox Institute of Water Technology, Lenox, MA, and Kofta Engineering Corp., Lenox, MA*

CLINT WILLIFORD, PhD • *Department of Chemical Engineering, University of Mississippi, University, MS*

ZUCHENG WU, PhD • *Department of Environmental Engineering, Zhejiang University, Hangzhou, People's Republic of China*

Toshiaki Yamamoto, PhD • *Department of Energy Systems Engineering, Osaka Prefecture University, Sakai, Osaka, Japan*

James T. Yeh, PhD • *National Energy Technology Laboratory, US Department of Energy, Pittsburgh, PA*

Ruihong Zhang, PhD • *Biological and Agricultural Engineering Department, University of California, Davis, CA*

Atmospheric Modeling and Dispersion

Lawrence K. Wang and Chein-Chi Chang

CONTENTS

1. AIR QUALITY MANAGEMENT

Air pollution is the appearance of air contaminants in the atmosphere that can create a harmful environment to human health or welfare, animal or plant life, or property (1). In the United States, air pollution is mainly the result of industrialization and urbanization. In 1970, the Federal Clean Act was passed as Public Law 91-604. The objective of the act was to protect and enhance the quality of the US air resources so as to promote public health and welfare and the productive capacity of its population. The Act required that the administrator of the US Environmental Protection Agency (EPA) promulgate primary and secondary National Ambient Air Quality Standards (NAAQS) for six common pollutants. NAAQS are those that, in the judgment of the EPA administrator, based on the air quality criteria, are requisite to protect the public health (Primary), including the health of sensitive populations such as asthmatics, children, and the elderly, and the public welfare (Secondary), including protection against decreased visibility, damage to animals, crops, vegetation, and buildings. These pollutants were photochemical oxidants, particulate matter, carbon monoxide, nitrogen dioxides, sulfur dioxide, and hydrocarbons.

1. Photochemical oxidants are those substances in the atmosphere that are produced when reactive organic substances, principally hydrocarbons, and nitrogen oxides are exposed to sunlight. For the purpose of air quality control, they shall include ozone, peroxyacyl nitrates, organic peroxides, and other oxidants. Photochemical oxidants cause irritation of the mucous membranes, damage to vegetation, and deterioration of materials. They affect the clearance mechanism of the lungs and, subsequntly, resistance to bacterial infection. The objective of photochemical oxidants' control is to prevent such effects.

2. A particulate is matter dispersed in the atmosphere, where solid or liquid individual particles are larger than single molecules (about 2×10^{-10} m in diameter), but smaller than about 5×10^{-4} m. Settleable particulates, or dustfall, are normally in the size range greater than

From: *Handbook of Environmental Engineering, Volume 2: Advanced Air and Noise Pollution Control*
Edited by: L. K. Wang, N. C. Pereira and Y.-T. Hung © The Humana Press, Inc., Totowa, NJ

10^{-5} m, and suspended particulates range below 10^{-5} m in diameter. The objective of suspended particulate control is the protection from adverse health effects, taking into consideration its synergistic effects.

3. Carbon monoxide is a colorless, odorless gas, produced by the incomplete combustion of carbonaceous material, having an effect that is predominantly one that causes asphyxia.

4. Nitrogen dioxide is a reddish-orange-brown gas with a characteristic pungent odor. The partial pressure of nitrogen dioxide in the atmosphere restricts it to the gas phase at usual atmospheric temperatures. It is corrosive and highly oxidizing and may be physiologically irritating. The presence of the gas in ambient air has been associated with a variety of respiratory diseases. Nitrogen dioxide gas is essential for the production of photochemical smog. At higher concentrations, its presence has been implicated in the corrosion of electrical components, as well as vegetation damage.

5. Sulfur dioxide is a nonflammable, nonexplosive, colorless gas that has a pungent, irritating odor. It has been associated with an increase in chronic respiratory disease on long-term exposure and alteration in lung and other physiological functions on short-term exposure.

6. Hydrocarbons are organic compounds consisting only of hydrogen and carbon. However, for the purpose of air quality control, hydrocarbons (nonmethane) shall refer to the total airborne hydrocarbons of gaseous hydrocarbons as a group that have not been associated with health effects. It has been demonstrated that ambient levels of photochemical oxidant, which do have adverse effects on health, are associated with the occurrence of concentrations of nonmethane hydrocarbons.

In 1990, the US Congress passed an amendment to the Clean Air Act of 1970. Under its requirements, the US EPA is to revise national-health-based standards—National Ambient Air Quality Standards (NAAQS) as shown in Table 1 (2)—and set the Significant Harm Levels (SHLs). The Standards, which control pollutants harmful to people and the environment, were established for six criteria pollutants. These criteria pollutants are ozone, particulate matter, carbon monoxide, nitrogen dioxides, sulfur dioxide, heavy metals (especially lead), and various hazardous air pollutants (HAPs). Descriptions for additional pollutants are described as follows.

Ozone (O_3) is a gas composed of three oxygen atoms. It is not usually emitted directly into the air, but at ground level it is created by a chemical reaction between oxides of nitrogen (NO_x) and volatile organic compounds (VOCs) in the presence of heat and sunlight. Ozone has the same chemical structure whether it occurs miles above the Earth or at ground level and can be "good" or "bad," depending on its location in the atmosphere. "Good" ozone occurs naturally in the stratosphere approx 10–30 miles above the Earth's surface and forms a layer that protects life on Earth from the sun's harmful rays. In the Earth's lower atmosphere, ground-level ozone is considered "bad."

$$VOC + NO_x + Heat + Sunlight = Ozone$$

Motor vehicle exhaust and industrial emissions, gasoline vapors, and chemical solvents are some of the major sources of NO_x and VOCs that contribute to the formation of ozone. Sunlight and hot weather cause ground-level ozone to form in harmful concentrations in the air. As a result, it is known as a summer air pollutant. Many urban areas tend to have high levels of bad ozone, but even rural areas are also subjected to increased ozone levels because wind carries ozone and pollutants that form it hundreds of miles away from their original sources.

Lead is a metal found naturally in the environment as well as in manufactured products. The major sources of lead emissions have historically been motor vehicles

Table 1
National Ambient Air Quality Standards (NAAQS)

Pollutant	Standard value[a]	Standard type
Carbon monoxide (CO)		
8-h Average	9 ppm (10 mg/m^3)	Primary
1-h Average	35 ppm (40 mg/m^3)	Primary
Nitrogen dioxide (NO$_2$)		
Annual arithmetic mean	0.053 ppm (100 µg/m^3)	Primary and Secondary
Ozone (O$_3$)		
1-h Average	0.12 ppm (235 µg/m^3)	Primary and Secondary
8-h Average[b]	0.08 ppm (157 µg/m^3)	Primary and Secondary
Lead (Pb)		
Quarterly average	1.5 µg/m^3	Primary and Secondary
Particulate (PM 10)[c]		
Annual arithmetic mean	50 µg/m^3	Primary and Secondary
24-h Average	150 µg/m^3	Primary and Secondary
Particulate (PM 2.5)[c]		
Annual arithmetic mean[b]	15 µg/m^3	Primary and Secondary
24-h Average[b]	65 µg/m^3	Primary and Secondary
Sulfur dioxide (SO$_2$)		
Annual arithmetic mean	0.03 ppm (80 µg/m^3)	Primary
24-h Average	0.14 ppm (365 µg/m^3)	Primary
3-h Average	0.50 ppm (1300 µg/m^3)	Secondary

[a]Parenthetical value is an approximately equivalent concentration.

[b]The ozone 8-h standard and the PM 2.5 standards are included for information only. A 1999 federal court ruling blocked implementation of these standards, which the EPA proposed in 1997. The EPA has asked the US Supreme Court to reconsider that decision. The updated air quality standards can be found at the US EPA website (2).

[c]PM 10: particles with diameters of 10 µm or less; PM 2.5: particles with diameters of 2.5 µm or less.

(such as cars and trucks) and industrial sources. Because of the phase out of leaded gasoline, metals processing is the major source of lead emissions to the air today. The highest levels of lead in air are generally found near lead smelters. Other heavy metals in other stationary sources are waste incinerators, utilities, and lead-acid battery manufacturers (4–6).

The list of HAPs and their definitions can be found in ref. 7. New Source Review (NSR) reform and HAPs control likely will have the most immediate impact on industrial facilities. HAP control will be very active in the 21st century on several fronts— new regulations, the Maximum Achievable Control Technology (MACT) hammer, and residual risk. Each presents issues for industrial plant compliance at the present. The Clean Air Act's HAP requirements will be a major challenge for any facility that has the potential to emit major source quantities of HAPs (10 tons/yr of any one HAP or 25 tons/yr of all HAPs combined). It is important to realize that these thresholds apply to all HAP emissions from an industrial facility, not just the emissions from specific activities subject to a categorical MACT standard.

In addition to the air quality indices, air effluent dispersion is another air pollution topic worthy of discussion. In the past decade, there has been a rapid increase in the height of power plant stacks and in the volume of gas discharged per stack. Although interest in tall stacks has increased, there is still a lack of proven pollutant (such as sulfur dioxide) removal devices. Accordingly, air quality control, in part, should continue to rely on the high stacks for controlling the ground-level pollutant concentrations. The dispersion of such airborne pollutants, thus, must be monitored and/or predicted. Most of the mathematical models used for the control of airborne effluents are reported in a manual, *Recommended Guide for the Prediction of the Dispersion of Airborne Effluents*, published by the American Society of Mechanical Engineers (3). In addition to the models presented for calculating the effective stack height, pollutant dispersion, and pollutant deposition, the manual also describes meteorological fundamentals, experimental methods, and the behavior of airborne effluents.

2. AIR QUALITY INDICES

There have been several air quality indices proposed in the past. These indices are described in the following subsections.

2.1. US EPA Air Quality Index

Initially, the US EPA produced an air quality index known as the Pollutant Standards Index (PSI) to measure pollutant concentrations for five criteria pollutants (particulate matter, sulfur dioxide, carbon monoxide, nitrogen dioxide, and ground-level ozone). The measurements were converted to a scale of 0–500. An index value of 100 was ascribed to the numerical level of the short-term (i.e., averaging time of 24 h or less) primary NAAQS and a level of 500 to the SHLs. An index value of 50, which is half the value of the short-term standard, was assigned to the annual standard or a concentration. Other index values were described as follows: 0–100, good; 101–200, unhealthful; greater than 200, very unhealthy. Use of the index was mandated in all metropolitan areas with a population in excess of 250,000. The EPA advocated calculation of the index value on a daily basis for each of the four criteria pollutants and the reporting of the highest value and identification of the pollutant responsible. Where two or more pollutants exceeded the level of 100, although the PSI value released was the one pertaining to the pollutant with the highest level, information on the other pollutants was also released. Levels above 100 could be associated with progressive preventive action by state or local officials involving issuance of health advisories for citizens or susceptible groups to limit their activities and for industries to cut back on emissions. At a PSI level of 400, the EPA deemed that "emergency" conditions would exist and that this would require cessation of most industrial and commercial activity.

In July 1999, the EPA issued its new "Air Quality Index" (AQI) replacing the PSI. The principal differences between the two indices are that the new AQI does the following:

1. Incorporates revisions to the primary health-based national ambient air quality standards for ground-level ozone and particulate matter, issued by the EPA in 1977, incorporating separate values for particulate matter of 2.5 and 10.0 μg ($PM_{2.5}$ and PM_{10}), respectively.
2. Includes a new category in the index described as "unhealthy for sensitive groups" (index value of 101–150) and the addition of an optional cautionary statement, which can be used at the upper bounds of the "moderate" range of the 8-h ozone standard.
3. Incorporates color symbols to represent different ranges of AQI values ("scaled" in the manner of color topographical maps from green to maroon) that must be used if the index is reported in a color format.

4. Includes mandatory requirements for the authorities to supply information to the public on the health effects that may be encountered at the various levels, including a requirement to report a pollutant-specific sensitive group statement when the index is above 100.
5. Mandates that the AQI shall be routinely collected and that state and local authorities shall be required to report it, for all metropolitan areas with more than 350,000 people (previously the threshold was urban areas with populations of more than 200,000).
6. Incorporates a new matrix of index values and cautionary statements for each pollutant.
7. Calculates the AQI using a method similar to that of the PSI—using concentration data obtained daily from "population-oriented State/Local Air Monitoring Stations (SLAMS)" for all pollutants except particulate matter (PM).

2.2. The Mitre Air Quality Index (MAQI)

2.2.1. Mathematical Equations of the MAQI

The Mitre Air Quality Index (MAQI) was based on the 1970 Secondary Federal National Ambient Air Quality Standards (8). The index is the root-sum-square (RSS) value of individual pollutant indices (9), each based on one of the secondary air quality standards. This index is computed as follows:

$$\text{MAQI} = \left[I_s^2 + I_c^2 + I_p^2 + I_n^2 + I_o^2 \right]^{0.5} \tag{1}$$

where I_s is an index of pollution for sulfur dioxide, I_c is an index of pollution for carbon monoxide, I_p is an index of pollution for total suspended particulates, I_n is an index of pollution for nitrogen dioxide, and I_o is an index of pollution for photochemical oxidants. These subindices are explained below.

Sulfur Dioxide Index (I_s): The sulfur dioxide index is the RSS value of individual terms corresponding to each of the secondary standards. The RSS value is used to ensure that the index value will be greater than 1 if one of the standard values is exceeded. The index is defined as

$$I_s = \left[\left(C_{sa}/S_{sa} \right)^2 + K_1 \left(C_{s24}/S_{s24} \right)^2 + K_2 \left(C_{s3}/S_{s3} \right)^2 \right]^{0.5} \tag{2}$$

where C_{sa} is the annual arithmetic mean observed concentration of sulfur dioxide, S_{sa} is the annual secondary standard value (i.e., 0.02 ppm or 60 µg/m³) consistent with the unit of measure of C_{sa}, C_{s24} is the maximum observed 24-h concentration of sulfur dioxide, S_{s24} is the 24-h secondary standard value (i.e., 0.1 ppm or 260 µg/m³) consistent with the unit of measure of C_{s24}, C_{s3} is the maximum observed 3-h concentration of sulfur dioxide, S_{s3} is the 3-h secondary standard value (i.e., 0.5 ppm or 1300 µg/m³) consistent with the unit of measure of C_{s3}, K_1 is 1 if $C_{s24} \geq S_{s24}$ and is 0 otherwise, and K_2 is 1 if $C_{s3} \geq S_{s3}$ and is 0 otherwise.

Carbon Monoxide Index (I_c): The carbon monoxide index component of the MAQI is computed in a fashion similar to the sulfur dioxide index:

$$I_c = \left[\left(C_{c8}/S_{c8} \right)^2 + K \left(C_{c1}/S_{c1} \right)^2 \right]^{0.5} \tag{3}$$

where C_{c8} is the maximum observed 8-h concentration of carbon monoxide, S_{c8} is the 8-h secondary standard value (i.e., 9 ppm or 10,000 µg/m³) consistent with the unit of measure of C_{c8}, C_{c1} is the maximum observed 1-h concentration of carbon monoxide, S_{c1} is the 1-h secondary standard value (i.e., 35 ppm or 40,000 µg/m³) consistent with the unit of measure of C_{c1}, and K is 1 if $C_{c1} \geq S_{c1}$ and is 0 otherwise.

Total Suspended Particulates Index (I_p): Total suspended particulate concentrations are always measured in micrograms per cubic meter. The index of total suspended particulates is computed as

$$I_p = \left[\left(C_{pa}/S_{pa} \right)^2 + K \left(C_{p24}/S_{p24} \right)^2 \right]^{0.5} \tag{4}$$

where C_{pa} is the annual geometric mean observed concentration of total suspended particulate matter. The geometric mean is defined as

$$g = \left[\prod_{i=1}^{n} X_i \right]^{1/n} \tag{4a}$$

Because of the nature of a geometric mean, a single 24-h reading of 0 would result in an annual geometric mean of 0. The EPA recommends that one-half of the measurement method's minimum detectable value be substituted (in this case, 0.5 µg/m³) when a "zero" value occurs. S_{pa} is the annual secondary standard value (i.e., 60 µg/m³), C_{p24} is the maximum observed 24-h concentration of total suspended particulate matter, S_{p24} is the 24-h secondary standard value (i.e., 150 µg/m³), and K is 1 if $C_{p24} \geq S_{p24}$ and is 0 otherwise.

Nitrogen Dioxide Index (I_n): The index of nitrogen dioxide does not require the RSS technique because only a single annual federal standard has been promulgated. The index is

$$I_n = C_{na}/S_{na} \tag{5}$$

where C_{na} is the annual arithmetic mean observed concentration of nitrogen dioxide and S_{na} is the annual secondary standard value (i.e., 0.05 ppm or 100 µg/m³) consistent with the unit of measure of C_{na}.

Photochemical Oxidants Index (I_o): The index is computed in a manner similar to the nitrogen dioxide index. A single standard value is used as the basis of the index, which is

$$I_o = \left[C_{o1}/S_{o1} \right] \tag{6}$$

where C_{o1} is the maximum observed 1-h concentration of photochemical oxidants and S_{o1} is the 1-h secondary standard value (i.e., 0.08 ppm or 160 µg/m³) consistent with the unit of measure of C_{o1}.

2.2.2. Application of the MAQI

A MAQI value of less than 1 indicates that all standards are being met for those pollutants in the MAQI computations. Because nine standards for five pollutants are involved in computing MAQI, any MAQI value greater than 3 guarantees that at least one standard value has been exceeded. If the MAQI values to be estimated by Eq. (1) are based on only five standards for three pollutants, then, for these figures, any MAQI value greater than 2.24 guarantees that at least one standard has been exceeded.

2.3. Extreme Value Index (EVI)

2.3.1. Mathematical Equations of the EVI

The extreme value index (EVI) was developed by Mitre Corporation (9) for use in conjunction with the MAQI values. It is an accumulation of the ratio of the extreme values for each pollutant. The EVIs for individual pollutants are combined using the RSS method. Only those pollutants are included for which secondary "maximum values not to be exceeded more than once per year" are defined. The EVI is given by

$$\text{EVI} = \left[E_c^2 + E_s^2 + E_p^2 + E_o^2 \right]^{0.5} \tag{7}$$

where E_c is an extreme value index for carbon monoxide, E_s is an extreme value index for sulfur dioxide, E_p is an extreme value index for total suspended particulates, and E_o is an extreme value index for photochemical oxidants.

Carbon Monoxide Extreme Value Index (E_c): The carbon monoxide extreme value is the RSS of the accumulated extreme values divided by the secondary standard values. The index is defined as

$$E_c = \left[\left(A_{c8}/S_{c8} \right)^2 + \left(A_{c1}/S_{c1} \right)^2 \right]^{0.5} \tag{8}$$

where A_{c8} is the accumulation of values of those observed 8-h concentrations that exceed the secondary standard and is expressed mathematically as

$$A_{c8} = \sum_i K_i \left(C_{c8} \right)_i \tag{8a}$$

where K_i is 1 if $(C_{c8})_i \geq S_{c8}$ and is 0 otherwise, S_{c8} is the 8-h secondary standard value (i.e., 9 ppm or 10,000 µg/m³) consistent with the unit of measure of the $(C_{c8})_i$ values, A_{c1} is the accumulation of values of those observed 1-h concentrations that exceed the secondary standard and is expressed mathematically as

$$A_{c1} = \sum_i K_i \left(C_{c1} \right)_i$$

K_i is 1 if $(C_{c1})_i \geq S_{c1}$ and is 0 otherwise, and S_{c1} is the 1-h secondary standard value (i.e., 35 ppm or 40,000 µg/m³) consistent with the unit of measure of the $(C_{c1})_i$ values.

Sulfur Dioxide Extreme Value Index (E_s): The sulfur dioxide extreme value is computed in the same manner as the carbon monoxide EVI. This index also includes two terms, one for each of the secondary standards, which are maximum values, and to be expected more than once per year. It should be noted that no term is included for the annual standard. The index is computed as

$$E_s = \left[\left(A_{s24}/S_{s24} \right)^2 + \left(A_{s3}/S_{s3} \right)^2 \right]^{0.5} \tag{9}$$

where A_{s24} is the accumulation of those observed 24-h concentrations that exceed the secondary standard and is expressed mathematically as

$$A_{s24} = \sum_i K_i \left(C_{s24} \right)_i \tag{9a}$$

where K_i is 1 if $(C_{s24})_i \geq S_{s24}$ and is 0 otherwise, S_{s24} is the 24-h secondary standard value (i.e., 0.1 ppm or 260 mg/m³) consistent with the unit of measure of the $(C_{s24})_i$ values, A_{s3} is the accumulation of values of those observed 3-h concentration that exceed the secondary standard and is expressed mathematically as

$$A_{s3} = \sum_i K_i \left(C_{s3} \right)_i$$

where K_i is 1 if $(C_{s3})_i \geq S_{s3}$ and is 0 otherwise, and S_{s3} is the 3-h secondary standard value (i.e., 0.1 ppm or 260 µg/m³) consistent with the unit of measure of the $(C_{s3})_i$ values.

Total Suspended Particulates Extreme Value Index (E_p): A secondary standard single maximum value not to be exceeded more than once per year is defined for total suspended particulates. The total suspended particulates EVI has only one term; no annual term is included. This index is computed as

$$E_p = A_{p24}/S_{p24} \tag{10}$$

where A_{p24} is the accumulation of those observed 24-h concentrations that exceed the secondary standard and is expressed mathematically as

$$A_{p24} = \sum_i K_i \left(C_{p24} \right)_i$$

where K_i is 1 if $(C_{p24}) \geq S_{p24}$ and is 0 otherwise, and S_{p24} is the 24-h secondary standard value (i.e., 150 µg/m^3).

Photochemical Oxidants Extreme Value Index (E_o): The index, like the total suspended particulates index, consists of a single term. The index is calculated as

$$E_o = A_{o1}/S_{o1} \tag{11}$$

where A_{o1} is the accumulation of those observed 1-h concentrations that exceed the secondary standard and is expressed mathematically as

$$A_{o1} = \sum_i K_i \left(C_{o1} \right)_i$$

where K_i is 1 if $(C_{o1})_i \geq S_{o1}$ and is 0 otherwise, and S_{o1} is the 1-h secondary standard value (i.e., 0.08 ppm or 160 µg/m^3) consistent with the unit of measure of the $(C_{o1})_i$ values.

2.3.2. Application of the EVI

The number or percentage of extreme values provides a meaningful measure of the ambient air quality because extreme high air pollution values are mostly related to personal comfort and well-being and affect plants, animals, and property. The EVI and its component indices always indicate that all standards are not being attained if the index values are greater than 0. The index value will always be at least 1 if any standards based on a "maximum value not to be exceeded more than once per year" is surpassed.

It should be noted that the index truly depicts the ambient air quality only if observations are made for all periods of interest (i.e., 1 h, 3 h, 8 h, and 24 h) during the year for which secondary standards are defined. Trend analyses using EVI values based on differing numbers of observations may be inadequate and even misleading.

2.4. Oak Ridge Air Quality Index (ORAQI)

2.4.1. Mathematical Equations of the ORAQI

The Oak Ridge Air Quality Index (ORAQI), which was designed for use with all major pollutants recognized by the EPA (10), was based on the following formula:

$$\text{ORAQI} = \left[\text{COEF} \sum_{i=1}^{3} \left(\text{Concentration of Pollutant } i/\text{EPA Standard for Pollutant } i \right) \right]^{0.967} \tag{12}$$

COEF equals 39.02 when $n = 3$, and equals 23.4 when $n = 5$. The concentration of the pollutants was based on the annual mean as measured by the EPA National Air Sampling Network (NASN). These are the same data on which the MAQI was based.

The EPA standards used in the calculation were the EPA secondary standards normalized to a 24-h average basis. For SO_2, the standard used was 0.10 ppm; for NO_2, it was 0.20 ppm; and for particulates, it was 150–160 µg/m^3.

2.4.2. Application of the ORAQI

The coefficient and exponent values in the ORAQI formula mathematically adjust the ORAQI value so that a value of 10 describes the condition of naturally occurring unpolluted air. A value of 100 is the equivalent of all pollutant concentrations reaching the federally established standards.

2.5. Allowable Emission Rates

2.5.1. Allowable Emission Rate of Suspended Particulate Matter

The allowable emission rate of suspended particulate matter from an air contamination source can be calculated (10) by the following equation:

$$Q_a = 0.5\left(p\pi u C^2 X^{2-n}\right)\exp\left(H_e^2/C^2 X^{2-n}\right) \tag{13}$$

where Q_a is the allowable emission rate of suspended particulate matter, (g/s), p is the ground-level concentration (0.15×10^{-3} g/m^3) (note: Pennsylvania state regulation), u is the mean wind speed set at 3.8 m/s, C^2 is the isotropic diffusion coefficient, set at 0.010 for neutral conditions, with dimensions, m^n, X is the downwind distance from the source (horizontal distance from the stack to the nearest property) (m), n is the stability parameter, nondimensional, set at 0.25 for neutral stability conditions, H_e is the effective stack height (m), and $\pi = 3.14$.

Substituting the above values into Eq. (13), the equation for calculating the allowable emission rate becomes

$$Q_a = \left(8.95\times10^{-6}X^{1.75}\right)\exp\left(100\,H_e^2/X^{1.75}\right) \tag{14}$$

The effective stack height (H_e) is the stack height plus the height that the effluent plume initially rises above the stack owing to the stack draft velocity and/or the buoyancy of the effluent.

2.5.2. Allowable Emission Rate of Particle Fall

The allowable emission rate of particle fall from an air contamination source can be calculated (11) by the following equation:

$$Q_a = \left(f\pi u C^2 X^{2-n}\right)\exp\left(Z^2/C^2 X^{2-n}\right) \tag{15}$$

where Q_a is the allowable emission rate of particle (dust) fall (g/s), f is the ground-level concentration (g/m^3) determined by dividing the ground-level particle (dust) fall rate (2.22×10^{-6} g/m^2/s, Pennsylvania state regulation) by the terminal setting velocity (0.03 m/s) for 25-μm particle size, quartz, u is the mean wind speed set at 3.8 m/s, C^2 is the isotropic diffusion coefficient, set at 0.010 for neutral conditions, with dimensions, m^n, X is the downwind distance from the source (m), n is the stability parameter, nondimensional, set at 0.25 for neutral stability conditions, and Z is the elevation of the plume above ground adjusted for dust fall (m),

$$Z = H_e - \left(XV/v\right) \tag{15a}$$

where H_e is the effective stack height (m) and v is the terminal settling velocity (0.03 m/s).

Substituting the above values into the Eq. (15), the equation for calculating the allowable emission rate becomes

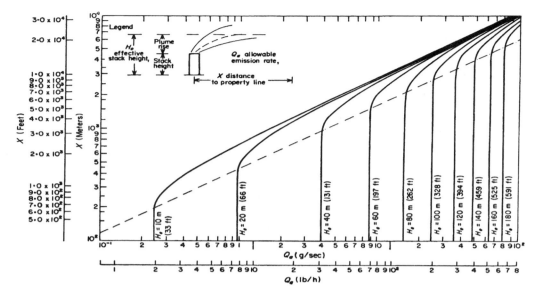

Fig. 1. Suspended particulate matter.

$$Q_a = \left(8.83 \times 10^{-6} X^{1.75}\right) \exp\left[100\left(H_e - 7.89 \times 10^{-3} X\right)^2 \Big/ X^{1.75}\right] \qquad (16)$$

2.6. Effective Stack Height

The effective stack height is the physical stack height plus the height that the effluent plume initially rises above the stack owing to the stack draft velocity and/or the buoyancy of the effluent (*see* Fig. 1).

2.6.1. Effective Stack Height for a Stack with Low Heat Emission

Unless it can be demonstrated otherwise, for a stack with low heat emission (the temperature of the flue gas equal to, or less than, 65(F) the effective stack height is calculated by the following equation:

$$H_e = H + d\left(V_s/u\right)\left(1 + \Delta T/T_s\right) \qquad (17)$$

where H_e is the effective stack height (m), H is the height of the stack (m), V_s is the stack gas ejection velocity (m/s), d is the internal diameter of the stack top (m), u is the wind speed (m/s) (assume 3.8 m/s unless other acceptable meteorological data are available for the stack locality), ΔT is the stack gas temperature minus ambient air temperature (K) (assume ambient air temperature is 283 K unless other acceptable meteorological data are available for stack locality), and T_s is the stack gas temperature (K).

2.6.2. Effective Stack Height for a Stack with High Heat Emission

Unless it can be demonstrated otherwise, for a stack with large heat emission (the temperature of the flue gas greater than 65°F) the effective stack height is calculated by the following equation:

$$H_e = H + (1.5V_s d + 4.09 \times 10^{-5} Q_h)/u \tag{18}$$

where H_e is the effective stack height (m), H is the height of the stack (m), V_s is the stack gas ejection velocity (m/s), d is the internal diameter of the stack top (m), u is the wind speed (m/s) (assume 3.8 m/s unless other acceptable meteorological data are available for the stack locality), and Q_h is the heat emission rate of the stack gas relative to the ambient atmosphere (cal/s),

$$Q_h = Q_m C_{ps} \Delta T \tag{18a}$$

where Q_m is the mass emission rate of the stack gas (g/s), C_{ps} is the specific heat of the stack gas at constant pressure (cal/g/k), $\Delta T = T_s - T$, T_s is the temperature of the stack gas at the stack top (K), T is the temperature of the ambient atmosphere (K) (assume ambient atmospheric temperature is 283 K unless other acceptable meteorological data are available for the stack locality).

2.7. Examples

2.7.1. Example 1

Problem

The observed values of atmospheric pollutants in 1965 at the Chicago CAMP (Continuous Air Monitoring Program) Station were as follows:

$C_{c8}=44$ ppm	$C_{pa}=194$ µg/m3
$C_{c1}=59$ ppm	$C_{p24}=414$ µg/m3
$C_{sa}=0.13$ ppm	$C_{na}=0.04$ ppm
$C_{s24}=0.55$ ppm	$C_{o1}=0.13$ ppm
$C_{s3}=0.94$ ppm	

Determine the carbon monoxide index, the sulfur dioxide index, the total suspended particulates index, the nitrogen dioxide index, the photochemical oxidants index, and the overall MAQI. Also discuss the calculated MAQI.

Solution

The sulfur dioxide index (I_s), carbon monoxide index (I_c), total suspended particulates index (I_p), nitrogen dioxide index (I_n), and photochemical oxidants index (I_o) can be computed by their respective equations. The results of these indices for the pollutants observed at the Chicago CAMP Station in 1965 are as follows:

$$I_s = \left[(0.13/0.02)^2 + 1(0.55/0.1)^2 + 1(0.94/0.5)^2\right]^{0.5}$$
$$= 8.72 > \sqrt{3}, \text{ standards exceeded,}$$
$$I_c = \left[(44/9)^2 + 1(59/35)^2\right]^{0.5}$$
$$= 5.17 > \sqrt{2}, \text{ standards exceeded,}$$
$$I_p = \left[(194/60)^2 + 1(414/150)^2\right]^{0.5}$$
$$= 4.25 > \sqrt{2}, \text{ standards exceeded,}$$
$$I_n = 0.04/0.05 = 0.80 < 1.0 \text{ OK,}$$
$$I_o = 0.13/0.08 = 1.62 > 1.0, \text{ standards exceeded.}$$

The above calculated individual pollutant indices are then used for the calculation of the overall MAQI. The corresponding value is

$$\text{MAQI} = \left[(8.72)^2 + (5.17)^2 + (4.25)^2 + (0.80)^2 + (1.62)^2 \right]^{0.5}$$

$$= 11.14 > \sqrt{9}, \text{ standards exceeded.}$$

If each of the individual pollutants had been at exactly the standard values, the MAQI would have been equal to √9, or 3. This value is arrived at by noting that nine standard values are defined: two for carbon monoxide, three for sulfur dioxide, two for total suspended particulates, and one each for nitrogen dioxide and photochemical oxidants. Hence, any MAQI value in excess of 3 guarantees that at least one pollutant component has exceeded the standards. It is apparent that the ambient air quality measured by the Chicago CAMP Station in 1965 was worse than the Federal Secondary Standard Values.

Interpretation of this index, as of any aggregate index, should be in terms of its relative (rather than absolute) magnitude with respect to a national or regional value of index. Cost of living and unemployment indices for a given location, for example, are frequently interpreted in this manner.

It is not apparent, by inspection of only the overall MAQI value, which standards were exceeded. It is recommended, therefore, that each of the individual pollutant indices be considered together with the MAQI in order to obtain a true picture of the actual situation. According to the individual pollutant indices derived, it is apparent that the standards of sulfur dioxide, carbon monoxide, total suspended particulates, and photochemical oxidants were exceeded.

2.7.2. Example 2

Problem

At the Chicago CAMP Station in 1965, the air quality was continuously monitored by the EPA and reported as follows:

1. About 1% of the measured 1-h carbon monoxide concentrations and 93.4% of the measured 8-h concentrations exceeded the respective secondary standards. From the raw EPA data, the accumulations of these values were $A_{c8} = 16,210$ ppm and $A_{c1} = 2893$ ppm.
2. The observed sulfur dioxide concentrations resulted in accumulated values of $A_{s24} = 37.52$ ppm and $A_{s3} = 38.63$ ppm, where 49.9% of the 24-h values and 2.5% of the 3-h values exceeded the secondary standards.
3. Sixty-six Hi-Volume Sampler 24-h measurements were taken. Of these, approx 74.2% exceeded the secondary standard value. The observed accumulated total suspended particulate concentrations in excess of the 24-h standard were $A_{p24} = 11535$ µg/m³.
4. Of the observed 1-h concentrations of photochemical oxidants, 1.8% exceeded the secondary standard. The accumulation of these values was $A_{o1} = 9.45$ ppm.

Determine the carbon monoxide extreme value index, the sulfur dioxide extreme value index, total suspended particulates extreme value index, the photochemical oxidants extreme value index, and the combined EVI. Also discuss the calculated EVI.

Solution

The extreme value indices of carbon monoxide (E_c), sulfur dioxide (E_s), total suspended particulates (E_p), and photochemical oxidants (E_o) are calculated by the equations in Section 2.3.1:

$$E_c = \left[(16210/9)^2 + (2893/35)^2\right]^{0.5}$$
$$= 1803.01,$$

$$E_s = \left[(37.52/0.10)^2 + (38.63/0.50)^2\right]^{0.5}$$
$$= 383.07,$$

$$E_p = 11535/150 = 76.90,$$
$$E_o = 9.45/0.08 = 118.12.$$

The individual pollutant EVIs are then combined and the overall EVI calculated by Eq. (7) for the Chicago CAMP Station:

$$\text{EVI} = \left[(1803.01)^2 + (383.07)^2 + (76.90)^2 + (118.12)^2\right]^{0.5}$$
$$= 1848.64.$$

The EVI and its component indices always indicate that all standards are not being attained if the index values are greater than 0. The index value will always be at least 1 if any standard based on a maximum value not to be exceeded more than once per year is surpassed.

The calculated EVI (i.e., 1848.64) tends to depict the degree to which the secondary standards have been exceeded. It is probably most useful as an indicator of the trend over time of the air quality in a particular locality. A characteristic of the EVI is its tendency to increase in magnitude as the number of observations in excess of standards increases. This growth of the index value is desirable. The EVI index truly depicts the ambient air quality because the observations were made for all periods of interest (i.e., 1 h, 3 h, 8 h, and 24 h) during the year for which secondary standards are defined.

The percentage of observed values exceeding the standard also helps to depict the situation, without having to inspect all of the available data. An analysis of available CAMP Station data reveals that the carbon monoxide 1-h secondary standard is rarely exceeded, even though the 8-h standard is exceeded as much as 93% of the time. As an option, this carbon monoxide EVI could be calculated strictly from the 8-h concentration values as

$$E_c = A_{c8}/S_{c8}$$

without under distortion of the true situation. For example, the Chicago CAMP Station data yield a value of $E_c = 1801.11$, compared with the previous value of 1803.01.

An inspection of CAMP sulfur dioxide data suggests that the 3-h standard is rarely exceeded, and when it is, the contribution of the 3-h extreme values to the sulfur dioxide EVI is negligible. The index, therefore, could optionally be calculated as

$$E_s = A_{s24}/S_{s24}$$

For example, computation in this manner using the Chicago CAMP data results in an index value of 375.20, a value that is 98% of the index value, which included the 3-h term.

2.7.3. Example 3

Problem

Calculate the ORAQI assuming that the three major pollution concentrations reach the US federally established standards (normalized to a 24-h average basis):

SO$_x$=0.10 ppm (by volume)
NO$_2$=0.20 ppm (by volume)
Particulates=150 μg/m^3

Solution

The ORAQI is calculated as

$$\text{ORAQI} = \left[39.02(0.10/0.10 + 0.20/0.20 + 150/150)\right]^{0.967}$$
$$= 100$$

A value of 100 is the equivalent of all three pollutant concentrations reaching the federally established standards. Note that the condition of naturally occurring unpolluted air will have a value of 10.

2.7.4. Example 4

Problem

What will be the formula for the calculation of the ORAQI if all five of the major pollutants recognized by the EPA are included?

Solution

The ORAQI can be calculated based on the following formula:

$$\text{ORAQI} = \left[23.40\sum_{i=1}^{5}(\text{Concentration of Pollutant } i/\text{EPA Standard for Pollutant } i)\right]^{0.967}$$

When all five pollutant concentrations (CO, SO$_2$, NO$_2$, particulates, and photochemical oxidants) reach the federally establishes standards, the index will be equal to 100:

$$\text{ORAQI} = (23.40 \times 5)^{0.967} = 100$$

2.7.5. Example 5

Problem

Determine the allowable emission rate of suspended particulate matter assuming that the following data are given:

Ground-level concentration (p)=150 μg/m^3
Mean wind speed (u)=3.8 m/s
Isotropic diffusion coefficient (C^2)=0.01
Stability parameter (n)=0.25
Effective stack height (H_e)= 10 m, 20 m, 40 m, 60 m, 80 m, 100 m, 120 m, 140 m, 160 m, and 180 m
Horizontal distance from the stack to the nearest property line (X)= 100 m, 150 m, 200 m, 250 m, 300 m, 500 m, 700 m, 1000 m, 3000 m, and 9000 m

Solution

Using either Eq. (13) or (14), one can calculate the allowable emission rate of suspended particulate matter (Q_a), because the values of p, u, C^2, and n are all identical to those recommended by the local government. If at least one of the four values is different from that recommended by the local government, only Eq. (13) could be used.

When $X=9000$ m and $H_e=20$ m, both Eqs. (13) and (14) indicate that $Q_a=74.79$ g/s:

$$Q_a = 0.5\left(p\pi u C^2 X^{2-n}\right)\exp\left(H_e^2/C^2 X^{2-n}\right)$$

$$= 0.5\left(150\times10^{-6}\times3.14\times3.8\times0.01\times9000^{2-0.25}\right)\times\exp\left(20^2/0.01\times9000^{2-0.25}\right)$$

$$= 74.79 \text{ g/s, calculated by Eq. (13).}$$

$$Q_a = \left(8.95\times10^{-6}X^{1.75}\right)\exp\left(100\,H_e^2/X^{1.75}\right)$$

$$= \left(8.95\times10^{-6}\times9000^{1.75}\right)\exp\left(100\times20^2/9000^{1.75}\right)$$

$$= 74.79\,\text{g/s, calculated by Eq. (14).}$$

The Q_a values for various X values (note: $H_e=20$ m) are also calculated and are as follows:

X (m)	Q_a (g/s)
100	8815.81
150	28.97
200	4.09
250	1.79
300	1.23
500	1.01
700	1.30
1000	1.99
3000	11.2
9000	74.79

It is important to note that Q_a decreases with increasing X from 100 to 500 m, then increases with increasing X from 500 to 9000 m. For air quality control, only the region where Q_a increases with X should be considered. Accordingly, Fig. 1 is prepared for air quality control and management. For Fig. 1, effective stack heights of 10, 20, 40, 60, 80, 100, 120, 140, 160, and 180 m were plotted while downwind distance ranged from 100 to 10,000 m. This graph shows the solution only for the region where Q_a increases with X. The region where Q_a decreases with X has been replaced by a vertical line. Figure 1 can be used only when p, u, C^2, and n are the same as stated in this problem.

2.7.6. Example 6

Problem

Determine the allowable emission rate of particle fall (or dust fall) from an air contamination source, assuming the following data are given:

Ground level particle fall rate $(q)=2.22\times10^{-6}$ g/m²/s
Terminal settling velocity for 25-µm quartz $(v)=0.03$ m/s
Ground-level particle concentration

$$(f) = \left(2.22\times10^{-6}\,\text{g/m}^2/\text{s}\right)/\left(0.03\,\text{m/s}\right) = 74\times10^{-6}\,\text{g/m}^3$$

Mean wind velocity $(u)=3.8$ m/s
Isotropic diffusion coefficient $(C^2)=0.01$
Stability parameter $(n)=0.25$
Effective stack heights $(H_e)=10$ m, 20 m, 40 m, 60 m, 80 m, 100 m, 120 m, 140 m, 160 m, and 180 m
Horizontal distance from the stack to the nearest property line $(X)=100$ m, 150 m, 200 m, 250 m, 300 m, 500 m, 700 m, 1000 m, 3000 m, and 9000 m

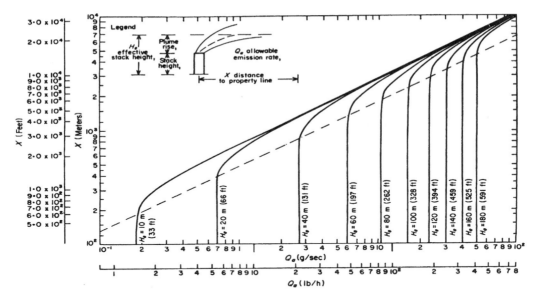

Fig. 2. Particle fall.

Solution

Using Eq. (15) or (16), one can calculate the allowable emission rate of particle fall (Q_a), because the values of q, v, f, u, C^2, and n are all identical to those recommended by the local government. If at least one of the six values is different from that recommended by the government, only Eq. (15) can be used.

When $X = 9000$ m and $H_e = 20$ m, both Eqs. (15) and (16) indicate that $Q_a = 75.77$ g/s:

$$Q_a = \left(f\pi u C^2 X^{2-n}\right)\exp\left[\left(H_e - \left(Xv/u\right)\right)^2 \Big/ C^2 X^{2-n}\right]$$
$$= \left(74 \times 10^{-6} \times 3.14 \times 3.8 \times 0.01 \times 9000^{2-0.25}\right)$$
$$\times \exp\left\{\left[20 - \left(9000 \times 0.03/3.8\right)\right]^2 \Big/ \left(0.01 \times 9000^{2-0.25}\right)\right\}$$
$$= 75.77\,\text{g/s, calculated by Eq. (15).}$$
$$Q_a = \left(8.83 \times 10^{-6} X^{1.75}\right)\exp\left[100\left(H_e - 7.89 \times 10^{-3} X\right)^2 \Big/ X^{1.75}\right]$$
$$= \left(8.83 \times 10^{-6}\ 9000^{1.75}\right)\exp\left[100\left(20 - 7.89 \times 10^{-3} \times 9000\right)^2 \Big/ 9000^{1.75}\right]$$
$$= 75.77\,\text{g/s, calculated by Eq. (16).}$$

The Q_a values for various X and H_e values can also be calculated. Finally, Fig. 2 was prepared. For the graph, stack heights of 10, 20, 40, 60, 80, 100, 120, 140, 160 and 180 m were plotted while distances downwind ranged from 100 to 10,000 m. Again, Fig. 2 shows the solution only for the region where Q_a increases with X. The region where Q_a decreases with X has been replaced by a vertical line.

3. DISPERSION OF AIRBORNE EFFLUENTS

3.1. Wind Speed Correction

It is necessary to adjust the wind speed and the standard deviations of the directional fluctuations for the difference in elevation when meteorological installations are not at the source height. The variation of wind speed with height can be estimated from the following equation:

$$UH = U(HS/H)^A \tag{19}$$

where UH is the mean wind speed at the stack height (m/s), U is the mean wind speed at the instrument height (m/s), HS is the stack height (m), H is the instrument height (m), and A is the a coefficient (0.5 for a stable condition and 0.25 for unstable, very unstable, and neutral conditions).

3.2. Wind Direction Standard Deviations

The standard deviations of the wind direction fluctuations must be adjusted for the difference between the height of measurement and the height of the stack. The following two equations are used:

$$SAH = SA(U/UH) \tag{20}$$

$$SEH = SE(U/UH) \tag{21}$$

where SAH is the standard deviation of the wind direction fluctuation in the horizontal direction (deg) at the stack height, SA is the standard deviation of the wind direction fluctuation in the horizontal direction (deg) at the instrument height, SEH is the standard deviation of the wind direction fluctuation in the vertical direction (deg) at the stack height, and SE is the standard deviation of the wind direction fluctuation in the vertical direction (deg) at the instrument height.

3.3. Plume Standard Deviations

When SAH and SEH are available from wind vanes, one can then determine the plume standard deviations:

$$SY = B(SAH)X^C \tag{22}$$

$$SZ = B(SEH)X^C \tag{23}$$

where SY is the standard deviation of the plume profile in the crosswind direction (m), SZ is the standard deviation of the plume profile in the vertical direction (m), X is the downwind distance from the source (m), B is a coefficient (0.15 for a stable case and 0.045 for neutral, unstable and very unstable cases), and C is a coefficient (0.71 for a stable case and 0.86 for neutral, unstable and very unstable cases).

3.4. Effective Stack Height

The effective stack height (HT) is the sum of two terms: (1) actual stack height (HS) and (2) the plume rise (HP) caused by the velocity of the stack gases and by the density difference between the stack gases and the atmosphere, as shown in Fig. 3:

$$HT = HS + HP \tag{24}$$

For small-volume sources having appreciable exit speeds (greater than or equal to 10 m/s) but little temperature excess (less than 50°C above ambient temperature), the height of plume rise (HP) can be determined by the following equation if VS is greater than UH:

$$HP = D(VS/UH)^{1.4} \tag{25}$$

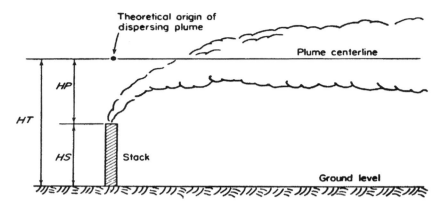

Fig. 3. Effective stack height (*HT*), actual stack height (*HS*), and plume rise (*HP*).

where D is the diameter of the stack (m) and VS is the vertical efflux velocity at release temperature (m/s).

For plumes having temperatures considerably above that of the ambient air (greater than or equal to 50°C), and a large-volume release (greater than or equal to 50 m³/s), the following equation can be used for calculating the HP, in meters. Under stable conditions,

$$HP = 2.9\left[F/(UH)\,G\right]^{1/3} \tag{26}$$

where F is the buoyance flux (m⁴/s³)$=g(VS)(0.5D)^2\,(RA-RS)/RA$, g is the acceleration of gravity (m/s²) ($=9.8$), VS is the vertical efflux velocity at release temperature (m/s), RS is the density of the stack at the stack top (g/m³), RA is the density of ambient air at the stack top (g/m³), G is the stability parameter (s⁻²) $(g/PT)(VLR)$, PT is the potential temperature at stack height (K) $[(TA)(P_0/P)^{0.29}]$, P is atmospheric pressure (mbar), $P_0 = 1013$ mbar (Standard), TA is the absolute ambient air temperature (K), VLR is the vertical potential temperature lapse rate (K/100 m)$=\Delta TA/\Delta Z + ALR = LR + ALR$, and ALR is the adiabatic lapse rate (0.98 K/100 m).

Under neutral and unstable conditions,

$$HP = \left[7.4(HS)^{2/3}\,F^{1/3}\right]/UH \tag{27}$$

3.5. Maximum Ground-Level Concentration

Based on the actual meteorological cases and effective stack heights, realistic maximum concentrations can be estimated. The maximum value occurs at the downwind distance (X), where

$$SZ = (HT)/2^{0.5} = SSZ \tag{28}$$

Using Eq. (23), one can calculate the downwind distance where the maximum ground-level concentration occurs. Then, using Eq. (22), one can calculate SY (or SSY). Finally, the maximum ground-level concentration (C_{max}, in mg/m³) can be determined with the following equation:

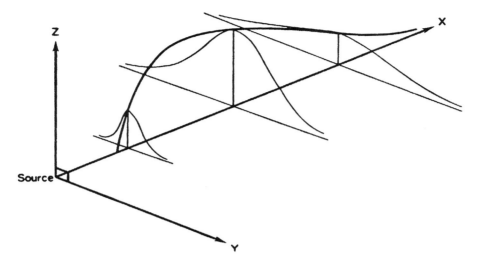

Fig. 4. Precise estimate of receptor concentration. X, downward distance; Y, crosswind distance; Z, vertical distance.

$$C_{max} = \left\{2Q\big/\left[2.718 \times 3.14(UH)(HT)^2\right]\right\}\big/(SSZ/SSY) \qquad (29)$$

where Q is the pollutant emission rate at the source (units/s), (e.g., g/s).

3.6. Steady-State Dispersion Model (Crosswind Pollutant Concentrations)

Steady-state models, that describe air transport by a diffusing plume convected by means of wind have been used by many scientists. The concentrations of atmospheric pollutants in the plume are generally assumed to be distributed in a Gaussian profile. The equation giving ground-level concentrations from an elevated point source (i.e., a typical stack) is

$$R(X, Y, Z = 0) = \left\{Q\big/[3.14(UH)(SY)(SZ)]\right\} \exp\left\{-\left[0.5HT^2\big/SZ^2 + 0.5Y^2\big/SY^2\right]\right\} \qquad (30)$$

where R is the pollutant concentration (units/m³) (e.g., mg/m³), X, Y, and Z are rectangular coordinates with X downwind, Y crosswind, and Z vertical (m). (Note: origin at source and ground level.)

Equation (30) is generally used for the computation of crosswind pollutant concentrations. Figure 4 shows a pattern of the distribution of pollutant concentrations at ground level derived from the steady-state dispersion model.

3.7. Centerline Pollutant Concentrations

The centerline pollutant concentrations can be estimated with Eq. (30) by letting $Y=0$, or

$$R(X, Y = 0, Z = 0) = \left\{Q\big/[3.14(UH)(SY)(SZ)]\right\} \exp\left\{-0.5\,HT^2\big/SZ^2\right\} \qquad (31)$$

Sometimes one wishes to examine the pollutant concentration pattern directly downwind assuming that the source is at ground level, and, in this case, $HT=Y=0$ in Eq. (30).

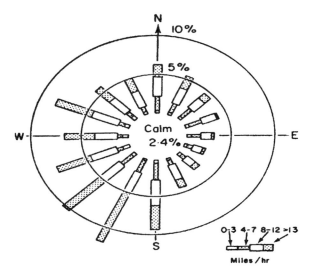

Fig. 5. A typical wind rose of Cincinnati in January.

3.8. Short-Term Pollutant Concentrations

Short-term peak concentrations (C_{peak}, in mg/m^3) may be calculated with Eq. (32):

$$C_{\text{peak}}/C_{\text{max}} = (3600/T)^E \qquad (32)$$

where T is the time (s) and E is a coefficient that varies with the dispersion conditions, as follows:

    $E = 0.65$ under very unstable conditions
    $E = 0.52$ under unstable conditions
    $E = 0.35$ under neutral conditions

No E value is given for stable conditions because elevated sources do not normally produce ground-level concentrations under such conditions. For practical applications, E is assigned to be zero for stable conditions in computer analysis.

3.9. Long-Term Pollutant Concentrations and Wind Rose

Over extremely long periods, such as 1 mo, there is a simple adaptation of the basic dispersion equation that can be used. Equation (33) is not a rigorous mathematical development, but it is satisfactory for rough approximations:

$$CW = \left\{(360WQ)/\left[100N(3.14)^{1.5}2^{0.5}(UH)(SZ)X\right]\right\}\exp\left\{-\left[0.5 \times HT^2/(SZ)^2\right]\right\} \quad (33)$$

where N is the angular width of a direction sector (deg), W is the frequency (%) with which a combination of meteorological condition of interest together with winds in that sector may be found, and CW is the long-term pollutant concentration (mg/m^3).

When W and N are 1% and 20°, respectively, Eq. (33) can be rewritten as

$$CW = \left\{0.07181\left[Q/[3.14(UH)]/[X(SZ)]\right]\right\}\exp\left\{-\left[0.5 \times HT^2/(SZ)^2\right]\right\} \qquad (34)$$

A typical wind rose documenting the necessary meteorological data is shown in Fig. 5. The monthly distribution of wind direction and wind speed of Cincinnati, Ohio in

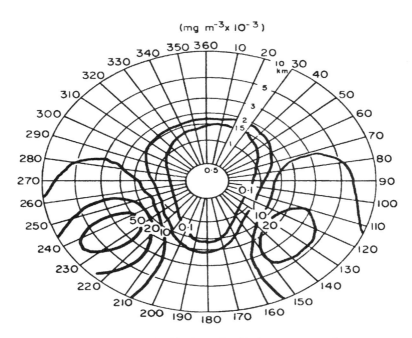

Fig. 6. Monthly distribution of pollution concentrations.

January are summarized on the polar diagram. The positions of the spokes show the direction from which the wind was blowing; the length of the segments indicate the percentage of the wind speeds in various groups.

Figure 6 shows a typical monthly distribution of long-term pollutant concentrations (3). It is seen that the isolines of pollutant concentration are drawn on a polar diagram for presenting the computed long-term pollutant concentrations surrounding an isolated plant stack. Note that the peak valve located 3 km to the southwest of the stack has a concentration about 1/100th of a typical hourly maximum concentration.

When there are four stability classes, it would be necessary to add the contributions of several classes to arrive at the final plot of concentrations. It is also advised (3) that such an analysis would normally be made for different seasons or months to show the variation throughout the year.

3.10. Stability and Environmental Conditions

Stability is related to both wind shear and temperature structure in the vertical of the atmosphere, although the latter is generally used as an indicator of the environmental condition. The "stability" of the atmosphere is defined as its tendency to resist or enhance vertical motion or, alternatively, to suppress or augment existing turbulence. Under stable conditions, the air is suppressed, and under unstable conditions, the air motion is enhanced.

In vertical motion, parcels of air are displaced. Because of the decrease of pressure with height, an air parcel displaced upward will encounter decreased pressure, expand, and increased volume. The rate of cooling with height is the dry adiabatic lapse rate and is approx $-1°C/100$ m ($-0.01°C/m$). If a parcel of dry air were brought adiabatically

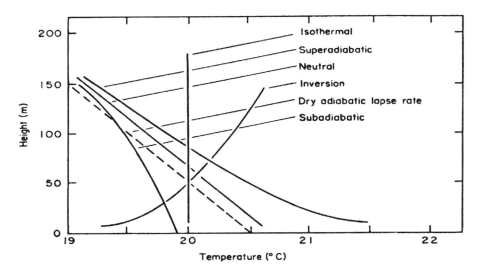

Fig. 7. Typical environmental lapse rates.

from its initial state to an arbitrarily selected standard pressure of 1000 mbars, it would assume a new temperature, known previously as the "potential temperature." This quantity is closely related to the dry adiabatic rate.

Similarly, if the displacement is downward so that an increase in pressure and compression is experienced, the parcel of air will be heated. The actual distribution of temperature in the vertical of the atmosphere is defined as the "environmental lapse rate" (LR). Typical examples are shown in Fig. 7, in comparison with the dry adiabatic lapse rate, which serves as a reference for distinguishing unstable from stable cases. The position of the dashed line in Fig. 7 representing the adiabatic lapse rate is not important; it is significant only as far as its slope is concerned. A superadiabatic condition favors strong convection, instability, and turbulence. It occurs on days when there is strong solar heating or when cold air is being transferred over a much warmer surface. The rate of decrease of temperature with height exceeds −1°C/100 m. Air parcels displaced upward will attain temperature higher than their surroundings, whereas air parcels displaced downward will attain lower temperatures than their surroundings. Because the displaced parcels will tend to continue in the direction of displacement, the vertical motions are enhanced and the layer of air is classified as "unstable."

If the environmental lapse rate is nearly identical to the dry adiabatic lapse rate, −1°C/100 m, the condition is classified as neutral, implying no tendency for a displaced parcel to gain or lose buoyancy.

A subadiabatic condition is classified as "stable" in which the lapse rate in the atmosphere is less than −1°C/100 m. Air parcels displaced upward attain temperature lower than their surroundings and will tend to return to their original levels. Air parcels displaced downward attain higher temperatures than their surroundings and also tend to return to their original levels. When the ambient temperature is constant with height, the layer is termed "isothermal," and, as in the subadiabatic condition, there is slight tendency for an air parcel to resist vertical motion; therefore, it is another "stable" condition.

Table 2
Meteorological Data

	Stable case	Unstable case
LR	0.8°K/100 m	−1.1°K/100 m
U	7.8 m/s	8.8 m/s
SA	2.0°	8.0°
SE	0.5°	5.5°

Under certain environmental conditions, the thermal distribution can be such that the temperature increases with height within a layer of air. This is termed "inversion" and constitutes an "extremely stable" condition (Fig. 7). The reader is referred to the recent literature (15–22) for updated information on quality management.

3.11. Air Dispersion Applications

3.11.1. Example 1

Problem and Tasks

There is a modern 700 MW (e.g., megawatt) coal-fired power plant having the following parameters, given in units:

Fuel consumption	750 lb of coal/MW/h
Sulfur content of coal	3%
Stack height	183 m (600 ft)
Stack diameter	6.08 m (19.95 ft)
Effluent temperature	275°F
Ambient air temperature at the stack top	50°F
Effluent density	9.92×10^{-4} g/cm^3
Ambient air density at the stack top	1.25×10^{-3} g/cm^3
Stack effluent velocity	15.54 m/s (51 ft/s)
Potential temperature	50°F
Atmospheric pressure	1013 mbars (standard)

Instruments and samplers for meteorological measurements are commercially available (12, 13). In this example, the meteorological measurements are also assumed to be available from a suitable tower at a height 108 m above ground, and, in this example, only two dispersion cases are considered and their surveyed data are presented in Table 2. The stable case represents a typical clear night with light low-level winds. The unstable case represents a typical sunny afternoon with moderate low-level winds. The wind rose for unstable conditions has been divided into 20° intervals. Table 3 lists the angular width of the direction sectors (deg) versus the frequency (%). The specific tasks of this project are as follows:

1. Document the given meteorological data.
2. Compute the pollutant emission rate at the source.
3. Compute the wind speed at the stack height.
4. Compute the standard deviation of the azimuth angle at the source height.
5. Compute the standard deviation of the elevation angle at the source height.
6. Print the actual stack height and compute the effective stack height.
7. Compute the centerline pollutant concentrations at the downwind distances of 100, 1000, 2000, 5000, 10,000, 50,000, and 100,000 m.
8. Compute the crosswind pollutant concentrations at the downwind distance of 4000 m (i.e., $X = 4000$ m) and at the crosswind distances (Y) of 0, 100, 200, 300, 500, and 1000 m.

Table 3
Wind Rose

Direction sector, K–J (deg)	Frequency, W(I) (%)
350–10	1
10–30	2
30–50	8
50–70	16
70–90	5
90–110	4
110–130	2
130–150	2
150–170	2
170–190	1
190–210	1
210–230	1
230–250	1
250–270	4
270–290	5
290–310	10
310–330	8
330–350	2
Total	75

9. Compute the short-term pollutant concentrations (i.e., the 1-min peak value and the 10-min peak value) at the centerline location where the maximum ground-level concentration occurs.
10. Compute the long-term pollutant concentrations for the completion of a wind rose analysis.

Solution

Initially, the given data must be converted to the desired units. Using the given plant description and conversion table (14), the input parameters become

TA, the ambient air temperature, $= 50°F = 283°K$
TR, the reference temperature, $= 273°K$
Q, the stack emission rate, $= (700\ MW)\ (750\ lb\ coal/MW/h)(0.03\ lb\ sulfur/lb\ coal)$
    $(2\ SO_2/S)\ (1\ h/3600\ s)(453{,}600\ mg/lb)\ (283°K/273°K) = 4.11×10^6\ mg\ SO_2/s$
$H = 108\ m$
VS, the stack effluent velocity, $= 51\ ft/s = 15.48\ m/s$
D, the stack diameter, $= 19.95\ ft = 6.08\ m$
HS, the stack height, $= 600\ ft = 183\ m$
$U = 7.8\ m/s$ for a stable condition and 8.8 m/s for an unstable condition (Table 2).

The second step is for wind speed correction. Data are available for instruments at $H = 108$ m and the stack height $HS = 182.88$ m. Equation (19) gives the mean wind speed at the stack height UH:

UH (stable) $= 7.8\ (183/108)^{0.5} = 10.15\ m/s$
UH (unstable) $= 8.8\ (183/108)^{0.25} = 10.04\ m/s$

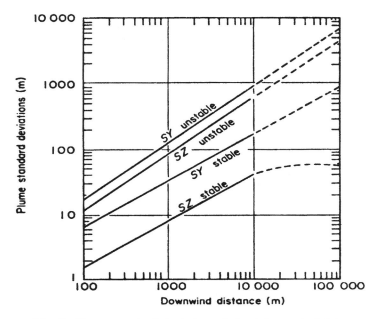

Fig. 8. Plume standard deviations derived from wind data.

The third step is for the determination of wind direction standard deviations. *SA* and *SE* values are given in Table 2. Two *UH* values have been calculated with Eq. (19). For the horizontal wind direction (or azimuth), Eq. (20) is used for standard deviation determinations:

SAH (stable)$= 2 \, (7.8/10.15) = 1.54°$
SAH (unstable)$= 8 \, (8.8/10.04) = 7.01°$

For the vertical wind direction (or elevation), Eq. (21) is used:

SEH (stable)$= 0.5 \, (7.8/10.15) = 0.38°$
SEH (unstable)$= 5.5 \, (8.8/10.04) = 4.82°$

The fourth step involves the determination of the plume standard deviation from wind direction standard deviations. Using Eq. (22), the standard deviation of plume profile in crosswind direction (*SY*) can be determined:

$$SY(\text{stable}) = 0.15(1.54)X^{0.71} = 0.24X^{0.71} \tag{35}$$

$$SY(\text{unstable}) = 0.045(7.01)X^{0.86} = 0.32X^{0.86} \tag{36}$$

Using Eq. (23), the standard deviation of plume profile in vertical direction (*SZ*) can be determined.

$$SZ(\text{stable}) = 0.15(0.38)X^{0.71} = 0.06X^{0.71} \tag{37}$$

$$SZ(\text{unstable}) = 0.045(4.82)X^{0.86} = 0.216X^{0.86} \tag{38}$$

Equations (35)–(38) are plotted in Fig. 8.

In Step 5, the effective stack heights (*HT*) are to be estimated for both stable and unstable conditions:

RA	$= 1.25 \times 10^{-3}$ g/cm^3
RS	$= 9.92 \times 10^{-4}$ g/cm^3
g	$= 9.8$ m/s^2
HS	$= 183$ m
F	$= 9.8\,(15.48)(0.5 \times 6.08)^2\,(1.25 \times 10^{-3} - 9.92 \times 10^{-4})/(1.25 \times 10^{-3})$
	$= 2.91 \times 10^2$ m^4/s^3
P	$= 1013$ mbars
P_0	$= 1013$ mbars
ALR	$= 0.98$ K/100 m
LR	$= 0.8$ K/100 m for a stable condition (Table 2)
VLR	$= 0.8 + 0.98 = 1.8$ K/100 m for a stable condition
PT	$= (283 \text{ K})(1013/1013)^{0.29} = 283$ K
G	$= (9.8/283)(1.8/100) = 6.16 \times 10^{-4}$ s^{-2}
HP (stable)	$= 2.9[2.91 \times 102/(10 \times 6.16 \times 10^{-4})] = 105$ m
HT (stable)	$= 183 + 105 = 288$ m
HP (unstable)	$= [7.4(183)^{2/3}(2.91 \times 10^2)^{1/3}]/10 = 158$ m
HT (unstable)	$= 183 + 158 = 341$ m

In Step 6, the maximum ground-level concentration is to be calculated. The stable case, of course, produces no maximum at ground level and, thus, only the unstable condition needs to be calculated. According to Section 3.5, the maximum value occurs at the downwind distance, where

$$SZ = SSZ = (H/T)/2^{0.5} = 341/1.41 = 242 \text{ m} \tag{39}$$

The downwind distance (X) is then estimated to be 3478 m with Eq. (38), and the SY value is estimated to be 340 m with Eq. (36). The maximum ground-level concentrations are estimated in two common units:

$$C_{\max} = \left\{ 2\left(4.11 \times 10^6\right) \big/ \left[2.718 \times 3.14(10)(341)^2\right] \right\} \big/ (242/340)$$

$$\text{PPM} = 0.58\,\text{mg/m}^3$$

$$\text{PPM}_{SO_2} = (22.4/64); \tag{40}$$

$$\text{PPM}_{\max} = C_{\max}\left(\text{PPM}_{SO_2}\right)$$
$$= 0.58(22.4/64) = 0.2 \text{ ppm of } SO_2 \tag{41}$$

which occurs at

$$X = X_{\max} = 3478 \text{ m.}$$

It should be noted that several wind speeds must be tried to determine the maximum ground-level concentration. For this example, the greatest PPM$_{\max}$ is 0.20 ppm occurring at 8.6 m/s.

The centerline pollutant concentrations are computed using Eq. (31) in Step 7.

$$SQ1 = Q/(3.14 \times UH) = 4.11 \times 10^6/(3.14 \times 10) = 1.31 \times 10^5$$

$$HDS = -0.5(HT/SZ)^2 \tag{42}$$

$$R = [SQ1/(SY \times SZ)]\exp(HDS) \tag{43}$$

Table 4
Computation of Centerline Pollutant Concentrations

X (m) (Assigned)	SY (m) [Eq. (36)]	SZ (m) [Eq. (38)]	R (mg/m³) [Eq. (43)]	C_{PPM} (ppm) [Eq. (44)]
1,000	120.0	82.5	0.00	0.00
2,000	217.7	149.7	0.29	0.10
5,000	478.8	329.2	0.49	0.17
10,000	869.1	597.5	0.20	0.07
50,000	3,468.6	2,384.7	0.03	0.01
100,000	6,295.7	4,328.3	0.00	0.00

$X_{max} = 3478$ m $\qquad\qquad C_{max} = 0.58$ mg/m³ $\quad$ PPM$_{max} = 0.20$ ppm

$$C_{PPM} = PPM_{SO_2}(R). \tag{44}$$

A table of required values is then prepared for the computation.

Table 4 indicates the computed centerline pollutant concentrations for Example 1. With the computed values (X versus C_{PPM} and X_{max} versus PPM$_{max}$), the centerline pollutant concentrations at ground level can be graphically plotted if desired.

Computation of crosswind concentrations is accomplished in Step 8 using Eq. (30). The distance of

$$X = CROSS = 4000 \text{ m}$$

is simply selected as an example.

$$SCZ = B(SEH)(CROSS)C = SZ \tag{45}$$

$$SCY = SCZ(SAH)(SEH) = SY \tag{46}$$

$$SQ2 = \{Q/[3.14(UH)(SY)(SZ)]\}\exp(HDS) \tag{47}$$

$$SQ2 = \{SQ1/[(SCY)(SCZ)]\}\exp(HDS) \tag{48}$$

$$R = (SQ2)\exp\left[-0.5(Y/SCY)^2\right] \tag{49}$$

Equation (49) is a simplified version of Eq. (30), developed for saving the computation by a digital computer. Table 5 indicates the calculated results.

Table 5
Computation of Crosswind Pollutant Concentrations

Y (m) (Assigned)	R [Eqs. (45)–(47) and (49)]	C_{PPM} (ppm) [Eq. (44)]
0	0.54	0.19
±100	0.54	0.19
±200	0.69	0.17
±300	0.43	0.15
±500	0.26	0.09
±1000	0.03	0.01

All of the concentrations presented in this example have been representative of hourly means. Estimation of short-term concentrations can be accomplished in Step 9 with Eq. (32). For 1-min peak values,

$$T = 60 \text{ s}$$
$$C_{peak} = (PPM_{max})(3600/T)^E$$
$$C_{peak} = (0.2 \text{ ppm}) (3600/60)^{0.52} = 1.67 \text{ ppm}$$

Similarly for 10-min peak values, $C_{peak} = 0.50$ ppm.

Long-term pollutant concentrations are estimated in Step 10 using Eq. (33). The pollutant dispersion conditions, wind speeds, and wind directions at a given site vary continuously from hour to hour, thus must be taken into account in the estimation. For this analysis, it is assumed that the wind rose for unstable conditions has been divided into 20° intervals ($N = 20$), as indicated in Table 3. It is time-saving to set the computation so that the sector pollutant concentrations will correspond to a 1% ($W = 1\%$) direction frequency. The data can then be multiplied by the actual percentages given in Table 3. Therefore,

$$\text{FACTOR} = 360W / \left[100N(3.14)^{0.5} 2^{0.5} \right]$$
$$= 360(1) / \left[100 \times 20 \times (3.14)^{0.5} 2^{0.5} \right] \tag{50}$$
$$= 7.181 \times 10^{-2}$$

$$HOD = -0.5[HT/SZ]^2 \tag{51}$$

$$CW(\text{mg/m}^3) = \{\text{FACTOR}(SQ1) / [X(SZ)]\} \exp(HOD) \tag{52}$$

$$CW(10^{-3} \text{ mg/m}^3) = \{71.81(SQ1) / [X(SZ)]\} \exp(HOD) \tag{53}$$

$$SQ1 = 1.31 \times 10^5 \text{(determined previously)},$$

$$CWR = PPM_{SO_2}(CW) \tag{54}$$

For the assumed $N = 20°$ and $W = 1\%$, the CW (10^{-3} mg/m³) and CWR (10^{-3} ppm) values of each 20° sector can be calculated with Eqs. (53) and (54) for $X = 1000$ m, 2000 m, 5,000 m, and 10,000 m and would have the values shown in Table 6. Finally, the wind rose for the unstable cases indicated in Table 3 is considered; thus,

$$CW_{actual\%} = CW_{1\%} \times W(I) \tag{55}$$

where $CW_{actual\%}$ is the long-term pollution concentration (10^{-3} mg/m³) of each 20° sector considering actual percentage frequency W; $CW_{1\%}$ is the long-term pollution concentration (10^{-3} mg/m³) of an assumed 20° direction sector at 1% frequency [note: $CW_{1\%}$ is calculated using Eq. (53)]; $W(I)$ is the given frequency data of a wind rose (see Table 3, for example); and I is the number of the assigned direction sector. It should be noted that the meteorological records have shown in Table 3 that the unstable case occurs during 75% of all hours with a mean wind speed of 10 m/s, and the stable case is found during the remaining 25% of the hours, also with an approx 10-m/s mean wind.

In this example, the stable case has been completely eliminated from further consideration because it contributes nothing to the ground-level concentrations for the unstable case alone and distributes these concentrations radically according to the wind rose associated with the unstable case.

Table 6
Computation of Long-Term Pollutant Concentrations

X (m) (Assigned)	SZ (m) (Predetermined)	exp(HOD) [Eq. (51)]	CW (10^{-3} mg/m^3) [Eq. (53)]	CWR (10^{-3} ppm) [Eq. (54)]
1,000	82.5	2.0×10^{-4}	0.03	0.01
2,000	149.7	1.0×10^{-1}	2.34	0.82
5,000	329.2	6.0×10^{-1}	3.31	1.16
10,000	597.5	8.6×10^{-1}	1.34	0.47

It is important to point out that in the dispersion equations [Eqs. (35)–(38)], the horizontal standard deviations of a plume (*SY*) and the vertical standard deviations of a plume (*SZ*) are functions of downwind distance (*X*) and meteorological conditions. The stable plot of *SZ* approaches a constant beyond 10,000 m, because it is believed that vertical dispersion almost ceases in such conditions.

3.11.2. Example 2

Problem

Discuss the following issues:

1. The possibility of computer-aided air quality management and air dispersion analyses of various gaseous pollutants.
2. The availability of commercial software for air dispersion analysis.
3. The availability of a special training program.

Solution

1. The computer programs for air quality management and air dispersion analysis are available in the literature (15,16). The hand-calculated results in this chapter agree closely with the computer-calculated results. Although Example 1 in Section 3.11.1 predicts the dispersion of airborne sulfur dioxide from a stack, a slightly modified calculation procedure and computer program can predict the dispersion of other types of gaseous pollutants from a stack. Specifically, the stack emission rate *Q* and the parameter PPM_{SO_2} must be recalculated and replaced, respectively, considering the new physical data of another gaseous pollutant. Yang (17) and ASME (3) provided more background information.
2. Software for computer-aided air dispersion analysis is also commercially available (18,19).
3. Training for air dispersion analysis is available through the university continuing education programs and the training institutes (20,21).

NOMENCLATURE

A	A coefficient; 0.5 (stable), 0.25 (neutral), 0.25 (unstable), 0.25 (very unstable)
A_{c1}	Accumulation of values of those observed 1-h concentrations that exceed the secondary standard
A_{c8}	Accumulation of values of those observed 8-h concentrations that exceed the secondary standard

A_{o1}	Accumulation of those observed 1-h concentrations that exceed the secondary standard
A_{p24}	Accumulation of those observed 24-h concentrations that exceed the secondary standard
A_{s3}	Accumulation of values of those observed 3-h concentrations that exceed the secondary standard
A_{s24}	Accumulation of those observed 24-h concentrations that exceed the secondary standard
B	A coefficient; 0.15 (stable), 0.045 (neutral), 0.045 (unstable) or 0.045 (very unstable)
C	A coefficient; 0.71 (stable), 0.86 (neutral), 0.86 (unstable), or 0.86 (very unstable)
C^2	Isotropic diffusion coefficient, set at 0.010 for neutral conditions, with dimensions m^n
C_{an}	Annual arithmetic mean observed concentration of nitrogen dioxide
C_{c1}	Maximum observed 1-h concentration of carbon monoxide
C_{c8}	Maximum observed 8-h concentration of carbon monoxide
C_{o1}	Maximum observed 1-h concentration of photochemical oxidants
C_{p24}	Maximum observed 24-h concentration of total suspended particulate matter
C_{pa}	Annual geometric mean observed concentration of total suspended particulate matter
C_{ps}	Specific heat of stack gas at constant pressure (cal/g/k)
C_{s24}	Maximum observed 24-h concentration of sulfur dioxide
C_{s3}	Maximum observed 3-h concentration of sulfur dioxide
C_{sa}	Annual arithmetic mean observed concentration of sulfur dioxide
C_{max}	Maximum ground-level concentration (mg/m^3)
C_{peak}	Short-term peak concentration (mg/m^3)
C_{PPM}	Pollutant concentration at downwind distance $X(l)$ (ppm)
CROSS	Downwind distance (m) at which the crosswind concentrations are to be calculated
CW	Long-term pollutant concentration (mg/m^3)
CWR	Long-term pollutant concentration (ppm)
d	Internal diameter of the stack top (m)
D	Stack diameter (m)
E	A coefficient; 0.0 (stable), 0.35 (neutral), 0.52 (unstable), 0.65 (very unstable)
E_c	An extreme value index for carbon monoxide
E_s	An extreme value index for sulfur dioxide
E_p	An extreme value index for total suspended particulates
E_o	An extreme value index for photochemical oxidants
EVI	Extreme Value Index
F	Buoyance flux (m^4/s^3)
FUEL	Pounds of fuel per megawatt-hour
G	Stability parameters (s^{-2})

H	Instrument height, or the altitude at which data were taken (m)
H_e	Effective stack height (m)
HDS	$-(HT)^2/[2(SZ)^2]$
HOD	$-0.5[HT/SZ]^2$
HS	Stack height (m)
HT	Effective stack height (m)
I	Internal variable
I_c	Index of pollution for carbon monoxide
I_n	Index of pollution for nitrogen dioxide
I_o	Index of pollution for photochemical oxidants
I_p	Index of pollution for total suspended particulates
I_s	Index of pollution for sulfur dioxide
J	Direction section angles
JST	Stability parameter; 1 (stable), 2 (neutral), 3 (unstable), 4 (very unstable)
K	Direction section angles
L	Internal variable
LR	Temperature lapse rate ($\Delta T/\Delta Z$)
MAQI	Mitre Air Quality Index
n	Stability parameter, nondimensional, set at 0.25 for neutral stability conditions
N	Angular width of a direction sector (deg)
NAME	"STABLE," "NEUTRAL," "UNSTABLE," "VERY UNSTABLE"
NUM	Number of wind rose data (W) to be read in
Q_a	Allowable emission rate of suspended particulate matter (g/s)
Q_m	Mass emission rate of stack gas (g/s)
ORAQI	Oak Ridge Air Quality Index
p	Ground-level concentration, 0.15×10^{-3} g/m^3 (Note: Pennsylvania state regulation)
POWER	Capacity of plant (MW)
PPM_{max}	Maximum ground-level concentration (mg/m^3)
PPM_{SO_2}	Variable for converting SO$_2$ concentration to ppm from mg/m^3
Q	Emission rate (mg/s)
R	Pollutant concentration (mg/m^3)
RA	Ambient air density at stack top (g/cm^3)
RS	Effluent density (g/cm^3)
S_{c1}	1-h secondary standard value (i.e., 35 ppm or 40,000 µg/m^3) consistent with the unit of measure of C_{c1}
S_{c8}	8-h secondary standard value (i.e., 9 ppm or 10,000 µg/m^3) consistent with the unit of measure of C_{c8}
S_{na}	Annual secondary standard value (i.e., 0.05 ppm or 100 µg/m^3) consistent with the unit of measure of C_{an}
S_{o1}	1-h secondary standard value (i.e., 0.08 ppm or 160 µg/m^3) consistent with the unit of measure of C_{o1}
S_{p24}	24-h secondary standard value (i.e., 150 µg/m^3)
S_{pa}	Annual secondary standard value (i.e., 60 µg/m^3)

S_{sa}	Annual secondary standard value (i.e., 0.02 ppm or 60 μg/m³) consistent with the unit of measure of C_{sa}
S_{s3}	3-h secondary standard value (i.e., 0.5 ppm or 1300 μg/m³) consistent with the unit of measure of C_{s3}
S_{s24}	24-h secondary standard value (i.e., 0.1 ppm or 260 μg/m³) consistent with the unit of measure of C_{s24}
SA	Standard deviation of the wind direction fluctuation in the horizontal direction (deg) at the instrument height
SAH	Standard deviation of the wind direction fluctuation in the horizontal direction (deg) at the stack height
SE	Standard deviation of the wind direction fluctuation in the vertical direction (deg) at the instrument height
SEH	Standard deviation of the wind direction fluctuation in the vertical direction (deg) at the stack height
SCY	(SCZ) (SAH/SHE) (Note: SCY = SY when X = CROSS)
SCZ	B(SHE)(CROSS)C (Note: SCZ = SZ when X = CROSS)
SQ1	Internal variable = Q/(3.14159UH)
SQ2	Internal variable = {Q/[3.14159(UH)(SY)(SZ)}exp[−0.5(HT/SZ)²]
SSY	SSZ (SAH/SHE)
SSZ	HT/2⁰·⁵ (m)
STED	Number of runs to be made
SULFUR	Percentage sulfur in fuel
SY	Standard deviation of plume profile in the crosswind direction (m)
SZ	Standard deviation of plume profile in vertical direction (m) [B(SHE)(X)(C)]
t	Time interval (min)
T	Temperature of ambient atmosphere (K)
ΔT	Stack gas temperature minus ambient air temperature (K)
TA	Ambient air temperature (K)
TAF	Ambient air temperature at stack top (°F)
T_s	Temperature of stack gas at stack top (K)
TS	Effluent temperature (K)
TSF	Effluent temperature (°F)
u	Wind speed (m/s) (assume 3.8 m/s unless other acceptable meteorological data are available for the stack locality)
U	Wind velocity at instrument height (m/s)
UH	Wind speed at stack height (m/s)
V_s	Stack gas ejection velocity (m/s)
VS	Stack effluent velocity (m/s)
VSF	Stack effluent velocity (ft/s)
W	Wind rose data; frequency (%) with which a combination of meteorological condition of interest together with winds in that sector may be found
X	Downwind distance from source (horizontal distance from the stack to the nearest property) (m)
X_{max}	Distance at which C_{max} occurs (m)

Y	Crosswind distance (m)
YYY	Internal variable for determining SO_2 concentration
Z	Vertical distance (m)
ZZ	Internal variable

REFERENCES

1. L. K. Wang, *Environmental Engineering Glossary*, Calspan Corp., New York, 1974.
2. US Environmental Protection Agency, website, http://www.epa.gov/airs/criteria, 2003.
3. ASCE, *Recommended Guide for the Prediction of Airborne Effluents*, 3rd ed. American Society for Mechanical Engineers, New York, 1979.
4. C. V. Weilert, *Environ. Protect.* **13**(3), 54 (2002).
5. B. S. Forcade, *Environ. Protect.* **14**(1), 22–25 (2003).
6. B. Geiselman, *Waste News* 10–11 (2003).
7. L. K. Wang, N. C. Pereira, and Y. T. Hung, (eds.), *Air Pollution Control Engineering*, Humana, Totowa, NJ, 2004.
8. US Congress National Primary and Secondary Ambient Air Quality Standards, *Federal Register* **36**(84) 1971.
9. Mitre Corporation, *National Environmental Indices: Air Quality and Outdoor Recreation*, Mitre Corporation Technical Report MTR-6159, 1972.
10. US Government, *Environmental Quality, The Third Annual Report of the Council on Environmental Quality*, US Government Printing Office, Washington, DC, 1972, pp. 5–44.
11. US Department of Health, *A Compilation of Selected Air Pollution Emission Control Regulations and Ordinances*, US Government Printing Office, Washington, DC, 1968.
12. Editor, *Environ. Protect.* **14**(3), 100–101 (2003).
13. Editor, *Pollut. Eng.* **32**(12), 24–26 (2000).
14. M. H. Wang, L. K. Wang, and W. Y. W. Chan, *Technical Manual for Engineers and Scientists*, Manual No. PB 80-143266, US Department of Commerce, National Technology Information Service, Springfield, VA, 1980.
15. M. H. S. Wang, L. K. Wang, T. Simmons, and J. Bergenthal, *J. Environ. Manag.*, 61–87 (1979).
16. L. K. Wang, M. H. S. Wang, and J. Bergenthal, *J. Environ. Manag.*, 247–270 (1981).
17. M. Yang, in *Handbook of Environmental Engineering* (L. K. Wang, N. C. Pereira, and H. E. Hesketh, eds.), Humana, Totowa, NJ, 1979, Vol. 1, pp. 199–270.
18. US EPA, *Environmental Protection April 2002 Software Guide*, US Environmental Protection Agency, Dallas, TX, 2002; available at www.eponline.com.
19. A. Wiegand, *Environ. Protect.* **13**(10), 22 (2002).
20. Lakes Environmental, *Environ. Protect.* **13**(3), 80 (2002); available at www.lakes-environmental.com.
21. Editor, *Environ. Protect.* **14**(3), 21–29 (2003).
22. C. Wehland and L. Earl, *Environ. Protect.*, **15**(6), 20–23 (2004).

2

Desulfurization and Emissions Control

Lawrence K. Wang, Clint Williford, and Wei-Yin Chen

CONTENTS

1. INTRODUCTION

Desulfurization removes elemental sulfur and its compounds from solids, liquids, and gases. Predominantly, desulfurization involves the removal of sulfur oxides from flue gases, compounds of sulfur in petroleum refining, and pyritic sulfur in coal cleaning. This chapter discusses the following topics:

1. Sulfur pollution (sulfur oxides, hydrogen sulfide, and organic sulfur pollutants).
2. The US Air Quality Act.
3. Solid-phase desulfurization (coal cleaning, gasification, and liquefaction).
4. Liquid-phase desulfurization (acid-lake restoration for H_2SO_4 removal and groundwater decontamination for H_2S removal).
5. Gas-phase desulfurization (SO_x and H_2S removals from air emission streams).

From: *Handbook of Environmental Engineering, Volume 2: Advanced Air and Noise Pollution Control*
Edited by: L. K. Wang, N. C. Pereira and Y.-T. Hung © The Humana Press, Inc., Totowa, NJ

Special emphasis is placed on gas-phase desulfurization, introducing various technologies for the removal of SO_x and H_2S from air emission streams. Of these technologies, the most important is lime/limestone flue gas desulfurization (FGD). This chapter describes FGD process systems, facilities, chemistry, and technology demonstrations, design configurations, gas handling/treatments, reagent/feed preparation, waste handling/disposal, Operation and maintenance (O&M), and process control. The wet and dry scrubbing chapter of this handbook series (chapter 5, volume 1) presents additional technical information on the unit process and unit operation aspects of scrubbing/absorption (73).

1.1. Sulfur Oxides and Hydrogen Sulfide Emissions

Sulfur oxides (SO_x) and hydrogen sulfide are two major sulfur-containing air pollutants. Both cause great environmental concern.

Hydrogen sulfide gases are released from sanitary landfill sites, sanitary sewer systems, wastewater-treatment plants, reverse-osmosis drinking water plants, septic tank systems, and hydrogeothermal plants (1–8). However, H_2S releases are negligible in comparison with SO_x releases. Accordingly, only the quantitative information of SO_x emissions is presented in this section.

More than 25 million metric tons of sulfur oxides (SO_x) are emitted annually in the United States, and about 65 million metric tons are annually emitted by the entire world's industrialized nations. The US SO_x emission alone represents about a quarter of the releases from human activities and natural sources throughout the world.

Major air pollutants are as follows: (a) sulfur oxides 14%, (b) nitrogen oxides 12%, (c) carbon monoxide 53%, (d) hydrocarbons 15%, and (e) suspended particulate matter (PM) 6%. Most sulfur oxides are released in the form of sulfur dioxide, which reacts in the atmosphere to sulfates. These interfere with normal breathing patterns, reduce visibility, and contribute to the formation of acid rain.

Coal combustion for power generation accounts for more than two-thirds of SO_x emissions in the United States. Sulfur is a natural contaminant of coal and is almost completely converted to sulfur oxide when coal is burned. The substitution of oil and natural gas for coal reduces emissions. However, these fuels are more expensive.

In terms of SO_x emission sources, the following is a statistical breakdown:

1. Industrial boilers, 8%.
2. Electric generation stations, 69%.
3. Copper smelters, 8%.
4. Petroleum refining, 5%.
5. Transportation, 5%.
6. Residential, commercial, and institutional, 5%.

1.2. SO_x Emissions Control Technologies

Most SO_2 control systems contact a calcium-based compound with the sulfur dioxide to form calcium sulfite, $CaSO_3$. This is oxidized to $CaSO_4$. The first scrubbers were introduced in Great Britain in the 1920s and 1930s and demonstrated 90% removal of SO_2. In the 1960s, installations followed in Japan and Europe. Under the provisions of the Clean Air Act of 1970, installations were made at new power plants in the United Sates. However, older plants were exempt, until the 1990 Clean Air Act

Amendments (CAAA) required controls for older plants. The major technologies include the following.

1.2.1. Dry and Semidry Sorbent Injection

In this method, particles of limestone or a quickly drying slurry of lime are injected into the economizer or flue gas. This latter is called the semidry method and it dominates sorbent injection applications.

1.2.2. Sulfuric Acid Production

Less commonly performed, the SO_2 is oxidized over a catalyst to SO_3, which is dissolved in water.

1.2.3. Conventional Wet FDG Technology

Flue gas from a particulate collector flows to the SO_2 scrubber, and the flue gas is contacted with a slurry containing particulate limestone to form calcium sulfite. The gas flux is limited to prevent entrainment, and mass transfer determines the absorber height. The calcium sulfite oxidizes to calcium sulfate, which crystallizes to gypsum ($CaSO_4 \cdot 2H_2O$). A dewatering system concentrates the gypsum to 80–90% solids for disposal or fabrication of wallboard. Over its life, a 500-MWe coal-fired plant, using a conventional scrubber, produces enough gypsum sludge to fill a 500-acre pond, 40 ft deep. Early scrubber systems also had poor reliability, requiring installations of spare modules.

1.2.4. Innovative Wet FGD Technology

Innovative scrubbers incorporate better designs and materials. They are characterized by greater compactness, lower capital and operating costs, high reliability (eliminating the need for spares), and elimination of waste disposal problems by producing wallboard-quality gypsum. A number of these have been demonstrated at a commercial scale through the US Department of Energy Clean Coal Technology program. Sections 9–11 of this chapter describe these systems and summarize performance data.

2. SULFUR OXIDES AND HYDROGEN SULFIDE POLLUTION

Although SO_x is a symbol of all oxides of sulfur (e.g., SO_2 and SO_3), about 95% of all sulfur oxides are in the form of sulfur dioxide (SO_2). It is a colorless gas that when cooled and liquefied can be used as a bleach, disinfectant, refrigerant, or preservative. In the atmosphere, however, SO_2 is a precursor of highly destructive sulfates (SO_4^{2-}), formed by the chemical addition of oxygen (O_2). SO_3 is not a stable compound and may react with water (H_2O) to form sulfuric acid (H_2SO_4), a component of acid rain (9,10).

Hydrogen sulfide and organic sulfur-containing compounds cause odor pollution at low concentrations, but cause extreme public concern when their concentrations are high. SO_x in the atmosphere has been recognized as a major air pollution problem in the United States since the inception of clean air legislation. SO_x emissions cause acid rain, affect public health, corrode materials, and restrict visibility.

2.1. Acid Rain

Acid rain is composed primarily of two acids: sulfuric (H_2SO_4) and nitric (HNO_3). Sulfuric acid, resulting from sulfur oxide emissions, comprises from 40% to 60% of the acidity, depending on regional emission patterns. Acid rain is a major problem throughout the world, especially in Scandinavia, Canada, and the eastern United States.

Rain in the northeastern United States averages 10–100 times the acidity of normal rainwater. More than 90 lakes in the Adirondack Mountains of New York State no longer contain fish because the increased acidity of lake water has caused toxic metals in the lakebeds and surrounding soils to be released into the lakes. Similar effects are beginning to occur in other areas of the United States such as northern Minnesota and Wisconsin. Preliminary studies indicate that the direct effects of acids on foliage and the indirect effects resulting from the leaching of minerals from the soil can reduce the yield from some agricultural crops (11).

2.2. Public Health Effects

As the concentration of sulfur oxides in the air increases, breathing becomes more difficult, resulting in a choking effect known as pulmonary flow resistance. The degree of breathing difficulty is directly related to the amount of sulfur compounds in the air. The young, the elderly, and individuals with chronic lung or heart disease are most susceptible to the adverse effects of sulfur oxides. Sulfates and sulfur-containing acids are more toxic than sulfur dioxide gas. They interfere with normal functioning of the mucous membrane in respiratory passages, increasing susceptibility to infection. The toxicity of these compounds varies according to the nature of the metals and other chemicals that combine with sulfur oxides in the atmosphere.

H_2S and organic sulfur compounds only cause offensive odor at low concentrations (5–8). At high concentration levels, both H_2S and organic sulfur compounds are toxic. Fortunately, their offensive odors may serve as a warning sign for people to move to safety.

2.3. Materials Deterioration

Sulfur acids corrode normally durable materials, such as metals, limestone, marble, mortar, and roofing slate. As a result, acidic sulfates are destroying statuary and other archeological treasures that have resisted deterioration for thousands of years. These include such well-known structures as the Parthenon in Greece and the Taj Mahal in India, as well as lesser known bronze and stone statuary in US cities.

Corrosive destruction of statuary is most severe in areas where droplets of moisture collect sulfates and other atmospheric particles, forming a crust on the statuary that retains moisture and promotes the formation of sulfuric acid. This acid destroys the surface of the statuary, causing smooth metal sculptures to become pitted and resulting in such severe spalling of stone figures that the outlines of the features become blurred.

2.4. Visibility Restriction

Small particles suspended in a humid atmosphere are the major cause of reduced visibility in the eastern United States. Sulfates constitute 30–50% of the suspended particles.

3. US AIR QUALITY ACT AND SO$_x$ EMISSION CONTROL PLAN

The Air Quality Act of 1967 required that states develop ambient air quality standards for SO_2. The Clean Air Act (CAA) of 1970 mandated performance standards for new and significantly modified sources of SO_2. In 1971, the US Environmental Protection

Agency (EPA) issued the first such standards for fossil-fuel-fired boilers greater than 25 MWe. These source performance standards (NSPS) limited allowable emissions to 1.2 lb of SO_2 per million British thermal units (Btu) of heat input to a boiler, and essentially restricted operators of these boilers to two choices: use low-sulfur coal or apply FGD technology (12,13).

In accordance with the 1977 Clean Air Act Amendments, the EPA established regulations that require electric power companies and industries to take steps the reduce SO_x emissions. In 1979, the NSPS were revised for power plants, requiring a percentage reduction of SO_2. This mandate was intended to be technology forcing, essentially requiring all new power plants to add SO_2-removal equipment to the base design (13).

In the 1980s, the US Congress began debating the need for additional SO_2 control as a means of reducing damage from acid rain, culminating in the Clean Air Act Amendments (CAAA) of 1990. Two portions of the CAAA of 1990 are important for SO_2 emissions control. These are Title I and Title IV. Title I establishes the National Ambient Air Quality Standards (NAAQS) for six criteria on pollutants, including SO_2. National Ambient Air Quality standards for sulfur oxides establish a maximum safe level of the pollutant in the atmosphere. According to these standards, atmospheric concentrations of SO_x should not exceed 0.5 part per million (ppm) during a 3-h period, or 0.14 ppm during a 24-h period. The annual mean concentration should not exceed 0.03 ppm.

Title IV, sometimes called the Acid Rain Program, sets requirements for reducing SO_2 emissions in three distinct phases:

1. Phase I targeted specific large sources to reduce SO_2 emissions by 5 million tons by January 1, 1995, using a limit of 2.5 lb/10^6 Btu.
2. Phase II required, by January 1, 2000, reduction of all power plants to a nationwide emission level of 1.2 lb SO_2/10^6 Btu and a sliding scale percentage reduction of 70–90%, depending on input sulfur content. The SO_2 emission levels are generally 0.3 and 0.6 lb/10^6 Btu for low- and high-sulfur coals, respectively.
3. Phase III required that SO_2 emissions be capped beyond the year 2000.

Title IV was the first large-scale approach to regulating emissions by using marketable allowances. These can be bought and sold in units of 1 ton of SO_2 emitted (14). Until the late 1990s, prices for allowances had remained low, but then increased. This has provided the potential for a large-scale retrofit. To meet limits, especially with high-sulfur coals, requires high removal efficiencies of 90% or more. Higher efficiencies will generate salable credits. In addition, Title IV has an option whereby unregulated sources can reduce SO_2 emissions and receive credits. Regardless of credits held, Title I sets the limits for compliance. Through certification provisions, a Title IV Permit serves as the primary verification and documentation of a facility's compliance with all applicable requirements of the Clean Air Act.

As a result of Phase I, SO_2 emissions declined by 20% from 1990 to 1997. Advanced scrubbers essentially halved the cost of conventional scrubbers (prior to the CAAA). The Clean Coal Technology (CCT) Program has supported the development of a number of options for meeting the requirements: advanced scrubbers, low-cost absorbent injection, clean fuels, and advanced power generation systems (14). Under Title IV, utilities meeting limits by repowering with advanced technologies were allowed a 4-yr extension to December 31, 2003.

Table 1
Representative Sulfur Content of Coal: Speciation and Totals

Coal	Total wt%	Pyritic wt%	Sulfatic wt%	Organic wt%
Illinois	4.490	1.23	0.060	3.210
Kentucky	6.615	5.05	0.135	1.415
Martinka	2.200	1.48	0.120	0.600
Westland	2.600	1.05	0.070	1.480
Texas lignite	1.200	0.40	—	0.800

4. DESULFURIZATION THROUGH COAL CLEANING

Coal contains pyritic and organic sulfur, as well as some sulfatic (sulfate) forms. Pyritic sulfur is a mineral form, whereas the organic sulfur is chemically bound in the structure of the coal. Most mineral sulfur can be removed by mechanical coal cleaning processes, but removing organic sulfur requires chemical processing (1–4,10,15–22). Illustrative values appear in Table 1 (23).

4.1. Conventional Coal Cleaning Technologies

Conventional coal cleaning processes are physical and mechanical processes. Coal is crushed to < 50mm in diameter and screened into coarse, intermediate, and fine particle size fractions. Crushing to a smaller size liberates ash-forming minerals and nonorganically bound sulfur (e.g., pyrites, FeS_2). The mineral matter has a higher density than organic-rich coal particles and can be separated from the coarse and intermediate particles of coal by jigs, dense-medium baths, cyclone systems, and concentrating tables (Table 2) (24).

From 40% to 90% of the total sulfur content in coal can be removed by this physical cleaning process. Physical cleaning cannot remove organically bound sulfur, which requires chemical or biological methods. Cleaning effectiveness depends on the size of

Table 2
Conventional Coal Cleaning Technologies

Technology type	Process
Crushing	Grinders pulverize coal, which is then screened into coarse (>50 mm in diameter), intermediate, and fine (<0.5 mm) particles. The crushing liberates the inorganic, bound mineral particles from the coal. Because these mineral particles are denser than the organically rich coal, they can be separated from the coal by further processing (see next items).
Jigs (G)	For coarse to intermediate particles.
Dense-medium baths (G)	For coarse to intermediate particles.
Cyclones (G)	For coarse to intermediate particles.
Froth flotation (G)	For fines; relies on the different surface properties of ash (hydrophilic) versus coal (hydrophobic); high potential, but current technologies do not handle the small particles efficiently.

Note: G = gravity(density)-based separation.

Table 3
Advantages and Disadvantages of Physical Coal Cleaning

Advantages	Disadvantages
10–40% lower SO_2	Coal grinding is energy intensive.
Higher pulverizer and boiler availability (estimated: 1% improvement in availability for every 1% decrease in ash content)	Water-based coal cleaning methods add moisture to the coal, which reduces boiler and power plant efficiency.
Lower maintenance costs (less wear and tear ing. on coal preparation equipment and boiler)	A 2–15% energy loss occurs during clean-
Less boiler slagging and fouling	
Lower dust loading of ESP/bag filter	
Lower transportation costs (applicable to cleaning at the mine only)	

pyritic sulfur particles and the proportion of sulfur in pyritic form. Thus, the larger the percentage of organically bound sulfur in the coal, the lower the percentage of sulfur that can be removed by physical methods.

Nearly half of the sulfur in coals from eastern Kentucky, Tennessee, Georgia, and Alabama and most of the sulfur in coal from the western mountain states is in the pyritic form and is relatively easy to remove by mechanical cleaning. The combination of physical coal cleaning and partial FGD enables many generating stations to meet SO_x emission standards at less expense than using FGD alone. The advantages and disadvantages of coal cleaning are summarized in Table 3.

The cost of physical cleaning varies from US$1 to US$10/ton, depending on the coal quality, the cleaning process used, and the degree of cleaning desired. In most cases, cleaning costs range from US$1 to US$5/ton (25).

4.2. Advanced Coal Cleaning Technologies

A number of advanced cleaning technologies have been developed to improve performance and economics and to exploit currently underutilized coal fines (as discussed below) (26,27). These technologies, including advanced physical, aqueous, and organic phase pretreatment and selective agglomeration, are at an earlier stage of commercialization and development with less well-established effectiveness and economics. Advanced coal cleaning technologies are listed in Table 4 (24).

The potential for cleaning the coal fines is greater than that for the coarse and intermediate coal particles, but conventional technologies do not handle the fines efficiently. Fines <0.5 mm in diameter can be separated by froth flotation exploiting surface differences between coal and ash. Coal's surface is hydrophobic. Ash's surface is hydrophilic. One of the more widely commercialized flotation methods is Microcel Flotation Technology. As of the mid-1990s, there were 44 Microcel installations around the world. The process diagram appears in Fig. 1. Ash removal can reach 60%; total sulfur removal is 10–40%, increasing in tandem with a rising percentage of pyritic (mineral) sulfur in the coal. Weight recovery (the percentage of coal retained) is 60–90%, and thermal recovery (percentage of heating value retained) is 85–98% (25). The Peak Downs plant near Queensland, Australia (Fig. 2) represents the largest installation of

Table 4
Advanced Coal Cleaning Technologies

Technology type	Process
Advanced physical cleaning	Advanced froth flotation (S)
	Electrostatic (S)
	Heavy liquid cycloning (G)
Aqueous phase pretreatment	Bioprocessing
	Hydrothermal
	Ion exchange
Selective agglomeration	Otisca
	LICADO
	Spherical Agglomeration Aglofloat
Organic phase pretreatment	Depolymerization
	Alkylation
	Solvent swelling
	Catalyst addition (e.g., carbonyl)
	Organic sulfur removal

Note: G = gravity(density)-based separation; S = surface-effect-based separation.

Microcel coal columns in the world. Sixteen 3-m-diameter Microcel columns replaced a traditional split-feed flotation circuit. The ash content of the froth product was reduced from about 9.5% to 6%. This allowed the operating gravities in the coarse circuit to be raised, increasing total plant yield by 4% (28).

Advanced coal cleaning has found particular application for coal fines because they are already ground fine and are often an unused byproduct (waste). Thus, advanced coal cleaning provides coal and energy recovery and waste minimization. However, the treatment of fine coal is the least efficient and most costly step in coal preparation. Based on data in the literature (25), for a typical preparation plant in the eastern United

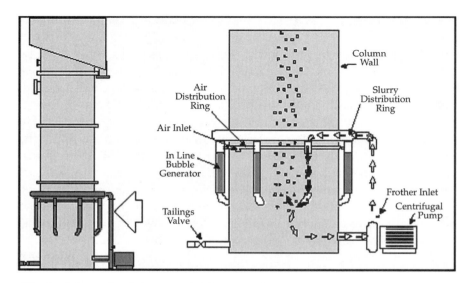

Fig. 1. Schematic of the essential features of the Microcel Flotation Technology.

Fig. 2. Microcel installation at the Peak Downs preparation plant, Queensland, Australia.

States, as particle size decreases, the ash content of the clean coal increases from 7.5% to 10.8%. Significantly, the moisture content of the fine fraction is nearly five times that of the coarse fraction (i.e., 25.1% vs 5.1%). This higher moisture content requires drying to meet increasingly stringent moisture constraints imposed by utility contracts. Field surveys at Virginia Tech suggest that the average cost to treat fine coal is three to four times higher than that for coarse coal. Consequently, the US coal producers currently discard between 27 and 36 million metric tons of fresh fine coal to refuse ponds each year. The US Department of Energy indicates that approx 1.8 billion metric tons of fine coal have been discarded in abandoned ponds and 450–725 million tons are in active ponds.

Low-ash products can be recovered from fine-coal streams using technologies such as horizontal belt filters. However, the coal has high water content, on the order of 36–42% for flotation froth (25). This restricts the use of this coal and requires further drying. Reagents have been developed that significantly improve the dewatering. Results indicate addition of 1.5 kg/ton of a novel dewatering aid can reduce filtering time by a factor of 4–5 and reduce water content by 40–60% or more, depending on time. Economic estimates show that dewatering (at US$1–5/ton) can improve profitability (25).

The economics of conventional and advanced coal cleaning for refuse pond recovery have been compared (29). Conventional circuits could not produce satisfactory ash reduction or, if they did, could not maintain profitability. The advanced fine-coal cleaning circuit, however, did achieve satisfactory ash reduction and demonstrated return on investment of 50% for both water-only and dense-medium systems, without tailings management. The dense-medium system appeared better suited to producing quality products. Tailings management was found to enhance economics. The Electric Power Research Institute has also compared economics for strategies of applying advanced coal cleaning (ACC) and FGD. Cost estimates suggested that combining ACC with FGD was more expensive than FGD alone. However, the estimates also suggested better economics for combining ACC and dry FGD versus ACC and wet FGD. Cleaned coal also offers other operational advantages not reflected in the estimates for desulfurization alone (27).

Several advanced processes are now available for improving the performance of fine coal cleaning and dewatering circuits. Any economic evaluation of these technologies must, however, consider the important interactions between the coarse- and fine-coal circuits on overall plant performance. A reduction in the ash and/or moisture content of the fine-coal product often allows higher cutpoints to be employed in the coarser-coal circuits without diminishing the quality of the overall plant product. This trade-off generally results in a substantial increase in total plant production. In many cases, the improved profitability can be used to justify improvements to the fine-coal circuit, although the apparent benefit to the fines circuit alone may be relatively small (25).

4.3. Innovative Hydrothermal Desulfurization for Coal Cleaning

Hydrothermal treatment removes organic sulfur (as well as inorganic), reduces the water-holding capacity, and removes other hazardous pollutants, such as mercury. It is a cleaning and upgrading treatment, particularly suited for coals with high organic sulfur or mercury, or for low-rank coals with high ash and water content. In general, it involves contacting the coal with hot water and, in some cases, at high pressure. Catalyst or inorganic reactants may be added.

An example of this technology uses supercritical water and a catalyst. The Energy and Environmental Research Center (EERC) at the University of North Dakota, Center for Air Toxic Metals (CATM) is investigating the hydrothermal pretreatment of coal to remove heavy metals. Pyritic and organic sulfur are also removed. An advantage is that it uses water to remove the sulfur and no costly or hazardous chemicals are needed for this process, such as would be needed for caustic leaching.

Raw materials include coal, water, and a cobalt/molybdenum catalyst. Estimates indicate the production of 1 million tons of product would require 1.3 million tons of raw coal and 240 gal water/ton coal. The cobalt/molybdenum catalyst has a service life of approx 6 yr.

In the first stage, crushed coal is slurried with water and pumped to system pressure, eliminating the need for a lock hopper. Supercritical water (>3200 psi and >374°C) is then forced into the bottom of the vessel and through the coal. The water removes the tars, oils, and impurities (including mercury and sulfur) from the coal. The overhead stream flows into a pair of cyclones, where the ash is removed. From the cyclones, the stream passes through a catalyst bed for further desulfurization of the tars and oils carried in the supercritical water.

The supercritical stream flows to a flash tank where the pressure is reduced. Hydrogen sulfide and some water vaporize. In an industrial system, the sour gas and sulfur-contaminated water are pumped to conventional sulfur-recovery systems. The liquid in the flash tank gravity separates into an organic layer and a water layer. The tars and oils are recombined with the cleaned product and pressed into briquettes for the finished product. The tars and oils increase the energy density of the product and serve as a binder. Feed coal and product composition are shown in Table 5.

Processing costs are estimated at US$0.57/$10^6$ Btu, or US$16.50/ton of clean coal. The processing cost would be partially offset by SO_2 allowances (US$12/ton coal at

Table 5
Comparison of Feed and Hydrothermally Treated Composition

	Feed coal	Treated coal
Moisture (wt%)	13.8	0.1
Ash (wt%)	10.3	3.7
Volatile matter (wt%)	34.6	29.5
Fixed carbon, (wt%)	41.3	66.8
HHV[a] (Btu/lb)	10,778	14,475
Total sulfur (wt%)	2.9	0.8
Hg (ppb)	X	Y

[a]Higher heating value.

US$200/ton SO_2 allowance), efficiency savings from moisture and ash reduction (US$2.74/ton coal). The reduced ash content should produce other savings from reduced maintenance. A rate-of-return (ROR) calculation indicated a 15% ROR could be achieved at a selling price as low as US$43/ton (US$1.48/10^6 Btu), making this product competitive for both utility and commercial applications.

The experimental work and plant design proposed have been completed. The results of the effort indicate that hydrothermal treatment of high-sulfur coals is technically and economically viable. The EERC is currently seeking funding opportunities to bring this technology through the pilot scale to demonstration (30).

5. DESULFURIZATION THROUGH VEHICULAR FUEL CLEANING

Gasoline, diesel fuel, and jet fuel all contain sulfur that is emitted in the form of sulfur oxides after combustion. Although motor vehicle emissions currently account for only about 3% of the total national sulfur oxide emissions, the EPA is concerned about them for two reasons.

First, the catalytic converter being installed in cars to control hydrocarbon and carbon monoxide emissions can convert exhaust sulfur dioxide to the more toxic compound sulfuric acid. These acid fumes could adversely affect the health of people driving in heavy traffic.

Second, diesel fuel and gasoline demand continue to rise, with projections for further increases of 1 and 2 million barrels per day, respectively, from 2002 to 2020 (DOE, 2001, EIA Annual Energy Review). As of 2002, diesel fuel is being produced with 350 ppm sulfur. Likewise, fluid catalytic cracker naphtha is a major component in blended gasoline (35% by a 1999 estimate) (31).

Major changes in vehicle fleet composition are anticipated over the next two decades: Most significantly, there will be more diesel-powered vehicles and fuel-cell-powered vehicles with onboard fuel reformers. Even with more highly fuel efficient vehicles, the US Energy Information Administration predicts that fuel demand will increase by about 50% by the year 2020 (32).

United States fuel regulations call for further reductions in sulfur content. By 2004, gasoline is to attain an average of 30 ppm, with a maximum of 80 ppm. Diesel will be required to meet the "80/20" rule, with production of 80% ultralow-sulfur diesel (ULSD) with 15 ppm maximum 7–10 ppm average) and 20% 500-ppm highway diesel between June 2006 and June 2010, and a 100% requirement for ULSD after June 2010 (33).

Sulfur can be removed from diesel fuel, gasoline, and jet fuels during the refining process. When the sulfur in petroleum is exposed to hydrogen in the presence of a catalyst, hydrogen sulfide gas is formed. This compound can be commercially marketed.

Meeting the sulfur standard for gasoline only requires technologies already developed: low-pressure hydrodesulfurization (HDS) and catalytic adsorption. Capital costs have been estimated at US$8 billion, and an incremental cost estimated at US$0.045 per gallon. Adsorption technologies have been, and continue to be, developed that may improve on these cost estimates by avoiding hydrogen use costs. These adsorbents (transitional metals in some cases) preferentially remove the organic sulfur-containing compounds without removing other aromatic molecules. Sulfur-containing molecules such as benzothiophene can be altered to remove the sulfur and allow the remaining aromatic ring to continue through the process S Zorb (34).

Meeting the diesel standard will likely require all of the hydrocracked stock and the entire straight-run material to go to high-pressure, high-temperature, two-stage distillate desulfurization units. This uses proven technology, but will probably require two or more hydrogenation units in series to achieve the desired sulfur levels with heavier crude oils. Cracked stock and coker distillate will have to be diverted to other markets.

Other options for processing diesel include sulfur adsorption on zeolite and selective partial oxidation. These offer significant advantages through a lower hydrogen requirement. However, they are under pilot- and laboratory-scale development. Based on the more conventional technologies, capital costs have been estimated at US$8 billion, and an incremental cost estimated at US$0.07 to US$0.15 per gallon. The implementation time line produces scheduling conflicts that will complicate meeting the goals. One third of US diesel fuel is from cracked stock—more difficult to process to low-sulfur fuel (32).

Although not discussed here at length, sulfur content in jet fuel is also a concern. The US Air Force has proposed halving the maximum sulfur content from 3000 to 1500 ppmw. This becomes more problematic because the Air Force data suggest a trend with time of rising sulfur content in the fuels used (35).

6. DESULFURIZATION THROUGH COAL LIQUEFACTION, GASIFICATION, AND PYROLYSIS

Alternatives to combustion use thermal conversion, gasification, liquefaction, and pyrolysis of coal to produce gas, liquid, and solid fuels respectively. These fuels have much reduced sulfur content, allowing combustion with little on no emissions controls. These fuels may also be used as feedstock for other chemical processes, e.g., synthesis gas to methanol (14). In this section, we describe coal gasification, liquefaction, and pyrolytic conversion.

6.1. Coal Gasification

A very elementary process of coal gasification was designed in the late 1700s to fuel the gas lights that illuminated cities. Since that time, approx 70 different coal gasification processes have been used commercially or are currently under development.

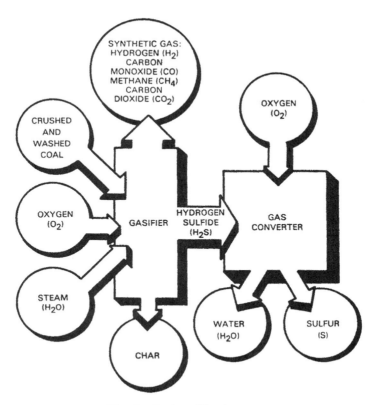

Fig. 3. Coal gasification.

Three basic steps are common to all coal gasification processes: coal pretreatment, gasification, and gas cleaning. Coal pretreatment includes various stages of coal washing and pulverization. Gasification produces either a low- or high-heat content gas by applying heat and pressure or using a catalyst to break down the components of coal. Coal is gasified in an atmosphere of limited oxygen. Generally, oxidation of the coal provides a gas containing carbon monoxide (CO), hydrogen (H_2), carbon dioxide (CO_2), water (H_2O), methane (CH_4), and contaminants such as hydrogen sulfide (H_2S) and char (*see* Fig. 3).

This "synthesis gas" is composed primarily of carbon monoxide and hydrogen. Variations in the process may increase the quantity of methane formed, producing a gas that releases more heat when it is burned.

The sulfur in coal is converted primarily to hydrogen sulfide (H_2S) during the gasification process. It exits from the gasifier with the methane and synthesis gas and is subsequently removed during the gas cleaning process. After removal, the hydrogen sulfide is then converted to elemental sulfur (S) through partial oxidation and catalytic conversion.

Four systems for gasification and combined cycle production of electricity have been demonstrated through the US DOE's Clean Coal Technology Program. These demonstration projects were performed from 1994 to 2001 and demonstrated a variety of gasifier types, cleanup systems, and applications:

- PSI Energy's Wabash River Coal Gasification Repowering Project began in 1995 and ran to 2000. It employed a two-stage entrained flow gasifier. Carbonyl sulfide was catalytically converted to hydrogen sulfide. This was removed using Methyldiethanolamine (MDEA)-based absorption/stripper columns. A Claus unit produced salable sulfur. SO_2 capture was greater than 99%, with emissions consistently below 0.1 lb/10^6 Btu (14).
- The Tampa Electric Integrated Gasification Combined-Cycle Project used a Texaco gasifier. Operations began in 1996. A COS hydrolysis reactor converted one of the sulfur species to a more easily removed form. The further cooled syngas then entered a conventional amine-based sulfur-removal system. SO_2 emissions were kept below 0.15 lb/10^6 Btu (97% reduction). In 2000, there were 10 domestic and international projects planned or under construction using the Texaco gasifier technology (14).
- The Sierra Pacific Power Company Pinon Pine IGCC Power Process was operated as a demonstration from 1998 to 2001. The gasifier used dry injection of limestone with the coal, with calcium sulfate being removed with the coal ash in the form of agglomerated particles suitable for landfilling. Final traces of sulfur were removed with a metal oxide absorbent (14).
- The fourth demonstration project by Kentucky Pioneer Energy was constructed in 2003. Operations will involve injection of limestone into the combustor. Conventional gas cleanup will be used to remove hydrogen sulfide emissions. SO_2 emissions are expected to be less than 0.1 lb/10^6 Btu (99% reduction) (14).

6.2. Coal Liquefaction

There are two basic approaches in converting coal to oil. One involves using a gasifier to convert coal to carbon monoxide, hydrogen, and methane, followed by a condensation process that converts the gases to oils. The second approach (*see* Fig. 4) involves using a solvent or slurry to liquefy pulverized coal and then processing this liquid into a fuel similar to heavy oil. The first approach, known as the Fischer–Tropsch process, was developed by Fischer and Tropsch in Europe in the 1930s. Tests and demonstrations of processes for producing synthetic oil from coal were initiated in the United States in the early 1960s, and a major plant was built in South Africa using the Fischer–Tropsch process.

Most recent research has involved the latter approach. Solvents and slurries used in these processes are usually produced from the coal and recycled in the system. Recently developed liquefaction processes have combined the use of solvents and distillation techniques to produce hydrocarbon gas and various hydrocarbon liquids. The Advanced Concepts for Direct Liquefaction Program was begun in 1991 by the DOE. The advanced two-stage liquefaction technology developed at Wilsonville, Alabama brought the estimated costs of liquefaction to US$33/bbl (1990 cost), with an ultimate goal of US$25 (36). Reference 36 describes research focused on improving process economics—coal cleaning, distillate hydrotreating, and dispersed catalyst development, among others. International research has continued in China, Germany, and Japan, where, in 1996, a 150 tonne/d plant was built (37).

These processes involve solvents, and slurries commonly remove sulfur from the liquefied coal by using hydrogen (H_2) to convert the sulfur to hydrogen sulfide gas. As in the gasification processes, this hydrogen sulfide is then partially oxidized to form elemental sulfur and water. More than 85% of sulfur in coal is removed during the liquefaction process. EPA research efforts currently focus on determining the sulfur content in synthetic

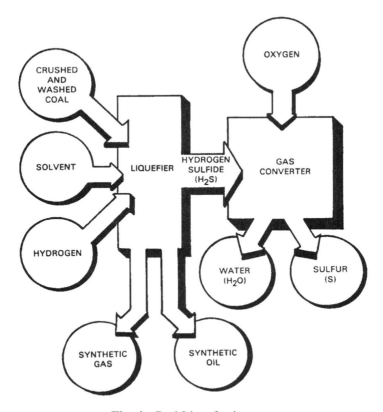

Fig. 4. Coal Liquefaction.

oils produced by different liquefaction processes and on identifying ways to improve their sulfur-removal efficiencies. EPA is also initiating programs to develop improved systems for preventing the escape of hydrogen sulfide (H_2S) and sulfur dioxide (SO_2) from the gas converter into the atmosphere.

6.3. Pyrolysis

A third approach involves thermal treatment of coal to produce a high-quality, low-sulfur fuel. Two technologies investigated under the Clean Coal Technology program, and described here, are referred to as ENCOAL and SynCoal.

The ENCOAL Mild Gasification Project was performed by the ENCOAL Corporation (a subsidiary of Bluegrass Coal Development Company) near Gillette, Wyoming using SGI International's "Liquids-From-Coal" (LFC) process. Work was performed from 1992 to 1997. The process consists of a drying step, followed by pyrolysis at 1000°F. The solid process-derived fuel (PDF) is then cooled, rehydrated, contacted with oxygen to reduce the potential for spontaneous combustion, and mixed with a dust suppressant. The process gases are cooled to condense coal-derived liquid (CDL). The PDF had a sulfur content of 0.36% versus 0.45% for the feed coal. The CDL had a sulfur content of 0.6% versus 0.8% for No. 6 fuel oil.

The SynCoal process involves the upgrading of low-rank coal to a high-quality, low-sulfur, solid fuel. The process was demonstrated under the Clean Coal Technology

Program by Western SynCoal LLC. The project was performed at Colstrip, Montana from 1992 to 2001. High-moisture, low-rankcoal is fed to a vibratory fluidized-bed dryer, where it is heated by a combustion gas. Water is driven off, and the coal is transferred to a second vibratory reactor where it is heated to nearly 600°F, driving off chemically bound water, carboxylic groups, and volatile sulfur compounds. A small amount of tar is released that seals the coal. The coal shrinks, fractures, and releases ash-forming minerals. Deep-bed stratifiers using air pressure and vibration are used to separate mineral matter, including pyrite.

SynCoal has been used by electric utilities and industries, primarily by cement and lime plants. Reduction in sulfur emissions has been demonstrated for electrical generation (14).

7. DESULFURIZATION THROUGH COAL-LIMESTONE COMBUSTION

7.1. Fluidized-Bed Combustion

Because sulfur oxides are emitted from the stacks of electrical-generating stations and industries, SO_x control efforts initially focused on flue gases. An alternative is to combine the sulfur dioxide absorbent and coal directly at the point of combustion. This is accomplished by the injection of limestone or mixing of fuel and limestone before their injection into the boiler. Fluidized-bed combustors provide the contact time needed for fuel–limestone interaction and offer other advantages. The following paragraphs describe atmospheric and pressurized fluidized-bed combustors, as well as coal–limestone pelletization.

In fluidized-bed combustion (FBC), a grid supporting a bed of crushed limestone or dolomite is set in the firebox (*see* Fig. 5). Air forced upward through the grid creates turbulence, causing the bed of limestone or dolomite to become suspended and move in a fluidlike motion. Natural gas is injected into the firebox, ignited, and then followed by pulverized coal. Once the coal has started to burn well, the natural gas is shut off and the fire is maintained by burning coal. Sulfur oxidized during combustion reacts with the limestone or dolomite in the firebox, forming calcium sulfate. Calcium sulfate and residual limestone or dolomite from fluidized-bed combustion can be disposed of in landfills or used in construction materials. Fluidized-bed combustion eliminates the need for FGD because the bed of limestone or dolomite can remove more than 90% of the sulfur oxides created during combustion.

Basic FBC concepts have been proven, and the US DOE and its industrial partners are now working on implementation and system design, scale-up, reliability, and control issues. Goals include achieving efficiency of 45% for high-efficiency, domestic, greenfield applications after the year 2005 (5% more than a modern pulverized coal unit with flue gas scrubbing). The target cost is US$750/kW for new FBC systems (38).

FBC in boilers can be particularly useful for high-ash coals and/or those with variable characteristics. A variation on FBC is the pressurized fluidized-bed combustor (PFBC). PFBCs have undergone significant development during the 1990s, and demonstration units have been built in Germany, Spain, and the United States.

Advantages of PFBCs include compact units, high heat transfer, potential usefulness for low-grade coals and for those coals with variable characteristics. As for atmospheric FBC, bubbling and circulating beds may be used. All commercial-scale operating units

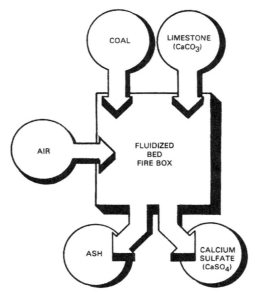

Fig. 5. Fluidized-bed combustion.

use bubbling beds, and PFBC normally refers to pressurized bubbling bed units. A pressurized circulating FBC was also constructed for demonstration.

In the PFBC, the combustor and hot gas cyclones are all enclosed in a pressure vessel. Coal and sorbent are fed and ash removal is performed across the pressure boundary. Hard coal and limestone can be crushed together and then fed as a paste. Units operate at pressures of 1–1.5 MPa, with combustion temperatures of 800–900°C. NO_x formation is less than in pulverized coal combustors (PCCs). SO_2 emissions are lowered by the injection of sorbent (limestone or dolomite) and its subsequent removal with the ash.

The residues consist of the original mineral matter, most of which does not melt at the combustion temperatures used. Where sorbent is added for SO_2 removal, there will be additional CaO/MgO, $CaSO_4$, and $CaCO_3$ present. There may be a high free-lime content, and leachates will be strongly alkaline. Carbon-in-ash levels are higher in FBC residues that in those from PCCs (39).

7.2. Lime–Coal Pellets

Burning pellets composed of a limestone and coal mixture is another way of eliminating the need for FGD. The US EPA's Office of Research and Devlopment (ORD) research has shown that the combustion of these pellets in conventional stoker boilers not only reduces sulfur oxide emissions but also enhances boiler performance. The pellets are made by pulverizing coal and limestone and adding a binder material to form small cylinders. As the pellet burns, the calcium in the limestone absorbs the SO_2 generated from burning the coal, resulting in the formation of calcium sulfate ($CaSO_4$).

The ability of the pellet to control sulfur emissions depends on the ratio of limestone to coal, pellet size, binder material, and types of coal and limestone used. For example, ORD has developed a binder material that enables as much as 87% of the SO_2 to be absorbed by the limestone when a pellet composed of two-thirds coal and one-third limestone is used.

The expense of preparing fuel with pellets would add more than US$15 per ton to the cost of coal, which is substantially less than the cost of installing and operating wet scrubber systems for industrial boilers. In the future, fuel pellets will be developed for a greater range of coal and boiler types. This research could enable users of high-sulfur coal from eastern US mines to meet SO_2 pollution control requirements.

8. HYDROGEN SULFIDE REDUCTION BY EMERGING TECHNOLOGIES

Hydrogen sulfide emissions derive from numerous sources including petroleum hydrodesulfurization, anaerobic wastewater treatment, and landfills. Major approaches to control include absorption into amine solution, catalytic oxidation to elemental sulfur, and biological oxidation. Absorbed hydrogen sulfide can be stripped from solution with steam, and sent to a Claus plant for partial oxidation to sulfur.

8.1. Innovative Wet Scrubbing Using a Nontoxic Chelated Iron Catalyst

In an innovative wet scrubber—in this case, for H_2S reduction—the process involves mass transfer from the gas to liquid phases. The offending specie, H_2S (or some other malodorous gas), is present in an emission stream or gas phase. The liquid phase is the scrubbing solution, distributed as a flowing bulk liquid into the scrubber. The scrubber is controlled by dispersing the gas phase (i.e., air emission stream with target pollutant, H_2S) as small gas bubbles into the passing liquid phase (i.e., scrubbing solution with scrubbing chemicals). The flow pattern can be either counterflow or cross-flow. The process equipment of an innovative wet scrubber resembles that for the aeration basin of an activated sludge system and is described in detail in Chapter 5, volume 1, on wet and dry scrubbing.

The innovative wet scrubber can achieve very high efficiencies (99+%) and has very high turndown capabilities. The liquid redox system is considered by some to be the best available control technology for geothermal power plants. The process employs a nontoxic, chelated iron catalyst, which accelerates the oxidation reaction between H_2S and oxygen to form elemental sulfur (1).

The air emission stream is contacted with the aqueous, chelated iron solution, where the H_2S is absorbed and ionized into sulfide and hydrogen ions as follows:

$$H_2S \text{ (vapor)} + H_2O \rightarrow 2H^+ + S^{2-} \tag{1}$$

$$S^{2-} + 2Fe^{3+} \rightarrow S \text{ (elemental sulfur)} + 2Fe^{2+} \tag{2}$$

$$0.5\ O_2 \text{ (vapor)} + H_2O + 2Fe^{2+} \rightarrow 2Fe^{3+} + 2OH^- \tag{3}$$

$$H_2S + 0.5\ O_2 \rightarrow S \text{ (elemental sulfur)} + H_2O \tag{4}$$

where S^{2-} is the sulfide ion, Fe^{3+} is the trivalent iron ion, S is elemental sulfur, Fe^{2+} is the divalent iron ion, O_2 is oxygen vapor, H_2O is water, and OH^- is the hydroxide ion.

The final chemical reaction presented in Eq. (4) is the summary of the three chemical reactions preceding it. The nontoxic, chelated iron catalyst allows the hydrogen sulfide to be oxidized to elemental sulfur, for recovery and reuse. The readers are referred Chapter 5, volume 1 (73).

8.2. Conventional Wet Scrubbing Using Alkaline and Oxidative Scrubbing Solution

In the conventional wet scrubber, mass transfer of contaminant occurs from the gas to the liquid phase. The process is controlled by dispersing the liquid phase (i.e., scrubbing solution containing scrubbing chemicals) as liquid droplets or thin films into a passing gas phase (i.e., air emission stream with target pollutant, H_2S). The flow pattern can be either counterflow or cross-flow. The conventional wet scrubber has shown a hydrogen sulfide removal efficiency of 99.9% from a contaminated airstream at various flow rates, superficial gas velocities, liquid flux rates, tower diameters, and HTU values.

The following reactions are for a single-stage scrubbing system using 0.1% caustic and 0.3% sodium hypochlorite to control hydrogen sulfide emissions:

$$H_2S + 2NaOH \rightarrow Na_2S + 2H_2O \tag{5}$$

$$NaOCl + H_2O \rightarrow HOCl + NaOH \tag{6}$$

$$4HOCl + Na_2S \rightarrow Na_2SO_4 + 4HCl \tag{7}$$

$$HCl + NaOH \rightarrow NaCl + H_2O \tag{8}$$

where H_2S is hydrogen sulfide, NaOH is caustic soda, or sodium hydroxide, Na_2S is sodium sulfide, NaOCl is sodium hypochlorite, Na_2SO_4 is sodium sulfate, HCl is hydrochloric acid, and NaCl is sodium chloride.

8.3. Scavenger Adsorption

Geothermal power plants are environmentally attractive because they employ a renewable energy source. Geothermal steam, however, contains varying amounts of noncondensible gases (NCG), such as carbon dioxide and hydrogen sulfide, which cause serious environmental, health, and safety problems (1,2). If the removal rate of hydrogen sulfide from the NCG is less than approx 140 kg/pd, it is generally economical to employ an H_2S scavenger such as Sulfur-Rite™ manufactured by US Filter/Gas Technology Products. This system is a fixed-bed process consisting of an iron-based solid material, which reacts with H_2S to form innocuous iron pyrite:

$$H_2S + \text{Sulfur-Rite} + \text{Iron} \rightarrow H_2O + FeS_2 \tag{9}$$

where H_2S is hydrogen sulfide, H_2O is water, FeS_2 is iron pyrite, and Sulfur-Rite is a H_2S scavenger.

The process is a relatively simple, batch-type system consisting of a carbon steel vessel(s), which hold the iron-based media. The NCG pass through the vessel(s) until all of the iron has been converted to pyrite. The vessel is then shut down, emptied, and refilled with fresh media. A "Lead-Lag" arrangement can be employed, which permits continuous treatment of the NCG even during changeouts. In this processing scheme, the NCG flows through two Sulfur-Rite vessels in series. When the outlet H_2S concentration in the first vessel is the same as the inlet concentration, the vessel is shutin and the medium is replaced. During the changeout, the NCG is processed through the second vessel only. When the change out is complete, the flow direction is reversed.

The medium is nonregenerable, resulting in a relatively high operating cost of approx US$12.00 per kilogram of H_2S removed. However, the system is simple and noncorrosive, which results in a relatively low capital investment. The major difficulty involved in using solid-based scavengers in Europe is disposal of the spent material. In other parts of the world, the spent material is simply landfilled in nonhazardous facilities. However, in Europe, landfilling is discouraged and expensive. Consequently, means of using the material as raw material for bricks and so forth are being investigated.

The solid-type scavengers can be replaced with liquid scavengers, which are generally triazine-based. The advantage of liquid-based scavengers is that the spent material can be injected down-hole for disposal. The disadvantage of liquid scavengers is that they are very expensive, having a relative cost of approx US$33.00 per kilogram of H_2S removed (1).

8.4. Selective Oxidation of Hydrogen Sulfide in Gasifier Synthesis Gas

In this process, oxygen is directly injected into the synthesis gas, where a selective catalytic oxidation converts the hydrogen sulfide to elemental sulfur. The process has the advantage of converting and removing the sulfur in one stage. In the case of the design for the Tampa Electric Company integrated gasification combined cycle plant, the process allowed elimination of sour gas coolers, the amine absorption unit, Claus plant, and tail gas treatment and incinerator (40,41).

8.5. Biological Oxidation of Hydrogen Sulfide

Biological oxidation has been used for odor control in hydrogen sulfide–containing airstreams. Bacteria convert the H_2S to sulfate, water, and carbon dioxide. The airstream is first humidified and warmed as needed. It then passes though a packed-bed biofilter where the H_2S is absorbed into a liquid film and oxidized there by bacteria. Collected water is removed to a sanitary drain. Hydrogen sulfide removal of 99% or greater can be achieved with inlet concentrations of up to 1000 ppm. Industrial-scale and smaller package units are in wide application (42).

9. "WET" FLUE GAS DESULFURIZATION USING LIME AND LIMESTONE

Flue gas desulfurization is the most commonly used method of removing sulfur oxides resulting from the combustion of fossil fuels. FGD processes result in SO_x removal by inducing exhaust gases to react with a chemical absorbent as they move through a long vertical or horizontal chamber, known as a wet scrubber (43–54).

A typical, no-frills FGD system is shown in Fig. 6. The efforts of research engineers to bring wet FGD to commercial acceptance resulted in the following innovations researched and developed at various demonstration facilities: (1) use of high liquid-to-gas ratios (enhanced scrubber internal recirculation) to prevent scaling, (2) use of forced oxidation to avoid scaling and improve disposal/salability of solids, (3) use of thiosulfate-forming additives to inhibit scaling, and (4) use of organic acid buffers to increase SO_2 removal and improve sorbent utilization (55).

Many different FGD processes have been developed, but only a few have received widespread use. Of the systems currently in operation, 90% use lime or limestone as the chemical absorbent.

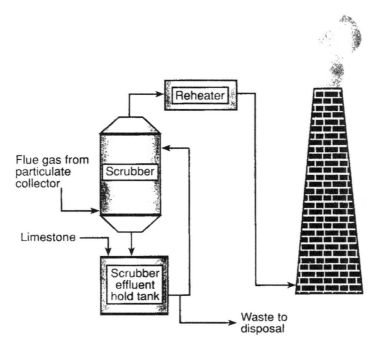

Fig. 6. Basic lime/limestone FGD process flow diagram.

9.1. FGD Process Description

The basic lime/limestone FGD process appears schematically in Fig. 6. Some systems produce a salable byproduct (i.e., gypsum for wallboard construction), but most use the throwaway process configuration. As shown in Fig. 7, flue gas, from which fly ash has been removed in a particulate collection device such as an electrostatic precipitator (ESP) or a fabric filter, is brought into contact with the lime/limestone slurry in the absorber, where SO_2 is removed. The chemical reaction of lime/limestone with SO_2 from the flue gas produces waste solids, which must be removed continuously from the slurry loop. These waste solids are concentrated in a thickener and then dewatered in a vacuum filter to produce a filter "cake" that is mixed with fly ash. The resulting stabilized mixture is then transported to a landfill. This lime/limestone FGD system is called a "throwaway" process because it produces a waste byproduct for disposal rather than for processing to recover salable gypsum.

The principal chemical reactions for the lime/limestone FGD process are presented below according to SO_2 absorption, limestone dissolution, and lime dissolution.

9.2. FGD Process Chemistry

9.2.1. Sulfur Dioxide Absorption

Chemical reactions for SO_2 absorption in a scrubber/absorber are as follows.

$$SO_2(g) \rightarrow SO_2(aq) \tag{10}$$

$$SO_2(aq) + H_2O \rightarrow H_2SO_3(aq) \tag{11}$$

$$H_2SO_3(aq) \rightarrow HSO_3^-(aq) + H^+(aq) \tag{12}$$

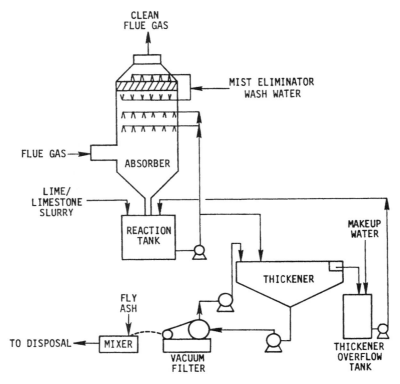

Fig. 7. Complete lime/limestone FGD process flow diagram.

$$HSO_3^- + (aq) \rightarrow SO_3^{2-}(aq) + H^+(aq) \tag{13}$$

$$SO_3^{2-}(aq) + \frac{1}{2}O_2(aq) \rightarrow SO_4^{2-}(aq) \tag{14}$$

$$HSO_3^-(aq) + \frac{1}{2}O_2(aq) \rightarrow SO_4^{2-}(aq) + H^+(aq) \tag{15}$$

where g is the gas phase, aq is the aqueous phase, HSO_3^- is the bisulfite ion, SO_3^{2-} is the sulfite ion, SO_4^{2-} is the sulfate ion, O_2 is oxygen, H^+ is the hydrogen ion, and SO_2 is sulfur dioxide.

9.2.2. Lime Dissolution and Lime FGD Chemical Reaction

Chemical reactions for lime dissolution in a scrubber/absorber (*see* Fig. 8) are as follows:

$$CaO(s) + H_2O \rightarrow Ca(OH)_2 \ (aq) \tag{16}$$

$$Ca(OH)_2(aq) \rightarrow Ca^{2+}(aq) + 2OH^-(aq) \tag{17}$$

$$OH^- (aq) + H^+ (aq) \rightarrow H_2O \tag{18}$$

$$SO_3^{2-}(aq) + H^+(aq) \rightarrow H_2SO_3^-(aq) \tag{19}$$

$$Ca^{2+}(aq) + SO_3^{2-}(aq) + 0.5H_2O \rightarrow CaSO_3 \bullet 0.5H_2O(s) \tag{20}$$

$$Ca^{2+}(aq) + SO_4^{2-}(aq) + 2H_2O \rightarrow CaSO_4 \bullet 2H_2O(s) \tag{21}$$

where S is the Solid phase.

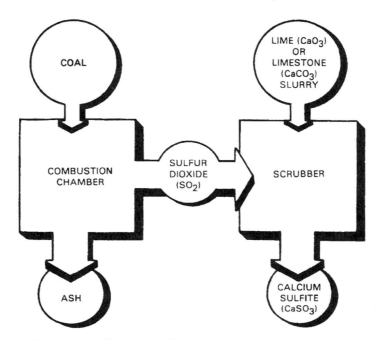

Fig. 8. Lime/limestone FGD process and coal combustion.

In the FGD system (*see* Figs. 6 and 7), the sulfur dioxide reacts with lime to form calcium sulfite and water, in accordance with the following overall chemical reaction:

$$SO_x + CaO + H_2O \rightarrow CaSO_3 + H_2O \tag{22}$$

where SO_x is the sulfur oxides (SO_2 or SO_3), CaO is lime (calcium oxide), H_2O is water, $CaSO_3$ is calcium sulfite, $Ca(OH)_2$ is calcium hydroxide, Ca^{2+} is the calcium ion, OH^- is the hydroxide ion, H^+ is the hydrogen ion, SO_3^{2-} is the sulfite ion, HSO_3^- is the bisulfite ion, $CaSO_4$ is calcium sulfate, and SO_4^{2-} is the sulfate ion. Calcium sulfite is the final product from the scrubber.

9.2.3. Limestone Dissolution and Limestone FGD Chemical Reactions

Chemical reactions for limestone dissolution in a scrubber/absorber (*see* Fig. 8) are as follows:

$$CaCO_3(s) \rightarrow CaCO_3 \,(aq) \tag{23}$$

$$CaCO_3 \,(aq) \rightarrow Ca^{2+} \,(aq) + CO_3^{2-} \,(aq) \tag{24}$$

$$CO_3^{2-} \,(aq) + H^+ \,(aq) \rightarrow HCO_3^- \,(aq) \tag{25}$$

$$SO_3^{2-} \,(aq) + H^+ \,(aq) \rightarrow HSO_3^- \,(aq) \tag{26}$$

$$Ca^{2+} \,(aq) + SO_3^{2-} \,(aq) + 0.5H_2O \rightarrow CaSO_3 \cdot 0.5H_2O(s) \tag{27}$$

$$Ca^{2+} \,(aq) + SO_4^{2-} \,(aq) + 2H_2O \rightarrow CaSO_4 \cdot 2H_2O(s) \tag{28}$$

The use of limestone in a FGD process system (*see* Figs. 6 and 7) results in a similar chemical reaction, but also yields carbon dioxide:

$$SO_2 + CaCO_3 + H_2O \rightarrow CaSO_3 + H_2O + CO_2 \tag{29}$$

where SO_2 is sulfur dioxide, $CaCO_3$ is calcium carbonate (limestone), and CO_2 is carbon dioxide.

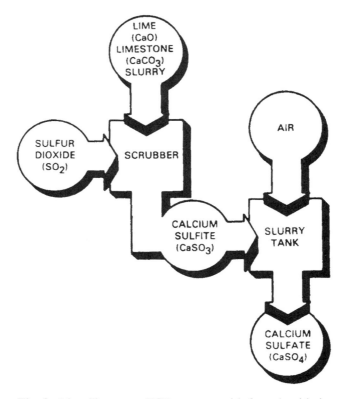

Fig. 9. Lime/limestone FGD process with forced oxidation.

Figure 8 shows that the final product from the limestone FGD chemical reactions is calcium sulfite from the scrubber/absorber.

9.2.4. Forced Oxidation Chemistry

As shown in Fig. 9, calcium sulfite is formed during the scrubbing/absorbing process. The substance presents a serious operational problem because it settles and filters poorly and can be removed from the scrubber/absorber slurry only in a semiliquid, or pastelike, form that must be stored in lined ponds. A solution to this problem involves forced oxidation in which air is blown into the tank that holds the used scrubber slurry, which is composed primarily of calcium sulfite and water. Dissolved oxygen then oxidizes the calcium sulfite to calcium sulfate (Fig. 9).

The following is the process chemistry of forced oxidation:

$$CaSO_3 + H_2O + \frac{1}{2}O_2 \rightarrow CaSO_4 + H_2O \tag{30}$$

where $CaSO_3$ is calcium sulfite and $CaSO_4$ is calcium sulfate.

9.3. FGD Process Design and Operation Considerations

Although not shown in the process diagrams (Figs. 6 and 7), the major equipment design difference between the lime and limestone processes is reagent feed preparation.

In the lime process, the reagent is slaked. In the limestone process, limestone is ground in a ball mill.

The basic operational factors one should be acquainted with when inspecting lime/limestone slurry FGD systems are discussed in the following Subsections. Knowing these factors and how they are interrelated with the process chemistry of each system will provide an understanding of how each process functions, in addition to providing a set of guidelines to be used during an inspection.

9.3.1. Stoichiometric Ratio

The stoichiometric ratio (SR) is defined as the ratio of the actual amount of SO_2 reagent, calcium oxide (CaO), or calcium carbonate ($CaCO_3$) in the lime or limestone fed to the absorber, to the theoretical amount required to neutralize the SO_2 and other acidic species absorbed from the flue gas. Theoretically, 1 mol of CaO or $CaCO_3$ is required per mole of SO_2 removed (SR = 1.0). In practice, however, it is usually necessary to feed more than the stoichiometric amount of reagent in order to attain the degree of SO_2 removal required. This is because of mass transfer limitations that prevent complete reaction of the absorbent.

If a high SO_2 removal efficiency is required, the absorber may not be able to achieve such removal unless extra alkalinity is provided by feeding excess reagent. The amount of excess reagent required depends on the SO_2 concentration in the inlet gas, gas flow, percentage SO_2 removal required, and absorber design. For lime reagent, the SR employed in commercial FGD systems is 1.05 for newer designs; it is up to 1.2 for older designs. For limestone reagent, a SR of 1.1 is used in newer designs, but can be as high as 1.4 in older designs (43,44).

If the reagent feed is too much in excess, the results are wasted reagent and increased sludge volume. Excessive overloading can also result in scaling in the form of $CaCO_3$ in the upper part of the absorber for lime systems, and calcium sulfite ($CaSO_3 \cdot \frac{1}{2}H_2O$), sometimes referred to as soft scale, in the lower part of the absorber for limestone systems. Excess reagent can also be carried up into the mist eliminator by entrainment, where it can accumulate, react with SO_2, and form a hard calcium sulfate ($CaSO_4 \cdot 2H_2O$) scale (by sulfite oxidation). This is particularly a problem with limestone systems. Calcium sulfate (or gypsum) scale is especially undesirable because it is very difficult to remove. Once formed, the scale provides a site for continued precipitation. Calcium sulfite scale can generally be easily removed by reducing the operating slurry pH or rinsing manually with water.

Scale formation is usually more prominent in limestone systems than lime systems, particularly for high-sulfur coal applications. Lime systems have a greater sensitivity to pH control because lime is a more reactive reagent. The change in pH across lime systems is more pronounced than in limestone systems partly because limestone dissolves more slowly.

9.3.2. Liquid/Gas Ratio

The ratio of slurry flow in the absorber to the quenched flue gas flow, usually expressed in units of gal/1000 ft^3 is termed the liquid-to-gas (L/G) ratio. Normal L/G values are typically 30–50 gal/1000 ft^3 for lime systems (44) and 60–100 gal/1000 ft^3 for limestone systems (49). Lime systems require lower L/G ratios because of the higher reactivity of lime. A high L/G ratio is an effective way to achieve high SO_2 removal. This also tends to reduce the potential for scaling, because the spent slurry from the absorber is more dilute with respect to absorbed SO_2. Increasing the L/G ratio can also increase system capital and operating costs, because of greater capacity requirements of the reaction tank and associated hold tanks, dewatering equipment, greater pumping requirements, slurry preparation and storage requirements, and reagent and utility necessities.

9.3.3. Slurry pH

Commercial experience has shown that fresh slurry pH as it enters the absorber should be in the range 8.0–8.5 for lime systems and 5.5–6.0 for limestone systems (43,44,48). In both FGD processes, as the SO_2 is absorbed from the flue gas, the slurry becomes more acidic and the pH drops. The pH of the spent slurry as it leaves the absorber is in the range 6.0–6.5 for lime systems and 4.0–5.0 for limestone systems. In the reaction tank of the absorber, the acidic species react with the reagent, and the pH returns to its original fresh slurry value. Slurry pH is controlled by adjusting the feed stoichiometry. Operation of lime/limestone FGD systems at low pH levels, approaching 4.5, will improve reagent utilization but will also lower SO_2 removal efficiency and also increase the danger of hard scale (gypsum) formation because of increased oxidation at lower pH levels. Operation of lime/limestone FGD systems at high pH levels, above 8.5 and 6.0, respectively, will tend to improve removal efficiency but will also increase the danger of soft scale (calcium sulfite) formation. Hence, control of slurry pH is essential to reliable operation. The inability to maintain sensitive control of the slurry pH can lead to both lowered SO_2 removal efficiencies and hard/soft scale formation.

9.3.4. Relative Saturation

In lime/limestone FGD processes, the term "relative saturation" (RS) pertains to the degree of saturation (or approach to the solubility limit) of calcium sulfite and sulfate in the slurry. RS is important as an indicator of scaling potential, especially of hard scale, which can present severe maintenance problems. Relative saturation is defined as the ratio of the product of calcium and sulfate ion activities (measured in terms of concentrations) to the solubility product constant. The solution is subsaturated when RS in less than 1.0, saturated when RS equals 1.0, and supersaturated when RS is greater than 1.0. Generally, lime/limestone processes will operate in a scale-free mode when the RS of calcium sulfate is maintained below a level of 1.4 and the RS of calcium sulfite is maintained below a level of approx 6.0. Operation below these levels provides a margin of safety to ensure scale-free operation. This is achieved through proper design and control of process variables (e.g., L/G, pH).

9.3.5. Overall FGD System Parameters

Important overall FGD system parameters include reagent type, water loop, solids dewatering, absorber parameters, reheat, reagent preparation, and fan location. A brief summary for each FGD system consideration is provided.

9.3.5.1. REAGENT TYPE

The gas handling and treatment subsystem, ductwork, and stack show a strong relation between lime systems and unreliability. This is probably because lime FGD systems are predominantly used for higher-sulfur coal applications. Limestone shows a high correlation with unreliability in the slurry circuit (limestone slurry is more abrasive than lime slurry).

9.3.5.2. WATER LOOP

There are two variations: open and closed water-loop FGD systems. There are some expectations that closed water loops, higher in chloride, will be less reliable. However, open water-loop systems appear less reliable. One explanation for this observation is that virtually all of the early generation commercial lime/limestone FGD systems were originally designed for closed water-loop (no discharge) operation. Because of a variety of problems (e.g., buildup of dissolved salts), the water loop was eventually opened up as one of the first measures to relieve these problems. (In other words, the water-loop variable is an "effect" rather than a "cause.")

9.3.5.3. SOLIDS DEWATERING

Results confirmed the expectation that FGD systems without dewatering were more reliable than systems with dewatering. They have less equipment to cause downtime and lower concentrations of dissolved salts that build up in the liquor loop.

9.3.5.4. ABSORBER PARAMETERS

Results indicated that towers with internals (packed, tray) have a high correlation with unreliability. The type of absorber exhibiting the highest unreliability is the packed tower. Spray tower absorbers exhibited the highest reliability. However, mist eliminators showed a high correlation of unreliability with spray tower absorbers. This is to be expected when considering the open structure of a spray tower, the high L/G ratio, and the upward flow of the gas without impediment or a change in direction. Absorbers with internals have been associated with a high degree of unreliability and are generally excluded from new designs. Another consideration in absorbers is the use of "prescrubbers." Prescrubbers include upstream scrubbers, presaturators, and quench towers. A number of systems are equipped with one of these devices to remove particulates, effect initial SO_2 absorption, and/or condition the gas stream prior to the absorber. Systems without prescrubbers appear to be more reliable than systems with prescrubbers. This is an expected result because systems with prescrubbers have an additional subsystem that may fail. However, the presence of a prescrubber shows a high correlation with reliability for SO_2 absorbers in contrast to their effect on the total system. A possible explanation is that the combination of flue gas quenching and chloride, particulate, and initial SO_2 removal that occurs in a prescrubber serves to protect the SO_2 absorber from failures.

9.3.5.5. REHEAT

The order of decreasing reliability for type of reheat is no reheat, bypass reheat, inline reheat, and indirect reheat. Figure 6 shows the position of a reheat unit; Fig. 10 shows the FGD system reheat schematic diagrams. Reheaters are described at length in Section 9.6.1.4.

9.3.5.6. REAGENT PREPARATION

Reagent preparation in a ball mill (limestone) is associated with considerably higher costs for slurry circuit equipment (e.g., pipes, valves) than is reagent preparation in a slaker (lime).

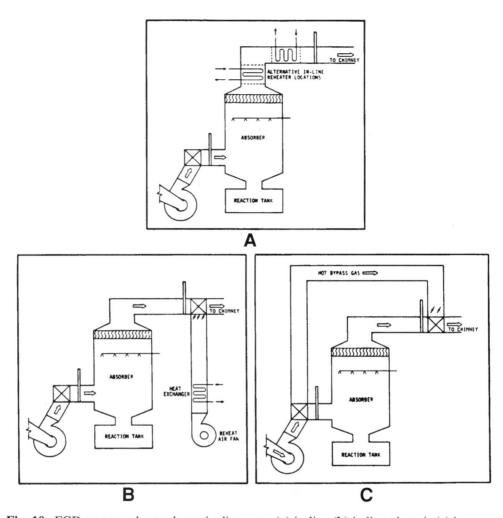

Fig. 10. FGD system reheat schematic diagrams: **(a)** in-line **(b)** indirect hot air **(c)** bypass.

9.3.5.7. FAN LOCATION

Fan unreliability was affected by fan location between the scrubber and the absorber. This location means that the fan operates completely wet, and more downtime is expected. There was little difference between downtime for fans located either upstream (operating on hot, particulate-cleaned gas) or downstream (operating on reheated gas) from the FGD system.

9.4. FGD Process Modifications and Additives

9.4.1. Forced Oxidation Modifications

The most important chemical consideration in lime/limestone processes is the oxidation of sulfite to sulfate. (*see* Fig. 9) Uncontrolled oxidation across the absorber leads to sulfate formation and resultant hard scaling problems on the absorber internals. Sulfite oxidation can occur either naturally or can be artificially promoted (i.e., forced oxidation). Natural oxidation occurs when sulfite in the slurry reacts with dissolved oxygen (O_2), which has been absorbed either from the flue gas or from the atmosphere (e.g., during agitation in the reaction tank). With forced oxidation, air is bubbled into the absorber reaction tank to further promote oxidation. This prevents the dissolved sulfite in the slurry from returning to the absorber, which minimizes the potential for the oxidation of the sulfite to sulfate in the absorber and resultant hard scaling problems. Forced oxidation has additional advantages of reducing the total volume of waste generated because of improved dewatering characteristics of the sulfate solids and improved characteristics of the final solid-waste product. Oxidation tends to increase with decreasing slurry pH. For this reason, forced oxidation is normally employed only with limestone systems.

The process chemistry of forced oxidation has been presented above. It should be noted that the calcium sulfate formed by this reaction grows to a larger crystal size than does calcium sulfite. As a result, the calcium sulfate can easily be filtered to a much drier and more stable material that can be disposed of as landfill. In some areas, the material may be useful for cement or wallboard manufacture or as a fertilizer additive.

Another problem associated with limestone scrubbing is the clogging of equipment by calcium sulfate scale. Forced oxidation can help control scale by removing calcium sulfite from the slurry and by providing an abundance of pure gypsum (calcium sulfate) to rapidly dissipate the supersaturation normally present. The scrubber operation also requires less freshwater, which is scarce in many western locations.

9.4.2. Chemical Additives

9.4.2.1. ADIPIC ACID ADDITIVE

The recent discovery that the addition of adipic solid to FGD limestone can increase the level of SO_2 removal from 85% to 95–97% represented a major breakthrough in SO_2 removal technology. Adipic acid, a crystalline powder derived from petroleum, is available in large quantities.

EPA experiments have shown that when limestone slurry reacts with SO_2 in the scrubber, the slurry becomes very acidic. This acidity limits SO_2 absorption. Dicarboxylic acids, in the form of adipic acid or dibasic acids, have been used commercially. Dibasic acids enhance SO_2 removal in a special manner. Acting as buffers, they tend to neutralize acid-generated hydrogen ions (H^+), which, in turn, prevents the decrease of the system pH and SO_2 removal.

Adding adipic acid to the slurry slightly increases the slurry's initial acidity, but prevents it from becoming highly acidic during the absorption of SO_2. The net result is an improvement in scrubbing efficiency. Adipic acid can reduce total limestone consumption by as much as 15%. Furthermore, the additive is nontoxic (it is used as a food additive) and does not degrade calcium sulfite sludge ($CaSO_3$) and gypsum ($CaSO_4$), the FGD wastes. In addition, high liquid-phase calcium concentrations permitted by the dibasic acids leads to a reduced potential for scaling tendencies in the absorber (47).

9.4.2.2. MAGNESIUM OXIDE ADDITIVE

In recent years, inorganic additives have been used to improve SO_2 removal efficiency, increase reagent utilization, decrease solid-waste volume, and decrease scaling potential of lime/limestone FGD systems. In lime/limestone FGD systems, inorganic additives enhance utilization by improving dissolution. This allows a lower stoichiometric ratio, which reduces limestone addition and the resulting volume of solid waste.

Magnesium oxide additives permit a higher SO_2 removal rate per unit volume of slurry. This is because the salts formed by the reaction of magnesium-based additives with the acid species in the slurry liquor are more soluble with respect to those of the calcium-based salts. This, in turn, increases the available alkalinity of the scrubbing liquor, which promotes a higher SO_2 removal rate.

9.4.3. Limestone Utilization

Adding adipic acid is one way to increase limestone utilization in the scrubber system. Researchers are studying other factors that affect SO_2 absorption and limestone utilization, including the limestone's particle size, impurities, and geological structure.

Limestone used in a scrubber system is crushed into small particles to allow more calcium carbonate ($CaCO_3$) molecules on the surface of the particles to react with the sulfur dioxide (SO_2) gas. ORD scientists are testing two sizes of limestone particles: a coarse grind, similar to that of sugar or salt, and a fine grind, similar in consistency to flour. Various types of limestone, crushed to the same particle size, are currently being compared for their effectiveness in removing sulfur oxides from exhaust gases.

These tests have shown that different limestones of equal particle size vary in their absorption effectiveness. Impurities in the limestone account for part of this difference. Recent experiments have shown that the presence of magnesium carbonate, the main impurity in limestone, inhibits calcium carbonate from reacting with the sulfur dioxide.

The presence of such impurities, however, cannot fully account for variations in the efficiencies of various limestones. Researchers are investigating such geological factors as crystal size and pore size to determine why some kinds of limestone work better than others. These data can then be used to improve the utilization of all limestones employed in FGD systems.

9.5. Technologies for Smelters

9.5.1. Water as a Scrubbing Solution

In a copper scrubber, if an air emission stream contains a high enough concentration of SO_x, water alone can be used, at least at an initial stage as a scrubbing solution. The final product will be sulfuric acid, which can be reused or sold. Specifically, copper ore contains large amounts of sulfur that are converted to sulfur oxides when the ore is processed. About 2 tons of sulfur dioxide (SO_2) is generated for each ton of copper produced.

Smelters produce two streams of gases containing sulfur oxides: a strong stream containing a 4% or greater concentration of SO_x, and a weak stream normally containing less than 2% SO_x. The strong stream is usually treated by a chemical process that converts SO_2 to sulfuric acid (H_2SO_4). In this process, SO_2 is cleaned and converted to SO_3, which reacts with water, producing H_2SO_4. In 1980, 13 of the 16 copper smelters in the United States operated sulfuric acid plants (9). The sulfuric acid can be used in ore processing operations or sold to other industries.

Most of the SO_2 emissions from copper smelters come from reverberatory furnaces, which burn gas, oil, or coal. When copper is heated, sulfur is released and mixes with gases from the burning fuel and with large quantities of air and is converted to SO_2. The concentration of SO_2 ranges from 0.5% to 3.5%, but rarely exceeds 2.5%. This SO_2 concentration is lower than the 4% or more required to process SO_2 into sulfuric acid (H_2SO_4), so the furnace exhaust gases are vented to the atmosphere. As of the early 1990s, none of the reverberatory furnaces operating in this country were equipped with controls for SO_2 emissions (45,46).

9.5.2. Wet Scrubbing Using Citrate Solution

The Industrial Environmental Laboratory in Cincinnati, Ohio investigated a citrate process for copper smelters, which concentrates SO_2 gas from the smelter furnace to allow the production of sulfuric acid.

In the citrate process, sulfur dioxide is dissolved in water and thus removed from the exhaust system:

$$SO_2 + H_2O \rightarrow HSO_3^- + H^+ \qquad (31)$$

where HSO_3^- is the bisulfite ion, H^+ is the hydrogen ion, H_2O is water, and SO_2 is sulfur dioxide.

Adding citrate to the water increases the amount of SO_2 that the water will absorb because the citrate ion (CIT) chemically bonds with the hydrogen ions (H^+). Sulfur can then be removed from the citrate solution in the form of an SO_2 stream strong enough to be used in the acid plant and converted to marketable sulfuric acid.

The citrate process was demonstrated in a copper smelter in Sweden and in a zinc smelter in Pennsylvania. The demonstrations have shown the citrate process to have at least 90% removal efficiency for SO_2 from an air emission stream.

9.5.3. Wet Scrubbing Using Magnesium Oxide Slurry

In the second SO_x control process, magnesium oxide is mixed with water to form a slurry. Washing the smelter gas with this slurry causes the SO_2 in the gases to combine with the magnesium and form magnesium sulfite.

The magnesium sulfite is collected, dried, and heated to temperatures of from 670°C to 1000°C (1250°F to 1800°F). The heat causes the magnesium sulfite molecules to break apart, regenerating magnesium oxide that can be reused and a highly concentrated SO_2 gas that can be converted to sulfuric acid.

The magnesium oxide process was tested for its effectiveness in removing SO_2 from the exhaust gases of industrial boilers and electric generating plants (9,14). The magnesium oxide process was also demonstrated in a smelter in Japan. The process was shown to be at least 90% effective in removing SO_2 from exhaust systems. Adapting them to the US smelting industry would be a major step in reducing national sulfur oxide emissions (52).

9.6. FGD Process Design Configurations

This Subsection briefly describes important equipment items one is likely to encounter when inspecting a conventional lime/limestone FGD system. Descriptions and diagrams are provided for each of the equipment items discussed. Operation and maintenance considerations for the equipment described here are presented later.

The equipment is organized by three major equipment areas: (1) gas handling and treatment, (2) reagent preparation and feed, and (3) waste solids handling and disposal.

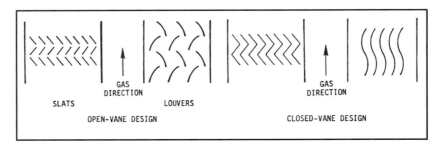

Fig. 11. Baffle-type impingement mist eliminators.

The gas handling and treatment facilities include the following:

1. Fans
2. Scrubbers/absorbers (*see* Figs. 6–10)
3. Mist eliminators (*see* Fig. 11)
4. Reheaters (*see* Figs. 6 and 10)
5. Ductwork and dampers
6. Stack

The reagent preparation and feed facilities include the following:

1. Reagent conveyors and storage (*see* Figs. 12 and 13)
2. Ball mills (*see* Fig. 14)
3. Slakers (*see* Fig. 15)
4. Tanks

The waste solids handling and disposal facilities include the following:

1. Thickeners
2. Vacuum filters
3. Centrifuges
4. Waste processing
5. Waste disposal
6. Pumps and valves

9.6.1. Gas Handling and Treatment Facilities

9.6.1.1. Fans

Fans move gas by creating a pressure differential by mechanical means. Fans are used to draw or push flue gas from the boiler furnace through the FGD system.

Fans used in FGD systems are either centrifugal or axial. Most fans used in FGD systems are of the centrifugal variety. Both fan designs may be equipped with variable-pitch vanes (or blades), which provide more efficient fan operation and better gas flow control.

9.6.1.2. Scrubbers/Absorbers

Strictly speaking, the term "scrubber" (*see* Fig. 6) applies to first-generation systems that remove both particulate and SO_2. "Absorber" (*see* Figs. 7 and 10) applies to the second- and third- generation systems (*see* Table 6) that remove SO_2 only, although the term "scrubber" is also used by some for this application. The basic scrubber/absorber types are described in Chapter 5, Volume 1 (73). There are various gas/slurry contacting devices used in the FGD systems.

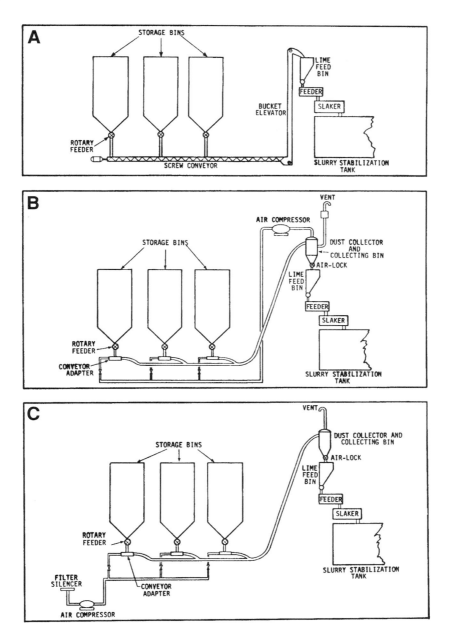

Fig. 12. Three types of conveying equipment used to transport lime: (**a**) mechanical conveyor (**b**) closed-loop conveyor (**c**) positive-pressure pneumatic conveyor.

9.6.1.3. MIST ELIMINATORS

A mist eliminator (*see* Fig. 11) removes entrained material introduced into the gas stream by the scrubbing slurry. These materials include liquid droplets, slurry solids, and/or condensed mist.

There are two basic types of mist eliminator used in FGD systems: the precollector and the primary collector. A precollector precedes the primary collector and is designed

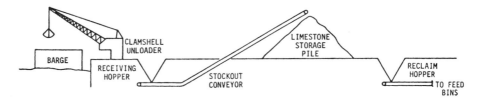

Fig. 13. Barge-based limestone handling and storage system.

to remove the larger entrained particles from the gas stream before it passes through the primary collector. A primary collector typically sees the heaviest duty with respect to entrainment loading and required removal efficiency.

Precollectors are of the bulk separation or knockout type. Bulk separation is effected by baffle slats, perforated trays, or a gas direction change (90° to 180°). Bulk separation devices are characterized by a low potential for solids deposition, a low gas-side pressure drop, and simplicity. Knockout-type precollectors are either the wash tray or trap-out tray design. Knockout devices remove large solid and liquid particles, they also provide a means to recycle the mist eliminator wash water. By recirculating the relatively clean wash water, the flow rate of the wash water to the mist eliminator can be significantly

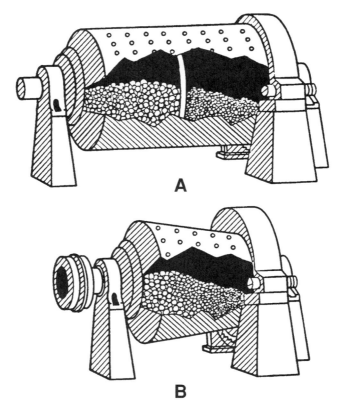

Fig. 14. Two types of ball mill used in limestone slurry FGD systems: **(a)** compartmented ball mill **(b)** Hardinge ball mill.

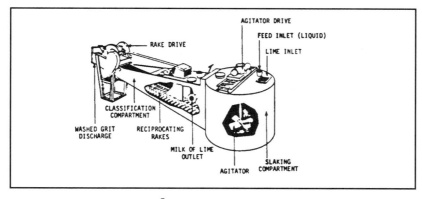

A Detention slaker

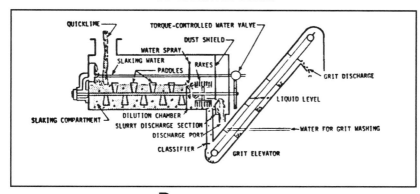

B Paste slaker

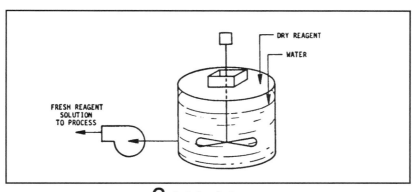

C Batch slaker

Fig. 15. Basic types of slaker.

increased, which allows greater flexibility in washing operations, wash water treatment, and the addition of scaling inhibitors. Despite all of these advantages, knockout-type precollectors are not used at most installations primarily because of plugging, high pressuredrop (approx 3 in. H_2O), increased complexity, and operating problems.

Impingement (or inertial impaction) removes mist by collection on surfaces placed in the gas streams. Entrained mist is collected in such devices by forcing the gas to make

Table 6
Typical Characteristics of First-, Second-, and Third-Generation Lime/Limestone Slurry FGD Systems

	Generation		
Characteristics	First	Second	Third
Duty	SO_2/fly ash	SO_2	SO_2
Absorber design	Venturi Tower	Tower with internals[a]	Open spray tower
Chemistry	High stoichiometric ratio	Moderate stoichiometric ratio	Low stoichiometric ratio
Water loop	Open	Closed	Closed with integrated water inventory
Waste processing	Ponding (no dewatering)	Primary dewatering and waste treatment	Primary and secondary dewatering and solid-waste physical/chemical treatment
Redundancy	None[b]	None[b]	Sparing of a number of system components

[a]Some spray towers are also included in late second-generation systems.
[b]Most of these systems incorporate minimal redundancy (e.g., pumps); however, spares are usually not provided for major components (e.g., absorbers).

changes in flow direction as it passes through the slats. The liquid droplets thus collected coalesce and fall by gravity back into the scrubbing slurry. Impingement-type mist eliminators used widely in lime/limestone slurry systems include baffle configurations. Baffle-type mist eliminators include the conventional open-vane (slat) and closed-vane chevron designs. The baffle-design mist eliminators are most common and constitute the simplest method of mist elimination.

9.6.1.4. REHEATERS

Reheaters (*see* Figs. 6 and 10) raise the temperature of the scrubbed gas stream in order to prevent condensation of acidic moisture and subsequent corrosion in the downstream equipment (ducts, fans, and stack). FGD systems that do not use reheaters must be equipped with specially lined stacks and exit ductwork to prevent corrosion. Such liners require special attention, and FGD systems using them must be equipped with emergency deluge sprays in the event of a temperature excursion.

The generic reheat strategies discussed in this Subsection include in-line, indirect hot air, and flue gas bypass (*see* Fig. 10). In-line reheat involves the use of a heat exchanger in the gas stream downstream of the mist eliminator (*see* Fig. 10a). The heat exchanger is a set of tube bundles through which the heating medium of steam or hot water is circulated. When steam is used, the inlet steam temperatures and pressures range from 350°F to 720°F and 115 to 200 psig, respectively. Saturated steam is preferred because the heat transfer coefficients of condensing steam are much higher than those of superheated steam. When hot water is used, inlet temperature of the hot water typically ranges from 250°F to 350°F and the temperature drop (water) over the heat exchanger is 70°F to 80°F.

Indirect hot-air reheat systems inject hot air into the gas stream (*see* Fig. 10b). There are two types of indirect hot-air reheater: the external heat exchanger and the boiler preheater design. In the external-heat-exchanger design, reheat is achieved by heating ambient air with an external heat exchanger using steam at temperatures of 350°F–450°F. The heating tubes are usually arranged in two to three banks in the heat exchanger. Hot air and flue gas may be mixed by use of a device such as a set of nozzles or a manifold in the reheater mix chamber section. In the boiler preheater design, reheat is achieved through the use of the boiler combustion air preheater to provide hot air. In this case, part of the heat that would have been used to heat the combustion air is used to reheat the stack gas. As a consequence, the temperature of the combustion air entering the boiler is lowered, thus somewhat reducing boiler efficiency.

In the bypass reheat system (*see* Fig. 10c) a portion of the hot flue gas from the boiler bypasses the absorber(s) and is mixed with scrubbed flue gas. Two variations of this method are "hot-side" bypass, in which the flue gas is taken upstream of the boiler air preheater, and "cold-side" bypass, in which flue gas in taken downstream of the boiler air preheater. In the former, a separate particulate-removal device (ESP or fabric filter) specifically for the bypass gas stream is required for fly ash control when an upstream (i.e., hot-side) particulate collector is not used.

9.6.1.5. DUCTWORK, DAMPERS, AND STACKS

Ductwork is used to channel the flow of gas within the FGD system. Ductwork in an FGD system in usually made of carbon steel plates ³⁄₁₆ or ¼ in thick, welded in a circular or rectangular cross setion. It is supported by angle frames that are stiffened at uniform intervals. The following design factors are considered for ductwork in lime/limestone slurry systems:

1. Pressure and temperature
2. Velocity
3. Configuration (cylindrical or rectangular)
4. Flow distribution
5. Variations in operating conditions
6. Materials of construction
7. Material thicknesses
8. Pressure drop

The ductwork must be designed to withstand the pressures and temperatures that occur during normal operation and also those that occur during emergency conditions. Ductwork is subject to a variety of conditions, depending on location within the system. The following list identifies the basic variants:

1. Inlet ductwork
2. Bypass ductwork (all or part of the flue gas)
3. Outlet ductwork (with reheat and without bypass)
4. Outlet ductwork (with reheat and with bypass for start-up)
5. Outlet ductwork (without reheat and without bypass)
6. Outlet ductwork (without reheat and with bypass for startup)

Dampers are used to regulate the flow of gas through the system by control or isolation functions. The entire system or subsystems may be regulated by the use of dampers. They are mainly used at the inlet duct to the module, the outlet duct from the module, and the bypass duct. Dampers may be used individually or in combination. A variety of damper designs are in use in lime/limestone slurry systems, including louver, guillotine, butterfly, and blanking plates.

Readers are referred to another chapter of this handbook for the stack design.

9.6.2. *Reagent Preparation and Feed Facilities*
9.6.2.1. REAGENT CONVEYORS AND STORAGE

Conveying equipment (*see* Fig. 12) used to transport limestone from unloading to storage includes dozing equipment, belt conveyors, and bucket elevators. Limestone is transported to feed bins by conveyors and bucket elevators. Limestone can be stored in silos, piles, or a combination of both. Short-term storage feed bins are used with both systems to feed limestone to the additive preparation system. Storage piles require more land to store a given quantity of limestone than silos. However, silos are more expensive and can experience flow problems such as plugging and jamming. Covered piles are sometimes used for limestone storage. The covers keep precipitation off the limestone pile and prevent freezing or limestone mud from developing. The primary design criterion of a limestone storage system is capacity. The storage facilities must have sufficient capacity so that the storage system does not limit the availability of the overall FGD system. There should be enough storage capacity to account for disruptions in the normal shipping schedule. Figure 13 shows an example of a limestone handling and storage system.

Conveying equipment used to transport lime can be of three basic types, as shown in Fig. 12. Most in-plant lime conveying involves simple elevation of the lime from a storage bin into a smaller feed bin. A simple combination of mechanical devices can move lime from storage at less than the initial cost and with less power consumption than a pneumatic conveyor. Mechanical conveying requires careful arrangement of bins and equipment. Alignment in a single straight row is preferable because each change of direction usually requires another conveyor.

9.6.2.2. BALL MILLS

A ball mill consists of a rotating drum loaded with steel balls that crush the limestone by the action of the tumbling balls as the cylindrical chamber rotates. Ball mills used in FGD systems fall into two categories. The long drum or tube mill variety is a compartmented type (*see* Fig. 14a) and the Hardings ball mill is noncompartmented and somewhat conical in shape (*see* Fig. 14b).

9.6.2.3. SLAKERS

A slaker is used in lime systems to convert dry calcium oxide to calcium hydroxide. The objective of lime slaking is to produce a smooth, creamy mixture of water and very small particles of alkali. Depending on the type of slaker used, the slurry produced contains 20–50% solids. A lime slaker combines regulated streams of lime, water under agitation, and temperature conditions needed to disperse soft hydrated particles. Dispersion must be rapid enough to prevent localized overheating and rapid crystal growth of the calcium hydroxide from occurring in the exothermic reaction. However, the mixture must be held in the slaker long enough to permit complete reaction.

Three basic types of slaker are presently used in lime slurry systems: detention, paste, and batch. A simplified diagram of each type is presented in Fig. 15.

9.6.3. Waste Solids Handling and Disposal Facilities

9.6.3.1. THICKENERS

The function of a thickener (*see* Fig. 7) is to concentrate solids in the slurry bleed stream in order to improve waste solids handling and disposal characteristics and recover clarified water. The slurry bleed stream usually enters a thickener at a solids level of about 5–15% and exits at a concentration of 25–40% solids. A thickener is a sedimentation device that concentrates the slurry by gravity. There are two basic types of thickener: gravity and plate. Only the gravity type will be described here, because plate thickeners are rarely used on utility FGD systems. A typical gravity thickener consists of a large circular holding tank with a central vertical shaft for settling and thickening of waste solids.

9.6.3.2. VACUUM FILTERS

Vacuum filters are widely used as secondary dewatering devices because they can be operated successfully at relatively high turndown ratios over a broad range of solids concentrations. A vacuum filter also provides more operating flexibility than other types of dewatering device as well as producing a drier product. Because a vacuum filter will not yield an acceptable filter cake if the feed solids content is too low, it is usually preceded by a thickener. A vacuum filter produces a filter cake of 45–75% solids from feed slurries containing 25–40% solids. The filtrate, typically containing 0.5–1.5% solids, is recycled to the thickener.

Two types of vacuum filter are used in conventional FGD system designs: drum and horizontal belt. Each has different characteristics and applicability. The drum type is the most widely applied.

9.6.3.3. CENTRIFUGES

Centrifuges are used to a lesser extent than vacuum filters in solids dewatering operations. The centrifuge product is consistent and uniform and can be handled easily. Centrifuges effectively create high centrifugal forces, about 4000 times that of gravity. The equipment in relatively small and can separate bulk solids rapidly with a short residence time.

There are two types of centrifuge: those that settle and those that filter. The settling centrifuge, which is the only kind used in commercial lime/limestone slurry FGD systems, uses centrifugal force to increase the settling rate over that obtainable by gravity settling.

9.6.3.4. WASTE PROCESSING AND DISPOSAL FACILITIES

Readers are referred to other volumes of this handbook series and elsewhere (10) for the details of various waste processing and disposal facilities. Only the following three, The most common waste processing processes, and three disposal processes are introduced and discussed in this Subsection: (a) forced oxidatio, (b) fixation, (c) stabilization, (d) ponding, (e) landfilling, and (f) stacking.

Forced oxidation supplements the natural oxidation of sulfite to sulfate by forcing air through the material. The advantages of a calcium sulfate (gypsum)–bearing material include better settling and filtering properties, less disposal space required, improved structural properties of the disposed waste, potential for utilization of the gypsum (e.g., wallboard production), and minimal chemical oxygen demand of the disposed material.

Forced oxidation, unlike fixation and stabilization, is not typically a tail-end operation; in many systems, this operation often occurs in the reaction tank.

Fixation increases the stability of the waste through chemical means. This may be accomplished by the addition of alkali, alkaline fly ash, or proprietary additives along with inert solids to produce a chemically stable solid. Examples of commercial processes of this type are those marketed by Conversion Systems, Inc. (e.g., Poz-O-Tec) and Dravo Corporation (e.g., Calcilox).

Stabilization is accomplished by the addition of nonalkaline fly ash, soil, or other dry additive. The purpose of stabilization is to enable the placement of the maximum quantity of material in a given disposal area, to improve shear strength, and to reduce permeability. Disadvantages are that the stabilized material is subject to erosion and rapid saturation and has residual leachability potential.

Waste disposal refers to operations at the disposal site for FGD waste following all handling and/or treatment stages. There are three basic FGD disposal site types: (1) ponding, (2) landfilling, and (3) stacking (10).

The most common waste disposal type is ponding. Ponds are either lined or unlined; lined ponds used for conventional FGD processes are typically clay lined. Landfilling is another waste disposal method. Wastes that have been fixated or stabilized are usually (although not always) landfilled. Stacking is only used for FGD systems designed to produce gypsum.

9.7. FGD Process O&M Practices

This section introduces the various types of operation and maintenance (O&M) practices for lime/limestone slurry FGD processes, the conditions under which the practices are implemented, and specific activities involved in each. More thorough treatment of the subject can be found elsewhere (56–59). This Subsection introduces the O&M requirements for these standard operating practices.

Increasingly stringent SO_x limits require a strong commitment from the owner/operator utility to FGD operation, including adequate staffing. Operators should be assigned specifically and solely to the FGD system during each shift. FGD system operation must be coordinated with the unit's power generation schedule and even into the purchasing of coal (i.e., sulfur, ash, and chlorine characteristics). Some of the current difficulties with lime/limestone FGD systems relate to poor operating practices, including overly complex procedures. In some cases, even properly installed equipment rapidly deteriorates and fails because of improper O&M practices. The operating characteristics of the FGD system can be established during the initial start-up period, which is also a time for finalizing operating procedures and staff training. Once steady-state operating conditions are reached, the system must be closely monitored and controlled to ensure proper performance.

During periods of changing load or variation of any system parameter, additional monitoring is required. Some standard O&M procedures (55,60) are described in the following Subsections.

9.7.1. Varying Inlet SO_2 and Boiler Load

As boiler load increases or decreases, modules are respectively placed in or removed from service. With each change in load, the operator must check the system to verify that all in-service modules are operating in a balanced condition. As the SO_2 concentration in the inlet flue gas changes, the FGD system performance changes. To maintain proper system response, slurry recirculation pumps can be added and removed from service as the SO_2 concentration increases or decreases.

9.7.2. Verification of Flow Rates

The easiest method of verifying liquid flow rates is for an operator to determine the discharge pressure in the slurry recirculation spray header with a hand-held pressure gage (permanently mounted pressure gages frequently plug in slurry service). Flow in slurry piping can be checked by touching the pipe. If the piping is cold to the touch at the normal operating temperature of 125–130°F, the line may be plugged.

9.7.3. Routine Surveillance of Operation

Visual inspection of the absorbers and reaction tanks can identify scaling, corrosion, or erosion before they seriously impact the operation of the system. Visual observation can identify leaks, accumulation of liquid or scale around process piping, or discoloration on the ductwork surface resulting from inadequate or deteriorated lining material.

9.7.4. Mist Eliminators

Many techniques have been used to improve mist collection and minimize operational problems. The mist eliminator (*see* Fig. 11) can be washed with process makeup water or a mixture of makeup and thickener overflow water. Successful, long-term operation without mist eliminator plugging generally requires continuous operator surveillance, both to check the differential pressure across the mist eliminator section and to visually inspect the appearance of blade surface during shutdown periods.

9.7.5. Reheaters

In-line reheaters (*see* Figs. 6 and 10) are frequently subject to corrosion by chlorides and sulfates. Plugging and deposition can also occur, but are rarer. Usually, proper use of soot blowers prevents these problems. The reliability of various reheater configurations is discussed in Section 9.3.5.5.

9.7.6. Reagent Preparation

Operational procedures associated with handling and storage of reagent are similar to those of coal handling. Operation of pumps, valves, and piping in the slurry preparation equipment is similar to that in other slurry service.

9.7.7. Pumps, Pipes, and Valves

Operating experience has shown that pumps, pipes, and valves can be significant sources of trouble in the abrasive and corrosive environments of a lime/limestone FGD system. The flow streams of greatest concern are the reagent feed slurry, the slurry recirculation loop, and the slurry bleed streams. When equipment is temporarily removed from slurry service, it must be thoroughly flushed.

9.7.8. Thickeners

Considerable operator surveillance is required to minimize the suspended solids in the thickener (*see* Fig. 7) overflow so that this liquid can be recycled to the system as supplementary pump seal water, mist eliminator wash water, or slurry preparation water. For optimum performance, the operator must maintain surveillance of such parameters as underflow slurry density, flocculent feed rate, inlet slurry characteristics, and turbidity of the overflow.

9.7.9. Waste Disposal

For untreated waste slurry disposal (*see* Fig. 7), operation of both the discharge to the pond and the return water equipment requires attention of the operating staff. In addition to normal operations, the pond site must be monitored periodically for proper water level, embankment damage, and security for protection of the public. Landfill disposal involves the operation of secondary dewatering equipment. Again, when any of the process equipment is temporarily removed from service, it must be flushed and cleaned to prevent deposition of waste solids. For waste treatment (stabilization or fixation), personnel are required to operate the equipment and to maintain proper process chemistry.

9.7.10. Process Instrumentation and Controls

Operation of the FGD system requires more of the operating staff than monitoring automated control loops and attention to indicator readouts on a control panel. Manual control and operator response to manual data indication may be more reliable than automatic control systems and are often needed to prevent failure of the control system. Many problems can be prevented when an operator can effectively integrate manual with automated control techniques.

10. EMERGING "WET" SULFUR OXIDE REDUCTION TECHNOLOGIES

Current efforts are directed toward further converting calcium sulfate by forced oxidation, using the limestone more efficiently, removing more SO_2 from the exhaust gas, improving equipment reliability, and altering the composition of the calcium sulfate to allow use in wallboard.

Although the meaning of "generation" is somewhat subjective, FGD systems may be distinguished in accordance with the evolution of technology per the following guidelines:

1. First generation: Designs that remove SO_2, and possibly fly ash, with gas contactors developed for or based on particulate matter scrubbing concepts. Included are lime/limestone slurry processes that use gas contactors with Venturi or packing-type internals.
2. Second generation: Designs that remove SO_2 primarily in gas contractors, developed specifically for SO_2 absorption, which utilize features to improve the chemical or physical means. Included are lime/limestone slurry processes using additives or spray towers, combination towers, or special reactors.
3. Third generation: Improved second-generation designs that encompass additional process refinements and are currently under demonstration or early commercial operation. Included are spray tower designs with spare absorbers, closed water-loop operations, and gypsum production.

Table 6 summarizes the basic characteristics of the system within the three generations.

As FGD technology evolved, more effective measures were adopted and modifications were made to earlier systems to upgrade performance.

The following Subsections summarize demonstration tests of three emerging wet FGD technologies. These have in common innovative reactor schemes for contacting flue gas with calcium-based sorbents in a slurry or solution.

10.1. Advanced Flue Gas Desulfurization Process

The Advanced Flue Gas Desulfurization (AFGD) Demonstration Project was performed by Pure Air on the Lake, L.P. (a subsidiary of Pure Air) at the North Indiana Public Service Company's Bailey Generating Station. Operational tests ran from 1992 to 1995. The process uses one absorber vessel to perform three functions: prequenching the flue gas, absorbing SO_2, and oxidation of the resulting calcium sulfite to wallboard-grade gypsum. The flue gas contacts two tiers of fountainlike sprays and passes through a gas–liquid disengagement zone, over the slurry reservoir, and through a mist eliminator. Variables studied included the sulfur content of the coal, slurry recirculation rate, Ca/S ratio, and the liquid/gas ratio in the absorber.

The process achieved SO_2 removal efficiencies of 95% and higher at Ca/S ratios of 1.07–1.10 with coal sulfur contents of 2.25–4.5%. The system had 99.5% availability and produced wallboard-grade gypsum with an average purity of 97.2%. The system effectively captured acid gases and trace elements associated with particulates. Some boron, selenium, and mercury passed to the stack gas as vapor.

Efficient operation and high reliability eliminated the need for a spare absorber. These advantages, with compactness, reduced space requirements. Concurrent flow allowed high flue gas velocities (up to 20 ft/s). A nonpressurized slurry distribution system reduced recirculation pump power requirements by 30%. The fountainlike flow of absorber reduces mist loading by as much as 95%. Use of dry pulverized limestone eliminates the need for sorbent preparation equipment. An air rotary sparger combines agitation with oxidation to enhance performance. A novel wastewater evaporation system controls chlorides without creating a new waste stream. A compression mill system (PowerChip™) modifies the physical structure of the gypsum.

Cost estimates were made for a 500-MWe power plant firing a 3% sulfur coal and achieving 90% SO_2 removal. The capital costs were estimated at US$94/kW, with a 15-yr levelized cost of US$6.5mil/kWh, equivalent to US$302/ton of SO_2 removed. This costs is about half that for conventional wet FGD.

High efficiency, compactness, elimination or use of byproduct streams, and costs of about one-half those of a conventional wet FGD process make the system highly applicable. As of 1999, there were no sales yet, but the system remains in operation with commercial gypsum sales (14).

10.2. CT-121 FGD Process

The CT-121 FGD system was demonstrated by Southern Companies Services, Inc. at the Georgia Power Company's Plant Yates, No.1 in Coweta County, Georgia from 1992 to 1994. The system uses a unique absorber called the Jet Bubbling reactor (JBR). In one vessel (*see* Fig. 16), the JBR combines limestone AFGD, forced oxidation, and gypsum crystallization. Flue gas is quenched with water injection and is bubbled into the scrubbing solution. SO_2 is absorbed and forms calcium sulfite. Air bubbled into the bottom of the reactor oxidizes the calcium sulfite to gypsum. The gypsum is removed from the slurry in a settling pond.

SO_2 removal efficiency was over 90% at SO_2 inlet concentrations of 1000–3500 ppm. Limestone utilization was over 97%. Particulate removal efficiencies of 97.7–99.3% were achieved. Hazardous air pollutant capture was greater than 95% for

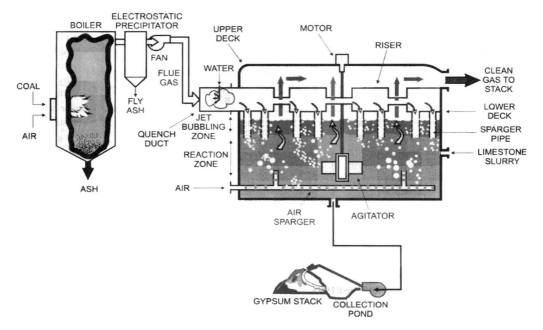

Fig. 16. Diagram for innovative applications of the CT-121 FGD process.

hydrogen chloride; 80–98% for most trace metals, but less than 50% for mercury and less than 70% for selenium. The gypsum was suitable for making wallboard, although requiring washing to remove chloride.

Availability was 95–97%, eliminating the need for a spare absorber. Simultaneous SO_2 and particulate removal were achieved at ash loadings for which an electrostatic precipitator has marginal performance. The fiberglass-reinforced plastic equipment proved durable and eliminated the need for a flue gas prescrubber and reheater.

The technology is applicable to new construction and retrofitting. Capital costs were estimated at US\$80–95/kW, with operating costs at US\$34–64/ton (1994 US\$) of SO_2 depending on specific conditions. Elimination of the need for a flue gas prescrubber, gas preheater, and a spare absorber should significantly reduce capital costs compared to conventional FGD. This technology is sold internationally (14).

10.3. Milliken Clean Coal Technology Demonstration Project

The Milliken Clean Coal Technology Demonstration Project was carried out by the New York State Electric & Gas Corporation (NYSEG) and other team members, at the NYSEG Milliken Station in Tomkins County, New York from 1995 to 1999. The demonstration used the Saarberg–Holter–Umwelttechnik (S-H-U) FGD process. This uses a space-saving concurrent/countercurrent absorber vessel. The vessel is Stebbins tile lined and constructed of reinforced concrete. The process is specifically designed to benefit from the use of formic acid to buffer the slurry to low pH, improving the rate of limestone dissolution and calcium solubility. This enhances SO_2 absorption efficiency and reduces limestone consumption. Energy efficiency and byproduct quality are improved. Formic acid use improved SO_2 removal efficiency to 98%, versus 95% without it.

Grinding the limestone finer, from 90% −325 mesh to 90% −170 mesh, improved SO_2 removal by 2.6%. The capital costs of the FGD system were estimated at US$300/kW (1998 US$), with operating costs at US$412/ton of SO_2 removed (1998 US$) (14).

11. EMERGING "DRY" SULFUR OXIDES REDUCTION TECHNOLOGIES AND OTHERS

11.1. Dry Scrubbing Using Lime or Sodium Carbonate

Chapter 5, volume 1, Wet and Dry Scrubbing, introduces the dry scrubbing process in detail. Dry scrubbing is a modification of wet scrubbing flue gas desulfurization technology. As in other FGD systems, the exhaust gases combine with a fine slurry mist of lime or sodium carbonate. This system, however, takes advantage of the heat in the exhaust gases to dry the reacted slurry into particles of calcium sulfite or sodium sulfite, depending on lime or sodium carbonate being the scrubbing slurry. The following is a chemical reaction if lime is used:

$$SO_2 + CaO \rightarrow CaSO_3 \tag{32}$$

where CaO is lime and $CaSO_3$ is calcium sulfite.

If a fine slurry mist of sodium carbonate is used for SO_2 removal, the following will be the chemical reaction:

$$SO_2 + Na_2CO_3 \rightarrow Na_2SO_3 + CO_2 \tag{33}$$

where Na_2CO_3 is sodium carbonate, Na_2SO_3 is sodium sulfite, and CO_2 is carbon dioxide. Dry scrubbing normally removes 70% of the dioxide in an air emission stream.

11.2. LIMB and Coolside Technologies

LIMB stands for the "Lime/Limestone Injection Multistage Burners" process. The LIMB and Coolside demonstrations were performed at Ohio Edison's Edgewater Station, Unit No. 4 during 1989 to 1992. The LIMB process (*see* Fig. 17) involves injection of a calcium-based sorbent into the boiler, above the burners, near a temperature of 2300°F. The sorbent calcines to calcium oxide, reacts with SO_2 and oxygen, and is removed with the fly ash in an electrostatic precipitator (ESP). The process produces particulates that are difficult to remove, but this is overcome by humidifying the stream prior to the ESP. SO_2 removal efficiencies at a Ca/S ratio of 2.0, minimal humidification, and the four sorbents tested ranged from 22% to 63%. Grinding the limestone sorbent to finer particle size ranges (100% < 44 and 10 µm) improved SO_2 removal efficiencies another 10–17%, respectively. SO_2 removal efficiencies were improved by about another 10% with humidification to a 20°F approach-to-saturation. Incorporating low-NO_x burners reduced NO_x emissions 40–50%. Availability was 95%, and humidifier operation in the vertical mode (versus. horizontal) was indicated to reduce floor deposits. Capital costs were US$31–102/kW (1992 US$) for plants ranging from 100 to 500 MWe, coals with 1.5–3.5% sulfur, and a 60% SO_2 reduction target. Commercialization includes sale of the LIMB technology to an independent power plant in Canada and multiple sales of the low-NO_x burners.

In the Coolside technology, the sorbent (hydrated lime) is injected into the flue gas downstream of the air preheater. Injection is followed by humidification with a mist

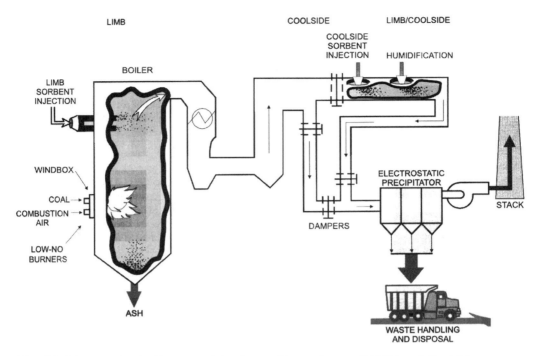

Fig. 17. Diagram for the limestone injection multistage burner (LIMB) and Coolside systems.

containing sodium hydroxide or sodium carbonate. The sorbent reacts with SO_2 in the presence of the sodium compounds to desulfurize the stream. The mist cools the flue gas from about 300°F to 140–145°F, with an approach to saturation of 20–25°F, to maximize sulfur capture. The sorbent and sodium compounds retain high reactivity and are recycled with fresh hydrated lime.

Coolside technology achieved an SO_2 removal efficiency of 70% at a Ca/S ratio of 2.0, a Na/Ca ratio of 0.2, and a 20°F approach-to-saturation temperature, using commercial hydrated lime and a 2.8–3.0% sulfur coal. Recycling sorbent, reduced sorbent and additive usage by up to 30% and improved SO_2 removal efficiency by 20+%.

The LIMB and Coolside technologies are applicable to most utility and industrial coal-fired units. They provide alternatives to conventional wet flue gas desulfurization, and retrofits require modest capital investment and downtime. Space requirements are also substantially less than for conventional processes (14).

11.3. Integration of Processes for Combined SO_x and NO_x Reduction

The reduction of SO_x emissions can be integrated with measures to reduce NO_x emissions. The latter can be achieved through burner design (61), natural gas injection, and reburning (62). NO_x reduction using natural gas enhances SO_2 reduction because natural gas displaces coal and its sulfur content. Gas reburning and sorbent injection (GR-SI) were demonstrated at Illinois Power's Hennepin No. 1 power plant, beginning in 1990. Tests indicated successful integration of the technologies. SO_2 emissions were reduced by 18% through displacement of coal with sulfur-free natural gas. NO_x emissions

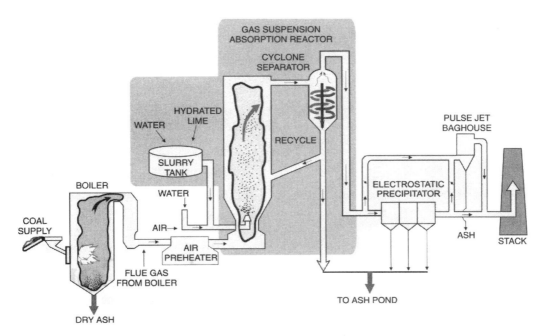

Fig. 18. Diagram of gas suspension absorption system.

control technologies are compatible with a wide variety of SO_2 emissions control methods: furnace sorbent injection, dust sorbent injection, wet scrubbers, dry scrubbers, and coal switching (62). SO_2 emissions credits can also help defray NO_x emission control costs, and in some scenarios, they could cancel that cost.

11.4. Gas Suspension Absorbent Process

The gas suspension adsorption (GSA) 10-MWe demonstration was performed at the Tennessee Valley Authority's Shawnee Fossil Power Plant near Paducah, Kentucky from 1992 to 1995. The GSA system (*see* Fig. 18) can be described as a semidry FGD technology. Flue gas passes upward through a vertical reactor. Solids coated with hydrated lime are injected into the bottom of the reactor. A major feature is that about 99% of solids are recycled to the reactor by a cyclone. The heat and mass transfer characteristics are superior to those in conventional semidry technology using a lime slurry directly sprayed into a duct or spray dryer.

Two sets of tests were performed using an electrostatic precipitator (ESP) and a pulse jet baghouse (PJBH). With ESP, an SO_2 removal efficiency of 90% was achieved at a Ca/S ratio of 1.3–1.4 and an approach-to-saturation temperature of 8–18°F. With a PJBH, an SO_2 removal efficiency of 96% was achieved at a Ca/S ratio of 1.4 and an approach-to-saturation temperature of 18°F. Both methods removed 99.9+% of particles, 98% of hydrogen chloride, 96% of hydrogen fluoride, and 99% or more of trace metals except cadmium, antimony, mercury, and selenium. GSA/PJBH removed 99+% of the selenium.

The Ca/S ratio, approach-to-saturation temperature, and chloride content significantly affected SO_2 removal efficiency. As the Ca/S ratio increased from 1.0 to 1.3, SO_2

Table 7
Cost Estimate for Gas Suspension Adsorption of SO$_2$

	Capital cost (1990 \$/kW)	Levelized cost (mils/kWh)
GSA: three units at 50% capacity	149	10.35
WLFO	216	13.04
Spray dryer	172	—

Note: Assumes 90% SO$_2$ removal at a Ca/S ratio of 1.3 and uses EPRI TAG™ method.
Source: U.S. DOE (2001).

removal increased from 76% to 93%. As the approach-to-saturation temperature decreased from 30°F to 12°F, SO$_2$ removal increased from 81% to 95% (at Ca/S of 1.3). As the chloride content (percentage of lime feed) increased from 0.5% to 2.0%, the SO$_2$ removal increased from 85% to 99%.

Lime utilization was better than for spray drying systems: with ESP, 66.1%; with PJBH, 70.5%. Because of improved heat and mass transfer, the same performance was achieved in one-fourth to one-third the size of a spray dryer, facilitating retrofitting in space-limited plants and reducing installation costs. The GSA system achieved lower particulate loading, 2–5 gr/ft^3 versus 6–10 gr/ft^3 for a spray dryer, allowing compliance with a lower ESP efficiency. The direct recycling of solids eliminates the need for multiple or complex nozzles as well as the need for abrasion-resistant materials. Thus, special steels are not required for construction and only a single spray nozzle is needed. The system demonstrated high availability and reliability similar to that for other commercial applications.

Cost estimates using EPRI's TAG™ method were prepared for a moderately difficult GSA retrofit of a 300-MWe boiler burning 2.6% sulfur coal. A SO$_2$ removal of 90% at a Ca/S was specified. Capital and levelized costs were compared for those of a wet limestone scrubber with forced oxidation, and a spray dryer (*see* Table 7) (14).

11.5. Specialized Processes for Smelter Emissions: Advanced Calcium Silicate Injection Technology

The Advanced Calcium Silicate Injection (ADVACATE) technology (*see* Fig. 19) is perhaps the most competitive with conventional technology, offering comparable (90+%) SO$_2$ control and annualized costs in comparison with the competing lime/limestone forced oxidation FGD technology. ADVACATE was evaluated on a 10-MWe prototype in the early 1990s, and demonstrations on a commercial scale were planned in the United States and overseas. The ADVACATE process was codeveloped by APPCD with the University of Texas and is currently licensed for worldwide use (63,64).

12. PRACTICAL EXAMPLES

Example 1

In FGD lime/limestone process operation for SO$_x$ removal, the inlet SO$_2$ concentration is largely dependent on the sulfur content of the coal fired in the boiler. To estimate SO$_2$ emissions (in units of lb SO$_2$/10^6 Btu), a field inspector usually uses the following equation:

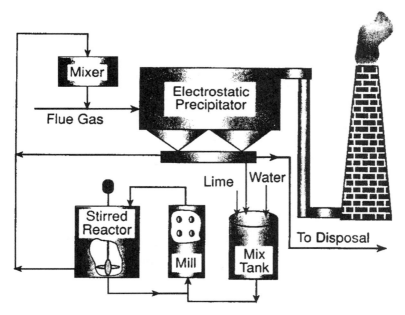

Fig. 19. ADVACATE process.

$$Q_{SO_2} = S_{\%W}\left(2\times10^4\right)C_f/GCV \tag{34}$$

where Q_{SO_2} is the SO_2 emission rate (lb $SO_2/10^6$ Btu), $S_{\%W}$ is the percentage sulfur in coal by weight, C_f is the fractional conversion of sulfur in coal to SO_2, and GCV is the gross caloric value (heating value of coal), (Btu/lb).

Answer the following:

1. What should an engineer do if the C_f value is unknown?
2. What are the GCV numbers for various coals?

Solution

1. If the fractional conversion C_f value is unknown, use the US EPA AP-42 emission factors that assign SO_2 conversion factors as follows (65):
   $C_f = 0.97$ for bituminous coal
   $C_f = 0.88$ for subbituminous coal
   $C_f = 0.75$ for lignite coal

2. The following heating values, or GCV, for various coals may be assumed if the actual GCV is unknown:
   GCV = 10,680 Btu/lb for bituminous coal
   GCV = 11,500 Btu/lb for subbituminous coal
   GCV = 12,000 Btu/lb for lignite coal

Example 2

What is the SO_2 emission rate (lb $SO_2/10^6$ Btu) if $S_{\%W}$ is known to be 3.5% S and C_f is known to be 0.92 for a high-quality subbituminous coal?

Solution

$$Q_{SO_2} = (3.5) \times (2 \times 10^4)(0.92)/(11,500 \text{ Btu/lb})$$
$$= 3.5 \times 2 \times 10^4 \times 0.92/11,500$$
$$= 5.6 \text{ lb SO}_2/10^6 \text{ Btu}$$

Example 3

Derive an engineering equation for converting normal air pollutant concentrations (lb/ft³) to the US government required units (lb/10⁶ Btu) for coal-fired electrical generation plants.

Solution

$$F_W = \frac{10^6[5.56(\%H)+1.53(\%C)+0.57(\%S)+0.14(\%N)-0.46(\%O_2)+0.21(\%H_2O)]}{\text{GCV}} \tag{35}$$

$$E = C_{WS} F_W \frac{20.9}{20.9(1-B_{WA})-\%O_{2W}} \tag{36}$$

where F_W is the coal analysis factor on a wet basis (std. ft³/10⁶Btu), GCV (gross caloric value) is the high heating value of coal (Btu/lb), E is the pollutant emission rate (lb/10⁶ Btu), C_{WS} is the pollutant concentration given as a wet basis (lb/ft³), B_{WA} is the ambient air moisture fraction, and O_{2W} is the percent oxygen in flue gas on a wet basis. The standard GCV for coal is given in Example 1.

Example 4

Energy consumption of a FGD system using lime or limestone is presented in this example. FGD energy consumption is attributed to reheat, flue gas flow, slurry preparation, and slurry recirculation. Other energy-consuming operations include slurry transfer (pumping), tank agitation, solids dewatering (thickeners, vacuum filters, centrifuges), steam tracing, electrical instrumentation, and air supply. An increase in energy consumption in any of these areas usually indicates a problem. Please answer the following:

1. Why is reheating one of the three major energy-consuming items in an FGD lime/limestone process system? What is a quick approximation method to determine the reheat energy consumption?
2. Why is forcing flue gas through the FGD one of the three major energy-consuming items in an FGD lime/limestone process system? What are the quick approximation methods to determine the forced draft energy consumption?
3. How important is the slurry preparation and recirculation system? How can the slurry recirculation pumping requirements be determined?

Solution

1. Reheating the saturated flue gas consumes more energy than any other part of the FGD system (assuming reheat is used). Reheat provides buoyancy to the flue gas, reducing nearby ground-level concentrations of pollutants. Reheat also prevents condensation of acidic, saturated gas from the absorber in the induced draft fan, outlet ductwork, or stack. Further 2 more, reheat minimizes the settling of mist droplets (as localized fallout) and the formation of a heavy steam plume with resultant high opacity. An increase in reheater energy consumption is generally indicative of plugged or scaled in-line reheater tube bundles. Energy consumption is increased because the

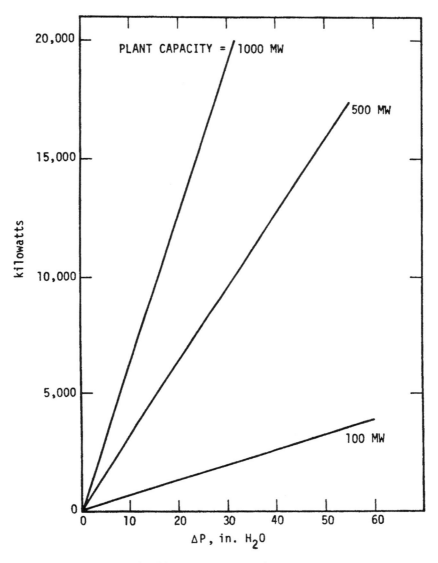

Fig. 20. Fan power requirements.

heat transfer efficiency of the reheater tubes is lowered. The following equation provides a quick approximation method to determine reheat energy consumption:

$$H_e = 0.01757 Q_{air} C_p \Delta T \tag{37}$$

where: H_e is heat energy (Btu), Q_{air} is the air flow rate at the inlet of reheat sections (lb/min), C_p is the specific heat [Btu/(lb)(°F)], and ΔT is the degree of reheat (°F).

2. Forcing flue gas through the FGD system consumes energy. Forced or induced draft fans use energy to overcome the gas-side pressure drop of the FGD system. An increase in fan energy consumption usually indicates either a mechanical problem with the fan and/or an increase in the pressure drop somewhere in the FGD system. The following equation provides a quick approximation method to determine FGD fan

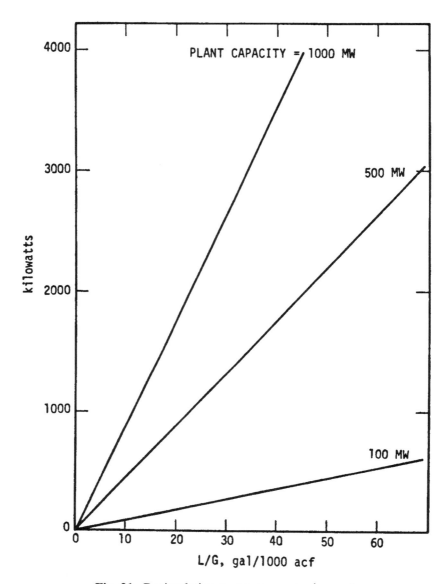

Fig. 21. Recirculation pump power requirements.

power requirements. Figure 20 provides a quick determination method if only plant size and gas-side pressure drop are known.

$$P = 0.0002617(\Delta P)Q_s \tag{38}$$

(assuming 80% fan efficiency), where P is power required [kw (for fan)], ΔP is the pressure drop through the FGD system (in. H_2O), and, Q_s is the gas flow rate at the outlet of scrubber/absorber (scfm).

3. Grinding limestone and slaking lime consume relatively small amounts of energy, as compared to other energy-consuming equipment. Any increases are usually the result of poor quality makeup water or mechanical problems with the slaker or ball mill.

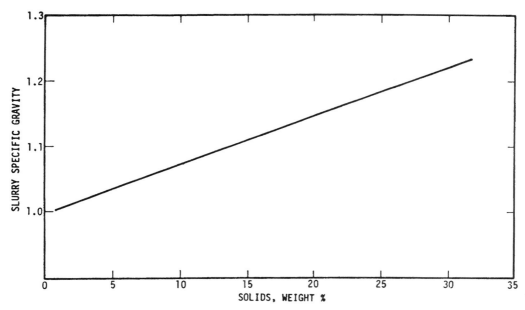

Fig. 22. Typical specific gravity of absorber recirculation slurry for lime/limestone FGD systems.

Energy is mainly consumed to recirculate the slurry to the absorber, to transfer water and slurry streams to various parts of the FGD system, and to treat and dispose of the solid-waste material. An increase in pumping energy consumption usually indicates either a mechanical problem or an increase in slurry side pressure drop in the system. The following equation provides a quick approximation method to determine recirculation pumping requirements. Figure 21 provides a quick determination for slurry recirculation pumping requirements if the plant size and *L/G* are known.

$$P = 0.000269 \times H_S \times (L/G) \frac{Q_S}{1000}$$
$$= H_S \times (L/G) Q_S \times (2.69 \times 10^{-7})$$

(39)

(assuming 90% pump efficiency), where P is the power required (KW) (for slurry recirculation pumps), Q_s is the gas flow rate at the outlet of scrubber/absorber (scfm), H_s is the head (ft), and *L/G* is the ratio of slurry flow to flue gas rate (gal/1000 scf, or gal/1000 scfm) at the outlet of the scrubber/absorber.

Example 5

A graphical representation of specific gravity as a function of the solids content of the slurry in lime/limestone FGD systems is presented in Fig. 22. What is the specific gravity of the recirculating slurry for the lime/limestone FGD system if the slurry's percentage solid (by weight) is 20%? Discuss its feasibility for the FGD system.

Solution

From Fig. 22, the slurry specific gravity should be equal to 1.14 if the slurry's solid content weight is 20%. Operation at consistent solids content in the various slurry process streams can improve the reliability of the absorber and slurry-handling equipment and improve process control. Specific gravity is a commonly used measure for determining

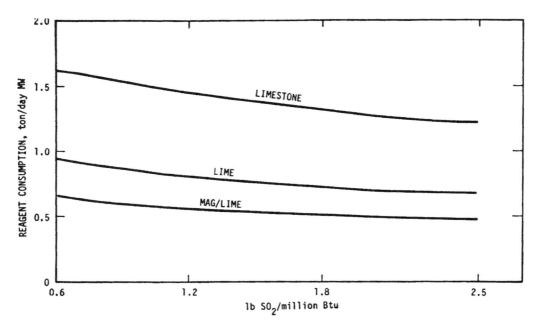

Fig. 23. Reagent requirement calculation.

slurry solids content. The design specific gravity of the recirculating slurry for lime/limestone FGD systems is usually between 1.05 and 1.14 (approx 7–20% solids).

Example 6

What are the three most common reagents for the lime/limestone FGD process? Show how to calculate approximate reagent requirements (ton reagent/day-MW)?

Solution

1. The three most common reagents for the lime/limestone FGD process are limestone ($CaCO_3$), lime (CaO), and magnesium lime (MgO). Dolomite (dolomitic lime or dolomitic limestone) is a crystallized mineral consisting of calcium magnesium carbonate, $CaMg(CO_3)_2$.
2. Reagent consumption is set by the stoichiometry of the process. As noted previously, it is necessary to feed more than the stoichiometric amount of reagent in order to attain the degree of SO_2 removal required (stoichiometric ratio). However, excessive reagent can lead to several operating problems, including wasted reagent, scale formation, and erosion of slurry-handling equipment. Figure 23 is a graphic representation of reagent consumption as a function of the SO_2 emission limitation and boiler size (i.e., equivalent FGD capacity in megawatts). This figure can be used by the field inspector to estimate reagent feed rates.
3. The megawatt is a unit used to describe gross or net power generation of a facility. One watt equals one joule per second (1 MW = 10^6 W).

Example 7

The solid-waste (sludge) production rate is one of the most important operational parameters for a lime/limestone FGD process system.

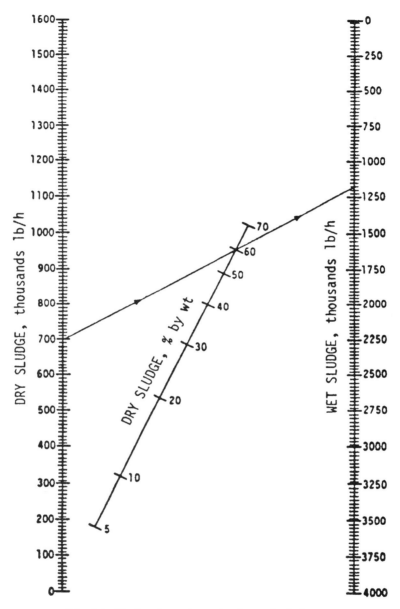

Fig. 24. Sludge (waste) production calculation.

1. Discuss the importance of solid-waste (sludge) production rate.
2. How does one convert between dry and wet sludge production?

Solution

1. Solid-waste (sludge) production will vary as a function of the inlet flue gas characteristics and FGD system design and operating characteristics. The constituents usually include solid-phase SO_2 reaction products, unreacted reagent, fly ash, and adherent liquor.

Increases in solid waste increase the burden on solids handling and disposal. This can mean higher energy consumption, possible deviation from closed water-loop operation because of excessive amounts of wastewater effluent, and reduced land area available for disposal. Variations in the quality of the slurry bleed stream to the thickener can either overload (high or "rich" solids content) or underutilize (low or "lean" solids content) the primary dewatering subsystem. The ratio of sulfite to sulfate contained in the spent slurry stream is also important because of the size differences between gypsum (1–100 μm in length) and calcium sulfite crystals (0.5–2.0μm in length). These differences can have a significant impact on the dewatering of the solid-waste material. Generally, as the ratio of sulfite to sulfate increases, the liquor content of the dewatered solid waste also increases.

2. Figure 24 provides a nomograph to convert between dry and wet sludge production.

Example 8

Acid rain is caused by SO_x and NO_x emissions release. Discuss possible engineering solutions to lake restoration assuming that the damage is done.

Solution

Both SO_x and NO_x emissions pollute lakes, usually at high elevations. SO_x and NO_x mix with normal rains, producing acid rain, in turn acidifying the lake water. In serious situations, the pH of lake water is too low to be habitable to many species of aquatic animals or plants.

Lakes polluted by acid rains (caused by SO_x and NO_x dissolution) can usually be restored by a neutralization process. Although any kind of alkaline chemicals can be used as a neutralizing agent, usually inexpensive lime is used.

In the above case of lime treatment, calcium in lime will react with sulfate ions in lake water, forming calcium sulfate. Calcium sulfate will precipitate from the lake water only when it exceeds its solubility. Most of sulfate ions will still remain in lake water.

It has been known that the ion-exchange process is technically feasible for removing sulfate ions from lake water, but it is not economically feasible.

Example 9

Removal of hydrogen sulfide from the gas phase has been introduced extensively in this chapter. It has been shown that sometimes the condensate (i.e., liquid phase) may contain high concentrations of H_2S at geothermal power plants (1). Introduce and discuss the engineering solutions to this case.

Solution

1. **First Solution: Gas Stripping.** The easiest process method for removing H_2S from condensate (i.e., a liquid phase) is simply to direct the condensate to the cooling tower where the H_2S will be stripped from the condensate and be exhausted with the effluent air from the cooling tower. In this situation, the cooling tower becomes a stripping tower for removing H_2S from condensate. This is an inexpensive process method but will create an odor nuisance or even a hazardous situation when H_2S reaches high concentration levels.

2. **Second Solution: EDTA Treatment and Filtration.** A safer process method of treating H_2S is to direct the condensate to the cooling tower but add chelated iron upstream of the cooling tower. The chelator is usually ethylenediaminetetraacetic acid (EDTA), whose only purpose is to increase the solubility of iron in water. The iron reacts with the dissolved H_2S as follows:

$$2Fe^{3+} + H_2S \rightarrow 2Fe^{2+} + S^0 + 2H^+ \tag{40}$$

where Fe^{3+} is the chelated trivalent ferric ion, H_2S is the target S-containing pollutant, Fe^{2+} is the divalent ferrous ion, S^0 is the insoluble elemental sulfur, and H^+ is the hydrogen ion.

Fe^{2+} is not very effective for reacting with H_2S, so it must be reoxidized to Fe^{3+} to be reused. This is accomplished when the cooling water is circulated in the cooling tower and comes into contact with oxygen (air) as follows:

$$2Fe^{2+} + 0.5O_2 + 2H^+ \rightarrow 2Fe^{3+} + H_2O \tag{41}$$

where O_2 is oxygen. The insoluble element sulfur, S^0, can be filtered out from the system in order to accomplish the goal of desulfurization.

3. **Third Solution: Combined EDTA and Sodium Sulfite Treatment.** The element sulfur, which is formed in Eq. (43), must be removed by filtration. The insoluble solid sulfur, left as is, will eventually plug the cooling tower. Instead of filtration removal of sulfur, sodium sulfite may be added to the cooling tower for dissolution of sulfur, forming soluble sodium thiosulfate, which can be reinjected into the geothermal formation:

$$S^0 + Na_2SO_3 \rightarrow Na_2S_2O_3 \tag{42}$$

where Na_2SO_3 is sodium sulfite and $Na_2S_2O_3$ is sodium thiosulfate.

13. SUMMARY

A recent perspective (66) summarizes energy use and emissions trends as well as key issues and developments for control of SO_x, NO_x, and particulate emissions. First, world energy use is projected to increase 50% by 2020 (67). Coal represents 80% of the world's fossil fuel proven recoverable reserves. The chief concerns for coal use involve emissions of SO_x, NO_x, particulates, and carbon dioxide. As noted earlier, the Clean Air Act and its amendments have lead to significant decreases in emission. From 1980 to 1999, SO_2 emissions have been reduced from 17.3 to 13.5 million tons. From 1970 to 1996, particulate matter emissions (<10 μm) have been reduced from 1.6 million tons to 260,000 tons. In the US State Implementation Plan, the goal is to reduce NO_x emissions 85% from 1990 levels.

Strategies for reducing SO_x emissions include fuel switching, blending, coal cleaning, and postcombustion FGD. Only about 25% of US power plants have FGD equipment, but this is expected to rise as existing plants retrofit.

In the above sections, we reviewed the two main technologies for FGD using calcium-based absorbents (wet and dry, using respectively, a lime/limestone slurry or a dry/semidry sorbent injection). The wet systems can achieve greater than 95% SO_x removal, but require ancillary equipment that can amount to 20% of overall plant costs. Dry methods involve injection of sorbent into the ductwork after the boiler, or injection directly into the furnace [furnace sorbent injection (FSI)]. The lime or limestone ($CaCO_3$) calcines, and sulfation reactions occur rapidly at the high temperatures. Injection of dry/semidry sorbents offers advantages of simplicity and ease of retrofit to existing plants. A concern relates to rate and percentage utilization of the sorbent because of the buildup of calcium sulfate that blocks pores and reduces transport and reaction. Recent research addresses this concern by producing a highly reactive absorbent by synthesis of $CaCO_3$

particles in aqueous slurry in the presence of small quantities of anionic surfactants or surface modifiers (68). This sorbent achieves more than 70% sulfation within 500 ms, in contrast to 20% for unmodified calcium carbonate. The sorbent can be regenerated to ultimately achieve almost complete sorbent utilization. The process, named "OSCAR," is being demonstrated at the Ohio McCracken Power Plant on the campus of Ohio State University.

Although this chapter focuses on SO_x and H_2S emissions control, the technologies discussed have important implications for other pollutants. For example, Ca-based sorbents, injected in the FSI process, can capture significant amounts of trace elements such as mercury, arsenic, and selenium.

Future research should be directed at the urgent needs: novel and efficient pollution control technologies, improved efficiency, advanced pulverized combustors, pressurized fluidized-bed combustors, and integrated gasification combined cycle systems. Future work will also see investigation of CO_2 sequestration technologies (eg., using the sorbent from the OSCAR process) (69). Hydrogen sulfide gas from different sources (1–7,70) can also be effectively removed by many FGD processes introduced in this chapter. It should also be noted that many systems are commercially available for flue gas conditioning, flue gas desulfurization, odor control (H_2S), waste disposal, process monitoring, and emission anlyses (71–73).

NOMENCLATURE

B_{WA}	Ambient air moisture fraction
C_f	Fractional conversion of sulfur in coal to SO_2
C_p	Specific heat [Btu/(lb)(°F)]
C_{WS}	Pollutant concentration given as a wet basis (lb/ft^3)
E	Pollutant emission rate (lb/10^6 Btu)
F_W	Coal analysis factor on a wet basis (std. ft^3/10^{6}Btu)
GCV	Gross caloric value (heating value of coal) (Btu/lb)
H_e	Heat energy (Btu)
H_s	Head (ft)
L/G	Ratio of slurry flow to flue gas rate, gal/1000 scf (or gal/1000 scfm)at the outlet of the scrubber/absorber
O_{2W}	Percent oxygen in flue gas on a wet basis
P	Power required (KW)
ΔP	Pressure drop through the FGD system (in. H_2O)
Q_{air}	Airflow rate at the inlet of reheat sections (lb/min)
Q_s	Gas flow rate at the outlet of scrubber/absorber (scfm)
Q_{SO_2}	SO_2 emission rate (lb SO_2/10^6 Btu)
$S_{\%W}$	Percentage sulfur in coal by weight
ΔT	Degree of reheat (°F)

REFERENCES

1. G. J. Nagl, *Environ. Technol.* **9**(7), 18–22 (1999).
2. T. R. Mason, *GRC Bull.* **1996**(6), 233–235 (June 1996).

3. J. Devinny, and M. Webster, *Biofiltration for Air Pollution Control*. Lewis Boca Raton, FL, 1999.
4. M. Lutz, and G. Farmer, *Water Environ. Fed. Oper. Forum* **16**(7), 10–17 (1999).
5. L. K. Wang, *Emissions and Control of Offensive Odor in Wastewater Treatment Plants*, Technical Report No. PB88-168042/AS, US Department of Commerce, National Technical Information Service, Springfield, VA, 1985.
6. L. K. Wang, *Identification and Control of Odor at Ferro Brothers Sanitary Landfill Site*, Technical Report No. P902FB-3-89-1, Zorex Corporation, Newtonville, NY, 1989.
7. L. K. Wang, *City of Cape Coral Reverse Osmosis Water Treatment Facility*, Report No. PB97-139547, US Department of Commerce, National Technical Information Service, Springfield, VA, 1997.
8. US EPA, *Septage Treatment and Disposal* US Environmental Protection Agency, Cincinnati, OH, Report No. EPA-625/6-84/009, 1984.
9. US EPA, *Controlling Sulfur Oxides,* Report No. EPA-600/8-80/029, US Environmental Protection Agency, Washington, DC, 1980.
10. US EPA, *Sulfur Emission: Control Technology and Waste Management*, US Environmental Protection Agency, Washington, DC, 1979.
11. F. T. Princotta, and C.B. Sedman, *Technological Options for Acid Rain Control*, The Electric Utility Business Environment Conference, 1993.
12. Federal Register US Congress, *Standards of Performance for New Stationary Sources*, 36:247, Part II, December 23, 1971.
13. Federal Register US Congress, *Standards of Performance for New Stationary Sources*, 44:133, Part II, June 11, 1979.
14. US DOE, *Clean Coal Technology Demonstration Program: Program Update 2000*, Report No. DOE/FE-0437, US Department of Energy, Washington, DC, 2001.
15. US EPA, *Utility FGD Study: October–December 1979*, Report No. EPA-600/7-80-029a, US Environmental Protection Agency, Washington, DC, 1980.
16. US EPA, *Survey of Dry SO_2 Control Systems*, Report No. EPA-600/7-80-030, US Environmental Protection Agency, Washington, DC,1980.
17. US EPA, *Field Tests of Industrial Stoker Coal-Fired Boilers for Emissions Control and Efficiency Improvement*, Report No. EPA-600/7-80-065a, US Environmental Protection Agency, Washington, DC, 1980.
18. US EPA, *Environmental Considerations of Energy-Conserving Industrial Process Changes. Executive Briefing;* Report No. EPA-625/9-77-001 US Environmental Protection Agency, Washington, DC, 1977.
19. US EPA, *Methods of Development for Assessing Air Pollution Control Benefits. Volume V, Executive Summary,* Report No. EPA-600/5-79-001e, US Environmental Protection Agency, Washington, DC, 1979.
20. US EPA, *Review of New Source Performance Standards for Coal-Fired Utility Boilers. Volume II, Economic and Financial Impacts*, Report No. EPA-600/7-78-155b, US Environmental Protection Agency, Washington, DC, 1978.
21. B. Sims, *Environ. Protect.* **12**(11), 34–36 (2001).
22. Editor, Clean air challenges. *Times Union*, p. A10. March 8, 2002.
23. A. Attar, *Hydrocarbon Process.* **58**, 175–79 (1979).
24. World Bank, *Coal Cleaning*, World Bank, Washington, DC, 2002; available at world bank.org/html/fpd/em/power/EA/mitigation/aqsocc.stm.
25. R. H. Yoon, D. I. Phillips, and G. H. Luttrell. *Applications of Advanced Fine Coal Cleaning and Dewatering Technologies*, Center for Coal and Minerals Processing, Virginia Polytechnic Institute and State University, Blacksburg, VA, 2002; available at egcfe.fossil.energy.gov/7thtech/p246.pdf

26. Electric Power Research Institute. *Coal Pretreatment to Enhance Conversion*, Report No. EPRI TR-101810, Electric Power Research Institute, Palo Alto, CA, 1994.

27. Electric Power Research Institute, *Proceedings: Advanced Physical Coal Cleaning—A State-of-the-Art Review*, Report No. EPRI TR-102635, Electric Power Research Institute, Palo Alto, CA, 1993.

28. I. R. Blake and G. Eldrige, *Proceedings of the 13th International Coal Preparation Exhibition and Conference*, 1996, pp. 87–106.

29. A. Patwardhan, Y. P. Chugh, M. K. Mohanty, and H. Sevim, *Comparative Economics of Advanced Fine Coal Cleaning in Refuse Pond Recovery and Active Mine Applications*, The Society for Mining, Metallurgy, and Exploration, Littleton, CO, 2002, SME Paper 02-043.

30. M. D. Mann, R. C. Timpe, C. M. Anderson, et al., *Environmental Aspects of Hydrothermal Treatment*, Center for Air Toxic Metals, Energy and Environmental Research Center, 2002; available at www.eerc.und.nodak.edu/catm/program_summaries/env_aspects_t.htm.

31. J. T. Capps, *Refiner Outlook for Clean Fuels, Hart World Fuels Conference*, 2001; available at www.chemweek.com/worldfuels/dc01/pdf/JTravis_Capps_Valero.pdf.

32. W. F. Lawson, Third US–China Oil and Gas Industry Forum, 2001.

33. US DOE, *The Transition to Ultra-Low-Sulfur Diesel Fuel Effects on Process and Supply*, Energy Information Administration, US Department of Energy, Washington, DC, 2001; available at www.eia.doe.gov/oiaf/servicerpt/ulsd/chapter6.html.

34. X. Ma, *Ultra Clean Transportation Fuels by Deep Desulfurization*, Pennsylvania State University Energy Institute, 2002; available at spacedaily.com/news/fuel-02e.html.

35. P. Pearce, *Sulfur in U.S. Military Jet Fuels, Fuels Branch*, US Air Force Research Laboratory, 2002; available at www.crcao.com/aviation/2002%20Aviation%20Meetings/ Wednesday/Pearce,%20P.%20-%20Operational%20Problems.ppt.

36. SRI International, *Highly Dispersed Catalysts for Coal Liquefaction*. Final Report to U.S. DOE, Pittsburgh Energy Technology Center, 1995; available at www.lanl.gov/cgi-bin/byte server.pl/projects/cctc/resources/pdfsmisc/coalliqf/M96002425.pdf.

37. Clean Coal Engineering & Research Center of Coal Industry (CCERC), Coal liquefaction; available at www.cct.org.cn/cct/ENG/7/content/0701-9.htm (accessed 2000).

38. US DOE, FETC's (Federal Energy Technology Center) Fluidized-Bed Combustion Program for 1998, presented by D.L. Bonk, 1998; available at www.netl.doe.gov/publications/pro ceedings/98/98ps/ps1-3.pdf.

39. IEA Coal Research, Pressurized fluidized bed combustion (PFBC); available at www.iea coal.org.uk/CCTdatabase/pfbc.htm (accessed 2002).

40. US DOE, *Selective Catalytic Oxidation of Hydrogen Sulfide for Simultaneous Coal Gas Desulfurization and Direct Sulfur Production*, National Energy Technology Laboratory, 2002.

41. US DOE, *Clean Coal Reference Plants: Integrated Gasification Combined Cycle*, Texaco, Draft Report, National Energy Technology Laboratory, 2001.

42. Biocube, Inc., www.biocube.com/hydrogen_sulfide_removal.htm (accessed 2002).

43. D. G. Jones, *Lime/Limestone Scrubber Operation and Control Study*, Research Project 630-2. Prepared for Electric Power Research Institute, Palo Alto, CA, 1978.

44. E. O. Smith, *Lime FGD Systems Data Book*, 2nd ed., Research Project 982-23, prepared for Electric Power Research Institute, Palo Alto, CA, 1983.

45. US DOE, *Utility FGD Survey January–December 1988*, Report No. ORNL/Sub/86-57949, US Department of Energy, Washington, DC, 1991.

46. IEA, *FGD Installations on Coal-Fired Plants*, IEA Coal Research Limited, London, 1994, pp. 36–42.

47. E. O. Smith, *Lime FGD Systems Data Book*, Research Project 1857-1, Prepared for Electric Power Research Institute, Palo Alto, CA, 1983.

48. H. Rosenberg, *Lime FGD Systems Data Book*, Research Project 982-1, prepared for the Electric Power Research Institute, Palo Alto, CA, 1979.

49. US EPA, *EPA Research Outlook*, Report No. EPA-600/9-80-005, US Environmental Protection Agency, Washington, DC, 1980.

50. US EPA, *EPA Research Highlights*, Report No. EPA-600/9-80-005, US Environmental Protection Agency, Washington, DC, 1980.

51. US EPA, *EPA/ORD Program Guide*, Report No. EPA-600/9-79-038, US Environmental Protection Agency, Washington, DC, 1979.

52. US EPA, *Sulfur Oxides Control in Japan* EPA Decision Series Report No. EPA-600/9-79-043, US Environmental Protection Agency, Washington, DC, 1979.

53. US EPA, *Coal Cleaning with Scrubbing for Sulfur Control: An Engineering/Economic Summary*, EPA Decision Series, Report No. EPA-600/9-77-017, US Environmental Protection Agency, Washington, DC, 1977.

54. US EPA, *Limestone FGD Scrubbers Users Handbook*, US Environmental Protection Agency, Industrial Environmental Research Laboratory, Research Triangle Park, NC, 1981.

55. US EPA, *Flue Gas Desulfurization Technologies for Control of Sulfur Oxides*, Report No. EPA/600/F-95/013, US Environmental Protection Agency, Washington, DC, 1995.

56. US DOE, *Failure Mode Analysis for Lime/Limestone FGD Systems. Volume I—Description of Study and Analysis of Results*, prepared for US Department of Energy, Morgantown, WV, 1984. Report No. DOE/METC/84-26 (DE84011958).

57. R. D. Delleny, and P. K. Beekley, *Process Instrumentation and Control in SO_2 Scrubbers*, Research Project 2249-1, Electric Power Research Institute, Palo Alto, CA, 1984.

58. NCA, *Steam Electric Plant Factors*. National Coal Association, Washington, DC, 1983.

59. M. A. Vuchetich, and R. J. Savoi, *Proceedings of Conference on Electrostatic Precipitator Technology for Coal-Fired Power Plants*, Report No. EPRI CS-2908, Electric Power Research Institute, Palo Alto, CA, 1983.

60. US EPA, *Flue Gas Desulfurization Inspection and Performance Evaluation*, Report No. EPA/625/1-95/019, US Environmental Protection Agency, Washington, DC, 1985.

61. US DOE, *Coolside and LIMB: Sorbent Injection Demonstrations Nearing Completion. Clean Coal Technology* Topical Report No. 2, Office of Communication, Washington, DC, 1990.

62. US DOE, *Reduction of NO_x and SO_2 Using Gas Reburning, Sorbent Injection and Integrated Technologies*. Clean Coal Technology Topical Report No. 3, Office of Communication, Washington, DC, 1993.

63. B. W. Hall, *JAWMA* **42**, 103 (1992).

64. US EPA, *Conference on Environmental Commerce, CONEC '93,*1993.

65. US EPA, *Compilation of Air Pollutant Emission Factor*, 3rd ed. (including Supplements 1–13), US Enviromental Protection Agency, Research Triangle Park, NC, 1977.

66. L.-S. Fan, and R. A. Jadhav, *AICHE J.* **48**(10), 2115–2123 (2002).

67. EIA (Energy Information Administration), *International Energy Outlook—2002*, Report No. Office of Integrated Analysis and Forecasting, US Department of Energy, Washington, DC, 2002. DOE/EIA-0484.

68. L.-S. Fan, A. Ghosh-Dastidar, and S. Mahuli, US Patent 5779464, 1998.

69. H. Gupta, and L.-S. Fan, *Ind. Eng. Chem. Res.* **41**, 4035 (2002).

70. D. H. Vanderholm, D. L. Day, A. J. Muehling, and Y. T. Hung, *Management of Livestock Wastes*, Technical Report No. PB2000-100255, US Department of Commerce, National Technical Information Service, Springfield, VA, 1999.

71. Editor, *Environ. Protect.* **14**(2) (2003).

72. B. Buecker and P. Dyer, *Environ. Protect.* **15**(6), 52–58 (2004).

73. L. K. Wang, J. Taricska, Y. T. Hung, J. E. Eldridge, and K. H. Li, Wet and dry scrubbing, Chapter 5 in *Air Pollution Control Engineering*, Humana Press, Totowa, NJ, 2004, pp. 197–306.

3
Carbon Sequestration

Robert L. Kane and Daniel E. Klein

CONTENTS

1. INTRODUCTION

1.1. General Description

"Carbon sequestration" refers to a portfolio of activities for the capture, separation and storage or reuse of carbon or CO_2. Carbon sequestration technologies encompass both the prevention of CO_2 emissions into the atmosphere as well as the removal of CO_2 already in the atmosphere.

Along with energy efficiency and lower carbon fuels, carbon sequestration represents a third pathway to reduce GHG emissions. This chapter presents an overview of the US Department of Energy's (DOE's) Carbon Sequestration Research and Development (R&D) Program. It outlines the road map developed by the DOE for government, industry, and academia to begin setting R&D directions related to carbon sequestration. DOE's R&D program concentrates on innovative sequestration concepts for longer-term solutions. This program covers research on (1) CO_2 capture and separation processes, (2) long-term storage options, especially in a variety of geologic formations, (3) enhancement of natural processes in terrestrial sinks, and (4) chemical or biological fixation for storage and/or beneficial use. With continued research, these technologies have the potential to provide a cost-effective mitigation option for limiting greenhouse gas (GHG)

From: *Handbook of Environmental Engineering, Volume 2: Advanced Air and Noise Pollution Control*
Edited by: L. K. Wang, N. C. Pereira and Y.-T. Hung © The Humana Press, Inc., Totowa, NJ

emissions and ultimately stabilizing GHG concentrations in the atmosphere. Consistent with the objectives of the Framework Convention on Climate Change, it is judicious to explore all potential GHG mitigation options in a comprehensive way, so that a broad range of strategies is available to help meet future policy goals.

1.2. Carbon Sequestration Process Description

Fossil fuels have been a major contributor to the high standard of living enjoyed by the industrialized world. However, possible requirements to reduce GHG emissions may limit or alter their use in the future. The major GHG is carbon dioxide, and fossil energy combustion accounts for the vast majority of anthropogenic (human-induced) CO_2.

Climate change is one of the primary environmental concerns of the 21st century. No single issue is as complex or holds as many potential implications for the world's inhabitants. Our response to this issue could dictate fundamental changes in how we generate and use energy.

By 2020, the world's appetite for energy is likely to be about 75% higher than what it was in 1990 (1). Without changes in energy policies, environmental policies, and/or technologies, global carbon emissions are also forecast to increase nearly 70% from 1990 levels. Atmospheric concentrations of CO_2 are currently about 30% above preindustrial levels and are rising. These rising concentrations are the focus of the United Nations Framework Convention on Climate Change (FCCC), which was ratified in 1992 and entered into force in 1994. The FCCC sets an "ultimate objective" of stabilizing "GHG concentrations in the atmosphere at a level that would prevent dangerous anthropogenic interference with the climate system" (2).

Many nations have begun taking actions to reduce or limit the growth of emissions of GHGs. Yet, most reductions in emissions would still lead to increasing concentrations of CO_2 in the atmosphere, because we are emitting faster than our terrestrial and ocean sinks can absorb them. To stabilize atmospheric CO_2 concentrations, even at double their current level, would require cutting global emissions by almost 70% relative to their 1990 levels.

The enormity of this challenge underscores the opportunities for new ideas and new technologies. Under assumptions of "business as usual," even factoring in anticipated trends in technologies and efficiencies, the direct costs of meeting long-term atmospheric concentration goals in the United States alone are measured in the hundreds of billions of dollars, and the worldwide costs are several times that amount. However, if advanced technologies can be developed and deployed, the total costs could be reduced by about half (3). Hence, the opportunities—like the challenges—are also enormous.

There are three primary means to reduce CO_2 emissions associated with energy production without reducing economic output:

1. Improve the efficiency of energy conversion and end-use processes.
2. Shift to lower-carbon-content fuels (including noncarbon sources, such as renewable energy and nuclear power).
3. Sequester the carbon released in energy production.

To reduce GHG emissions effectively and economically, we must be prepared to use all three of these methods. To date, most CO_2 mitigation strategies have focused on the first two, and these are considered by many to be the best and most cost-effective first

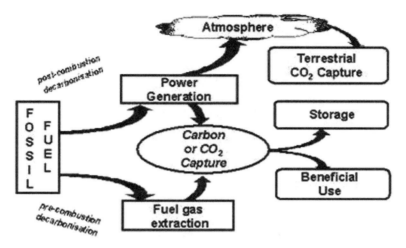

Fig. 1. What is "carbon sequestration"?

steps in managing GHG emissions. However, although energy efficiency measures and low-carbon fuels can reduce emissions, it is questionable whether they are sufficient to stabilize CO_2 concentrations, given continuing economic and population growth and the extent of emissions reductions that stabilizing CO_2 concentrations would entail. Therefore, it is prudent to investigate the role that can be played by the third pathway: carbon sequestration.

Many people consider the term "carbon sequestration" to be limited to the uptake of CO_2 by trees and other plants through photosynthesis and their storing it as carbon. However, carbon sequestration can encompass much more. Another form of sequestration—injecting CO_2 into partially depleted oil reservoirs—is already underway to enhance oil production. CO_2 could also be injected into unmineable coal seams, thus enhancing the recovery of the coal-bed methane. However, these are only current examples of the many sequestration options that may someday be technologically and economically available. More broadly, carbon sequestration is the removal of GHGs, either directly from the exhaust streams of industrial or utility plants or indirectly from the atmosphere, and storing them long term so that they cannot interact with the climate system (*see* Fig. 1).

The US DOE believes there may be new, innovative concepts for sequestration. The question is whether any of these ideas can be developed into practical, low-cost approaches. Efforts to date have shown both progress and promise, garnering important support from political leaders and the business community.

Importantly, carbon sequestration is a concept that is both compatible with the current energy infrastructure and a bridge to future energy systems. The major advantage of our present fossil-fuel-based energy system is, quite simply, that it works. It is relatively low cost. It uses low-cost and globally abundant resources, and it represents a huge capital investment in a global infrastructure. It will not be - nor should it be— discarded overnight (4).

In March 2001, in a letter to four senators, President George W. Bush of the United States defined his Administration's policy regarding the reduction of CO_2 emissions from US power plants. He cited "the lack of commercially available technologies for removing and storing carbon dioxide" and he said that the Administration would "continue to fully examine global climate change issues," with the goal of developing technologies and other creative ways to address global climate change. The president's letter describes the difficulties inherent in balancing energy, environmental, and economic goals. At the same time, it highlights the opportunities that new technologies—such as carbon sequestration—can play in a comprehensive and balanced national energy policy.

These views were expanded in remarks made by President Bush on global climate change in June 2001. The president stated that "there are only two ways to stabilize concentration of GHGs. One is to avoid emitting them in the first place; the other is to try to capture them after they're created..... America is the leader in technology and innovation. We all believe technology offers great promise to significantly reduce emissions—especially carbon capture, storage and sequestration technologies" (5).

As part of his June 11, 2001 announcement, President Bush directed the Secretaries of Energy and Commerce, along with the Administrator of the EPA, to develop a National Climate Change Technology Initiative (NCCTI). The NCCTI will develop innovative approaches in accordance with several basic principles, as outlined by the president. The approaches will (1) be consistent with the long-term goal of stabilizing greenhouse gas concentrations in the atmosphere, (2) be measured, as we learn more from science and build on it, (3) be flexible to adjust to new information and take advantage of new technology, (4) ensure continued economic growth and prosperity, and (5) pursue market-based incentives and spur technological innovation.

2. DEVELOPMENT OF A CARBON SEQUESTRATION ROAD MAP

Climate change is not a 10-yr or even a 20-yr challenge. It is a challenge best measured in generations rather than years or even decades (6). That is why long-term options must be considered. We must include—perhaps even concentrate on—options that offer progress toward stabilizing GHG concentrations on a global scale. A carbon sequestration strategy represents a long-term R&D approach that, if successfully developed, could offer a set of new options for dealing with GHGs, most likely in the post-2015 time frame.

Present technologies for carbon capture are not currently affordable, entail high energy penalties, and are limited in scope. To be viable, carbon sequestration will need to be less expensive, more efficient, and of higher capacity. Accordingly, the DOE has established the following program goals to guide its activities (7):

1. Provide economically competitive and environmentally safe options to offset all projected growth in baseline emissions of GHGs by the United States after 2010, with offsets starting in 2015:
2. Achieve a long-term cost goal for carbon sequestration in the range of US$10/ton of avoided net costs.
3. Offset at least one-half the required reductions in global GHG emissions, measured as the difference between a business-as-usual baseline and the emissions level corresponding to a concentration of 550 ppm CO_2 beginning in the year 2025.

The DOE has been developing and refining a road map for setting R&D directions related to carbon sequestration. As part of this road-mapping effort, the DOE drafted a report detailing the emerging science and technology of carbon sequestration (8), describing the research that the DOE is examining in the following areas:

1. System studies and assessments
2. Enhanced natural sinks
3. Capture and separation technology
4. Geologic storage
5. Ocean sequestration
6. semical and biological fixation and reuse

In the near term, value-added sequestration applications will provide a cost-effective means of reducing emissions and provide collateral benefits in terms of increased domestic production of oil and gas. In the mid-term and long term, advanced CO_2 capture technology and integrated CO_2 capture, storage, and conversion systems will provide cost-effective options for deep reductions in GHG emissions.

Private public partnerships and cost-shared R&D are a critical part of technology development for carbon sequestration. These relationships draw on pertinent capabilities that the petroleum and chemical industries have built up over decades, important knowledge and capabilities of the US Geological Survey, academia, and the National Laboratories, and collaborative efforts for the identification and selection of promising research pathways. Through partnerships, the program has been able to harness these capabilities and direct them to the goal of carbon sequestration technology development.

Recognizing that the need for new science and technologies to reduce GHG emissions is a global concern, the DOE's Sequestration Program is deeply engaged in building international partnerships throughout the world. These international partnerships take various forms, ranging from science and technology information exchanges to formal coordinated funding of cooperative research worldwide. As global interest and funding of carbon sequestration research grows, these collaborations will expand.

3. TERRESTRIAL SEQUESTRATION

Terrestrial ecosystems include both vegetation and soils containing microbial and invertebrate communities. They are widely recognized as major biological "scrubbers" for CO_2. Terrestrial sequestration is defined as either the net removal of CO_2 from the atmosphere or the prevention of CO_2 emissions from leaving terrestrial ecosystems.

The terrestrial biosphere is estimated to sequester large amounts of carbon—about 2 billion metric tons annually. The total amount of carbon stored in soils and vegetation throughout the world is estimated to be roughly 2 trillion metric tons (8). Hence, even a small change in the CO_2 flows could amount to large additional amounts sequestered.

Enhancing the natural processes that remove CO_2 from the atmosphere is thought to be one of the most cost-effective means of reducing atmospheric levels of CO_2, and forestation and deforestation-abatement efforts are already under way. Terrestrial sequestration can be enhanced in four ways:

1. Reversing land-use patterns
2. Reducing the decomposition of organic matter
3. Increasing the photosynthetic carbon fixation of trees and other vegetation
4. Creating energy offsets using biomass for fuels and other products

Table 1
Selected US DOE Projects in Terrestrial Sequestration

Research partners	Location	Research objective
Ohio State University	Columbus, OH, USA	Use of solid-waste-derived soil
Virginia Polytechnic Institute	Blacksburg, VA, USA	enhancers to improve carbon
		upadate of disturbed lands
Stephen F. Austin State University	Nacogdoches, TX, USA	Reclamation/reforestation program to sequester carbon on Appalachain
Texas Utilities (TXU)	Dallas, TX, USA	abandoned mine lands
Tennessee Valley Authority	Drakesboro, KY, USA	Amend soils using the Paradise Plant scrubber's gypsum byproducts and clarified wastewater for irrigation, both enhancing nature's "biological scrubbers" for CO_2
The Nature Conservancy	Arlington, VA, USA	Develop and implement various forestry sequestration projects and refine the tools and methods for measuring their long-term carbon storage potential

Sources: U.S. Department of Energy, DOE Fossil Energy Techline (9–12).

This program area is focused on integrating measures for improving the full life-cycle carbon uptake of terrestrial ecosystems, including farmlands and forests, with fossil fuel production and use. The efforts are being conducted in collaboration with the DOE Office of Science and the US Forest Service.

Research is already underway. Table 1 describes some of the recent project awards in these research areas. These projects also demonstrate the multiple benefits that often accompany terrestrial sequestration in the form of improved soil and water quality, better wildlife habitats, increased water conservation, and the like.

4. CO_2 SEPARATION AND CAPTURE

The idea of capturing CO_2 from the flue gases of power plants did not originate with GHG concerns. Rather, it initially gained attention as a possible source of supply of CO_2 especially for use in enhanced oil recovery (EOR) operations. In EOR, CO_2 is injected into oil reservoirs to increase the mobility of the oil and, thus, the productivity of the reservoir.

Similar opportunities for CO_2 sequestration may exist in the production of hydrogen-rich fuels (e.g., hydrogen or methanol) from carbon-rich feedstocks (e.g., natural gas, coal, or biomass). Such fuels could be used in low-temperature fuel cells for transport or for combined heat and power. Relatively pure CO_2 would result as a byproduct (13).

Roughly one-third of the United States' anthropogenic CO_2 emissions come from power plants. Given the enormous infrastructure of combustion-based power-generation assets in the United States, developing technologies to capture CO_2 from flue gas (post-combustion capture) is a major thrust and a wide range of options are being pursued.

To date, all commercial plants to capture CO_2 from power plant flue gas use processes based on chemical absorption with a monoethanolamine (MEA) solvent. MEA was

Table 2
Typical Energy Penalties Resulting from CO_2 Capture

Power plant type	Today	Future
Conventional coal	27–37% (15)	15% (14)
Gas	15–24% (15)	10–11% (14)
Advanced coal	13–17% (15)	9% (15)

developed over 60 yr ago as a general, nonselective solvent to remove acid gases such as CO_2 and H_2S from natural gas streams. The process was modified to incorporate inhibitors to resist solvent degradation and equipment corrosion when applied to CO_2 capture from flue gas. Also, the solvent strength was kept relatively low, resulting in large equipment and high regeneration energy requirements (14).

Therefore, CO_2 capture processes have required significant amounts of energy, which reduces the power plant's net power output. For example, the output of a 500-MWe (net) coal-fired power plant may be reduced to 400 MWe (net) after CO_2 capture, an "energy penalty" of 20%. The energy penalty has a major effect on the overall costs. Table 2 shows typical energy penalties associated with CO_2 capture, both as the technology exists today and as it is projected to evolve in the next 10–20 yr (14,15).

There are numerous options for the separation and capture of CO_2, many of which are commercially available. Many advanced methods are also under development, such as adsorbing CO_2 on zeolites or carbon-bonded activated fibers and then separating it using inorganic membranes (16). However, none have been applied at the scale required as part of a CO_2 emissions mitigation strategy, nor has any method been demonstrated for a broad range of anthropogenic CO_2 sources. Additional research into advanced processes seeks to improve the potential of these options.

Several major studies have analyzed the economics of capturing CO_2 from the flue gas of coal-fired power plants. These studies looked at CO_2 capture from pulverized coal (PC) power plants and from integrated gasification combined cycle (IGCC) power plants. MEA scrubbing was used in the PC plants, whereas IGCC plants allowed the use of more energy-efficient scrubbing processes involving physical absorption. All studies used commercially available technology and included the cost of compressing the captured CO_2 to about 2000 psia for pipeline transportation (17–22).

For PC plants, the cost of reducing CO_2 emissions is in the range of about US\$30–70/metric ton. of CO_2 avoided. For IGCC plants, the cost of reducing CO_2 emissions is about US\$20–30/metric ton. These costs exclude storage, which might add an additional US\$5–15/metric ton.

In general, these findings are very encouraging. Already, in Norway, under an existing carbon tax, a power plant with CO_2 capture for sequestration is scheduled to be built within the next few years. However, the viability of CO_2 capture from power plants should not be judged based on today's relatively expensive technology. There is great potential for technological improvements that can significantly lower costs. For example, improving the heat rate of fossil plants or reducing the energy penalty for CO_2 capture could significantly reduce costs. For coal-fired power plants, achieving a 50% thermal

Table 3
Selected US DOE Projects in CO_2 Separation and Advanced Boiler Design

Research partners	Location	Research objective
BP Corporation	Anchorage, AK, USA	Demonstrate the feasibility of capturing CO_2 from a variety of fuel types and combustion sources and storing it in unmineable coal seams and saline formations
Los Alamos National Laboratory	Los Alamos, NM, USA	Develop an improved high-temperature polymer membrane for separating CO_2 from methane and nitrogen gas streams
Idaho National Engineering and Environmental laboratory	Idaho Falls, ID, USA	
University of Colorado	Boulder, CO, USA	
Pall Corp.	East Hills, NY, USA	
Shell Oil Co.	Houston, TX, USA	
Media and Process Technology Co.	Pittsburgh, PA, USA	Develop a high-temperature CO_2-selective membrane as a reactor, which can enhance the water–gas shift reaction efficiency while recovering CO_2 simultaneously
Research Triangle Institute	Research Triangle Park, NC, USA	Develop a simple, low-cost CO_2-separation technology with a reusable, sodium-based sorbent to capture CO_2 from existing source flue gas
Alstom Power, Inc.	Windsor, CT, USA	Produce concentrated CO_2 by firing oxygen (rather than air) in an advanced boiler
Praxair, Inc.	Tonawanda, NY, USA	Develop a novel "oxy-fuel" boiler design using a membrane to separate oxygen from combustion intake air

Sources: U.S. Department of Energy, DOE Fossil Energy/Techline (9,10,12).

efficiency (i.e., a 30% decrease in heat rates) would result in a 30% decrease in capture costs alone. As Table 2 suggests, evolutionary developments could halve the energy penalties, thereby cutting capture costs about 40%.

Research and development is required to help fill the many gaps in the science and technology. Challenges exist in the areas of chemical and physical absorption and adsorption, low-temperature distillation, gas-separation membranes, and product treatment and conversion. Research will be needed to achieve the desired breakthroughs in cost, efficiency, and safety.

Several DOE projects are seeking technologies to lower the costs and improve the separation of CO_2 from the gas streams of energy facilities and other sources. Other research efforts are focusing on new boiler designs that utilize higher ratios of oxygen in the input airstream; among other effects, the CO_2 in the exhaust air is more concentrated, lending it to more economical capture and/or reuse. Table 3 describes some of the recent project awards in these research areas.

Table 4
Geology CO$_2$ Storage: Worldwide Potential

Storage option	Capacity (billion tons CO$_2$)
Deep saline reservoirs	400–10,000
Depleted oil and gas reservoirs	920
Coal seams	150++

Sources: IEA Greenhouse Gas R&D Programme (23) and Advanced Resources International estimates for coal seams (24).

5. GEOLOGIC SEQUESTRATION OPTIONS

Once captured, CO$_2$ needs to be sequestered. There are a variety of potential geologic sequestration options for long-term storage. This is not really a new concept. For example, CO$_2$ is currently injected into more than 70 operating oil fields to enhance oil production. What is new is the idea that storage of CO$_2$ is a desirable goal in and of itself.

Estimates of global storage potential span a broad range, as research into this topic is still at an early stage. However as depicted in Table 4 (23,24), the storage potential of these geologic options is enormous, possibly measured in trillions of metric tons. This is many times larger than total worldwide energy-related CO$_2$ emissions, estimated at about 22.3 billion metric tons of CO$_2$ in 1999 (1).

Deep saline reservoirs may be the best long-term underground storage option. Such reservoirs normally are too salty to provide potable water supplies and are generally hydraulically separate from more shallow reservoirs and surface water. Depending on the reservoir, injected CO$_2$ would displace the saline water, with some of the CO$_2$ dissolving, some reacting with the solids, and some remaining as pure CO$_2$. Deep saline reservoirs are located throughout much of the United States, so perhaps 65% of the CO$_2$ emitted from US power plants could be injected into them without the need for long pipelines. In the United States alone, the estimated storage potential of deep saline reservoirs ranges from 5 to 500 billion metric tons of CO$_2$.

There is already considerable experience with the use of deep saline reservoirs for storing large quantities of fluids. Over 167,000 oil and gas injection wells inject over about 2.5 billion metric tons./yr (25) and over 9 billion gal/yr, or 33 million metric tons./yr, of hazardous wastes are injected into wells.

The first commercial CO$_2$ capture and sequestration facility began in September 1996, when Statoil of Norway began storing CO$_2$ from the Sleipner West gas field in a sandstone formation 1000 m beneath the North Sea. This CO$_2$ derives from natural gas processing, where the CO$_2$ content of the extracted natural gas needs to be reduced from 9.5% to 2.5%. The economic incentive for this project is the Norwegian carbon tax (currently US$38/metric ton CO$_2$).

Instead of being vented, the captured CO$_2$ at the Sleipner Project is injected from a floating rig through five pipes at a rate of 20,000 metric tons/wk, as shown in Fig. 2 (26). (This corresponds to the rate of CO$_2$ produced from a 140-MWe coal-fired power plant.) Earlier pilot studies showed that most of the CO$_2$ will react to form solid calcite, with some dissolving in the groundwater and some remaining as a separate phase.

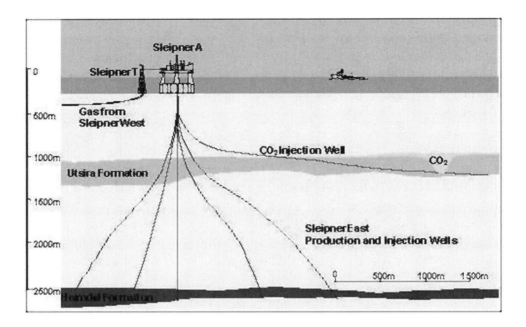

Fig. 2. Sleipner CO_2 injection project (courtesy of Tore A. Torp, Statoil).

Preliminary hydrogeologic and geochemical modeling showed that there is enormous potential for CO_2 sequestration in the midwestern United States, especially in the Mt. Simon Sandstone in the Illinois Basin. This capacity appears to be sufficient for storing emissions for several decades or more. However, the modeling also indicated that local factors, such as formation thickness, permeability, injectivity, and geochemistry, significantly influence the technical feasibility and cost-effectiveness of this technology. Accordingly, site-specific assessments will be an important step in better understanding this regional capacity. A realistic pilot test over a year or more would enable this concept to move beyond the laboratory and computer simulation stage.

Depleting oil and gas reservoirs also appear to be a promising land storage option. Because they have previously contained hydrocarbon gases for thousands of years, their geologic integrity is likely to be good. In the United States, currently abandoned oil and gas reservoirs could hold about 3 billion metric tons of CO_2. Over time, as the United States draws more oil and gas from its reservoirs, the inventory of depleted reservoirs will increase. The ultimate CO_2 storage capacity for the United States could be around 100 billion metric tons of CO_2. Several issues must be resolved, however, before these can be considered viable candidates for CO_2 storage.

In southern Saskatchewan, PanCanadian Petroleum is developing the Weyburn CO_2 EOR project, which began operation in September 2000. CO_2—96% pure—from the Dakota Gasification facility in North Dakota is injected into the Weyburn oil reservoir at a rate of 50 million cubic feet per day. In addition to extending the life of this field by another 25 yr, the project provides an excellent opportunity for developing an understanding of how CO_2 is stored underground. The performers will employ seismic methods and geochemical sampling to evaluate the distribution of CO_2 in the reservoir and the chemical reactions that occur between the CO_2 and the reservoir rock and fluids. Because the field is relatively shallow and has been extensively developed, there are numerous observation wells and a comprehensive understanding of the geology of the field. It is estimated that about half of the injected CO_2 will be locked up in the oil that remains in the ground and that over the 20-yr lifetime of the project, about 19 million metric tons of CO_2 will be stored (27). By developing an understanding of the fate of the CO_2 in the oil reservoir, confidence in EOR as a CO_2 storage option can grow.

Abandoned or uneconomic coal seams could also become CO_2 storage sites, because CO_2 injected into coal adheres to the coal surface and remains within the seam. Some of the early research regarding CO_2 storage in deep coal seams is particularly intriguing because many coal deposits contain methane—itself a potent GHG. The injected CO_2 displaces the sorbed methane from the coal surface, with two molecules of CO_2 being trapped for each molecule of methane released (28). By injecting CO_2 into methane-rich unmineable coal deposits, more methane is produced for energy.

United States coal-bed methane production is largely concentrated in the San Juan Basin in the southwestern United States and in Alabama's Black Warrior Basin. About 8 billion metric tons of CO_2 sequestration potential is contained in the San Juan Basin, along with enhanced recovery of the methane (24). Worldwide, storage potential is estimated to be upward of 150 billion metric tons.

CO_2-enhanced coal-bed methane recovery is being tested in the vast, deep coal beds in Alberta, Canada. In a process called enhanced gas recovery (EGR), CO_2 is injected into deep, unmineable coal beds to recover methane (29). The initial phases of the project, led by the Alberta Research Council, have been favorable, and the effort is now moving into the design and implementation of a full-scale pilot project. Injection began in 2000 and will continue for 12 mo; if successful, full-scale development can begin in 2002.

Several DOE projects are directed at identifying and resolving technical and environmental issues in sequestering CO_2 in a variety of geological formations. Table 5 describes some of the recent project awards in these research areas.

6. OCEAN SEQUESTRATION

The world's oceans may be a large potential sink for anthropogenic CO_2 emissions. However, because the oceans play such an important role in sustaining the biosphere, any potential changes to these ecosystem functions must be carefully and thoroughly considered. We currently have very little knowledge of how the potential pH change or other impacts resulting from CO_2 injection would affect the biogeochemistry and ecosystems in the deep ocean. Improvements in understanding marine systems are needed before marine sequestration could be implemented on a large scale.

An initial test of deep-sea CO_2 release has provided promising findings (30,31). A remote-operated vehicle was used to drip a small amount of liquid CO_2 into a 4-L laboratory beaker on the ocean floor, roughly 2 miles below the surface. At this depth, the CO_2 has a higher density than seawater and sinks rather than rises. The CO_2 quickly combined with water to form a block of icelike hydrate. It is expected that the hydrate

Table 5
Selected US DOE Projects in Geologic Sequestration

Research partners	Location	Research objective
Advanced Resources International	Houston, TX, USA	Demonstrate CO_2 sequestration in deep unmineable coal seams in the San Juan Basin using enhanced coal-bed methane recovery
Texas Tech University	Lubbock, TX, USA	Create a novel well-logging technique using nuclear magnetic resonance to characterize the geologic formation for long-term CO_2 storage, including the integrity and quality of the reservoir seal
University of Utah	Salt Lake City, UT, USA	Identify geochemical reactions in natural CO_2 fields in deep saline reservoirs, as analogs for repositories of CO_2 separated from the fuel gases of power plants

Source: U.S. Department of Energy, DOE Fossil Energy Techline (10).

would dissolve very slowly, which would minimize local concentration effects. Observations of nearby fish activity suggested that biological impacts would be small.

More tests of deep-sea carbon sequestration are scheduled, including one off the coast of Kona, Hawaii. In the proposed CO_2 Ocean Sequestration Field Experiment, a multinational group of researchers plans to inject about 40–60 tons of pure liquid CO_2 into ocean waters nearly 3000 ft (1 ft = 0.3048 m) deep over a period of about 2 wk. (32). The goals of this experiment are to gather data in the vicinity of the CO_2 injection point to improve our understanding of the basic physical phenomena and to use this in refining computer models to estimate potential environmental impacts.

To implement ocean CO_2 sequestration on a larger scale, several methods of injection have been proposed. One method is to transport the liquid CO_2 from shore via a pipeline and to discharge it from a manifold lying on the ocean bottom, forming a rising droplet plume. Another method is to transport the liquid CO_2 by tanker and then discharge it from a pipe towed by the moving ship. Still another approach is to inject the CO_2 as deeply as possible in order to maximize the sequestration efficiency. One such idea is to inject the liquid CO_2 to a sea-floor depression, forming a stable "deep lake" at a depth of about 4000 m.

7. CHEMICAL AND BIOLOGICAL FIXATION AND REUSE

The goal of CO_2 utilization is to design chemical processes that can convert CO_2 to useful and durable products that have reasonable lifetimes. Whereas storing CO_2 can mitigate the GHG problem, converting CO_2 to useful products can create additional economic and environmental benefits.

Research in advanced chemical and biological sequestration is aimed at permanent, stable sequestration and at recycling carbon to create new fuels, chemical feedstocks,

Table 6
Selected US DOE Projects in Advanced Processes

Research partners	Location	Research objective
Idaho National Engineering and Environmental Laboratory	Idaho Falls, ID, USA	Grow microorganisms known as cyanobacteria as "biofilms" that could capture and convert CO_2
Montana State University	Bozeman, MT, USA	through photosynthesis
University of Memphis	Memphis, TN, USA	
Ohio University	Athens, OH, USA	Attach photosynthetic organisms to specially designed growth surfaces in a bioreactor to minimize pressure drop and create a near-optimal enhanced photosynthesis process
Physical Science, Inc.	Andover, MA, USA	Developing photosynthesis technologies for selected species of microalgae to effective fix carbon from typical power plant exhaust gases
Yolo Country Planning and Public Works Department	Yolo Country, CA, USA	Full-scale demonstration of a new waste landfill "bioreactor" that traps landfill methane using special membrane on the surface and transports the gas to collection points

Source: U.S. Department of Energy, DOE Fossil Energy Techline (9,10).

and other products. Three possible end uses include particulate carbon in composite materials and construction materials, CO_2 as a feedstock for production of plastics, and carbon to create soil amendments.

Advanced chemical processes might lead to unique sequestration technologies or to improvements in our understanding of chemistry that will enhance the performance of other sequestration approaches. Advanced chemical technologies envisioned for the future would work with the technologies now being developed to convert recovered CO_2 economically into benign, inert, long-lived materials that can be contained and/or have commercial value.

Advanced biological processes can augment or improve natural biological processes for carbon sequestration from the atmosphere in terrestrial plants, aquatic photosynthetic species, and soil and other microbial communities. These technologies encompass the use of novel organisms, designed biological systems, and genetic improvements in metabolic networks in terrestrial and marine microbial, plant, and animal species.

All concepts for these technologies are at an early research stage. A better understanding of the basic processes and new chemistry and bioprocessing approaches is needed before practical, achievable technology performance or cost levels can be estimated.

The DOE projects in this area are directed at exploring novel chemical or biological methods for converting CO_2 into either commercial products or into inert, long-lived stable compounds. Table 6 describes some of the recent project awards in these research areas.

8. CONCLUDING THOUGHTS

To limit the costs of reducing emissions of GHGs, many technological options will be needed. To have these technological options available when we need them in the future, we need to do research today.

Carbon capture and sequestration is an area with great potential. Its potential benefits for our energy systems and our global environment are too great to ignore, and they warrant our best efforts to get it started right.

The US DOE has established an international leadership role in sequestration science and technology development through the Carbon Sequestration R&D Program. The program is managed by the Office of Fossil Energy and implemented by the National Energy Technology Laboratory. R&D management is strongly focused on public/private partnerships with industry, academia, state and local governments, and international entities.

Research and development in science and engineering are key to achieving breakthroughs in carbon sequestration. Advances in chemical sciences and the resulting technologies will enable CO_2 to be captured in greater quantities and at lower cost. Understanding the chemical interactions of CO_2 in underground and ocean environments will help us assess the longer-term fate and ultimate acceptability of these promising storage options. The transformation of CO_2 into materials with potential commercial value is a field in its infancy, but it may ultimately lead to the best uses of our planet's rich fossil-fuel resources.

The challenges may be great, but the rewards could be vastly greater.

NOMENCLATURE

CO_2	carbon dioxide
DOE	(United States) Department of Energy
EGR	enhanced gas recovery
EOR	enhanced oil recovery
FCCC	United Nations Framework Convention on Climate Change
GHG	greenhouse gas
IGCC	integrated gasification combined cycle (IGCC) power plants
MEA	monoethanolamine
MWe	megawatt equivalent
NCCTI	National Climate Change Technology Initiative
PC	pulverized coal
ppm	parts per million
R&D	research and development

ACKNOWLEDGMENT

Parts of this chapter are adapted from *Chemical Engineering Progress*, June 2001, with permission of the American Institute of Chemical Engineers.

REFERENCES

1. US DOE, *International Energy Outlook 2001*, Report No. DOE/EIA–0484 (2001), US DOE, Energy Information Administration, Washington, DC, 2001, Table 1; available at www.eia.doe.gov/oiaf/ieo/tbl_1.html (accessed March 2001).

2. United Nations Framework convention on climate change, available at www.unfccc.de/ resource/ conv/ conv_004.html (accessed May 1992).

3. S. H. Kim and J. A. Edmonds, *Potential for Advanced Carbon Capture and Sequestration Technologies in a Climate Constrained World,* Report No. PNNL-13095, Pacific Northwest National Laboratory, Washington, DC, 2000, p. 24.

4. R. S. Kripowicz, remarks given at the *Fourth International Conference on GHG Control Technologies,* 1998; available at www.fe.doe.gov/remarks/98_carbon.html.

5. President George W. Bush, President Bush discusses global climate change, The White House, Office of the Press Secretary, June 11, 2001; availble at http://www.whitehouse.gov/ news/releases/2001/06/20010611-2.html.

6. R. S. Kripowicz, *Carbon Sequestration Workshop*, US DOE, Gaithersburg, MD, 1999; available at www.fe.doe.gov/remarks/99_krip_seqconf.html.

7. US DOE, *Carbon Sequestration R&D Program Plan:* FY1999–FY2000, US DOE, National Energy Technology Lab, Pittsburgh, PA, 1999; p. 2; available at www.fe.doe.gov/ coal_power/sequestration/reports/programplans/99/index/shtml.

8. US DOE, *Carbon Sequestration Research and Development* Report No. DOE/SC/FE-1, US DOE, Office of Science and Office of Fossil Energy, Washington, DC, 1999; available at www.fe.doe.gov/coal_power/sequestration.

9. US DOE, DOE selects eight national lab projects as research to capture, store greenhouse gases expands, *DOE Fossil Energy Techline,* 2000; available at www.fe.doe.gov/techline/ tl_seqnatlb1.html.

10. US DOE, Energy Dept. launches thirteen new research projects to capture, store greenhouse gases, *DOE Fossil Energy Techline*, 2000; available at www.fe.doe.gov/techline/ tl_seq_ind1.html.

11. US DOE, Terrestrial carbon sequestration test underway at reclaimed mine site, *DOE Fossil Energy Techline,* 2001; available at www.fe.doe.gov/techline/tl_sequestration_kyproj.html.

12. US DOE, Energy department to study new ways to capture, store greenhouse gases, *DOE Fossil Energy Techline*, 2001; available at www.fe.doe.gov/techline/tl_sequestration_sel2.html.

13. R. Socolow, (ed.), *Fuels Decarbonization and Carbon Sequestration: Report of a Workshop,* Princeton University, Press, NJ, 1997; available at www.princeton.edu/~ceesdoe.

14. C. L. Leci. *Energ. Convers.Manag.* **38(Suppl.)**, S45–S50 (1997).

15. H. J. Herzog and E. M. Drake presented at the *23rd International Technical Conference on Coal Utilization and Fuel Systems,* 1998.

16. R. T. Watson, M. C. Zinyowera, and R. H. Moss, (eds.), *Climate Change 1995—Impacts, Adaptations, and Mitigation of Climate Change: Scientific-Technical Analyses, Contribution of Working Group II to the Second Assessment Report of the Intergovernmental Panel on Climate Change,* Cambridge University Press, New York, 1996.

17. R. D. Doctor, J. C. Molberg, and P. R. Thimmapuram, *KRW Oxygen-Blown Gasification Combined Cycle: Carbon Dioxide Recovery, Transport, and Disposal*, Report No. ANL/ESD-34, Argonne National Laboratory, Argonne, IL, 1996, as cited in ref. 15.

18. P. Condorelli, S. C. Smelser, and G. J. McCleary, *Engineering and Economic Evaluation of CO_2 Removal from Fossil-Fuel-Fired Power Plants, Volume 2: Coal Gasification Combined-Cycle Power Plants*, Report No. IE–7365, Electric Power Research Institute, Palo Alto, CA, 1991, as cited ref. 15.

19. C. A. Hendriks, *Carbon Dioxide Removal from Coal-fired Power Plants,* Kluwer Academic, Dordrecht, 1994, as cited in ref. 15.

20. H. Audus, P. Freund, and A. Smith, Global Warming Damage and the Benefits of Mitigation, IEA Greenhouse Gas R&D Programme, Cheltenham, UK, 1995, as cited ref. 15.

21. S. C. Smelser, R. M. Stock, and G. J. McCleary, *Engineering and Economic Evaluation of CO_2 Removal from Fossil-Fuel-Fired Power Plants, Volume 1: Pulverized-Coal-Fired* Power Plants, Report No. IE–7365, Electric Power Research Institute, Palo Alto, CA, 1991, as cited in ref. 15.

22. C. L. Mariz, *Sixth Petroleum Conference of the South Saskatchewan Section*, 1995, as cited ref. 15.

23. Putting Carbon Back in the Ground, IEA, Greenhouse Gas R&D Programme, Cheltenham, UK, 2001.

24. S. H. Stevens, et al., Presented at *Fourth International Conference on GHG Control Technologies* 1998.

25. US Environmental Protection Agency, *Underground Injection Control (UIC) Program*, US EPA, Washington, DC, 2001; available at www.epa.gov/safewater/uic.html.

26. T. A. Torp, presented at *5th International Conference on GHG Control Technologies (GHGT-5)*, 2000.

27. IEA, *CO$_2$ Enhanced Recovery in the Weyburn Oil Field: Report of a Workshop to Discuss Monitoring*, Report No. PH3/20, IEA Greenhouse Gas R&D Programme, Cheltenham, UK, 1999.

28. W. D. Gunter, et al., presented at the *Third International Conference on Carbon Dioxide Removal,* 1996.

29. S. Wong, and B. Gunter, Testing CO$_2$ enhanced coalbed methane recovery, *Greenhouse Issues*, IEA Greenhouse Gas R&D Programme, Cheltenham, UK 1999; available at www.ieagreen.org.uk/doc8.htm.

30. P. Brewer, presented as the Roger Revelle Commemorative Lecture, National Academy of Sciences, Washington, DC, 1999.

31. D. Whitman. and P. Brewer, *U.S. News and World Report*, 3 January 2000, p. 66; available at www.usnews.com/usnews/issue/000103/brewer.htm.

32. Pacific International Center for High Technology Research, CO$_2$ ocean sequestration field experiment: fact sheet, Pacific International Center for High Technology Research, Honolulu, HI, 2001; www.co2xperiment.org/facts.htm.

4

Control of NO$_x$ During Stationary Combustion

James T. Yeh and Wei-Yin Chen

CONTENTS

1. INTRODUCTION

Nitrogen oxides (NO$_x$) and sulfur oxides (SO$_x$) emissions are primary contributors to acid rain, which is associated with a number of effects including acidification of lakes and streams, accelerated corrosion of buildings, and visibility impairment. Among the various nitrogen oxides emitted from stationary combustion; nitrogen oxide (NO), nitrous oxide (N$_2$O), and nitrogen dioxide (NO$_2$) are stable, and NO predominates (over 90%). In health effects, NO$_2$ can irritate the lungs and lower resistance to respiratory infection. In the area of ozone nonattainment, NO$_x$ and volatile organic compounds (VOCs) react in the atmosphere to form ozone, a photochemical oxidant and a major component of smog. Atmospheric ozone can cause respiratory problems by damaging lung tissue and reducing lung function (1).

It is generally believed that over 80% of the total NO$_x$ emitted to the atmosphere originate at sources where fossil fuels and industrial wastes are burned. About one-half of the emissions are produced during combustion of fossil fuels in the utility industries. The rate of NO$_x$ formation is affected by fuel nitrogen content and by combustor design parameters. Higher firing temperature and combustor pressure increase NO$_x$ emissions. Nitric acid plants also produce large amounts of NO$_x$ as waste gas, but in much higher concentration than emissions from utility boiler flue gas.

Formation of nitrogen oxides during stationary combustion has been discussed in many reviews (2). Nitrogenous compounds in the fuel, or fuel nitrogen, are usually a main source, at least 75%, of NO$_x$ emissions in stationary combustion, such as pulverized coal combustion. The majority of the nitrogen in the volatile species converts sequentially to

From: *Handbook of Environmental Engineering, Volume 2: Advanced Air and Noise Pollution Control*
Edited by: L. K. Wang, N. C. Pereira and Y.-T. Hung © The Humana Press, Inc., Totowa, NJ

HCN and amine radicals, NH_i ($i = 0, 1, 2, 3$); oxidation of amine radicals leads to the formation of NO and NO_2. Char nitrogen is also a major source of NO emissions. A recent study of single particle char combustion suggests that nearly all nitrogen in the char convert to NO in bench-scale flame environments (3).

Oxidation of nitrogen in the air at high temperature is also a primary source of NO_x, or the "thermal" NO_x. It is generally believed that thermal NO_x formation follows the Zeldovich mechanism in which the rate-controlling step is

$$N_2 + O \leftrightarrow NO + N$$

The activation energy of this reaction, 314 kJ/mol, is high; thus, NO formation through this route becomes important at high temperatures.

Collisions of hydrocarbon radicals and molecular nitrogen in the fuel-rich region of the flames consist the third major NO_x formation pathway:

$$CH + N_2 \rightarrow HCN + N$$
$$C + N_2 \rightarrow CN + N$$

The cyanides are oxidized to NO after the primary flame; NO from this route is called "prompt NO."

In this chapter, we discuss the present regulations concerning the NO_x emission allowance, in-furnace and postcombustion NO_x control technologies, and issues facing the regulators and industries.

2. THE 1990 CLEAN AIR ACT

The 1990 Clean Air Act Amendment (CAAA) was enacted to address various air pollutant concerns: nonattainment for criteria pollutants (Title I), mobile source emissions (Title II), air toxic emissions (Title III), and acid rain (Title IV). Title IV is related to SO_x and NO_x criteria control. The 1990 CAAA's NO_x reduction target is 2 million tons/yr over 1980 levels by the year 2000. The specific NO_x standard was that a site must meet depends on several different factors, including application type, fuel, new or retrofit site status, and plant location. New sources must often meet more stringent New Source Performance Standards (NSPS). State Implementation Plans (SIP) can vary from state to state and may call for more stringent controls than required by federal regulations. Many of the 1990 CAAA regulations did not take effect fully until 1995 or beyond.

The estimated NO_x reduction achieved through the Acid Rain NO_x Program and the compliance options selected by the affected Phase I sources in the first 3 yr (1996–1998) of implementation are encouraging, as summarized by Krolewski et al. (4). The NO_x reductions from the 1990 level were averaging about 31% from about 250 sources recorded that were affected. The average NO_x emission rate was approx 0.41 lb/MMBtu.

Recently, the US Environment Protection Agency (EPA) has finalized a Federal NO_x Budget Trading Program based on the application of a populationwide 0.15-lb/MMBtu NO_x emission rate for large electricity-generating units (EGUs) and a 60% reduction from uncontrolled emissions for large non-EGU boilers and turbines for compliance on May 31, 2004 (5).

3. NO$_x$ CONTROL TECHNOLOGIES

One method to reduce NO$_x$ emissions is to minimize its formation via combustion modification, such as optimizing burner design for best aerodynamic distribution of air and fuel (low-NO$_x$ burner), controlling flame stoichiometry by regulating the overall fuel/air ratio supplied (low excess air), or other gross staging of combustion (overfire air ports, removing burners from service, derating, reburn). In addition to burner design changes and flame temperature control, another control measure is to reduce NO$_x$ after it is formed.

3.1. In-Furnace NO$_x$ Control

3.1.1. Low-NO$_x$ Burner

Low NO$_x$ burners control NO$_x$ formation by carrying out the combustion in stages. These burners control the combustion staging at and within the burner rather than in the firebox. Low NO$_x$ burners are designed to control both the stoichiometric and temperature histories of the fuel and air locally within each individual burner flame envelope. This control is achieved through design features that regulate the aerodynamic distribution and mixing of the fuel and air. There are two distinct types of design for low NO$_x$ burners: staged-air burners and staged-fuel burners.

3.1.2. Other In-Furnace NO$_x$ Control Techniques

There are a number of methods to reduce NO$_x$ formation during the combustion. In the low excess air (LEA) approach, less excess air is supplied to the combustor than normal, thus reducing the flame temperature and the formation of thermal NO$_x$. Fuel NO$_x$ is also reduced in the starved-air flame zone.

The overfire air (OFA) system uses conventional burners to introduce the fuel and substoichiometric quantities of combustion air (primary air). The remaining combustion air is introduced at some distance down the firebox through overfire air ports, thereby staging the combustion to cool the flame and suppress both thermal NO$_x$ and fuel NO$_x$ formation. Burners out of service (BOOS) is a variation of the staged combustion technique for reduction of NO$_x$ formation. It is a low-cost retrofit NO$_x$ control measure for existing fireboxes. Flue gas recirculation (FGR) is based on recycling a portion of cooled flue gas back to the primary combustion zone. This lowers the peak flame temperature and thus reduces thermal NO$_x$ formation.

Reburn, also referred to as in-furnace NO$_x$ reduction or staged-fuel injection, is the only NO$_x$ control approach implemented in the furnace zone but may also be defined as the postcombustion, preconvection section. A more detailed discussion of reburning follows.

3.1.3. Reburning

3.1.3.1. BACKGROUND

Engle (6) first demonstrated the reductive powers of hydrocarbons in the conversion of NO$_x$ to N$_2$ almost four decades ago. Wendt et al. (7) proposed application of this knowledge to the concept of fuel-staged combustion, or reburning, as an alternative to air-staged combustion to ensure the stability of the primary flame. Reburning is a three-stage combustion technology designed for the reduction of NO by introducing a small amount of reburning fuel after the primary flame, where the majority of NO is chemically reduced to nitrogen. The reburning zone usually has a stoichiometry at 0.85–0.95. The unburned hydrocarbons produced in the reburning zone are burned out in a third stage by introducing additional air. Tests on a full-scale boiler at Mitsubishi Heavy Industries (8) resulted in over a 50% NO_x reduction. Their results promoted significant interests in the United States because the regulations established by the Clean Air Act Amendments of 1990 mean that a single NO_x control technology is not likely to be sufficient for boilers in the ozone nonattainment areas.

Research in the last 20 yr has revealed much about the reburning process. Smoot et al. (9) critically reviewed the reaction mechanisms of nitrogen during reburning and the influences of fluid dynamics on NO reduction. Hill and Smoot reviewed the nitrogen oxide formation and destruction in combustion systems (10). The technological status of reburning has recently been summarized in a topical report of the US Department of Energy (11).

3.1.3.2. Homogeneous Chemistry

In the early development of reburning, volatile fuels were considered the primary candidates as reburning fuels. Detailed kinetic analysis of homogeneous, gas-phase NO reduction in a fuel-rich environment by simulation with CHEMKIN indicates that the majority of NO is reduced by hydrocarbon radicals C, CH, and CH_2 to HCN and, then, to amine radicals, NH_i, in a fuel-rich environment, such as the reburning zone (10,12). The following reaction steps seem to dominate the mechanisms:

$$NO + C \rightarrow CN$$
$$CN + H_2 \rightarrow HCN + H$$
$$HCN + O \rightarrow NCO + H$$
$$NCO + H \rightarrow NH + CO$$
$$NH + H \rightarrow N + H_2$$
$$N + NO \rightarrow N_2 + O$$

In addition, the nitrogen radicals may recycle back to NO through reactions with oxidants, such as OH and O_2,

$$N + OH \rightarrow NO + H$$
$$N + O_2 \rightarrow NO + O$$

or to HCN through the reducing agent CH_3,

$$N + CH_3 \rightarrow H_2CN + H$$
$$H_2CN \rightarrow HCN + H$$

In the burnout stage, HCN and amine radicals convert to either N_2 or NO, but conversion of HCN to NO is higher than that of amine radicals.

3.1.3.3. Heterogeneous Chemistry

The rate of reaction between NO and carbonaceous materials, such as coal-derived chars, soot, and fly ash, has been a subject of several fundamental studies. The reduction of nitric acid was critically reviewed by De Soete (13) and Aarna and Suuberg (14). When these carbonaceous materials were tested in laboratory reactors free from secondary reactions between char and product gas, the rates of NO and chars were usually low and seem to be less effective than natural gas in reburning. However, recent research suggested that lignites have NO reduction efficiencies comparable to that of methane in a simulated reburning environment (15). Moreover, it has been demonstrated that, in contrast to the findings related to bituminous coals, heterogeneous reactions on the lignite char surface contribute higher NO reduction than the corresponding gas-phase NO reactions in reburning (16). The effectiveness of lignite during reburning has also been demonstrated in a 1.0×10^6-Btu/h pilot-scale combustion facilities (17). Chen and Tang (18) recently reported their study on the variables, kinetics, and mechanisms of heterogeneous reburning. In addition to its large surface area, another factor reflecting effectiveness of lignite char appears to be its capability to minimize a number of reaction barriers desirable for reburning. First, it comprises catalytic components for chemisorption of both NO and O_2; moreover, the catalysts effectively carry the adsorbed oxygen atoms to carbon active sites (19). Second, it contains a large amount of carbon reactive sites for the formation and desorption of relatively weak surface oxides and, therefore, the production of CO (20). Third, CO, in turn, serves as a scavenger of surface CO and liberates the reactive site in a fuel-rich environment, such as reburning (18). These reaction steps are potentially rate controlling for reburning with many carbonaceous materials, such as chars derived from bituminous coals. However, the above-discussed catalytic features seem to render lignite char uniquely effective in reburning.

3.1.3.4. REBURNING TECHNOLOGIES

In addition to its capability of reducing NO, reburning is attractive to utilities and industries because of its ability to retrofit old boilers at costs much lower than those for the postcombustion technologies, such as selective catalytic reduction. The DOE has selected and cofunded four reburning projects for demonstration in the last 15 yr (11,21). Two of these projects involved natural gas and two others involved coal as reburning fuels; they are briefly discussed below.

Energy and Environmental Research Corporation's natural gas reburning and sorbent injection (GR-SI) technology (22) has been demonstrated at Illinois Power Company's Hennipin Plant with a 80-MWe (gross) tangentially fired boiler and at City Water, Light and Power's Lakeside Station in Springfield, Illinois, with a 40-MWe (gross) cyclone-fired boiler. A calcium compound (sorbent) is injected above the reburning zone for sulfur removal; hydrated lime [$Ca(OH)_2$] served as the baseline sorbent. To improve the gas penetration and therefore the mixing of the natural gas and flue gas from the primary combustion zone, a small amount of flue gas is recirculated and serves as the natural gas carrier in reburning. A 66% NO reduction is achieved in both locations. The capital cost of a natural gas reburning system on a 500-MW boiler is estimated to be around $30 per kilowatt (in 1996 dollars). Illinois Power has retained the gas reburning system, and City Water, Light and Power has retained the full technology for commercial use.

Babcock and Wilcox's coal reburning system (11) has been demonstrated at Wisconsin Power and Light Company's Nelson Dewey Plant in Cassville, Wisconsin, with an 100-MWe (gross) cyclone-fired boiler. Both bituminous coal and subbituminous coal were used as reburning fuels. A 50% or greater NO reduction is achieved. The capital cost of a natural gas reburning system on a 110-MW boiler is estimated to be around $66 per kilowatt, and a 605-MW boiler is estimated to be around $43 per kilowatt (in 1990 dollars). Wisconsin Power and Light has retained the full technology for commercial use.

Energy and Environmental Research Corporation's natural gas reburning (GR) and Foster Wheeler Energy Corporation's low NO_x burner (LNB) technology (11) have been demonstrated at Public Service Company of Colorado's Cherokee Station with a 158-MWe (net) wall-fired boiler. A 66% NO reduction was achieved. The capital cost of a natural gas reburning system on a 300-MW boiler is estimated to be around $26 per kilowatt (in 1996 dollars). Public Service Company of Colorado has retained the LNB-GR system. Energy and Environmental Research Corporation has been awarded two contracts to provide gas reburning systems for five cyclone coal-fired boilers: TVA's Allen Unit No. 1 with options for Unit Nos. 2 and 3 (identical 330-MWe units); and Baltimore Gas and Electric's C.P. Crane, Unit No. 2, with an option for Unit No. 1 (similar 200-MWe units). Use of the technology also extended to overseas markets. One of the first installations of the technology took place at the Ladyzkin State Power Station in Ladyzkin, Ukraine.

D.B. Riley's MPS mill and Fuller's MicroMill™ technology for producing micronized coal (11) has been demonstrated at New York State Electric and Gas Corporation's Milliken Station in Lansing, New York, with a 148-MWe (net) tangentially fired boiler and at Eastman Kodak Company's Kodak Park Power Plant in Rochester, New York, with a 60-MWe (net) cyclone boiler. Low volatile Pittsburgh seam bituminous coal was fired at both test sites. The project demonstrates micronized coal reburning, where the reburning coal is ground to 85% below 325 mesh. The small coal particles have greatly increased surface area, which, in turn, increases the rate of NO reduction. A 57% or greater NO reduction is achieved. The capital cost of a micronized coal reburning system on a 300-MW boiler is estimated to be around $14 per kilowatt (in 1999 dollars). The Eastman Kodak Company has retained the technology for commercial use.

3.1.4. Altered Reburning Processes

Yeh et al. (23,24) reported a modified gas reburning process, which was designed specifically for the destruction of nitrogen oxides in high concentration from industrial waste gas streams. Methane gas was injected into a combustor with a waste gas stream containing a high concentration of NO_x, which is to be destroyed. The combustor could be a furnace designed just for the purpose of NO_x destruction or an industrial boiler could be utilized for this auxiliary application without affecting normal boiler operation. The waste gas stream could be the off-gas from a sorbent regenerator in the NOXSO process (25) for simultaneous removal of SO_2 and NO_x from fossil fuel combustion flue gas, or the waste gas could be the off-gas from a nitric acid plant and fertilizer plant. In a DOE research facility, a simulated NOXSO process regenerator off-gas doped with methane gas was injected into a pulverized-coal-fired combustor and the efficiencies of injected NO_x reduction were recorded. At certain test conditions, no net increase of NO_x was detected in the combustor flue gas (23).

In a theoretical background search for in-furnace NO_x reduction, the thermodynamic equilibrium effect on NO_x concentrations was calculated using the DOE Multi-Phase Thermodynamic Equilibrium computer code and Sandia National Laboratory's CHEMKIN computer code. Equilibrium concentrations of NO_x at various flame temperatures were obtained. The thermodynamic model simulates pulverized-coal combustion at 20% excess air. Injection of additional NO (referred to as the recycle ratio, expressed as the ratio of moles of NO injected to mole of NO formed in the combustion process) was studied at levels of 0%, 90%, and 100%. The calculated data show that at NO recycle ratio as high as 1, virtually no net increase of NO occurs. However, although thermodynamics indicates nearly total destruction of injected NO is feasible (up to a recycle ratio equal 1), chemical kinetics reaction rates will dictate the final product mix.

METHANE de-NOX is another altered reburning process (26). In contrast to conventional reburning, where the reburn fuel is injected above the combustion zone to create a fuel-rich reburn zone, the METHANE de-NOX process involves direct injection of natural gas into the combustion zone above the grate; this results in a reduction of NO$_x$ formed in the coal bed and also limits its formation through the decomposition of the NO$_x$ precursors to form molecular nitrogen rather than nitrogen oxides.

3.2. Postcombustion NO$_x$ Control

Another major route for NO$_x$ control aims to reduce NO$_x$ after it is formed in the furnace. The methods include selective catalytic reduction (SCR), nonselective catalytic reduction (NSCR), and selective noncatalytic reduction (SNCR). A nitrogenous agent, such as ammonia (NH$_3$), urea (NH$_2$CONH$_2$), and cyanuric acid [(HNCO)$_3$], decomposes to amine radicals that, in turn, chemically reduces NO$_x$ to N$_2$ and H$_2$O in a postcombustion zone (27).

In the SCR process, NH$_3$, usually diluted with air, is injected into a flue gas stream upstream of a catalyst bed. NH3 reacts with NO$_x$ on the surface of the catalyst to form molecular nitrogen and water. Depending on system design, NO$_x$ removal of 80–90% and higher is available. The primary variable affecting NO$_x$ reduction is temperature. Three types of catalyst system have been deployed commercially: noble metal, base metal, and zeolites. Noble metals are typically washcoated on inert ceramic or metal monoliths and used for particulate-free low-sulfur exhausts. They function at the lower end of the SCR temperature range (460–520°F) and are susceptible to inhibition by SO$_x$ (1). Base metal vanadia–titania catalysts may either be washboard or extruded into honeycombs. Extruded monoliths are used in particulate-laden coal and oil-fired applications. Optimum NO$_x$ reduction occurs at a catalyst bed temperature between 570°F and 750°F. Zeolites may be loaded with metal cations (such as Fe and Cu) to broaden the temperature window (28).

The catalyst selected depends largely on the temperature of the flue gas being treated. A given catalyst exhibits optimum performance within a temperature range of ± 86°F. Below this optimum temperature range, the catalyst activity is greatly reduced, allowing unreacted NH$_3$ to slip through. Above 842°F, NH$_3$ begins to be oxidized to form additional NO$_x$. Depending on the catalyst substrate material, some catalyst may be damaged as a result of thermal stress (29).

Nonselective catalytic reduction is a special mix of catalyst, which promotes simultaneous reduction of NO$_x$, unburned hydrocarbon, and CO. The catalyst is generally a mixture of platinum and rhodium. It operates in the temperature range of 660–1470°F and typical NO$_x$ conversion range from 80% to 95%.

The SNCR process reduces NO$_x$ via NH$_3$ injection at a temperature between 1500°F and 1700°F (30). Urea and cyanuric acids have also been used for NO$_x$ reduction. The NO$_x$ molecule is broken up into nitrogen and water molecules during the SNCR process.

Temperature is the primary variable for controlling the NO_x reduction. Above 2200°F, NH_3 is oxidized to NO by oxygen in flue gas. Adequate gas residence in combustor is also important to achieve NO_x reduction at optimum temperature as well as good gas mixing of the reducing agent across the full furnace.

3.3. Hybrid Control Systems

Both SNCR and SCR can be used in conjunction with each other with some synergistic benefits. Also, both processes can be used in conjunction with LNBs. Some artificial intelligence software can also be incorporated into the above processes.

3.4. Simultaneous SO₂ and NOₓ Control

3.4.1. Dry Processes

The copper oxide process is a dry and regenerable process for the simultaneous removal of SO_2 and NO_x from flue gas. The sorbent is copper-impregnated (5–7% Cu) spheres (1/16 in. in diameter) of gamma alumina (31,32). Copper oxide can be operated in a fixed-bed, a fluidized-bed, or a moving-bed mold at 750°F absorber temperature. The SO_2 and O_2 in flue gas react with CuO to form $CuSO_4$. NO_x in flue gas is reduced by NH_3 injection into the absorber. The spent sorbent can be regenerated at 800°F with methane or other reducing gases, such as CO and H_2.

In a CuO moving-bed process test, a continuous-cycling long-term experiment produced greater than 90% removal for both SO_2 and NO_x (32). Ninety-five percent NO_x reduction was easily maintained during a long-term cycling test. The NH_3/NO_x molar ratio was maintained at near 1. A wet chemical sampling technique that bubbles the flue gas exiting the absorber through a hydrochloric acid solution was used to determine NH_3 slippage. The solution is further analyzed by using an ion electrode technique. All data indicated 0 to 5-ppm NH_3 slippage. Both CuO and $CuSO_4$ on the surface of alumina substrate act as very efficient catalysts for NO_x reduction by NH_3. Copper on alumina does not have shortcomings of SO_x and fly ash poisoning. Copper oxide-impregnated alumina can be used for catalyzing NO_x reduction by NH_3 alone in a fixed-bed setting.

The Ebara process (33), which was developed in Japan, uses high-energy irradiation to oxidize SO_2 and NO in the flue gas to form SO_3 and NO_2 while injecting NH_3 downstream of the irradiation chamber. The result is that $(NH_4)_2SO_4$ and NH_4NO_3 fertilizers are formed. The Ebara process for simultaneous removal of SO_2 and NO_x is currently operating in China. The desired level of NO_x reduction can be adjusted by the irradiation energy applied to the irradiation chamber. The percentage of power plant wattage that must be applied for a high level of SO_2 and NO_x removal is not known.

A general rule, which can be applied to any de-SOX and de-NOX processes, is that the cost of the process increases exponentially with the removal level of SO_2 and NO_x increase. Linear extrapolation of cost must never be used. The rule also applies to any similar processes (e.g., CO_2 removal in a scrubber).

3.4.2. Wet Scrubbing Processes

The wet limestone ($CaCO_3$) process is well known for removal of SO_2 in flue gas; however, the process is incapable of removing nitric oxide (NO) because of its low solubility in aqueous solutions. Research efforts to modify existing wet flue gas desulfurization

(FGD) processes for the simultaneous control of SO$_2$ and NO$_x$ emissions have led to several new approaches to enhance NO$_x$ absorption in scrubbing liquors. NO constitutes approx 95% of NO$_x$ in flue gas. The new approaches include the oxidation of NO to the more soluble NO$_2$ and the addition of various iron(II) chelates to the scrubbing liquors to bind and activate NO. These methods have not been found economical to operate.

The removal of SO$_3$ and NO$_2$ in flue gas by ammonium hydroxide solution was reported by Senjo et al. (34), and absorption of SO$_3$ and NO$_2$ by water was reported by Novoselov et al. (35). Oxidizing the flue gas with ozone or other strong oxidants, such as ClO$_2$ and H$_2$O$_2$, produced the SO$_3$ and NO$_2$ in flue gas.

Chang et al. (36–38) proposed using aqueous emulsions of yellow phosphorus (P$_4$) and an alkali in a simple wet scrubber, where NO is oxidized and absorbed in scrubber liquid together with SO$_2$ from the flue gas. The process byproducts are potentially valuable fertilizer chemicals, including phosphate, nitrate, and sulfate. NO oxidation through phosphorus chemistry was claimed to be more economical than other oxidants (36,39).

4. RESULTS OF RECENT DEMONSTRATION PLANTS ON NO$_x$ CONTROL

The results of several recent large-scale NO$_x$ control demonstrations are summarized in this section. These projects were cofunded by the DOE and industries under the Clean Coal Technology (CCT) Demonstration Program (21).

Foster Wheeler's LNB technology incorporating advanced overfire air (AOFA) in conjunction with EPRI's Generic NO$_x$ Control Intelligent System (GNOCIS) computer software was demonstrated in a 500-MWe power plant (Southern Company Services, Inc., Georgia Power Company's Plant Hammond, Unit No. 4). Using LNB/AOFA, long-term NO$_x$ emissions were 0.40 lb/MMBtu, representing a 68% reduction from base conditions. The capital cost for a 500-MWe wall-fired unit is $18.8 per kilowatt for LNB/AOFA and it is $0.5 per kilowatt for GNOCIS. The technology is applicable to the 411 existing pre-NSPS dry-bottom wall-fired boilers in the United States, which burn a variety of coals. The GNOCIS technology is applicable to all fossil-fuel-fired boilers, including multifuel boilers.

A SCR process combined with sorbent (calcium or sodium based) injection for combined SO$_2$, NO$_x$, and particulates removal from flue gas was promoted by the Babcock & Wilcox Company as SO$_x$-NO$_x$-Rox Box (SNRB™). It was demonstrated in a 5-MWe capacity slip stream in Ohio Edison Company's R.E. Burger Plant, Unit No. 5. It achieved greater than 70% SO$_2$ removal and 90% or higher reduction in NO$_x$ emissions while maintaining particulate emissions below 0.03 lb/MMBtu. The NO$_x$ removal is accomplished by injecting NH3 to selectively reduce NO$_x$ in the presence of a SCR catalyst. The NO$_x$ reduction of 90% was achieved with an NH$_3$/NO$_x$ molar ratio of 0.9 and temperature of 800–850°F. The capital cost in 1994 dollars for a 150-MWe retrofit was $253 per kilowatt, assuming 3.5% sulfur coal, baseline NO$_x$ emissions of 1.2 lb/MMBtu, 65% capacity factor, and 85% SO$_2$ and 90% NO$_x$ removal. Levelized cost over 15 yr in constant 1994 dollars was $553 per ton of SO$_2$ and NO$_x$.

A SNCR process combined with B&W's DRB-XCL low-NO$_x$ burners, in-duct sorbent injection, and furnace (urea) injection was demonstrated in a 100-MWe plant at Public Service Company of Colorado's Arapahoe Station, Unit No. 4. LNB with minimum

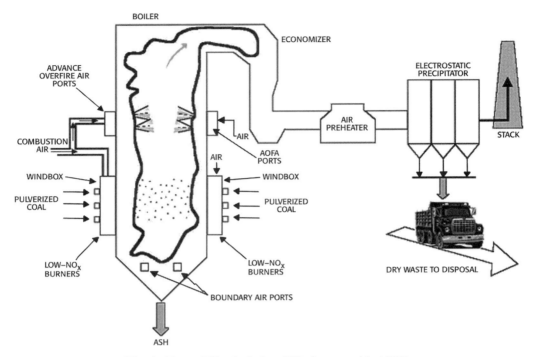

Fig. 1. Foster Wheeler's low-NO$_x$ burner with AOFA.

overfire air reduced NO$_x$ emission by more than 63% under steady-state conditions. With maximum overfire air (24% of total combustion air), a NO$_x$ reduction of 62–69% was achieved across the 50- to 110-MWe range. The SNCR system, using both stationary and retractable injection lances in the furnace, provided an additional 30–50% reduction of the remaining NO$_x$ in flue gas at an ammonia slip of 10 ppm. In other words, approx 10% of additional NO$_x$ was removed from the flue gas by the SNCR process, making a total final NO$_x$ reduction of 80%. The total capital cost for the technology ranges from $125 to $281 per kilowatt for 300- to 50-MWe plants, respectively.

The benefits derived from the CCT demonstration program in the area of NO$_x$ control technologies include (1) 75% of existing US coal-fired units have been or currently are being retrofitted with low-NO$_x$ burners, (2) highly effective commercial low-NO$_x$ burners were developed, (3) a 50% reduction in SCR cost has resulted since 1980, (4) an estimated 30% of US coal-fired generating capacity will incorporate SCR technology by 2004, and (5) over 60 million tons of NO$_x$ emissions has been caputered since 1970 based on average fleet emissions (21).

Eskinazi (40) provides a detailed engineering assessment on NO$_x$ controls for coal-fired boilers. Figures 1–3 are courtesy of the DOE (21,41). Figure 1 shows a schematic of Foster Wheeler's low-NO$_x$ burner with AOFA and EPRI's Generic NO$_x$ Control Intelligent System (GNOCIS) computer software. Figure 2 is the diagram of the Babcock & Wilcox Company's low-NO$_x$ cell-burner. Figure 3 shows a schematic of Energy and Environmental Research Corporation's combined NO$_x$ and SO$_2$ reduction using gas reburning and furnace sorbent injection technology.

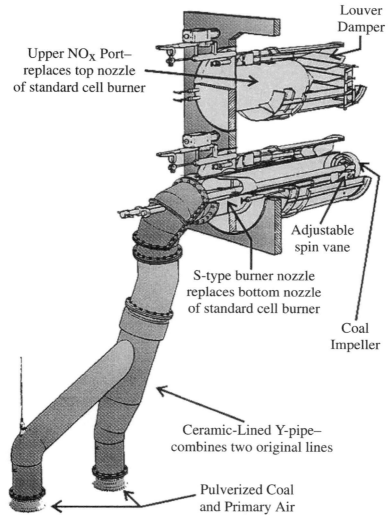

Upper NO$_X$ Port–
replaces top nozzle
of standard cell burner

Louver
Damper

Adjustable
spin vane

S-type burner nozzle
replaces bottom nozzle
of standard cell burner

Coal
Impeller

Ceramic-Lined Y-pipe–
combines two original lines

Pulverized Coal
and Primary Air

Fig. 2. Low-NO$_x$ cell burner (B&W).

5. FUTURE REGULATION CONSIDERATIONS

One potentially important development in the near future is the growing interest in developing three-pollutant (SO$_2$, NO$_x$, and toxics [e.g., mercury]) or four-pollutant (add CO2) emission control regulations in to reduce the total cost and uncertainty associated with improving the environmental performance of large stationary sources (42).

6. FUTURE TECHNOLOGY DEVELOPMENTS IN MULTIPOLLUTANT CONTROL

Aside from the SCR, SNCR, and reburning technology for NO$_x$ control, a multipollutant control process, which removes SO$_2$, NO$_x$, CO$_2$, and all other acid gases from coal-burning flue gas, is being tested in DOE's National Energy Technology Laboratory.

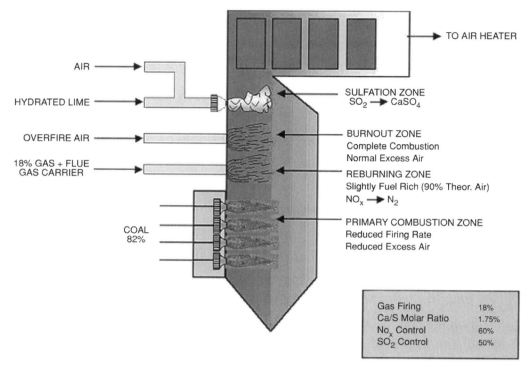

Fig. 3. Schematic of reburner technology.

The process uses ammonia solution to remove acid gases in a wet scrubber (43). The driving force in developing this aqua ammonia process is economy. CO_2 scrubbing by ammonia solution in a packed bed (with a structured packing) was highly successful (44,45). CO_2 removal efficiency easily exceeded 90% with 17% or higher ammonia solution at approx 2 s gas residence time.

REFERENCES

1. C. J. Pereira, and M. D. Amiridis, in *The ACS Symposium Series 587 on Reduction of Nitrogen Oxide Emission* (Ozkan et al., eds.), American Chemical Society, Washington, DC, 1995, pp. 1–13.
2. D. Smoot, *Fundamentals of Coal Combustion for Clean and Efficient Use,* Elsevier, Amsterdam, 1993.
3. L. S. Jensen, H. E. Jannerup, P. Glarborg, et al., *Experimental Investigation of NO from Pulverized Char Combustion*, the Combustion Institute, Pittsburgh, PA, 2000, pp. 2271–2278.
4. M. J. Krolewski, A. S. Mingst, and R. K. Srivastava, *Department of Energy 2000 Conference on Selective Catalytic and NonCatalytic Reduction for NO$_x$ Control*, US DOE, Pittsburgh, PA, 2000, pp. 1–4.
5. K. Culligan, and M. J. Krolewski, *DOE 2001 Conference on Selective Catalytic Reduction and Non-Catalytic Reduction for NO$_x$ Controls*. US DOE, Pittsburgh, PA, 2001.
6. G. Engel, *The Influence of Steam, Hydrogen and Oxygen, Compt. Rend.*, **231**, 1493–1500 (1950).

7. J. O. L. Wendt, C. V. Sternling, and M. A. Matovich, *14th Symposium (International) Combustion*, The Combustion Institute, Pittsburgh, PA, 1973, pp. 897–904.
8. Y. Takahashi, M. Sakai, T. Kunimoto, et al., *Proc. 1982 Joint Symp. Stationary Combustion NO$_x$ Control*, EPRI Report No. CS-3182, Palo, Alto, CA, 1983.
9. L. D. Smoot, S. C. Hill, and H. Xu, *Prog. Energy Combust.* **24**, 385–408 (1998).
10. S. C. Hill and L. D. Smoot, *Prog. Energy. Combust.* **26**, 417–458 (2000).
11. Department of Energy, *Reburning Technologies for the Control of Nitrogen Oxides Emissions from Coal-Fired Boilers*, Clean Coal Technology Topical Report No.14, US DOE, Washington, DC, 1999.
12. J. A. Miller and C. T. Bowman, *Prog. Energy Combust.* **15**, 287–337 (1989).
13. G. G. De Soete, in *Pulverized Coal Combustion: Pollutant Formation and Control*, 1970–1980, Report No. EPA-600/8-90-049, US Environmental Protection Agency, Cincinnati, Ohio, 1990, pp. 8-1-8-59.
14. I. Aarna and E. M. Suuberg, *Fuel* **76**, 475–491 (1997).
15. T. E. Burch, F. R. Tillman, W. Y. Chen, *Energy, Fuel* **5**, 231–237 (1991).
16. W. Y. Chen and L. Ma, *AIChE J.* **42**, 1968–1976 (1996).
17. R. Payne, D. K. Moyeda, P. Maly, et al., in *Proceedings of the EPRI/EPA 1995 Joint Symposium on Stationary Combustion NO$_x$ Control, Book 4*, 1995.
18. W. Y. Chen and L. Tang, *AIChE J.* **47**, 2781–2797 (2001).
19. M. J. Illan-Gomez, A. Linares-Solano, L. R. Radovic et al., *Energy Fuel* **10**, 158–168 (1996).
20. A. A. Lizzio, H. Jiang, and L. R. Radovic, *Carbon* **28**, 7–19 (1990).
21. Department of Energy, *Clean Coal Technology Demonstration Program: Program Update 2000*, US DOE, Washington, DC, 2001.
22. Department of Energy, *Reduction of NO$_x$ and SO$_2$ Using Gas Reburning, Sorbent Injection and Integrated Technologies*; Clean Coal Technology Topical Report No. 3, US DOE Washington, DC, 1993.
23. J. T. Yeh, J. M. Ekmann, H. W. Pennline, et al., in *Proceedings of the 194th ACS National Meeting*, 1987, pp. 471–483.
24. J. T. Yeh, J. M. Ekmann, H. W. Pennline, et al., US Patent 4,878,442, 1989.
25. J. T. Yeh, W. T. Ma, H. W. Pennline, et al., *Chem. Eng. Commun.* **114**, 65–88 (1992).
26. I. Rabovttser and M. Roberts, in *13th Annual International Pittsburgh Coal Conference*, 1996, pp. 1505–1510.
27. L. Muzio and G. Quartucy, *State-of-the-Art Assessment of SNCR Technology*, Report No. TR-102414, Electric Power Research Institute, Palo Alto, CA, 1993.
28. J. W. Byrne, J. M. Chen, and B. K. Speronello, *Catal. Today* **13**, 33 (1992).
29. L. M. Campbell, D. K. Stone, and G. S. Shareef, *Sourcebook: NO$_x$ Control Technology Data*, Report No. EPA/600/S2-91/029, US Environmental Protection Agency, Cincinnati, OH, 1991.
30. H. Bosch, and F. Janssen, *Catal. Today* **2**, 369 (1987).
31. J. T. Yeh, R. J. Demski, J. P. Strakey, et al., *Environ. Prog.* **4**(4), 223–228 (1985).
32. J. T. Yeh, H. W. Pennline, and J. S. Hoffman, in *89th AWMA Annual Meeting*, 1996.
33. R. R. Lunt, and J. D. Cunic, *Profiles in Flue Gas Desulfurization*. Center for Waste Reduction Technologies (CWRT), AIChE, New York, 2000.
34. T. Senjo, and M. Kobayashi, US Patent 4035470, 1977.
35. S. S. Novoselov, A. F. Gavrilov, V. A. Svetlichnyi, et al., *Therm. Eng.* **33**(9), 496–498 (1986).
36. S. G. Chang and D. K. Liu, *Nature*, **343**(6254), 151–153 (1990).
37. D. K. Liu, D. X. Shen, and S. G. Chang, *Environ. Sci. Technol.* **25**(1), 55–60 (1991).
38. S. G. Chang and G. C. Lee, *Environ. Prog.* **11**(1), 66–72 (1992).
39. G. C. Lee, D. X. Shen, D. Littlejohn, et al., in *SO$_2$ Control Symposium*, 1990.

40. D. Eskinazi, *Retrofit NO$_x$ Controls for Coal-Fired Utility Boilers: A Technology Assessment Guide for Meeting Requirements of the 1990 Clean Air Act Amendments*, EPRI, Palo Alto, CA, 1993.
41. US DOE *Clean Coal Technology Demonstration Program: Project Profiles*, National Energy Technology Laboratory, Pittsburgh, PA, 2001.
42. A. Farrell, in *2001 Conference on SCR and SNCR for NO$_x$ Control*, 2001, pp. 3–4.
43. J. T. Yeh and H. W. Pennline, FY2002 Research proposal to NETL/US DOE, 2001.
44. J. T. Yeh, NETL/US DOE, unpublished data.
45. J. T. Yeh, *Energy Fuel*, in press.

5

Control of Heavy Metals in Emission Streams

L. Yu Lin and Thomas C. Ho

Contents

INTRODUCTION
PRINCIPLE AND THEORY
CONTROL DEVICE OF HEAVY METALS
METAL EMISSION CONTROL EXAMPLES
NOMENCLATURE
REFERENCES

1. INTRODUCTION

Heavy metals are elements that are located in the periodic table from groups III to VI and periods 4 or greater. The elements have high atomic weight greater than sodium and high specific gravity (i.e., usually greater than 5.0 g/cm^3). In addition, the elements have high thermal conductivity and are characterized by malleability and ductility. There are 65 elements listed in the periodic table that can be defined as heavy metals based on the above definitions. Approximately 30 metals, either in elemental forms, in salts, or in organometallic compounds, have been used by industry. Heavy metals in the wastes will not be directly exposed to the atmosphere. Thermal or vaporization processes enhance heavy metal exposure into the atmosphere. Unlike organic compounds, metals cannot be completely destroyed by the thermal process. The thermal process can only oxidize the majority of metals to particulate matter. Only a small amount of volatile metals having a boiling point lower than the thermal/combustion operation temperature will be vaporized. Metals, such as arsenic, barium, beryllium, chromium, cadmium, lead, mercury, nickel, and zinc, are of great concern in waste incineration and coal combustion because of their presence in many wastes and because of possible adverse health effects from human exposure to emissions (1).

Metals in the feed to a combustor/incinerator primarily form metallic oxides in the combustion process or possibly metallic chlorides if chlorine is present. The metallic compounds that are formed will be distributed among three effluent streams from a combustion facility: combustor bottom ash, air pollution control device (APCD) effluent (or fly ash), and stack gas. The distribution among these effluents depends on a number of factors. More of the metallic compounds remain in the bottom ash if they are less volatile, if fewer particles are entrained by combustion gases, or if they are chemically bound in the bottom ash. APCDs, such as Venturi scrubbers, electrostatic precipitators, baghouses, and packed-bed scrubbers, will collect more metals if the metallic compounds are less volatile, if more of the compounds are on large particles, or if the solubility of the compounds in scrubbing liquids is higher. Metallic compounds that are not retained in the combustor bottom ash or are not collected by the control devices will be emitted in the stack gas (2). Because of the characteristics of metals, the incineration process may not be a good alternative for high-metal-content wastes. If the incineration needs to be used, appropriate pollution control devices should be applied in order to reduce the metals discharged to the environment.

As early as 1970, the emission standards in the Clean Air Act Amendments (CAAA) were established for regulating beryllium and mercury from industrial sources. Since then, more sophisticated operations and technologies have been developed, resulting in more contaminated pollutants, including heavy metals that were found in the atmosphere. The US Environmental Protection Agency (EPA) lists emission levels of 10 priority metals: arsenic, beryllium, cadmium, chromium, antimony, barium, lead, mercury, silver, and thallium. The emissions of these EPA priority heavy metals must also meet the risk standards on the Resource Conservation and Recovery Act (RCRA) guidelines. Under RCRA (1976), emissions of particulate matters, including metal particles, are limited to 0.08 grains/day standard cubic foot (gr/d-scf) for the stack gas corrected to 50% excess air. If excess air is greater or less than 50%, the particle concentration must be adjusted before it is compared with 0.08 gr/d-scf (3).

The CAAA of 1990 has promulgated more pollutants. Table 1 is the National Emission Standards for Hazardous Air Pollutants (NESHAPs). The standards are very specific to regulate the sources, mission standards, and control methods (5).

2. PRINCIPLE AND THEORY

This section is a review of basic theory related to heavy metals in incineration. The information consists of heavy metal reactions in an incinerator and control of metal emissions. This information is very important to understand metal compound formation and reaction behavior in the incinerator. It also provides the knowledge that will lead us to select the APCD for heavy metals. More information may be required for servicing the purposes of pollution control and incinerator operation, including metal stoichiometry, thermodynamics, and combustion.

2.1. Reactions in the Incinerator

2.1.1. Oxidation and Reduction Reactions

The term "oxidation" was originally used to describe reactions in which an element combines with oxygen. For example, the reaction between copper metal and oxygen to form copper oxide involves the oxidation of copper. Metals act as reducing agents in their chemical reactions. When copper is heated over a flame, the surface slowly turns black as the copper metal reduces oxygen in the atmosphere to form copper(II) oxide. If we turn off the flame and blow H_2 gas over the hot metal surface, the black CuO that forms on the surface of the metal is slowly converted back to copper metal. In the course of this reaction, CuO acts as an oxidizing agent and H_2 is the reducing agent. An impor-

Table 1
Summary of National Emission Standards for Hazardous Air Pollutants

Hazardous air pollutants	Original publication	Emission limits	References
Asbestos	36 FR[a] 5931, March 31, 1971	No visible emissions to the outside air from the wetting operation. However, the temperature must be recorded at least at hourly intervals, and records must be retained for at least 2 yr in a form suitable for inspection.	40 CFR[b] 60.01, 40 CFR 61, 149
Arsenic	45 FR 37886, June 5, 1984	Uncontrolled total arsenic emissions from the glass melting furnace shall be less than 2.5 Mg/yr or total arsenic emissions from the glass melting furnace shall be conveyed to a control device and reduced by at least 85%	40 CFR 60.01, 40 CFR 61.162
Beryllium	36 FR 5931, March 31, 1971	Emissions to the atmosphere from stationary sources shall not exceed 10 gr (0.022 lb) of beryllium over a 24-h period. Approval from the EPA, the ambient concentration limit in the vicinity of the stationary source is 0.01 μg/m^3 (4.37×10^{-6} g/ft^3) over a 30-d period.	40 CFR 60.01, 40 CFR 61.32, 40 CFR 61.42
Mercury	36 FR 5931, March 31, 1971	Emissions to the atmosphere from mercury ore processing facilities and mercury cell chlor-alkali plants shall not exceed 2.3 kg (5.1 lb) of mercury per 24-h period. Emissions to atmosphere from sludge incineration plants, sludge drying plants, or a combination of these that process wastewater-treatment plant sludges shall not exceed 3.2 kg (7.1 lb) of mercury per 24-h period.	40 CFR 60.01, 40 CFR 61.52

[a]Federal regulations.
[b]Code of Federal Regulations.
Source: ref. 4.

tant feature of oxidation–reduction reactions can be recognized by examining what happens to the copper in this pair of reactions. The first reaction converts copper metal into CuO, thereby transforming a reducing agent (Cu) into an oxidizing agent (CuO). The second reaction converts an oxidizing agent (CuO) into a reducing agent (Cu). Every reducing agent is therefore linked to a conjugate oxidizing agent and vice versa. Every time a reducing agent loses electrons, it forms an oxidizing agent that could gain electrons if the reaction were reversed.

$$2Cu(s) + O_2(g) \rightarrow 2\ CuO(s) \tag{1}$$

The two half-reactions of oxidation and reduction are expressed as

$$2\ Cu(s) \rightarrow 2\ Cu^{2+} + 4\ e^- \qquad \text{Oxidation Reaction}$$

$$O_2(g) + 4\ e^- \rightarrow 2\ O^{2-} \qquad \text{Reduction Reaction}$$

$$2\ Cu(s) + O_2(g) \rightarrow 2\ CuO(s) \qquad \text{Redox Reaction}$$

Oxidation states provide a compromise between oxidation and reduction reactions. Sometimes, there are uncertainties between electron donor or electron acceptor. The oxidation states can present as a hypothetical charge. Although this concept may not have chemical reality, the concept is useful and practical in explaining the stoichiometric reaction in combustion.

The oxidation processes in incineration are involved in the formation of particulate metals. The processes are so complex that the real reactions and formations are still unknown (2,6). For some, the knowledge may come from the burning oil or coal. Metal oxides after combustion become very small particles. APCDs may not easily collect them. Some researches tried to change physical or chemical properties in the incinerator in order to enhance metal oxidation process and to enlarge the particle size of metal oxides. However, these efforts are still being researched.

2.1.2. Hydrolysis Reaction

The hydrolysis reaction is another important reaction in incineration. The hydrolysis reaction is a chemical reaction of a compound with water and results in the formation of one or more new stable oxide or hydroxide compounds. The positive charge of metal ions takes electron density of the O–H bond from the water. This increases the bond's polarity, making it easier to break. When the O–H bond breaks, an aqueous proton is released to produce an acidic solution. The equilibrium for this reaction can be expressed as

$$\left[M(H_2O)_n \right]^{z+} + H_2O \rightarrow \left[M(H_2O)_{n-1}(OH) \right]^{(z-1)+} + H_3O^+ \tag{2}$$

Hydrolysis requires high temperature and high pressure, acid or alkali, and catalysts. In the incinerator, high temperature is presented in the combustion chamber. Much of the moisture and acidic solutions remain inside the chamber. As a result, heavy metal compounds undergo hydrolysis and become fly ashes or bottom ashes in the incinerator. Detailed hydrolysis reaction of metal in the incinerator is rarely understood. Currently, the formation of fly ashes or bottom ashes and the equilibrium of metal hydroxide can only be expressed using the solubility product.

2.1.3. Chlorination

Another important reaction is related to chloride in the incineration as a result of due to waste that contains plastic materials. The chloride compounds may convert to hydrochloric acid during the combustion process as well as organic or inorganic chlorides.

Because of hydrolytic reactions, inorganic chlorides may form HCl at high temperature. The reaction is

$$Cl_2 + H_2O \rightarrow 2\,HCl + \frac{1}{2}O_2 \tag{3}$$

The equilibrium constant can be expressed as

$$K = \frac{[P_{HCl}]^2 [P_{O_2}]^{\frac{1}{2}}}{[P_{H_2O}][P_{Cl_2}]} \tag{4}$$

where P is the partial pressures of each reactant. At 670°C, the value of K is approximately equal to 1.

The gaseous compounds and alkali chlorides ligand with metals, forming heavy metal chlorides in the incinerators. The experiments show that an increase in chlorine contents will enhance HCl formation as well as volatile metal formation in the incineration (7). However, the kinetics and reactions of these formations in the incinerator are also unknown. The experiments were also conducted at 900°C with the spiked chlorides, including ionic and molecular bond compounds, such as PVC, C_2Cl_4, NaCl, KCl, and FeCl. The chlorine contents vary from 0% to 1.6%. The results indicated that an increase in the organic and inorganic chloride contents enhances heavy metals, particularly Cd partitioned to the fly ash (8,9).

2.1.4. High-Temperature Metal Absorption

Chemical absorption reactions between metal vapors and a variety of sorbents at high temperatures have been observed both in a packed bed and in a fluidized bed (5,10–12). The following reactions between metals and sorbent constituents have been confirmed both theoretically and experimentally:

$$2\,PbO + SiO_2 \rightarrow Pb_2SiO_4(s) \tag{5}$$

$$CdO + SiO_2 \rightarrow CdSiO_3(s) \tag{6}$$

$$CdO + Al_2O_3 \rightarrow CdAl_2O_4(s) \tag{7}$$

$$PbCl_2 + Al_2O_3{:}2SiO_2 + H_2O \rightarrow PbO{:}Al_2O_3{:}2SiO_2(s) + 2HCl(g) \tag{8}$$

$$CdCl_2 + Al_2O_3 + H_2O \rightarrow CdAl_2O_4(s) + 2\,HCl(g) \tag{9}$$

As indicated, both SiO_2 and Al_2O_3 have been identified as potential sorbents for metal capture at high temperatures.

To further demonstrate metal–sorbent reactions, several selected simulation results are displayed in Figs. 1–3, where the equilibrium speciation of lead and cadmium at high temperatures in the presence of SiO_2 and Al_2O_3 is shown. These simulations involved a typical fuel with the following composition: carbon, 71.3 wt%; hydrogen, 5.2 wt%; nitrogen, 1.4 wt%; oxygen, 7.8 wt%; sulfur, 4.6 wt%; ash, 9.6 wt%, and metal, 0.1% (1000 ppm). The amount of sorbent used was 20% of that of the fuel, and the percentage excess air used was 50%.

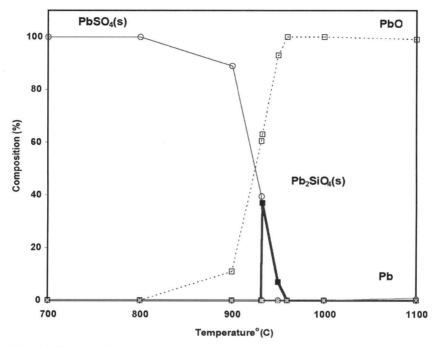

Fig. 1. Simulated lead speciation (System: Pb–S–S$_i$O$_2$). (From refs. 5 and 10.)

The results shown in Fig. 1 indicate that lead will react with both sulfur and silica during combustion. At a temperature below 900°C, PbSO$_4$(s) is the thermodynamically preferred lead compound; however, between 930°C and 960°C, Pb$_2$SiO$_4$(s) is the preferred compound; and above 960°C, PbO(g) is the dominating species. These results suggest that silica is thermodynamically capable of capturing lead. However, although not shown, the existence of sulfur will affect the capture process, especially at a temperature below 930°C.

The results shown in Figs. 2 and 3 for cadmium indicate that cadmium will react with Al$_2$O$_3$ and SiO$_2$ to form CdSiO$_3$(s) and CdAl$_2$O$_4$(s), respectively. The existence of sulfur does not seem to interfere with the reactions according to the equilibrium results shown. Once again, these simulation results suggest that silica and alumina have the potential to capture cadmium under combustion conditions.

2.2. Control of Metal Emissions

Among the 11 metals listed in the Clean Air Act Amendments of 1990 (i.e., Sb, As, Be, Cd, Cr, Co, Pb, Hg, Mn, Ni, and Se), mercury and selenium are volatile; arsenic, cadmium, and lead are semivolatile; and the remainder are refractory. The refractory metals are thermally stable and most of them stay in bottom ash or fly ash without being vaporized during high-temperature combustion. The effective control of fly ash emission therefore controls the emissions of these metals. The semivolatile metals (i.e., arsenic, cadmium, and lead), tend to volatilize during combustion and readily recondense during flue gas cooling to form metal fumes or to deposit fine particulate matter

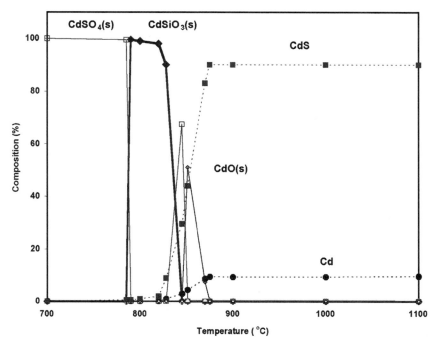

Fig. 2. Simulated cadmium speciation (System: Cd–S–S$_i$O$_2$). (From refs. 5 and 10.)

(PM). Because of the regulations, researchers pay more attention to volatile metal emissions from the incineration than the refractory metals remaining in the bottom or fly ashes. The boiling point and melting point of volatile metals are the key parameters of whether the metals stay in particulate forms or in vapor forms.

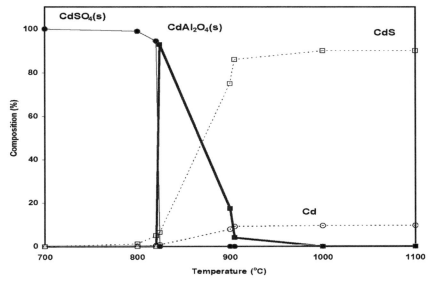

Fig. 3. Simulated cadmium speciation (System: Cd–S–Al$_2$O$_3$).

2.2.1. Control of Particulate Matters

The efficient collection of fine PM, either micron or submicron in size, is extremely crucial to control the emissions of these metals. These metals, however, are found to be relatively reactive and can be retained effectively by sorbents at high temperatures, as will be discussed below. The control of PM is an important aspect for air pollution control. The observed metal concentrations on particulate range from 0.24 to 9.0 μg/g for lead (μg lead/g particulate), 0.012 to 0.36 μg/g for Cd, 0.094 to 2.5 μg/g for Cr, and 0.024 to 2.7 μg/g for Ni. It is important to note, however, that it is difficult to extend the reported findings from these tests to any combustors/incinerators in general. This is because metal emissions are a function of the amount of metal input to the combustor (which is highly variable day-to-day and facility-to-facility) as well as the efficiency of the combustor/incinerator in controlling metal emissions.

Particles are collected by a combination of several mechanisms, such as gravity settling, centrifugal force, electrostatic impaction, inertial impaction, direct interception, and diffusion. The physical phenomenon involved in each mechanism is discussed in the following:

1. *Gravity Settling.* The gravity settling of a discrete particle is governed by three forces: gravity force, buoyant force, and the frictional drag force. Under the equilibrium and laminar-flow conditions, the particle settling velocity can be expressed by Stokes' law as follows:

$$V_s = \left[g \left(\rho_p - \rho \right) D^2 \right] / 18\mu \tag{10}$$

    where V_s is the settling velocity, D is the diameter of the particle, g is the acceleration resulting from gravity, ρ_p is the density of the particle, ρ is the density of the gas, and μ is the viscosity of the gas.

2. *Centrifugal Force.* Centrifugal force is a force exerted on a particle that depends on the curvature of the path or centrifugal velocity to the particle. According to Newton's third law of motion, centrifugal force is equal to centripetal force; that is, the magnitude of the centrifugal force is equal to the mass of the particle times its centrifugal velocity squared and divided by the radius of the circular motion:

$$\mathbf{F} = \frac{mv^2}{r} \tag{11}$$

    where $\mathbf{F}$ is the centrifugal force, m is the mass, v is velocity, and r is the radius.
    Only centripetal force acts on the body in motion. Centripetal force pulls on the mass to maintain the mass in a circular path. If the centripetal force is removed from the mass, the mass would travel in a straight tangential line as required by Newton's first law.

3. *Electrostatic Force.* Electrostatic force on a particle is the product of the charge on the particle multiplied by the collecting field. The particles in the gas stream pass an electrical field. The ionization of airflow between electrical fields is charged. That allows the particles to draft to oppositely charged plates. As described by Coulomb's law, the force between two charges is directly proportional to the product of the charges and inversely proportional to the square of the distance between them (13):

$$\mathbf{F}_e = K_e \frac{qq'}{r^2} \tag{12}$$

where $\mathbf{F}_e$ is the electrostatic force, and qq' are the charges, K_e is the proportionality constant, and r is the radius.

4. *Inertial Impact.* Inertia is the resistance force to a mass in changing its state of motion. The force could set a mass at rest into motion, or if in motion, to any change of speed. According to Newton's first law or the law of inertia, the inertial force is proportional to the mass and the acceleration. The greater the mass, the more it resists acceleration. In other words, the greater the inertia, the less acceleration force is applied to the mass. The large particles in the gas stream have too much inertia to follow the gas streamlines around the impactor. The large particles are impacted on the impactor surface, whereas the small particles and the gas tend to diverge and pass around the interceptor.
5. *Direct Interception.* The particles having less inertia can barely follow the gas streamlines. If the gas streams pass an opening media of filter, the openings have an opening section less than the particle radius. The particles in the gas streams will be directly intercepted by the filter media and remained in the surface of the filter. This mechanism is also called straining. In air pollution control, fabric filtration is a well-known method for separating dry particles from a gas stream. The open spaces among the fibers can be as large as 50 µm
6. *Diffusion.* Particles less than 0.1 µm have individual or random motion. They are governed by diffusion; that is, small particles are affected by collisions on a molecular level. The particles do not necessarily follow the gas streamlines, but move randomly. This is known as "Brownian motion." In thermal motion, the motion velocity of the particle is proportional to the square root of the temperature. Because of this random movement and high temperature in the APCD, the air pollution device may indiscriminately collect the particles.

2.2.2. *Control of Metal Vapor*

Vapor is only used for a condensable gas that is not far from a liquid or solid and can be condensed into a liquid or solid relatively easily. Major chemical compounds will change their phases from liquid to vapor as the temperature is raised. Some others, such as mercuric chloride, can directly sublimate from a solid phase to a vapor phase. Control of metal vapor depends on characteristics of the metal, operational condition in the incinerator, and APCDs. The volatile metals, (i.e., mercury and selenium), are the most troublesome because they tend to stay in the gas phase, which is difficult to control by APCDs, and are less reactive with sorbents at high temperatures. Additional control technologies such as low-temperature sorbent injection/carbon adsorption or acid scrubbing are required to effectively control their emissions. An alternative technology for metal emission control is to capture metal vapors at the hot end of the combustion process (i.e., in the combustion chamber). It is to employ suitable sorbents to chemically and/or physically absorb volatilized metal vapors in the combustion chamber. The sorbents can either be injected into the chamber or serve as fluidized particles in a fluidized-bed combustor.

2.2.2.1. MERCURY EMISSION CONTROL

The concern over mercury emissions from waste combustion/incineration systems and coal-burning utilities has been growing, especially since the passage of the Clean Air Act Amendments of 1990. Unlike most other trace elements, mercury is highly volatile and exists almost exclusively in the vapor phase of combustion flue gases, either in the form of elemental mercury or mercury salts such as $HgCl_2$, HgO, HgS, and $HgSO_4$. As flue gases cool, it is possible that a fraction of the mercury in the gas phase may nucleate or be adsorbed on residual carbon or other fly ash particles.

The performance standards imposed by federal and state agencies for mercury emissions have been tightening in recent years. Current proposed federal standards on mercury emission for hazardous waste incinerators, cement kilns burning hazardous waste, and lightweight aggregate kilns burning hazardous waste are 40, 72, and 47 g/dscm (gram per dry standard cubic meter), respectively. Similar standards are expected to be proposed for municipal waste incinerators and coal-fired utility combustors. It is generally reported (14) that without additional modifications, conventional flue gas desulfurization systems are capable of removing mercury to some extent. However, to achieve high removal efficiencies to meet the expected emission standards, modifications such as packed-bed absorption by effective absorbents or dry sorbent injection with activated-carbon-impregnated lime are required.

Activated carbons with or without chemical impregnation have been reported to be relatively effective for the sorption of mercury chloride and elemental mercury (15–17). It is generally observed that mercury chloride is more easily adsorbed by nonchemically impregnated activated carbons than elemental mercury (18). However, chemically impregnated activated carbons have been found to dramatically enhance elemental mercury sorption (19). Sulfur, iodine, and chlorine are commonly used as the chemical agents for chemical impregnation. Other chemical agents evaluated include hydrogen chloride and the chlorides of aluminum, zinc, iron, and copper (20).

The mechanisms of mercury sorption by various sorbents have also been investigated and both chemisorption and physisorption have been reported (17,20,21). It is generally reported that the rate of mercury sorption by sorbents vary with temperature, gas flow rate, mercury form, sorbent type, sorbent amount, and sorbent properties such as particle size, surface area, pore size distribution, and chemical impregnation procedures (22).

2.2.2.2. MERCURY SPECIATION

Figures 4–7 display the results from four mercury equilibrium simulations. The four systems shown are as follows: $HgCl_2$–air, $HgCl_2$–air–S, Hg–air, and Hg–air–S. As indicated, at low temperatures, the commonly encountered stable forms of mercury are $HgSO_4(s)$, $HgCl_2(s)$, $HgS(s)$, $HgO(s)$, and $Hg^0(l)$. As the temperature increases, these mercury species may vaporize, decompose, or react with chlorine or oxygen to form $HgO(g)$, $Hg^0(g)$, and $HgCl_2(g)$.

When chlorine is present, the $HgCl_2(g)$ is the dominant form of mercury at temperatures up to about 600°C. With a further increase in the temperature, $HgCl_2(g)$ may react with $H_2O(g)$ to form $HgO(g)$ according to the following equation (23):

$$HgCl_2(g) + H_2O \rightarrow HgO(g) + 2\,HCl \tag{13}$$

However, as indicated in Figs. 4 and 5, the formed $HgO(g)$ may decompose to form $Hg^\circ(g)$ according to

$$HgO(g) \rightarrow Hg^0(g) + \frac{1}{2}O_2(g) \tag{14}$$

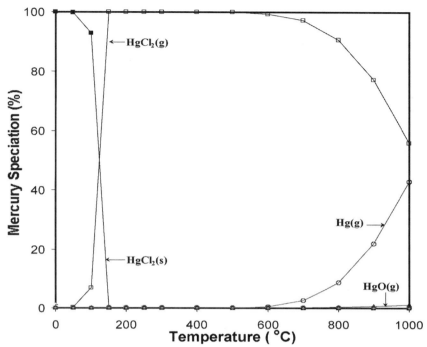

Fig. 4. Equilibrium mercury speciation in a HgCl$_2$–air system (HgCl$_2$: 1.6 wt%; air: 8.4%). (From refs. 5 and 10.)

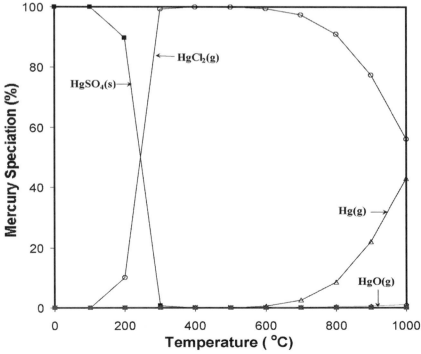

Fig. 5. Equilibrium mercury speciation in a HgCl$_2$–air–sulfur system (HgCl$_2$: 1.5 wt% air: 94.5 wt%; sulfur: 4.0 wt%). (From refs. 5 and 10.)

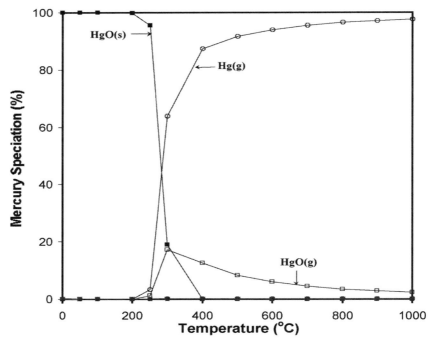

Fig. 6. Equilibrium Mercury Speciation in a Hg⁰–air system (Hg°: 1.6 wt%; air 98.4 wt%). (From refs. 5 and 10.)

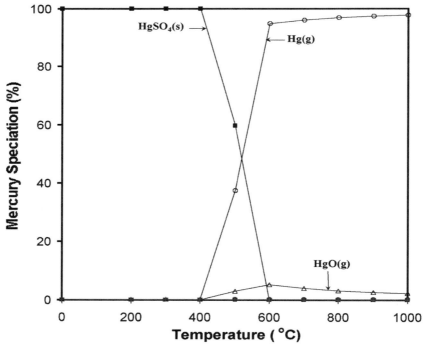

Fig. 7. Equilibrium mercury speciation in a Hg⁰–air–sulfur system (Hg⁰ 1.5 wt%; air: 94.5 wt%; sulfur: 4.0 wt%). (From refs. 5 and 10.)

Consequently, even with the existence of chlorine, elemental mercury is eventually the most dominant form of mercury at high temperatures of about 1000°C.

When chlorine is not present, either HgO(s) or $HgSO_4$(s) is the dominating species at a temperature below about 250°C for HgO(s) or 400°C for H_2SO_4(S) as indicated in Figs. 6 and 7. With an increase in the temperature, the two mercury species will decompose to form Hg^0(g), as indicated Figs. 6 and 7. Again, without the existence of chlorine, elemental mercury is the dominant form of mercury at high temperatures.

3. CONTROL DEVICE OF HEAVY METALS

Wastes are fed and burned in the incinerator. Partial metals can be destroyed or oxided, generating particulate matter, vapors, and ashes. These secondary pollutants need to be treated by an APCD prior to being discharged to the atmosphere. The most common APCDs among hazardous waste incinerators are the gravity settling chamber, cyclone, quench, baghouse, electrostatic precipitator (ESP), Venturi scrubber, wet scrubber, packed tower absorber, spray lower absorber, and so forth (2,6,24). More specifically, quench is used for gas cooling and PM removal. The settling chamber, cyclone, baghouse, ESP, or scrubbers are for PM removal. The absorber can eliminate vapor released from the burning. Most hazardous waste incinerator facilities use one of the following four combinations for air pollution control:

1. Baghouse (or ESP) and Venturi scrubber
2. Baghouse (or ESP) and dry scrubber
3. Quench (or Venturi scrubber) and packed tower absorber
4. Ionizing wet scrubber and a packed tower absorber

More than 30% of hazardous waste incinerators do not have any APCD because those facilities only handle a low concentration of pollutants (2).

The treatment of heavy metals in bottom or fly ashes from incinerators is very controversial. Bottom or fly ashes cannot be sent directly to a landfill or discharged into the environment. A toxicity characteristic leaching procedure (TCLP) needs to be conducted in order to determine the toxic level in the ashes. If the toxic level of metal exceeds the standards, further treatment of the ashes is required. Otherwise, ashes can be disposed of or treated in landfills. Four common ash treatments are (1) vitrification, (2) solidification by cement, (3) stabilization using chemical agents, and (4) extraction with acid or other solvent.

Detailed descriptions of APCDs have been discussed or stated in other chapters. In this section, only introductory information on these devices is provided.

3.1. Gravity Settling Chamber

This is a simple particulate collection device using the principle of gravity to remove the PM as a gas stream passes through a long settling chamber. Following Stoke's law in the settling chamber, the gas velocities must be sufficiently low in order for the particles to settle. Usually, the gas velocity is less than about 3 m/s and is needed to prevent re-entrainment of the settled particles. The gas velocity of less than 0.5 m/s will produce good settling results. Any baffles may reinforce settling. As the particles are suspended in the chamber, either more hydraulic detention time or less turbulence occurring in the chamber brings about more particles to settle. The pressure drop through the chamber is usually low mainly due to the entrance and exit head losses.

Settling chambers are commonly used as pretreatment devices. They are sometimes used in process industries, particularly in the food and metallurgical industries, as the first step in dust control. Use of gravity settling chambers can also reduce the maintenance cost of high-efficiency control equipment, which is subject to abrasive deterioration.

3.2. Cyclone

In addition to using gravity for particle removal, cyclones use centrifugal force for fine particle removal. Centrifugal force and inertia cause particles to move outwardly and then strike the cyclone's wall, resulting in the particle sliding down to the device. The cyclone consists of a vertically placed cylinder that has an inverted cone attached to its base. The particulate-laden gas stream enters tangentially at the inlet point to the cylinder. The velocity of this inlet gas stream is then transformed into a confined vortex, from which centrifugal forces tend to drive the suspended particles to the walls of the cyclone. The vortex turns upward after reaching the bottom of the cylinder in a narrower inner spiral. The clean gas is removed from a central cylindrical opening at the top while the dust particles are collected at the bottom.

3.3. Electrostatic Precipitator

The electrostatic precipitator is another particulate collection device that uses electrostatic force to remove particles less than 5 μm in diameter. Such a small particle size is difficult to remove effectively by gravity settlers and cyclones. However, electrostatic precipitators can easily remove those particles with approx 100% efficiency. For metal partition removal, the ESP serves as the primary air pollution device.

Electrostatic precipitators require the maintenance of a high potential difference between the two electrodes, one is a discharging electrode and the other is a collecting electrode. There are two types of ESP. In a single-stage ESP, gas ionization and particulate collection are processed in the single-stage ESP. It operates at ionizing voltages from 40,000 to 70,000 V DC. In the two-stage precipitator, it operates at DC voltages from 11,000 to 14,000 V for ionization. The metal particles are ionized in the first chamber and then collect in the second chamber.

The resistivity of metal particles is the dominant parameter affecting the performance of the ESP. Studies showed that decreased resistivity and increased cohesivity of the particles are the most effective approach in increasing ESP collection efficiency and decreasing the discharge emissions from ESP. In order to achieve these modifications of gas condition, such as additive agent and changing operation temperature, have provided a promising solution (24).

3.4. Quench

Air pollution control devices cannot operate at very high temperatures. Hot gas exiting from the incinerator usually has a temperature as high as 2000°F. When the hot gas from the combustion chamber directly discharges into the APCDs, it could damage the device and reduce the operation efficiency. A quencher is a cooling unit that is used to cool the waste temperature to approx 200°F. This cooling unit also provides a reduction of pollutants from the emission gases. There are two types of quench; air dilution and water dilution. For heavy metals, either in particulate forms or vapor forms, dilution with air is seldom used. Water dilution is a more popular method in cooling. When a large amount of water sprays into the gas stream from incineration combustion, the water takes the heat from the gas. During the quenching process, the amount of water sprayed into the stream exceeds the amount of water evaporated. As a result, some PMs in suspended forms and gases, such as HCl, Cl_2, SO_2, and others, could be removed.

3.5. Scrubber

Scrubbers are devices that remove PM by contacting the dirty gas stream with liquid drops. Generally, water is used as the scrubbing fluid. In a wet collector, the dust is agglomerated with water and then separated from the gas together with the water. The mechanism of particulate collection and removal by a scrubber can be described as a four-step process: (1) transport, (2) collision, (3) adhesion, and (4) precipitation. The physical principles involved in the operation of scrubbers are: (1) impingement, (2) interception, (3) diffusion, and (4) condensation.

The simpler types of scrubber with low-energy inputs are effective in collecting particles above 5–10 µm in diameter, whereas the more efficient, high-energy-input scrubbers will perform efficiently for collection of particles as small as 1–2 µm in diameter.

3.6. Fabric Filters

Fabric filtration is one of the most common techniques to collect PM from industrial waste gases. The use of fabric filters is based on the principle of filtration, which is a reliable, efficient, and economic method to remove PM from the gases. The air pollution control equipment using fabric filters are known as baghouses.

A baghouse or a bag filter consists of numerous long, vertically hanging bags approx 4–18 in. in diameter and 10–40 f long. The number of bags can vary from a few hundred to a thousand or more depending on the size of the baghouse. Baghouses are constructed as single or multiple compartmental units. For metal particle removal, the abrasion resistance, chemical resistance, tensile strength, and permeability, and the cost of the fabric should be considered. The fibers used for fabric filters can vary depending on the industrial application. Some filters are made from natural fibers, such as cotton or wool. Some use synthetic fibers, such as Nylon, Orlon, and polyester. In general, synthetic fibers can resist higher temperatures and corrosive chemicals.

Baghouses are widely used by industrial and waste incinerators for particle removal. The recommended filtering velocity varies from 1.5 to 4 ft/min. Once the baghouses have been used, resistance starts to build up inside the fabric filter. A typical pressure drop in the baghouses is from 4 to 8 in. H_2O.

3.7. Vitrification

The vitrification method has chemically been applied to process heavy metals and radioactive elements into a durable-resistant glass. In this method, electrodes are inserted into the soil or waste, allowing an electric current to pass between them. At approx 3600°F (or 1900°C), soil or waste melts, producing glasslike material. After the glasslike material cools to room temperature, it has a very low leaching capability. According to the study, the high cost of this process is of concern. The total cost of this operation is about $100 to $250 per ton of waste treatment.

3.8. Solidification

Solidification is a mechanical binding process that can encapsulate the waste in a monolithic solid and then preventing the waste from leaching or escaping from the surface of encapsulation. Several physical and chemical properties of ashes can affect solidification. The properties include pH, specific gravity, viscosity, temperature, solid/water contents, curing time, redox potential, ion exchange, absorption, and so forth. For metal solidification, two population processes have been used to treat metal in ashes: adding bulk agents and no bulk agents. Using bulk agents, the agents must be inert materials and quickly solidify. Typical bulk agents used in industry for solidification are Portland cement, lime, and polymer.

The advantages of solidification are as follows:

1. Low cost
2. Good long-term stability and low leachability
3. Low waste solubility and permeability
4. Nontoxic production
5. Easy to operate and in process

The disadvantages of solidification are as follows:

1. High volume increases
2 Resistance to biological degradation
3. Increased transportation cost

In a series of metal solidification tests conducted by the US EPA for electric arc furnace dust, the bulk agent-to-waste ratio, the water-to-waste ratio, mixture pH, and the cure time in days are 0.05, 0.5, 11.0, and 28, respectively. Table 2 shows the results of the solidification process of these tests (26).

3.9. Chemical Stabilization and Fixation

Stabilization uses either physical or chemical technologies to convert the toxic substances to the least soluble, leachable, and mobile compounds. Because ashes generated from hazardous waste incinerators have a relatively large surface area, the potential for leachability from the surface increases. The most popular operation of chemical stabilization and fixation is the addition of water and/or fix agents into the waste. The mixtures will go through a very slow dewatering and metamorphic process and become dry, hard, low-permeability clay or metamorphic rock. The selection criteria of fixation agents depend on pollutant concentrations, physical and chemical properties of pollutants, and the chemical reaction between pollutants and fixation agents. Possible agents for metal stabilization are cement, quicklime, clay, synthetic materials, and EDTA (ethylenediaminetetraacetic acid).

The advantages and disadvantages of stabilization and fixation are similar to metal solidification. Only the operation cost of stabilization and fixation is a little higher than the metal solidification.

Table 2
Metal Leaching Characteristics After the Solidification Process

Metal	Total concentration (ppm)	Untreated waste TCLP (mg/L)	Solidification TCLP (mg/L)
Arsenic	36	<0.01	<0.01
Beryllium	0.15	<0.001	<0.001
Cadmium	481	12.8	3.29
Chromium	1,370	<0.007	<0.017
Copper	2240	0.066	<0.02
Lead	20,300	45.1	1.1
Mercury	3.8	0.0026	0.0013
Nickel	243	0.027	0.020
Silver	59	0.021	<0.003
Zinc	244,000	445	22.6

Source: ref. 26.

3.10. Extraction

Extraction removes and recovers substances from waste using the extracting solvent and/or physical operations, such as stripping and distilling. This process has been widely used in recovering organic solutes, including phenols, oils, freons, and chlorinated hydrocarbons, from either aqueous or nonaqueous wastes. This extraction process has also been used in separating the metals from ore and mining. For metal extraction from fly ashes or bottom ashes in a hazardous waste incinerator, this process has seldom been used because the capital and operation and maintenance (O&M) costs are very high. In the meantime, the recovery metal values from the extraction are relatively low.

3.11. Fluidized-Bed Metal Capture

Metal capture experiments have been conducted in a fluidized bed during coal combustion at high temperatures (10). The fluidized-bed combustor was selected because it provides the most effective contact between metal vapors and sorbents. Typical experimental results indicating the effectiveness of metal capture by bauxite, zeolite, and lime are shown in Tables 3–5 for lead, cadmium, and chromium, respectively. Note that the results reported in these tables include only the amount captured by bed sorbents (i.e., the amount captured by fly ash sorbents is not included). It is worth pointing out here that the amount captured by fly ash sorbents can be significant especially for lime.

The results shown in Table 3 for lead capture indicate that all three sorbents tested are capable of in-bed capturing lead during fluidized-bed coal combustion, with the average capture efficiency ranging from 44%–69%. Bauxite and zeolite appear to have better "bed sorbent" capture efficiencies than lime. Lime, however, has greater "fly ash sorbent" capture efficiency as compared to zeolite and bauxite, which is not shown. As suggested by equilibrium simulations, the mechanism of lead capture by zeolite appears to be the result of the formation of $Pb_2SiO_4(s)$, and the mechanism of lead capture by bauxite could be the result of the formation of the same compound and/or alumino-silica

Table 3
Percentage Lead Capture by Bed Sorbents

Coal[a]/sorbent	Bauxite	Zeolite	Lime
IBC-101	64	74	75
IBC-102	80	68	16
IBC-106	77	57	67
IBC-109	62	47	49
IBC-110	73	51	36
IBC-111	49	62	32
IBC-112	74	44	22
Average	69	58	42

[a]IBC: Illinois Basin coal.
Source: refs. 5 and 10.

Table 4
Percentage Cadmium Capture by Bed Sorbents

Coal[a]/Sorbent	Bauxite	Zeolite	Lime
IBC-101	54	52	56
IBC-102	50	58	58
IBC-106	76	72	70
IBC-109	71	88	50
IBC-110	47	58	61
IBC-111	56	30	73
IBC-112	55	49	50
Average	58	58	60

[a]IBC: Illinois Basin coal.
Source: refs. 5 and 10.

Table 5
Percentage Chromium Capture by Sorbents

Coal[a]/Sorbent	Bauxite	Zeolite	Lime
IBC-101	0	4	2
IBC-102	0	26	30
IBC-106	0	22	9
IBC-109	0	37	26
IBC-110	0	66	44
IBC-111	0	10	47
IBC-112	0	54	51
Average	0	31	30

[a]IBC: Illinois Basin coal.
Source: refs. 5 and 10.

compounds. The mechanism of lead capture by lime, however, is suspected to be the result of the "melt capture", as suggested by Linak and Wendt (27).

For cadmium capture, the results shown in Table 4 indicate that the average "bed sorbent" capture efficiencies associated with the sorbents are similar to those of lead capture by bed sorbents. All three sorbents are seen to be relatively effective, with an average capture efficiency being around 60%. The effectiveness of cadmium capture by bauxite and zeolite appears to suggest the formation of $CdAl_2O_4(s)$ and $CdSiO_3(s)$, as revealed from equilibrium simulations. The formation of these compounds, however, could not be analytically confirmed because of their low concentrations in the sorbents.

The chromium capture results shown in Table 5 indicate that zeolite and lime are both capable of capturing the metal. The average capture efficiencies are seen to be 31% and 30%, respectively, which are much lower than those of lead and cadmium capture by the two sorbents. The mechanisms of chromium capture by these sorbents, however, are not clear at this time. Note that bauxite was not observed to capture any chromium because the original bauxite contained a high concentration of chromium, which continued to vaporize during combustion. The net result was that, in contrast to chromium capture, bauxite gave away chromium during the process.

4. METAL EMISSION CONTROL EXAMPLES

4.1. Municipal Solid-Waste Incineration

An incineration plant has a total annual capacity of 320,000 Mg/yr. The plant consists of a crane, a furnace chamber, a rotary kiln, an after-burning chamber, a boiler and recovery energy system, a semidry reactor, a bag filter, and a chimney. The plant was used for disposal of municipal solids from private households, institutions, commercial firms, offices, and so forth. The temperature of the incinerator was operated at over 900°C. Heavy metals such as Cd, Cr, Cu, Ni, Pb, and Zn in each control unit were measured and studied (28). Table 6 shows the mass balance of metals partitioning in the plant. As expected, Cu, Pb, and Zn had a higher quantity in the waste. Copper is one of the elements that forms a stable volatile compound in the combustion chamber, but very little is present in the vapor fractions. Lead and zinc are the other two major metals detected in the waste. Although lead and zinc are easily evaporated as chlorides present in the waste during the combustion, the study showed that a trace amount of these metals was released in the vapor phase. The majority of lead and zinc were still found in the bottom ash and the bag filter. According to this study, the baghouse for air pollution control provided a sufficient treatment and management for this solid-waste incineration plant.

4.2. Asphalt-Treatment Plants

There are approx 4500 asphalt plants in the United States. Firing hazardous waste and waste oil in asphalt plants is very economical because of the relatively low fuel cost. However, the plants' discharge of pollutants, including metals, is the primary concern of the US EPA. A study was conducted to determine the metal concentration in the burned gas and to determine metals removal efficiency by two APCDs: a baghouse and a scrubber. Two process plants were evaluated to determine metal concentrations in the aggregate and oil (29). Plant A is a drum mix plant with a 400 British

Table 6
Partitioning of Metals in the Solid-Waste Incineration Plant

Metal	Total amount (g/d)	Rotary kiln bottom (g/d)	Boiler bottom (g/d)	Bag Filter (Fly Ashes) (g/d)	Exhausted gas (g/d)
Cd	3,815 (100%)	576 (15.1%)	179 (4.68%)	3,060 (80.22%)	n.d.[a]
Cr	16,459 (100%)	16,070 (97.64%)	193 (1.16%)	196 (1.19%)	n.d.
Cu	140,679 (100%)	124,243 (88.32%)	1,564 (1.12%)	14,872 (10.57%)	n.d.
Ni	7,071 (100%)	6,739 (95.31%)	69 (0.97%)	263 (3.72%)	n.d.
Pb	203,939 (100%)	125,913 (61.74%)	798 (0.39%)	77,228 (37.87%)	n.d.
Zn	256,698 (100%)	188,928 (73.60%)	30,438 (11.86)	37,332 (14.54%)	n.d.

[a]n.d.: none detectable.
Source: ref. 28.

tons per hour production rate. The average fuel consumption rate was 500 gal of 100% waste oil per hour. Plant C had a lower production capacity at 76 British tons per hour production. The fuel burning rate was 165 gal of 100% waste oil per hour. Metals were measured in the aggregate feed, waste fuel oil, and inlet and outlet of the APCD. Table 7 shows the concentration of various metals in the aggregate feed and fuel oil. In general, less metals were present in the oil than in the aggregate. Both

Table 7
Metals Concentration in the Aggregate and Waste Oils

Metal	Plant A		Plant C	
	Aggregate (ppm)	Waste oil (ppm)	Aggregate (ppm)	Waste oil (ppm)
Arsenic	Not analyzed			
Aluminum	1,300	295	5	0–1
Boron	<1.0	<1.0	1,930	22
Barium	4.4	23.8	10	26
Cadmium	4.7	<0.3	0–1	1
Calcium	150,000	42.5	58,300	571
Chromium	2.7	287.5	11	3
Iron	5,900	570	8,120	188
Lead	11	71.8	7	416
Magnesium	48,000	11.5–14	3,590	286
Manganese	380	5.3	88	7
Molybdenum	5.1	8.1–9.6	4	7
Nickel	8.0	3.4	7	4
Silicon	270	69.3	665	150
Silver	<3.0	1.7–3.9	0–1	0–1
Sodium	160	0–20	23	385
Tin	<30	0–50	0–10	0–10
Titanium	42	1.8–2.3	73	0–5
Vanadium	7.3	10	10	12
Zinc	15	67	10	593

Source: ref. 29.

Table 8
Particulate Loading and Control Device Efficiencies

Plant	Sample location	Particulate concentration (gr/d-scf)	Particulate loading (lb/h)	Control efficiencies (%)
A	Inlet of baghouse	8.9	1536	99.75
	Outlet of baghouse	0.020	3.96	
C	Inlet of scrubber	2.32	271.9	99.24[a]
	Outlet of scrubber	0.015	2.07	

[a]The Venturi scrubber was operating at a pressure drop of 16–17.5 in. of water.
Source: ref. 29.

metal concentrations and feeding rate of the aggregate were over 100 times the feed rate of fuel oil. Table 8 shows that waste oil contributes significant amounts of aluminum, barium, chromium, iron, lead, molybdenum, nickel, silicon, and zinc to the PM at the APCD. Overall the efficiencies of the baghouse and scrubber are 99.75% and 99.24%, respectively. Most metals were effectively controlled by the baghouse. However, the scrubber was not so efficient for metals such as lead, sodium, and zinc. The study concluded that the baghouse can collect fine PM and the efficiency of the scrubber decreases as the particle size increases.

4.3. Hazardous Waste Incinerator Operation at Low-to-Moderate Temperature

The effectiveness of incineration at a low-to-moderate temperature in decontaminating soils was performed at the US EPA's Incineration Research Facility (19). The contaminated soils contained volatile organic compounds (VOCs) and toxic trace metals. With high-temperature incineration, the volatile metals in the combustion flue gas can pose a serious challenge to the air pollution control system. A series of tests was conducted

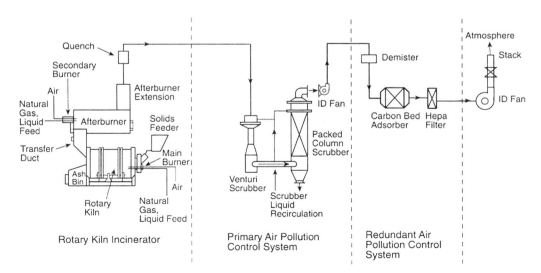

Fig. 8. Schematic of the IRF rotary kiln incineration system. (From ref. 30.)

in the rotary-kiln incineration system at the EPA (Fig. 8). Two objectives relative to trace metals are: (1) the distribution of trace metals in process discharges when a metal-contaminated soil is treated by thermal desorption and (2) whether thermal desorption treatment conditions affect a metal's leachability from the treated soil. The results showed arsenic, barium, cadmium, chromium, and lead were not volatile and remained in the soil. Mercury is a volatile metal, which tends to be equally distributed in the treated soil and the scrubber exit flue gas. At the testing temperature ranges (i.e., 320°C–650°C), mercury volatilization increased with an increase in treatment time. The effects of thermal treatment on metal leachabilities in the TCLP test varied. Arsenic leachabilities in the TCLP did not change significantly at a low-treatment temperature. As the operation temperature increased, less arsenic was leached from the soil. For cadmium, leachability in the TCLP may be reduced at a higher temperature. Compared to 13–16% leachable cadmium from the feed, only 2% was detected in the TCLP when the temperature was above 300°C. The data also showed a dramatic increase in the leachable fraction of chromium when soil treatment was increased from 225°C to 275°C. The mercury leachability data showed that the treated soil gave TCLP leachate mercury concentrations two or three times those of the feed samples.

NOMENCLATURE

P	Partial pressures of each reactant (kPa)
K	Equilibrium constant
V_s	Settling velocity (m/s)
D	Diameter of the particle (m)
g	Acceleration as a result of gravity (m/s^2)
ρ_p	Density of the particle (g/cm^3)
ρ	Density of the gas (g/cm^3)
μ	Viscosity of the gas (g/cm s)
$\mathbf{F}$	Centrifugal force (N)
m	Mass (g)
v	Velocity (m/s)
r	Radius (cm)
$\mathbf{F}_e$	Electrostatic force (N)
qq'	Charge (C)
K_e	Proportionality constant

REFERENCES

1. R. L. Davidson, D. F. S. Natush, J. R. Wallace, et al., *Environ. Sci. Technol.* **8**, 1107 (1974).
2. E. T. Oppelt, *JAPCA* **37**, 558–586 (1987).
3. The Resource Conservation and Recovery Act (1976), 42 USC 6901.
4. 40 CFR, Code of Federal Regulations, Title 40, Part 61. National Emission Standards for Hazardous Air Pollution, Washington, DC, 1990.
5. T. C. Ho, C. H. Chen, J. R. Hopper, et al., *Combust. Sci. Technol.* **85**, 101 (1992).
6. L. Theodore and J. Reynolds *Introduction to Hazardous Waste Incineration,* Wiley, New York (1987).
7. P. Gorman, A. Trenholm, D. Oberacker, et al., *Particulate and HCL Emissions from Hazardous Waste Incinerators* (H. M. Freeman, ed.), Technomic, Lancaster, PA, 1988.

8. K. S. Wang and K. Y. Chiang, in *Proceedings of the 90th Air & Waste Management Association*, 1997, paper 97-RP155.08.
9. K. S. Wang, K. Y. Chiang, S. M. Sin, et al,. in *Proceedings of the 91st Air & Waste Management Association*, (1998), paper 98-RA99.06.
10. T. C. Ho, H. T. Lee, H. W. Chu, et al., *Fuel Process. Technol.* **39**, 373 (1994).
11. W. A. Punjak, M. Uberoi, and F. Shadman, *J. AIChE* **35**, 1186 (1989).
12. M. Uberol, and F. Shadman, *J. AIChE* **36**, 307 (1990).
13. C. D. Cooper, and F. C. Alley, *Air Pollution Control: A Design Approach*, Waveland, Prospect Heights, IL (1994).
14. D. M. White, W. E. Kelly, M. J. Stucky, et al., *Emission Test Report: Field Test of Carbon Injection for Mercury Control, Camden County Municipal Waste Combustor*, Report No. EPA-600/R-93-181, US Environmental Protection Agency, Cincinnati, OH, 1993.
15. S. V. Krishnan, B. K. Gullett, and W. Jozewicz, *Envir. Prog.* **16**, 47 (1977).
16. S. V. Krishnan, B. K. Gullett, and W. Jozewicz, *ES&T* **28**, 1506 (1994).
17. Y. Otani, H. K. Emi, I. Uchijima, et al., *ES&T* **22**, 708 (1988).
18. K. Karjava, T. Laitinen, T. Vahlman, et al., *Int. J. Environ. Anal. Chem.* **49**, 73 (1992).
19. M. I. Guijarro, S. Mendioroz, and V. Munoz, *Ind. Eng. Chem. Res.* **33**, 375 (1994).
20. J. M. Quimby, in *Proceedings of the 86th Annual Meeting of the Air & Waste Management Association*, 1993, paper 93-MP-5.03.
21. Y. Otani, C. Kanaoka, S. M. Usui, and H. Emi, *ES&T* **20**, 735 (1986).
22. M. Babu, A. Licata, and L. P. Nethe, in *Proceedings of EPA Multipollutant Sorbent Reactivity Workshop*, 1996, p. 202.
23. B. Hall, O. Lindqvist, and E. Ljungstrom, *ES&T* **24**, p. 108 (1990).
24. M. D. LaGrega, P. L. Buckingham, and J. C. Evans, *Hazardous Waste Management*, McGraw-Hill, New York, 2001.
25. C. J. Bustard, K. E. Baldrey, M. D. Durham et al., 1997 in *Proceedings of the 90th Air & Waste Management Association*, 1997, paper 97-RP14.04.
26. Environmental Protection Agency, *Onsite Engineering Report for Waterways Experiment Station for K061*, EPA, Washington, DC, 1988.
27. W. P. Linak, and J. O. L. Wendt, *Prog. Energy Combust. Sci.* **19**, p. 145 (1993).
28. S. Binner, L. Galeotti, F. Lombardi, et al., in *Proceedings of 12th International Conference on Solid Waste Technology and Management* (1995).
29. L. E. Cottone, D. A. Falgout, and I. Licis, *Sampling and Analysis of Hazardous Waste and Waste Oil Burned in Three Asphalt Plants*, Technomic, Lancaster, PA, 1988.
30. J. Lee, D. Fournier, C. King, et al., *Evaluation of Rotary Kiln Incinerator Operation at Low-to-Moderate Temperature Conditions*, Report No. EPA/600/SR-96/105, US Environmental Protection Agency, Cincinnati, OH, 1997.

6

Ventilation and Air Conditioning

Zucheng Wu and Lawrence K. Wang

CONTENTS

AIR VENTILATION AND CIRCULATION
VENTILATION REQUIREMENTS
VENTILATION FANS
HOOD AND DUCT DESIGN
AIR CONDITIONING
DESIGN EXAMPLES
HEALTH CONCERN AND INDOOR POLLUTION CONTROL
HEATING, VENTILATING, AND AIR CONDITIONING
NOMENCLATURE
ACKNOWLEDGMENTS
REFERENCES
APPENDICES

1. AIR VENTILATION AND CIRCULATION

1.1. General Discussion

Ventilation is defined as the process of supplying air to, or removing it from, any enclosed space by natural or mechanical means. Such air may or may not be conditioned. There are at least five effects resulting from human occupancy of unventilated or poorly ventilated rooms: (1) a decrease in the oxygen content in air, (2) an increase in the carbon dioxide content in air, (3) a release of odor-causing organic compounds from the skin, clothing, and mouths of the occupants, (4) an increase in humidity owing to the moisture in the breath and evaporation from the skin, and (5) an increase in the room temperature owing to the heat generated in the body processes (1).

The first two effects are interrelated because the consumption of oxygen in breathing will reduce the oxygen content in the atmosphere of a closed room and proportionally increase the carbon dioxide content in the air. On the average, each person contributes about 1.699×10^{-2} m³/h (i.e., 0.6 ft³/h) of carbon dioxide to the atmosphere. According to the US Bureau of Mines Circular 33, 0.5 % carbon dioxide in the atmosphere, at the expense of oxygen, would require a slight increase in lung ventilation, whereas 10% cannot be endured for more than a few minutes. In order to keep the carbon dioxide content at a low level, many local codes specify that 0.28–0.89 m^3 per person per minute (i.e., 10–30 ft^3 per person per min) of fresh air should be provided in industrial and commercial areas to furnish necessary levels of oxygen and prevent odors.

The temperature of the normal human body is 37°C (98.6°F); however, people are most comfortable when the air temperature around them is in the 22–26°C (72–78°F) range. Sometimes, they can feel comfortable in a higher air temperature if there is air motion. Such man-made air motion is termed "ventilation," which causes a "wind chill effect" and provides evaporative cooling. The American Society of Heating, Refrigerating and Air-Conditioning Engineers (ASHRAE) guide (2) suggests air velocities of 6.1–15.2 m/min (20–50 ft/min), with the lower values applying to heating systems and the higher values to cooling.

Ventilation with heating as the only air treatment is generally used in schools, hospitals, commercial buildings, industrial plants, theaters, dwellings, and so forth. Originally, this consisted mainly of furnishing sufficient "fresh" air at as comfortable a level of temperature as was obtainable and in keeping out noticeable odors and drafts. Later, in some industrial plants, fumes and dusts were removed by exhaust systems and, when required by certain processes, the air was humidified.

Accordingly, two types of ventilation have evolved. The first is natural ventilation, which occurs through windows, doors, skylights, and roof ventilators and normally supplies an ample amount of air for a dwelling. The second is artificial ventilation, by which the air in a room is exhausted to the outside by a mechanical fan, thereby causing a lower pressure inside and a leakage inward through windows, doors, and walls. More specifically, a well-designed artificial ventilation system can control environmental conditions by effectively providing (1) air supply, introducing fresh outside air and distributing it where most needed in the proper volumetric flow rate and at the right velocities; (2) air exhaust, with rapid removal of fumes, or of overheated or contaminated air from working or storage areas; and (3) makeup air, elimination of negative pressures by replacing the air exhausted from a building.

The Occupational Safety and Health Act (OSHA) became law in April 1971 (3). The purpose of the Act is to ensure, as far as possible, safe and healthy working conditions for every worker in this country. There are seven areas having to do with air quality that are specifically mentioned in the OSHA: (1) exhaust ventilation for spray booths; (2) exhaust systems for blast-cleaning, grinding, polishing, and buffing operations; (3) mechanical ventilation where dangerous toxic or flammable liquids, vapors, and so forth occur or are stored; (4) mechanical exhaust systems for welding, cutting, and brazing operations; (5) special hood requirements and exhaust ventilation for plating, degreasing tanks, and similar applications; (6) exhaust ventilation for furnace rooms, drying, ovens, laundry, and washing operations; and (7) adequate makeup air systems. Complete details on these important safety and health standards are contained in the *Federal Register* (3). It should be noted that every industrial plant involved in at least one manufacturing operation requires artificial ventilation (i.e., mechanical ventilation) to conform to the OSHA requirements.

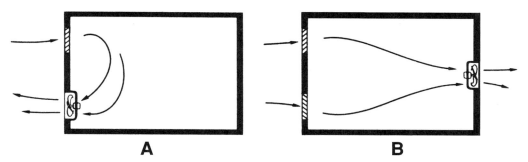

Fig. 1. (a) Improper and (b) proper positions of a fan in a room. (Courtesy of ILG Industries, 1972.)

1.2. Typical Applications

Selecting the proper location of fans in a room is extremely important for efficient mechanical ventilation. Air distribution, air velocities, and air directions all should be coordinated for optimum results. The improper positioning of a ventilation fan in a room where air "short circuits" to the fan and does not ventilate a section of the room to the right is shown in Fig. 1a when the inlet and outlet are at the same end of a large building. A duct system such as that shown in Fig. 1b should be used.

The basic systems shown in Fig. 2 are recommended for their economy and efficiency for both large and small buildings. In each case, the exhaust fan and/or power roof ventilator should be placed at points opposite to the intake, thus causing fresh air to be drawn across the entire room. The rate and direction of airflow is almost identical in each of the three cases.

Figures 3–7 illustrate typical examples of efficient ventilation systems using special arrangements to utilize the full benefits of the exhaust fans selected. Grilles and false ceilings can be used to eliminate the need for ductwork (*see* Fig. 3). Upblast or tube-axial fans can be used to exhaust fumes or vapors (*see* Figs. 4 and 5), and hoods or drop curtains are recommended over tanks or vats. The enclosure confines the fumes to be exhausted, enabling the fan to perform more effectively. Buildings with high ceilings and areas requiring localized air distribution can be ventilated more effectively by installing combined supply-and-exhaust systems (*see* Fig. 6) in which centrifugal, propeller, or roof fans may be used to circulate fresh air close to the workers. In areas in which several processes are producing fumes or emitting excessive heat (as shown in Fig. 7), hoods should be provided as close to the source as possible, and interconnected by means of ductwork to a remote ventilation fan.

Air circulators may be effectively used to boost air velocity through large buildings having flow patterns difficult to control. How three air circulators are used to redirect air into occupied areas near the floor level and to provide heat relief and cooling comfort to individuals in the area is shown in Fig. 8. It is generally desirable to position the fans 2.5–3.0 m above the floor and at approx 15.0-m intervals to obtain a continuously circulating column of air across a building.

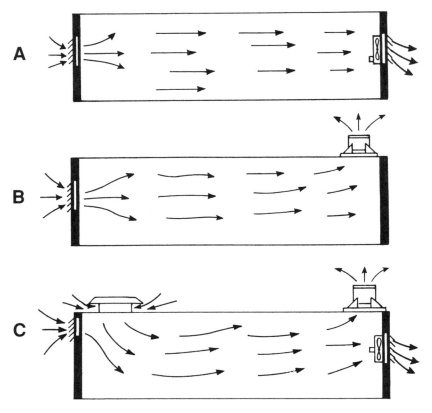

Fig. 2. Basic air distribution system for which the rate of minimum air velocity method can be applied: **(a)** wall fan system; **(b)** upblast power roof ventilation system; **(c)** combined system using a wall fan, an upblast power roof ventilator, and hooded-power roof ventilator. (Courtesy of American Coolair Corp., 1976.)

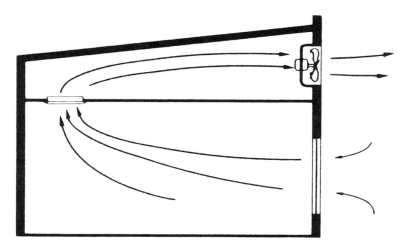

Fig. 3. Use of grilles and false ceilings, eliminating the need for ductwork. (Courtesy of ILG, Industries, 1972.)

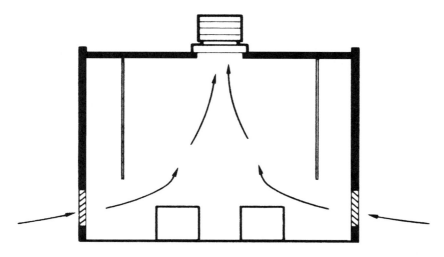

Fig. 4. Use of upblast fans to exhaust fumes and vapors. (Courtesy of ILG, Industries, 1972.)

When it is not practical to install a complete ventilation and cooling system, zone cooling may be used effectively in some buildings to provide comfort cooling to the small area of the building that is occupied. Such effective zone cooling may be accomplished by use of air circulators.

Figure 9 illustrates how deflectors and baffles are used for directing airflow. Deflectors can be installed near the air intake to direct air into occupied areas near the floor level. Baffles can be installed near the midpoint of the building to redirect and reconcentrate airflow into occupied areas, thus holding air velocity at a desired rate for effective cooling, as shown in Fig. 9a. Deflectors in front of a supply fan and under a supply power roof ventilator, shown in Fig. 9b, can diffuse intake air and prevent high-velocity air currents from flowing directly over occupants located near air intake areas.

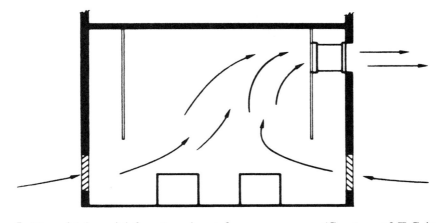

Fig. 5. Use of tube-axial fans to exhaust fumes or vapors. (Courtesy of ILG industries, 1972.)

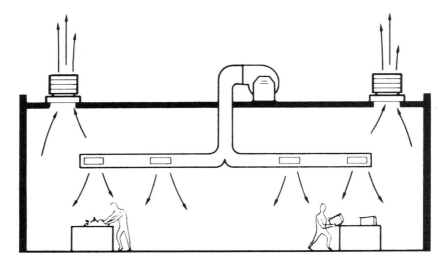

Fig. 6. Combined supply-and-exhaust systems. (Courtesy of ILG Industries, 1972.)

Baffles, deflectors, and diffusers are usually made from a variety of inexpensive materials such as polyethylene, Masonite, plywood, and sheet metal.

Buildings incorporating exhaust fans and/or power roof ventilators into their basic design sometimes may exhaust too large volumes of hot air, fumes, and other contaminants, thus causing severe negative pressures to develop in the buildings. This condition can reduce the efficiency of the exhaust system, create excessive air movement through doorways and other openings, and create downdrafts in flues, which may extinguish pilot lights and cause explosions and fire. The solution is to install supply fans or "makeup" air ventilators supplying outside air to "make up" the deficiency and thereby balance the internal pressure. The supply fans or "makeup" air ventilators should be located so that the airflow direction and air velocities created contribute to the total ventilation.

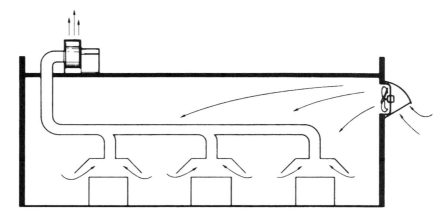

Fig. 7. Interconnected ventilation system using a remote exhaust fan. (Courtesy of ILG Industries, 1972.)

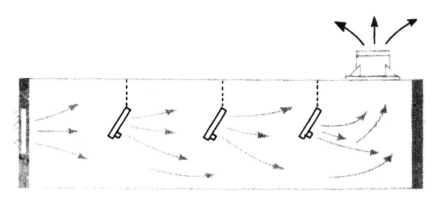

Fig. 8. Use of air circulators. (Courtesy of American Coolair Corp., 1976.)

2. VENTILATION REQUIREMENTS

The volume of air required to operate a ventilating and cooling system is not only the most fundamental but also the most important ventilation requirement. The method selected for determining air volume will relate closely to the objectives to be accomplished by the ventilation and cooling system. Generally, four methods are employed to calculate the required air volume: (1) rate of air change, (2) rate of minimum air velocity, (3) volumetric airflow rate per unit floor area, and (4) heat removal

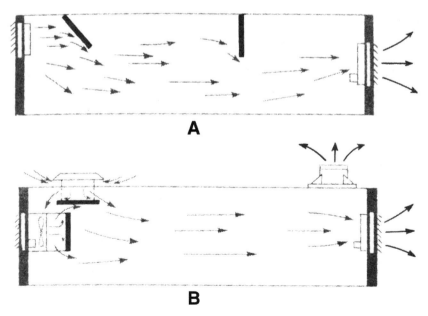

Fig. 9. Use of deflectors and baffles: **(a)** first case; **(b)** second case. (Courtesy of American Coolair Corp., 1976.)

Table 1
Recommended Rate of Air Change (min)

Class of buildings	Adequate ventilation	Sensible ventilation	Ref.
Auditoriums	5	2	5
Bakeries	3	1	5
Boiler rooms	4	2	5
Churches	5	1	6
Engine rooms	2	1	5
Factories	6	3	5
Foundries	4	1	5
Garages	6	3	5
Heat-treat rooms	2	1	5
Laundries	3	1	5
Locker rooms	5	2	5
Machine shops	5	3	5
Offices	5	2	5
Paper mills	3	2	5
Parking garages	5	2	5
Residences	5	1	6
Restaurants	6	4	6
Schools	5	1	6
Stores	8	5	5
Toilets	5	2	5
Transformer rooms	5	1	5
Warehouses	8	3	5

2.1. Rate of Air Change

This is a time-honored and the simplest approach to a calculation of air volume requirements. It is based on the assumption that a complete change of air in a given space should be made within a given time. The time, or rate of change, selected is based mainly on engineering experience with similar installations or may be established by a health or safety code. The formula for determining total ventilation requirements (4) by this method is

$$Q = C_V/R_a \tag{1}$$

(in metric or British units), where C_V is the volume of building to be ventilated (L^3), R_a is the recommended rate of air change, and Q is the measurement of volumetric airflow rate through a fan or fan system (L^3/t). (Note: Q is generally expressed in cubic feet of air per minute in British units and in cubic meters of air per minute in metric units.) Recommended rates of air change for typical installations are shown in Table 1. The "adequate ventilation" column in the table is based on average conditions. For more comfort, a greater rate of change may be utilized, as shown in the "sensible ventilation" column, where "sensible" means a rate of change of sufficient velocity such that the air motion can be "sensed," thus providing a cooling effect through moisture evaporation from the skin. This method has been used by engineers for many years and is considered to be satisfactory for small buildings or rooms. A typical layout of exhaust fans for the rate of air change method is shown in Fig. 10. For jobs that involve personnel

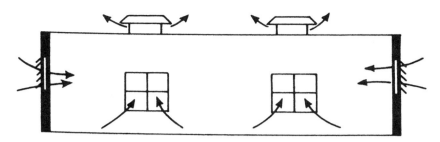

Fig. 10. Typical layout of fans for rate of air change method. (Courtesy of American Coolair Corp., 1976.)

comfort, this method is not recommended when the building is over 1416 m^3 in content or longer than 30.48 m (6). Separate exhaust systems should be provided where excessive heat or moisture is generated.

2.2. Rate of Minimum Air Velocity

It is possible that a high incidence of discomfort in workers may continue even though the exhaust and supply requirements of a large enclosure have been carefully determined and installed—a typical air circulation problem. As mentioned above, discomfort among workers can result in large buildings (i.e., over 1416 m^3 in content or longer than 30.48 m) when exhaust fans are roof-mounted and air is supplied through windows and wall openings, as shown in Fig. 10. The cooling effect on the workers in the area is negligible because airflow is generally from the window opening to the nearest roof exhauster. To be effective, air circulators should be installed at or near floor level in a large building to redirect the airflow, as shown in Fig. 8.

Three variations of the same basic system of air distribution to which the rate of minimum air velocity method can be applied are shown in Fig. 2. It has been determined from field experience that an average air velocity of 46–76 m/min (150–250 ft/min) is usually sufficient for personnel cooling under summertime conditions. The minimum air velocity recommended by the fan manufacturers (5,6) is indicated in Table 2. The volumetric airflow rate (m^3/min or ft^3/min) can be calculated easily by multiplying the cross section of an area through which the air is to move by the desired longitudinal air velocity. The typical layout of fans for the air velocity method is shown in Fig. 11. This method, although not always practical, is the most effective for comfort ventilation. When designing ventilation systems by this method, care must be taken to avoid air leakage or "short-circuiting" through unplanned openings.

2.3. Volumetric Airflow Rate per Unit Floor Area

This method of calculation is a modern adaptation of the rate of air change formula and is more suitable for large assembly areas, such as conference rooms, gymnasiums, or large dining rooms. The volumetric airflow rate (m^3/min or ft^3/min) is determined by multiplying the total floor area by an arbitrary figure ranging from 0.61 to 3.66 m^3/min/m^2 (2–12 ft^3/min/ft^2). The value 1.22 m^3/min/m^2 (4 ft^3/min/ft^2) has been recommended as a minimum for summer cooling of large assembly-type operations. This method is likely

Table 2
Recommended Air Velocity

Length of building		Air velocity	
ft	m	ft/min	m/min
Up to 100	Up to 30.5	150	45.7
100–200	30.5–61.0	200	61.0
200–300	61.0–91.4	250	76.2
300 and up	91.4 and up	250 plus booster fans	76.2 plus booster fans

Source: ref. 6.

to produce unsatisfactory results in many cases (4); therefore, for best results when using the method, air distribution and air velocity must be well controlled.

2.4. Heat Removal

When general ventilation involves a heat problem, the heat-removal method should also be considered. It is necessary to know the heat generated in Btu/h or g-cal/s, the average outside shade temperature, and the maximum inside temperature that can be tolerated. Equations (2) and (3) are used for calculating the volumetric airflow rate:

$$Q = 0.926q/(T_i - T_o) \tag{2}$$

(in British units), where Q is the ventilation rate (ft^3/min), q is the heat generated (Btu/h), T_i is the inside temperature (°F), and T_o is the outside temperature (°F), and

$$Q = 0.208q/(T_i - T_o) \tag{3}$$

(in metric units) where the units of Q, q, T_i, and T_o are m^3/min g-cal/s, degrees Celcius, and degrees Celcius, respectively.

3. VENTILATION FANS

3.1. Type

Mechanical fans are used to move air from one point to another for the purpose of air transport, circulation, cooling, or ventilation. There are two basic types of wheel used

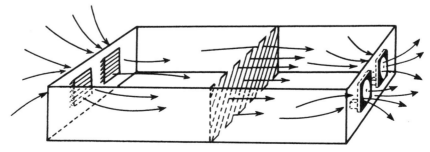

Fig. 11. Typical layout of fans for air velocity method. (Courtesy of American Coolair Corp., 1976.)

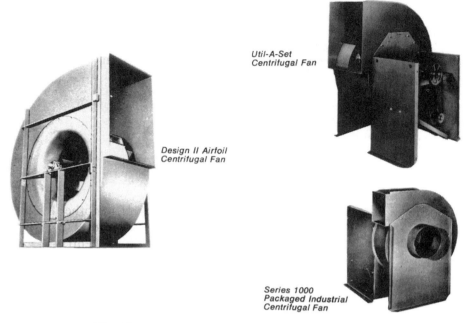

Util-A-Set Centrifugal Fan

Design II Airfoil Centrifugal Fan

Series 1000 Packaged Industrial Centrifugal Fan

Fig. 12. Centrifugal fans manufactured by ILG Industries, Inc.

in mechanical fans: (1) the centrifugal or radial-flow type, in which the airflow is at right angles to the axis of rotation of the rotor, and (2) the propeller or axial-flow type, in which the airflow is parallel to the axis of rotation of the rotor. Both are manufactured in several designs to suit various industrial usages. Only the most common types of mechanical fans are described in the following subsections.

3.1.1. Radial-Flow Fans

A radial-flow fan, sometimes referred to as a "blower" or a centrifugal fan, consists of a wheel or rotor mounted on a shaft that rotates in a scroll-shaped housing. Air enters at the eye of the rotor, makes a right-angle turn, and is forced through the blades of the rotor by centrifugal force into the scroll-shaped housing. A static pressure is thus imparted to the air by the centrifugal force. A portion of the velocity pressure is also converted into static pressure by the diverging shape of the scroll. Several centrifugal fans manufactured by the ILG Industries Inc., Chicago, Illinois are shown in Fig. 12. Centrifugal fans can be divided into three main classifications.

1. *Forward-curved-blade type*: A forward-curved-blade fan has about 20–64 blades that are shallow, with the leading edge curved toward the direction of rotation. The fan is normally referred to as a volume fan because it is designed to handle large volumes of air at low pressures. It rotates at relatively low speeds, which results in quiet operation. This type of fan can be used effectively in heating, ventilation, and air conditioning work, but should not be used for gases containing dusts or fumes because of the accumulation of deposits on the short curved blades.

2. *Backward-curved-blade type*: This type of fan has about 14–24 blades that are supported by a solid steel back plate and shroud ring and are inclined in a direction opposite to the

direction of rotation. It has a higher operating efficiency and higher initial cost than the forward-curved-blade type. Backward-curved, multiblade fans are used extensively in heating, ventilating, and air conditioning work but should not be used on gases containing dusts or fumes. They can be installed on the clean air discharge as an induced system in conjunction with other air pollution control devices.

3. *Straight-blade type*: This type of fan has a comparatively large rotor diameter. It has about 5–12 blades that are attached to the rotor by a solid steel back plate or a spider built up from the hub. The straight-blade fan is utilized for exhaust systems handling gas streams that are contaminated with dusts and fumes. Both its initial cost and its efficiency are less than that of the backward-curved-blade fan.

Centrifugal fans can be modified to form an air curtain. An air curtain fan substitutes as a door by discharging a blanket of high-velocity air over the doorway, parallel to the wall, as shown in Fig.13. It is used in restaurants, food markets, food processing plants, cold-storage plants, and many other places where frequent opening and closing of a door is not practical, because of continual traffic. The normal function of an air curtain fan is to create an air barrier that prevents entrance of insects, dust, etc., or to minimize infiltration, through the entrance of air that is heated, cooled, or air-conditioned. The effectiveness of the curtain depends upon many variables, such as air temperature, air pressure differential, and wind disturbance. It has been estimated that the effectiveness of an air curtain ranges from 75% to 90% compared with a normal door. Air curtain fans should be installed on the warm-air side of the doorway to guard against frosting and to eliminate chilled air blowing on personnel moving in and out of the cold-storage area.

3.1.2. Axial-Flow Fans

Axial-flow fans depend on the action of the revolving airfoil-type blades to pull the air in by the leading edge and discharge it from the trailing edge in a helical pattern of flow. Some axial-flow fans manufactured by American Coolair Corporation, Jacksonville. Florida, are shown in Fig. 14. They can be divided into three main classifications.

1. *Propeller type (Fig. 14a,b)*: This type of fan is mounted directly in a wall for ventilation against low resistance. It has from 2 to 16 blades that are either disklike or narrow, airfoil type. Its blades may be mounted on a large or small hub, depending on the use of the fan. The propeller fan is distinguished from the tube-axial and vane-axial fans in that it is equipped only with a mounting ring. Propeller fans are available in direct-drive and belt-drive versions. They are mostly used in free-air applications (without ductwork) for supply or exhaust ventilation. These fans provide maximum air volume with minimum horsepower requirements, at comparatively low initial costs; however, they are noisier than centrifugal fans for a given duty.

2. *Tube-axial type (Fig. 14c,d)*: The tube-axial fan is similar to the propeller fan except that it is mounted in a tube or cylinder. This design permits "straight-through" airflow and is particularly adaptable where space is limited. The tube-axial fan is more efficient than the propeller fan. Belt-drive types, with the motor out of the airstream, are most suitable for the exhaust of fumes, paint spray, and other similar applications. A two-stage, tube-axial fan, with one rotor revolving clockwise and the second counterclockwise, will recover a large portion of the centrifugal force as pressure and will approach vane-axial fans in efficiency.

3. *Vane-axial type*: The vane-axial fan is similar in design to a tube-axial fan except that air-straightening vanes are installed on the discharge side or suction side of the rotor. This type of

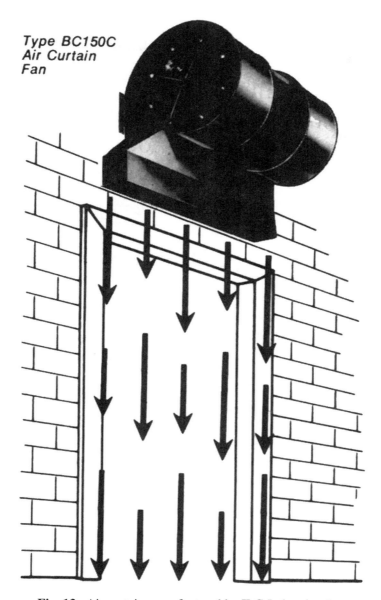

Type BC150C
Air Curtain
Fan

Fig. 13. Air curtain manufactured by ILG Industries, Inc.

fan can be readily adapted to multistage and be operated at a pressure of 16-in. water column at high volume and efficiency.

3.2. Fan Laws

The performance of mechanical fans and ventilation system characteristics are governed by the "fan laws" (4). The airflow through a mechanical fan is generally in the turbulent region, and the effect of viscosity is ignored. For homologous fans (or the same

Fig. 14. Axial-flow fans manufactured by American Coolair Corp: **(a)** direct-drive propeller fan (style C , A , E, and R); **(b)** belt-drive propeller fan (style Q and M); **(c)** direct-drive tube-axial fan (type TD duct fan); **(d)** belt-drive tube-axial fan (type TB duct fan).

fan) operating at the same point of rating, the volumetric airflow rate and the power will depend on the fan size, fan speed, and gas density:

$$Q = f D^m N^r d^s \qquad (4a)$$

where Q is the volumetric air flow rate (L³/t) and f is a coefficient dimensionless. D is the fan size (L), N is the fan speed (t⁻¹), and d is the gas density (ML⁻³). By substituting fundamental dimensional units (7), the values of m, r, and s are 3, 1, and 0, respectively. Hence,

$$Q = f D^3 N \qquad (4b)$$

Similarly, the following equation can be established for the system resistance developed:

$$H = fD^uN^vd^w \qquad (5a)$$

in which H is fundamental force per unit area mass times acceleration per area ($ML^{-1}t^{-2}$). The exponents u, v, and w then can be calculated from dimensional analysis as 2, 2, and 1, respectively. Hence,

$$H = fD^2N^2d \qquad (5b)$$

Repeating for the power required P (ML^2t^{-3}),

$$P = fD^aN^bd^c \qquad (6a)$$

The exponents a, b, and c are determined to be 5, 3, and 1, respectively, by dimensional analysis. Thus,

$$P = fD^5N^3d \qquad (6b)$$

Equations (4b), (5b), and (6b) can be simplified, combined, or modified to yield the so-called "fan laws," which enable users of fans to make certain necessary computations and also enable a manufacturer to calculate the operating characteristics for all the fans in homologous series from test data obtained from a single fan in the series. The "fan laws" are as follows (4):

1. Change in fan speed, but no change in size, gas density, and system.
   a. Q varies as fan speed (rpm).
   b. H varies as fan speed squared.
   c. P varies as fan speed cubed.

2. Change in fan size, but no change in fan speed and gas density.
   a. Q varies as the cube of the wheel diameter.
   b. H varies as the square of the wheel diameter.
   c. P varies as the fifth power of the wheel diameter.
   d. Tip speed varies as the wheel diameter.

3. Change in fan size, but tip speed and gas density constant.
   a. Fan speed varies inversely as the wheel diameter.
   b. Q varies as a square of the wheel diameter.
   c. H remains constant.
   d. P varies as a square of the wheel diameter.

4. Change in gas density, but system, fan speed, and fan size constant.
   a. Q is constant.
   b. H varies as the gas density.
   c. P varies as the gas density.

5. Change in gas density, but constant pressure and system, fixed fan size, and variable fan speed.
   a. Fan speed varies inversely as the square root of the gas density.
   b. Q varies inversely as the square root of the gas density.
   c. P varies inversely as the square root of the gas density.

6. Change in gas density, but constant weight of gas, constant system, fixed fan size, and variable fan speed.
   a. Fan speed varies inversely as the gas density.
   b. Q varies inversely as the gas density.

Table 3
Permissible Noise Exposure Limits[a]

Duration (h/d)	Sound level (dBA)
8	90
6	92
4	95
3	97
2	100

[a]Continuous operation.

    c. *H* varies inversely as the gas density.
    d. *P* varies inversely as the square of the gas density.

3.3. Fan Selection to Meet a Specific Sound Limit

3.3.1. Specification of Sound Level for Fans

Noise level in the industrial environment has become a critical matter. Limits on sound levels, under certain conditions in industry, have been established by the Occupational safety and Health Act in a section pertaining to occupational noise exposure. The specifications for fans and other mechanical equipment are expressed as "sound pressure" in decibels on the "A" scale (abbreviated dBA). These sound-levels are easily measured by a sound-level meter, reading directly in dBA. The meter, however, does not differentiate, but takes measures all noise-producing sources, whether fans or other machinery. The maximum sound levels that a worker may be exposed to are given in Table 3. For example, for a normal worker working 8 h/d, the permissible noise exposure limit is 90 dBA. Unfortunately, noise measurement and control are not that simple. Sound pressure in dBA can be measured in a laboratory at a given distance from the fan. However, the environment in which the fan is used can drastically affect the amount of sound heard from the fan. Therefore, specification of a dBA rating provides no assurance that a noise problem can be avoided (4,8).

The Air Moving and Conditioning Association (AMCA) has adopted a single number method of sound rating, known as "sones." The sones method of rating has the great advantage of being linear, rather than logarithmic (like the decibel system), and therefore is directly related to loudness as perceived by the human ear. Thirty sones, for example, is 3 times as loud as 10 sones. The practical limits of room loudness, in sones, for all types of building can be obtained from AMCA for selecting sone-rated fans. A correlation between dBA readings and sone levels has been established by the fan industry, through laboratory tests, making direct comparison possible. The sone levels compared with dBA readings are given in Table 4.

In order to avoid noise problems caused by fans and possible violation for OSHA noise standards, AMCA has established Standard 300, which is currently used by fan manufacturers as the official industry method of sound testing air-moving devices. For additional information on the sound characteristics of mechanical fans, the reader is referred to AMCA Publications 302 and 303, also AMCA Policy Statement dated

Table 4
Sone Levels Compared to dBA reading

Room sones	Approximate dBA
10	62
15	69
30	80
60	90

December 8, 1971. Appendix D of AMCA Publication 303 includes a recommended practice for the calculation of typical dBA sound pressure levels for ducted fan installations. According to the American Coolair Corporation, the procedure outlined in its Appendix D also applies with equal accuracy to conducted fans, and the procedure is recommended only when noise is not critical and a practical method for estimation results is needed (8). When noise is critical, sound power in octave bands should be specified for mechanical equipment.

3.3.2. Procedure for Fan Selection

Initially, ventilation fans should be selected according to their performance ratings, such as volumetric airflow rate (ft^3/min or m^3/min) at desired static pressure, blade diameter, motor horsepower, fan speed, brake horsepower, shape, weight, and so forth. Subsequently, the calculation for sound should be completed. It may be found that some changes in the initial selections are necessary to meet the noise criteria established by QSHA.

The American Coolair Corporation has developed a procedure for selecting mechanical fans using the same basic steps recommended by the AMCA. Because the suggested procedure is offered as a guideline only, neither ACC nor AMCA in any way accept responsibility for results obtained in using the procedure. The basic assumptions of the procedure are that a building or other facility exists and that additional ventilation is needed. The summary of the fan selection procedure and the explanation of step-by-step calculations are included in Appendix D to this chapter by permission of ACC.

The fans and blowers selection guides are available in the literature (9–13).

4. HOOD AND DUCT DESIGN

4.1. Theoretical Considerations

The actual flow contours and streamlines for airflow into a circular duct opening, representing a simple hood (14), are shown in Fig. 15. The contours in the figure are expressed as percentage of the velocity at the opening. The lines of constant velocity are contour lines, whereas those perpendicular to them are streamlines, which represent the direction of airflow. It should be noted that the hood in Fig.15b has a flange. At a distance of one or two hood diameters from the hood face, there is little difference in the centerline velocity of hoods of equal air volume; therefore, it is not necessary to distinguish between flanged and unflanged hoods. Accordingly, it is practical (16) to use one design equation for all shapes (i.e., square, circular, and rectangular) of up to about a 3 : 1 length-to-width ratio. The addition of a flange, however, improves the efficiency of the duct as a hood for

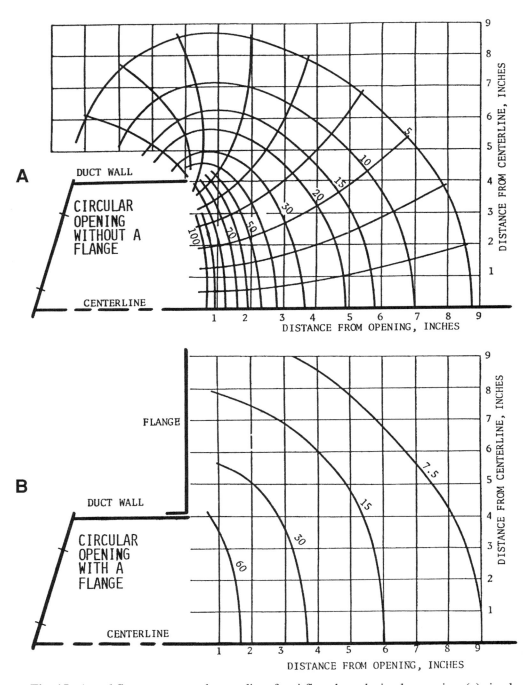

Fig. 15. Actual flow contours and streamlines for airflow through circular opening: **(a)** circular opening without a flange; **(b)** circular opening with a flange. (From ref. 15.)

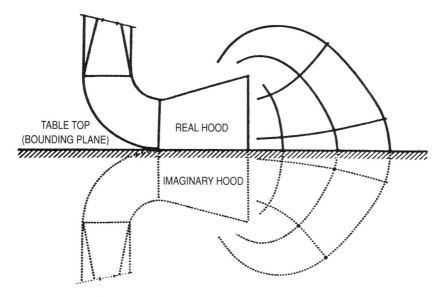

Fig. 16. Rectangular hood bounded by a plane surface. (Courtesy of the US EPA, 1973.)

a distance of about one diameter or less from the duct face. Dalla Valle (14) gives the following velocity equation for a freestanding or unobstructed circular opening:

$$p = 100 \, A/(A + 10X^2) \qquad (7)$$

(in metric or British units), where p is percentage of the velocity at the opening found at a point X on the axis of the duct (dimensionless), X is the distance outward along the axis from the opening (L), and A is the area of the opening (L^2).

The velocity at the opening can be computed from the familiar continuity equation

$$Q = AV \qquad (8)$$

(in metric or British units), where Q is the total air volume entering the opening (L^3/t) and V is the velocity of air at the opening (L/t).

By combining Eqs. (7) and (8) with appropriate units, the following is obtained for the freestanding or unobstructed circular opening:

$$Q = V_X \, (A + 10X^2) \qquad (9)$$

(in metric or British units), where V_X is the velocity of air at point X (L/t).

For a rectangular hood bounded on one side by a plane surface, as shown in Fig. 16, the following equation should be used:

$$Q = V_X(2A + 10X^2)/2 \qquad (10)$$

(in metric or British units), where the terms have the same meaning as above. The rectangular hood is considered to be twice its actual size, the additional portion being the mirror image of the actual hood and the bounding space being the bisector.

Analysis of both Eqs. (9) and (10) show that at the hood face $X = 0$ and $V_X = V$, and both equations reduce to Eq. (8). For large values of X, A becomes less significant.

Table 5
Exhaust Requirements for Various Operations

Operation	Exhaust arrangement	Remarks
Abrasive blast rooms	Tight enclosures with air inlets (generally in roof)	For 18.3- to 30.5-m/min downdraft or 30.5-m/min crossdraft in room
Abrasive blast cabinets	Tight enclosure	For 152.4 m/min through all openings, and a minimum of 20 air changes per minute
Bagging machines	Booth or enclosure	For 30.5 m/min through all openings for paper bags: 61 m/min for cloth bags
Belt conveyors	Hoods at transfer points enclosed as much as possible	For belt speeds less than 61 m/min, $V = 32.5 \text{m}^3/\text{min/m}$ belt width with at least 45.7 m/min through openings; for belt speeds greater than 61 m/min, $V = 46.5 \text{ m3/min/m}$ belt width with at least 61 m/min through remaining openings
Bucket elevator	Tight casing	For 30.5 $\text{m}^3/\text{min/m}^2$ of elevator casing cross-section (exhaust near elevator top and also vent at bottom if over 10.7 m high)
Foundry screens	Enclosure	Cylindrical: 122 m/min through openings and not less than 30.5 $\text{m}^3/\text{min/m}^2$ of cross-section; flat deck: 61 m/min through openings, and not less than 7.6 $\text{m}^3/\text{min/m}^2$ of screen area
Foundry shakeout	Enclosure	For 61 m/min through all openings, and not less than 61 $\text{m}^3/\text{min/m}^2$ of grate area with hot castings and 45, 7 $\text{m}^3/\text{min/m}^2$ with cool castings
Foundry shakeout	Side hood (with side shields when possible)	For 122- 152.4-$\text{m}^3/\text{min/m}^2$ grate area with hot castings and 06.7 to 122 $\text{m}^3/\text{min/m}^2$ with cool castings
Grinders, disk and portable	Downdraft grilles in bench or floor	For 61 to 122 m/min through open face, but at least 45.7 $\text{m}^3/\text{min/m}^2$ of plan working area
Grinders and crushers	Enclosure	For 61 m/min through openings
Mixer	Enclosure	For 30.5–61 m/min through openings
Packaging machines	Booth	For 15.2–30.5 m/min
	Downdraft	For 22.9–45.7 m/min
	Enclosure	For 30.5-122 m/min
Paint spray	Booth	For 30.5–61 m/min indraft, depending on size of work, depth of booth, and so forth.
Rubber roll (calendars)		For 22.9– 30.5 m/min through openings
Welding (arc)		For 30.5 m/min through openings

Source: ref. 15.

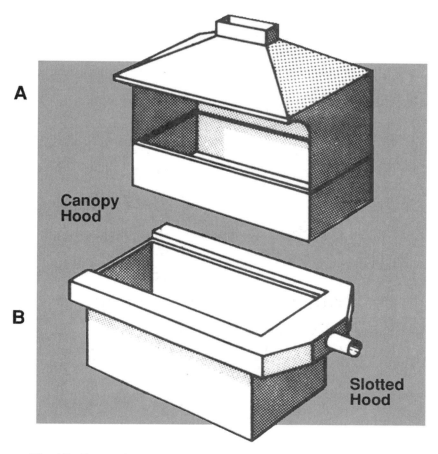

Fig. 17. Canopy hood and slot hood. (Courtesy of ILG Industries, 1972.)

For given values of A and X, the volume of air entering the hood can be determined by either Eq. (9) or (10), depending on the shape of the opening.

4.2. Hoods for Cold Processes

4.2.1. Hoods for Cold Operations Excluding Open-Surface Tanks

The recommended exhaust requirements for various cold operations, except using open-surface tanks, which were established over the years by various concerned organizations, are indicated in Table 5. The various cold operations are listed in the first column, and the second column specifies complete or nearly complete enclosures. The recommended specifications in the last column must be adjusted to specific applications.

4.2.2. Hoods for Open-Surface Tanks

The ventilation of open-surface tanks may be controlled by canopy hoods (*see* Fig. 17a) or by slot hoods (*see* Fig. 17b). The latter are commonly employed. The recommended minimum ventilation rates for open-surface tanks are indicated in Table 6. The usual practice is to provide a slot along each long side of the hood with a slot face velocity of 548.6–609.6 m/min (1800–2000 ft/min).

Table 6
Ventilation Rates for Open-Surface Tanks

| Process | Minimum ventilation rate (m³/min/m² hood opening) | | | | Minimum ventilation rate,[a] m³/min/m² of tank area, lateral exhaust W/L = (tank width)/(tank length) ratio | | | | | |
| | Enclosing hood | | Canopy hood | | W/L = 0.0–0.24 | | W/L = 0.24–0.49 | | W/L = 0.50–1.0 | |
	One open side	Two open side	Three open sides	Four open sides	A	B	A	B	A	B
Plating										
Chromium (chromic acid mist)	22.9	30.5	38.1	53.3	38.1	53.3	45.7	61.0	53.3	68.6
Arsenic (arsine)	19.8	27.4	30.5	45.7	27.4	39.6	35.5	45.7	39.6	51.8
Hydrogen cyanide	22.9	30.5	38.1	53.3	38.1	53.3	45.7	61.0	53.3	68.6
Cadmium	22.9	30.3	38.1	53.3	38.1	53.3	45.7	61.0	53.3	68.6
Anodizing	22.9	30.5	38.1	53.3	38.1	53.3	45.7	61.0	53.3	68.6
Metal cleaning (pickling)										
Cold acid	19.8	27.4	30.5	45.7	27.4	39.6	33.5	45.7	39.6	51.8
Hot acid	22.9	30.5	38.1	53.3	38.1	53.3	45.7	61.0	53.3	68.6
Nitric and sulfuric acids	22.9	30.5	38.1	53.3	38.1	53.3	45.7	61.0	53.3	68.6
Nitric and hydrofluoric acids	22.9	30.5	38.1	53.3	38.1	53.3	45.7	61.0	53.3	68.6
Metal cleaning (degreasing)										
Trichloroethylene	22.9	30.5	38.1	53.3	38.1	53.3	45.7	61.0	53.3	68.6
Ethylene dichloride	22.9	30.5	38.1	53.3	38.1	53.3	45.7	61.0	53.3	68.6
Carbon tetrachloride	22.9	30.5	38.1	53.3	38.1	53.3	45.7	61.0	53.3	68.6

Metal cleaning (caustic or electrolytic)										
Not boiling	19.8	27.4	30.5	45.7	27.4	39.6	33.5	45.7	39.6	51,8
Boiling	22.9	30.5	38.1	53.3	38.1	55.3	45.7	61.0	53.3	68.6
Bright dip (nitric acid)	22.9	30.5	38.1	53.3	38.1	53.3	45.7	61.0	53,3	68.6
Stripping										
Concentrated nitric acid	22.9	30.5	38.1	53.3	38.1	53.3	45.7	61.0	53.3	68.6
Concentrated nitric and sulfuric acids	22.9	30.5	38.1	53.3	38.1	53.3	45.7	61.0	53.3	68.6
Salt baths (molten salt)	15.2	22.9	22.9	38.1	18.3	27.4	22.9	30.5	27.4	33.5
Salt solution (Parkerise, Bonderise, etc.)										
Not boiling	27.4	27.4	30,5	45.7	27.4	39.6	33.5	45.7	39.6	51.8
Boiling	22.9	30.5	38.1	53.3	38.1	53.3	45.7	61.0	53.3	68.6
Hot water (if vent, desired)										
No boiling	15.2	22.9	22.9	38.1	18.3	27.4	22.9	30.5	27.4	33.5
Boiling	22.9	30.5	38.1	53.3	38.1	53.3	45.7	61.0	53.3	68.6

[a]Column A refers to the tank hood along one side or two parallel sides when one hood is against a wall or a baffle running length of tank and as high as tank is wide, also to tanks with exhaust manifold along center with W/w becoming tank with in W/L ratio. Column B regers to a freestanding tank with a hood along one side or two parallel sides.

Source: ref. 15.

173

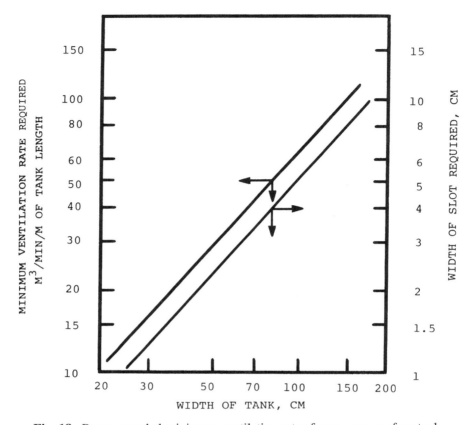

Fig. 18. Recommended minimum ventilation rates for an open-surface tank.

The recommended minimum ventilation rates for an open-surface tank with two parallel slot hoods are given in Fig. 18. Both Table 6 and Fig. 18 are modifications of the American Standards Association Code Z9.1, which makes no allowance for drafts. The effect of drafts can be minimized by the installation of baffles or an increase in ventilation rate.

4.3. Hoods for Hot Processes

4.3.1. Circular High-Canopy Hoods

A heated airstream, when rising from a hot surface, mixes turbulently with the surrounding air, as shown in Fig. 19. A hood may be considered a high-canopy hood when the distance between the hood and the hot source exceeds approximately the diameter of the source or 1 m, whichever is larger. The rising air column expands approximately according to the following empirical formula established by Sutton (17) and Hemeon (16):

$$D_c = 0.5X_f^{0.88} \tag{11a}$$

(in British units), where X_f is the distance from the hypothetical point source to the hood face (ft) (*see* Fig. 19) and D_c is the diameter of the hot column of air at the level of the hood face (ft).

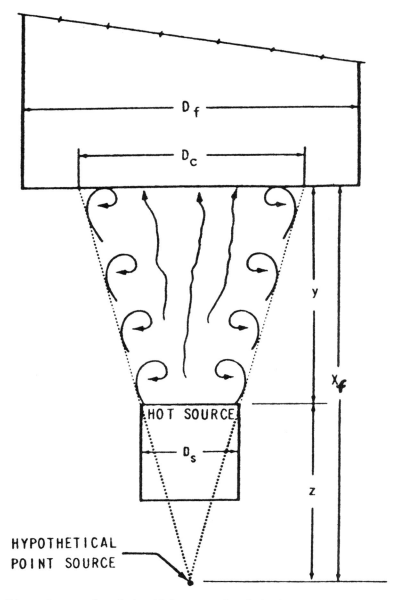

Fig. 19. Dimensions used to design high-canopy hoods for hot sources. (From ref. 16.)

The distance from the hypothetical point source to the hot source can be calculated using

$$z = (2D_s)^{1.138} = X_f - y \tag{12a}$$

(in British units), where z is the distance between the hypothetical point source and the hot source (ft), D_s is the diameter of hot source (ft), and y is the distance between the hot sources and the hood face (ft).

Table 7
Coefficients for Calculating Sensible Heat Loss by Natural Convection[a]

Shape or disposition of heat surface	Natural convection heat loss coefficient, $h^{b,c}$	
	Btu/h-ft2-°F	kg-cal/h-m2-°C
Vertical plates, over 2 ft (0.61 m) high	$0.3(T_d)^{0.25}$	$1.465(1.8T_d + 32)^{0.25}$
Vertical plates, less than 2 ft (0.61 m) high (X = height, ft or m)	$0.28(T_d/X)^{0.25}$	$1.016[(1.8T_d + 32)/X]^{0.25}$
Horizontal plates, facing upward	$0.38(T_d)0.25$	$1.855(1.8T_d + 32)^{0.25}$
Horizontal plates, facing downward	$0.2(T_d)^{0.25}$	$0.976(1.8T_d + 32)^{0.25}$
Single horizontal cylinders (D = diameter, in. or cm)	$0.42(T_d/D)^{0.25}$	$2.589[(1.8T_d + 32)/D]^{0.25}$
Vertical cylinders, over 2 ft (0.61 m) high (D = diameter, in. or cm)	$0.4(T_d/D)0.25$	$2.465[(1.8T_d + 32)/D]^{0.25}$

[a]Vertical cylinders less than 2 ft (0.61m) high.
[b]Multiply h value from the above formula by appropriate factor:

Height (ft)	Height (m)	Factor
0.1	0.03048	3.5
0.2	0.06096	2.5
0.3	0.09144	2.0
0.4	0.12192	1.7
0.5	0.15240	1.5
1.0	0.30480	1.1

[c]Metric units: T_d (°C); X (m); D (cm); k (kg-cal/h-m²-°C). British units T_d (°F); X (ft); D (in.); h (Btu/h-ft2°F).

The velocity of the rising column of air into the high-canopy hood may be calculated from

$$V_f = (37/X_f^{0.20})\, q^{0.333} \tag{13a}$$

(in British units), where V_f is the velocity of the hot-air jet at the level of the hood face (ft/min) and q is the rate at which the heat is transferred to the rising column of air (Btu/min).

The term q, in turn, can be calculated by

$$q = 0.0167hA_sT_d \tag{14a}$$

(in British units), where A_s is the area of the hot source (ft²), T_d is the temperature difference between the hot source and the ambient air (°F), and h is the natural convection heat loss coefficient listed in Table 7 (Btu/h-ft²-°F).

In general, a 15% safety factor should be applied in calculating the velocity of the rising column into a hood and the minimum ventilation rate to ensure complete capture of the emissions. Increasing the diameter of the hood by a factor of $0.8y$ is also recommended (15).

Although there are many different shapes or dispositions of heat surface, as indicated in Table 7, the high-canopy hoods usually control emissions arising from horizontal-plane surfaces. By combining Eqs. (13a) and (14a) with the heat transfer coefficient for horizontal-plane surfaces and allowing a 15% safety factor, the following equation for rising air velocity is derived:

$$V_f = 8A_a^{0.333}(T_d)^{0.417} X_f^{-0.25} \tag{15a}$$

(in British units), V_f, A_s, T_d, and X_f have been defined above.

Equations (11a)–(15a) can be rewritten using metric units:

$$D_c = 0.4336X_f^{0.88} \tag{11b}$$

$$z = 2.5929D_s^{1.138} = X_f - y \tag{12b}$$

$$V_f = (13.2612/X_f^{0.25}) \, q^{0.333} \tag{13b}$$

$$q = 0.00926hA_s(1.8T_d + 32) \tag{14b}$$

$$V_f = 3.9972A_s^{0.333}(1.8T_d + 32)^{0.417}(X_f)^{-0.25} \tag{15b}$$

where X_f is the distance from the hypothetical point source to the hood face (m), D_c is the diameter of the hot column of air at the level of the hood face (m), z is the distance between the hypothetical point source and the hot source (m), y is the distance between the hot source and the hood face (m), V_f is the velocity of the hot-air jet at the level of the hood face (m/min), q is the rate at which heat is transferred to the rising column of air (kg-cal/min), h is the natural convection heat loss coefficient listed in Table 7 (kg-cal/h-m^2-°C), A_s is the area of the hot source (m^2), and T_d is the temperature difference between the hot source and the ambient air (°C).

The volumetric airflow rate entering the hood can be calculated by

$$Q = V_r (A_s - A_c) + V_f A_c \tag{16}$$

(in metric or British units), where Q is the ventilation rate (L^3/t), V_f is the velocity of the rising air column at the hood face (L/t), A_c is the area of the rising column of contaminated air at the hood face (L^2), V_r is the required velocity through the remaining area of the hood $A_f - A_c$ (L/t), and A_f is the total area of the hood face (L^2).

4.3.2. Circular Low-Canopy Hoods

An important distinction between high- and low-canopy hoods is that the latter are so close to the hot source that there is little mixing between the rising air column and the surrounding atmosphere. In general, a hood is referred to as a low-canopy hood when the distance between the hot source and the hood face does not exceed approximately the diameter of the source or 1 m, whichever is smaller. The diameter of the air column may be considered equal to the diameter of the hot source.

For the practical design of low-canopy hoods, extending the hood 16 cm on all sides should be sufficient when drafts are not a serious problem. In other words, the hood face diameter must be taken as 32 cm greater than the diameter of the hot source. In addition, a 15%, or greater, safety factor is required.

By applying a 15% safety factor, the ventilation rate for the circular low-canopy hood can be calculated using Eqs. (17a) and (17b):

$$Q = 4.7 (D_f)^{2.33} (T_d)^{0.417} \tag{17a}$$

(in British units), where Q is the ventilation rate of the hood (ft^3/min), D_f is the diameter of the hood (ft), and T_d is the difference between the temperature of the hot source and the ambient atmosphere (°F), and

$$Q = 2.1203\ (D_f)^{2.33}\ (1.8T_d + 32)^{0.417} \tag{17b}$$

(in metric units) in which the units of Q, D_f, and T_d are m^3/min, m, and °C, respectively. To use the above two equations, select a hood diameter (D_f) 31 or 62 cm larger than the source. The ventilation rate required for a hood D_f in diameter then can be calculated for the actual temperature difference T_d between the atmosphere and the hot source.

4.3.3. Rectangular High-Canopy Hoods

Although a circular hood can be used to control a rectangular source of emission, the required volumetric ventilation rate would be excessive. Therefore, a rectangular source generally would require a rectangular hood in order to minimize the construction cost. General equations for designing a rectangular hood for a rectangular source are as follows:

$$z = (2W_s)^{1.138} = X_f - y \tag{18}$$

$$W_c = 0.5X_f^{0.88} \tag{19}$$

$$L_c = L_s + (W_c - W_s) \tag{20}$$

$$A_c = W_c L_c \tag{21}$$

$$W_f = W_c + 0.8y \tag{22}$$

$$L_f = L_c + 0.8y \tag{23}$$

$$A_f = W_f L_f \tag{24}$$

$$A_s = W_s L_s \tag{25}$$

$$V_f = 8A_s^{0.333}(T_d)^{0.417}X_f^{-0.25} \tag{26}$$

$$Q = V_f A_c + V_r(A_f - A_c) \tag{27}$$

(all in British units), where z is the distance between the hypothetical point source and the hot source (ft), X_f is the distance from the hypothetical point source to the hood face (ft), y is the distance between the hot source and the hood face (ft), W_s is the width of the hot source (ft), L_s is the length of the hot source (ft), W_c is the width of the rising air jet at the hood (ft), L_c is the length of the rising jet at the hood (ft), A_c is the area of the rising air jet at the hood (ft^2), W_f is the width of the rectangular hood (ft), L_f is the length of the rectangular hood (ft), A_r is the area of the rectangular hood (ft^2), V_f is the velocity of the hot-air jet at the level of the hood face, assuming that the hood controls emissions arising from horizontal-plane surfaces and allowing a 15% safety factor (ft/min), $A_s - L_s W_s$ is the area of the hot source (ft^2), T_d is the temperature difference between the hot source and the ambient air (°F), Q is the required ventilation rate (ft^3/min), and V_r is the required velocity through the remaining area of the hood, $A_f - A_c$ (ft/min).

The metric equations for designing a rectangular hood for a rectangular source are

$$z = 2.5929 W_s^{1.138} = X_f - y \tag{28}$$

$$W_c = 0.4336 X_f^{0.88} \tag{29}$$

$$L_c = L_s + (W_c - W_s) \tag{30}$$

$$A_c = W_c L_c \tag{31}$$

$$W_f = W_c + 0.8y \tag{32}$$

$$L_f = L_c + 0.8y \tag{33}$$

$$A_f = W_f L_f \tag{34}$$

$$A_s = W_s L_s \tag{35}$$

$$V_f = 3.9972 A_s^{0.333} (1.8T_d + 32)^{0.417} X_f^{-0.25} \tag{36}$$

$$Q = V_f A_c = V_r (A_f - A_c) \tag{37}$$

(in metric units); each term has been defined earlier. The metric units used in Eqs. (28)–(37) are shown in parentheses following each variable: z (m), X_f (m), y (m), W_s (m), L_s (m), W_c (m), L_c (m), A_c (m^2), W_f (m), L_r (m), A_r (m^2), A_s (m^2), V_r (m/min), T_d (°C), Q (m^3/min), and V_r (m/min).

4.3.4. Rectangular Low-Canopy Hoods

When the distance between the hood and the rectangular hot source does not approximately exceed the width of the rectangular hot source or 1 m, whichever is smaller, a rectangular low-canopy hood should be installed to control the source of emission. When a rectangular low-canopy hood is to be designed, the size of the air column can be considered essentially equal to the size of the hot source. The hood needs to be larger than the hot source by only a small amount to provide for the effects of waver and deflection resulting from drafts. When the problem of drafts is not serious, a rectangular low-canopy hood should be provided with dimensions 31 cm (1 ft) wider and 31 cm longer than the hot source. If drafts or toxic emission, or both, are a more serious problem, a safety factor greater than 15% is required; thus, the size of the hood should be increased an additional 31 cm or more, or a complete enclosure should be provided. The equations for designing the rectangular low-canopy hoods are

$$W_f = W_s + \Delta W \tag{38a}$$

$$L_f = L_s + \Delta L \tag{39a}$$

$$A_f = W_f L_f \tag{40a}$$

$$Q = 6.2(W_f)^{1.333}(T_d)^{0.417} L_f \tag{41a}$$

(in British units), where W_f is the width of the rectangular low-canopy hood (ft), L_f is the length of the hood (ft), W_s is the width of the rectangular hot source (ft), L_s is the length of the hot source (ft), ΔW is the increased width (usually 1 or 2 ft), ΔL is the increased length (usually 1 or 2 ft), A_f is the area of the rectangular hood (ft^2), T_d is the temperature difference between the hot source and the ambient air (°F), and Q is the required ventilation rate (ft^3/min).
The equivalent equations using metric units are

$$W_f = W_c + \Delta W \tag{38b}$$

$$L_f = L_s + \Delta L \tag{39b}$$

$$A_f = W_f L_f \tag{40b}$$

$$Q = 2.8069(W_f)^{1.333}(1.8T_d + 32)^{0.417} L_f \tag{41b}$$

in which all terms have been defined earlier. Their metric units are shown in parentheses following each variable: W_f (m), L_f (m), W_s (m), L_s (rn), ΔW (m), ΔL (m), A_f (m^2), T_d (°C), and Q (m^3/min).

4.3.5. Other Considerations

It is recommended (15) that the ventilation rate for an enclosure around a hot source be based on the same principles as that for a low-canopy hood. The thermal draft should be accommodated by the hood for hot processes. In addition to the determination of the ventilation rate required to accommodate the thermal draft, the hood face velocity or indraft through all openings should be calculated. The suggested minimum indraft velocities are as follows:

1. Any circumstances, 30.48 m/min (100 ft/min)
2. Air contaminants discharged with considerable force, 61.00 m/min (200 ft/min)
3. Air contaminants discharged with extremely great force, 152.40 m/min (500 ft/min)

Ideal ventilation hoods for hot processes should be airtight; however, openings may sometimes be unavoidable in the upper portions of an enclosure or canopy hood. Hemeon (16) proposes the use of Eq. (42) to calculate the volume of leakage from a sharp-edged orifice in a hood at a point above the hood face:

$$V_e = 200[\, Yq/A_o(460 + T)]^{0.333} \tag{42a}$$

(in British units), where V_e is the velocity of escape through orifices in the upper portions of a hood (ft/min), Y is the vertical distance above the hood face to the location of the orifice (ft), q is the rate at which heat is transferred from the hot source to the air in the hood (Btu/min), A_o is the area of the orifice (ft^2), and T is the average temperature of the air inside the hood (°F).

Based on Eq. (42a), the following equation using metric units is derived by the authors:

$$V_e = 53.3951[Yq/A_o(273 + T)]^{0.333} \tag{42b}$$

(in metric units), where the metric units to be used are given in parentheses after the variable: V_e (m/min), Y (m), q (kg-cal/min), A_o(m^2), and T (°C).

The approximate rate of heat generation can also be calculated using

$$q = QdST_d \tag{43}$$

(in metric or British units), where q is the approximate rate of heat generation (Btu/min or kg-cal/mm) Q is the ventilation rate or exhaust rate (ft^3/min or m^3/min) d is the average density of air mixture (0.075 lb/ft^3 or 1.2015 kg/m^3) S is the average specific heat of the air mixture, (0.24 Btu/lb-°F or 0.24 kg-cal/kg-°C) and T_d is the average hood temperature minus ambient air temperature (°F or °C).

4.4. Ducts

Ductwork is required to conduct the contaminants to a collection device. Ductwork may be either of rectangular or round construction and is usually made of galvanized sheet steel. When designing an air distribution system, the ventilation rate (m^3/min or

Table 8
Selection of Suitable Air Velocities

Operational units	Institutions and public buildings		Industrial buildings	
	Recommended air velocity (m/min)	Maximum air velocity (m/min)	Recommended air velocity (m/min)	Maximum air velocity (m/min)
Air filters	91.4	106.7	106.7	106.7
Air intakes	152.4	274.3	152.4	365.8
Air washers	152.4	152.4	152.4	152.4
Branch ducts	182.9–274.3	243.8–396.2	243.8–304.8	304.8–548.6
Branch risers	182.9–213.4	243.8–365.8	243.8	304.8–487.7
Cooling coils	152.4	152.4	182.9	182.9
Fan outlets	396.2–609.6	457.2–670.6	487.1–131.5	518.2–853.4
Heating coils	152.4	182.9	182.9	231.4
Main ducts	304.8–396.2	335.3–487.7	365.8–548.6	396.2–670.6

ft^3/min) should be determined first for each main and branch duct. After that, suitable air velocities (m/min or ft/min) should be selected according to the type of building and its intended application, as recommended in Table 8. Low air velocities or special acoustical treatment may be necessary for churches, libraries, and so forth, where quietness is a prime consideration.

The pressure drop, system resistance, system balance, and duct construction can be calculated when the ventilation rate and air velocities are known.

There are two general types of pressure loss encountered when air flows in ducts: (1) friction losses, which occur from the rubbing of the air along the surface of the duct, and (2) dynamic losses, which occur from air turbulence due to rapid changes in velocity or direction. More specifically, the losses in a ventilation system may include the inertia losses, orifice losses, straight-duct friction losses, elbow and branch entry losses, and contraction and expansion losses from the ductwork, and the pressure losses through the air pollution control equipment (64).

Inertia losses are defined as the energy required to accelerate the air from rest to the velocity in the duct and are expressed in terms of velocity pressure H_v:

$$H_v = (V_a/4005)^2 \qquad (44)$$

(in British units), where H_v is the velocity pressure or head [in. H_2O, $4005 = 1096.2$ (volume in ft^3 of 1 lb of air at 70°F and 14.7 psia)$^{0.5}$] and V_a is the velocity of air (ft/min)

Orifice losses may be defined as the pressure or energy losses at the hood or duct entrances, and they result mainly from the vena contract at the hood throat. The orifice losses (H_o) are usually expressed as a percentage of the velocity pressure (H_v) corresponding to the velocity (V_a) at the hood throat, as shown in Fig. 20.

Straight-duct friction losses (in in. H_2O per 100 ft of duct) are a function of duct diameter, velocity, and volumetric airflow rate. A resistance chart in which allowance has been made for moderate roughness of the duct is shown in Fig. 21.

The resistance of elbows and branch entries can be expressed in terms of the equivalent length of the straight duct of the same diameter that will have the same pressure

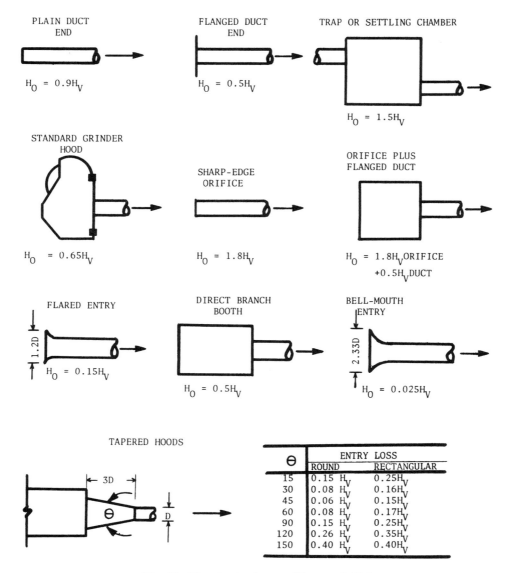

Fig. 20. Hood entry losses. (From ref. 18.)

loss as the fitting. In practical design, in order to determine the total resistance to air-flow, the equivalent lengths of the elbow and branch entry losses, as given in Tables 9 and 10, are added to the actual lengths of straight duct.

Contraction losses are encountered when the cross-sectional area contracts during transition from a larger duct area to one of a smaller area. The contraction pressure losses through decreasing duct transitions of various degrees of abruptness are given in Tables 11 and 12. Duct transitions, or reduction in duct size, should not be too abrupt or air turbulence and pressure losses will increase significantly. When the angle of the slope (θ) is 60° or less, the static pressure changes through a contracting duct transition can be calculated using

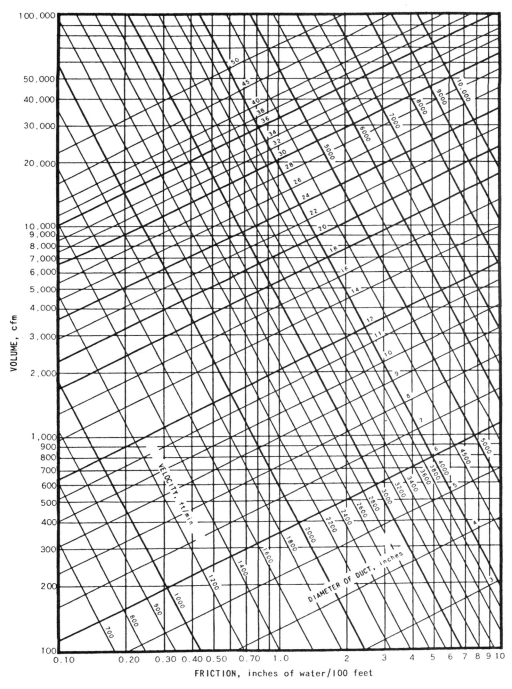

Fig. 21. Friction loss chart. (From the US EPA, 1973.)

Table 9
Airflow Resistance Caused by Elbows and Branch Entries Expressed as Equivalent Feet of Straight Duct

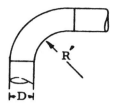 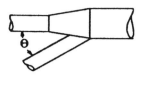

Diameter of round duct (in.)	Resistance (equivalent feet of straight duct)					
	Throat radius (R') of 90° elbow			Angle of branch entry		
	1.0D	1.5D	2.0D	15°	30°	45°
3	5	4	3	1	2	3
4	7	5	4	I	3	5
5	9	6	5	2	4	6
6	11	7	6	2	5	7
7	12	9	7	3	6	9
8	14	10	8	3	7	11
9	17	12	10	4	8	12
10	20	13	11	4	9	14
11	23	16	13	5	10	15
12	25	17	14	5	11	18
14	30	21	17	6	13	21
16	36	24	20	8	15	25
18	41	28	23	9	18	28
20	46	32	26	10	20	32
22	53	37	30	11	23	36
24	59	40	33	13	25	40
26	64	44	36	14	28	44
28	71	49	40	15	30	47
30	75	51	42	16	32	51
36	92	63	52	—	—	—
40	105	72	59	—	—	—
48	130	89	73	—	—	—

Source: ref. 15.

$$H_{s2} = H_{s1} - (H_{v2} - H_{v1}) - F_r(H_{v2} - H_{v1}) \tag{45}$$

in which H_{s1} and H_{s2} are the static pressures at points 1 and 2, respectively, H_{v1} and H_{v2} are the velocity pressures at points 1 and 2, respectively, and F_r is the loss fraction of velocity pressure difference.

When the angle of the slope (θ) is greater than 60°, the static pressure changes through a contracting duct transition should be calculated by

$$H_{s2} = H_{s1} - (H_{v2} - H_{v1}) - K(H_{v2}) \tag{46}$$

in which K is a simple factor indicated in Table 12.

Table 10
Airflow Resistance Caused by Elbows Expressed as Equivalent Diameters of Straight Round Duct

Aspect ratio (W/D) of rectangular duct	Resistance [equivalent diameter with of straight round duct, with throat radius (R') of 90° elbow]			
	0.25D	0.50D	1.00D	2.00D
4	13	8	6	6
2	17	10	8	7
1	18	10	8	7
0.5	24	14	9	9
0.25	30	19	14	14

Source: ref. 5.

Table 11
Contraction Pressure Losses Through Decreasing Duct Transitions at Taper Angle 60° or Below

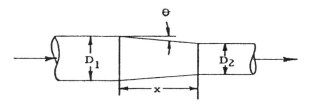

Taper angle, θ (deg)	$X/(D_1 - D_2)$	Loss fraction (F_r) of H_r difference
5	5.73	0.05
10	2.84	0.06
5	1.86	0.08
20	1.38	0.10
25	1.07	0.11
30	0.87	0.13
45	0.50	0.20
0	0.29	0.30

Source: ref. 15.

Table 12
Contraction Pressure Losses Through Decreasing Duct Transitions at Taper Angle Greater than 60°

D_2/D_2 ratio	Factor
0.1	0.48
0.2	0.46
0.3	0.42
0.4	0.37
0.5	0.32
0.6	0.26
0.7	0.20

Source: ref. 15.

Table 13
Static Pressure Losses and Regains Through Enlarging Duct Transitions

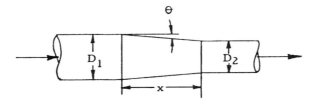

Taper angle θ (deg)	$X/(D_1 - D_2)$	Regain factor R_f	Loss factor, F_r
3.5	8.13	0.78	0.22
5.0	5.73	0.72	0.28
10.	2.84	0.56	0.44
15.0	1.86	0.42	0.58
20.0	1.38	0.28	0.72
25.0	1.07	0.13	0.87
30.0	0.87	0.00	1.00
>30.0	—	0.00	1.00

Notes: The regain and loss factors are expressed as a fraction of the velocity pressure difference between points 1 and 2. In calculating the static pressure changes through an enlarging duct transition, select from the table and substitute in Eq. (47).
Source: ref. 15.

When the cross-sectional area expands, a portion of the change in velocity pressure is converted into static pressure. The change in static pressure through an enlarging duct transition can be calculated by

$$H_{s2} = H_{s1} + R_f(H_{v1} - H_{v2}) \tag{47}$$

in which R_f is the regain factor expressed as a fraction of the velocity pressure difference between points 1 and 2 and can be selected from Table 13, H_v is a positive value, and H_s is positive in the discharge duct from the fan and is negative in the inlet duct to the fan.

The pressure drops through air pollution control equipment vary widely, but are often provided by the equipment manufacturers.

5. AIR CONDITIONING

5.1. General Discussion and Considerations

Complete air conditioning is defined as the total control of the air distribution and airborne dust, odors, toxic gases, and bacteria within any structure. Partial air conditioning is primarily the control of the air environment by filtration and by removal or addition of heat. Cooling, however, is defined as a heat transfer unit operation used for the removal of heat.

The human body is continually adjusting to the effective temperature of the air in order to maintain itself at the proper temperature. Effective temperature is defined by the American Society of Heating. Refrigerating, and Air Conditioning Engineers

(ASHRAE) as an arbitrary index of the degree of warmth or cold felt by the human body in response to the combined effects of temperature, humidity, and air movement. The numerical value of the effective temperature for any given air condition is fixed by the temperature of moisture-saturated air, which, at a velocity of 4.57–7.62 m/min (15–25 ft/min) or practically still air, induces a sensation of warmth or cold like that of the given condition (2). Accordingly, any air condition is assumed to have an effective temperature of 18.3°C (65°F) when it induces a sensation of warmth like that experienced in practically still air at 18.3°C and saturated with moisture (4).

When air temperature is hot, the mucous membranes of the nose normally show swelling, redness, and moisture, whereas cold temperature results in dry effects. The combination of low air temperature and high humidity usually causes discomfort (4). The amount of water vapor (i.e., moisture) in the air, expressed as a percentage of the maximum amount that the air could hold at the given temperature is termed "relative humidity," which is measured by a comparison of wet-bulb and dry-bulb temperatures (19). Regardless of the relative humidity, however, moisture on the human skin will cause chilliness. Figure 22 is the ASHRAE comfort chart (2) for air movement of 4.57–7.62 m/min (15–25 ft/min) with persons normally clothed and not strenuously active. Curves marked "summer" and "winter" indicate that near 100% of all persons will feel comfortable at an effective temperature of 21.7°C (71°F) in summer, and 20°C (68°F) in winter. Curves marked "slightly cool," "comfortable," and so on, represent sensations felt by the subjects after 3 h in the air conditioned space and adaptation to conditions, including moisture accumulations on skin and clothing. The ordinate of Fig. 22 is the "well-bulb temperature," which is the temperature of air as measured by a wet-bulb thermometer. The wet-bulb temperature is lower than that measured by the dry-bulb thermometer in inverse proportion to the humidity (12). Wet-bulb and dry-bulb temperatures are the same when the air is saturated.

In addition to the ASHRAE effort, a proposed heat stress standard for industry has been prepared by the Standards Advisory Committee on Heat Stress of the Occupation Safety and Health Administration and is known as the OSHA Heat Stress Standard. This standard establishes limits for wet-bulb globe temperature (WBGT) to which a worker can be exposed. The WBGT is a combination measured inside a 15.24-cm (6-in.) copper sphere painted black. The acceptable temperature (WBGT) for work in an area of low air movement is different from that of high air movement. With high air movement (91.44 m/min or greater), a higher WBGT is permitted. The OSHA Heat Stress Standard is given in Table 14. When many situations do not meet the standard, an air velocity of 91.44 m/min or greater should be provided over the worker. If this can be done, it will usually be the least expensive way of complying with the standard.

The quality of the atmospheric environment is governed by not only the temperature, humidity, and air movement but also by the concentration of gases, dusts, odors, and bacteria. Complete air conditioning should mean the total control of the air environment within a room or building; thus, various atmospheric pollutants should also be removed or reduced to permissible levels by air conditioning operation. It has been recognized (20) that many organic hydrocarbons occurring in the atmosphere are carcinogens (i.e., causative agents of cancer). Certain aromatic compounds have been associated with a significant increase in lung cancer. The inhalation of carcinogenic agents may also

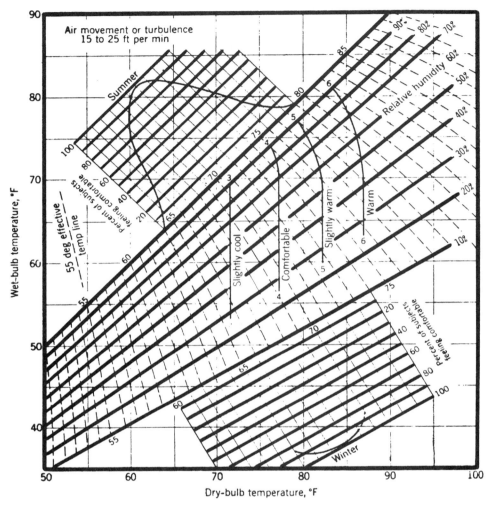

Fig. 22. A comfort chart established by American Society of Heating, Refrigerating and Air-Conditioning Engineers, Inc. (Courtesy of McGraw-Hill.)

cause cancers in parts of the human body other than the lungs. The threshold limit values (TLV) of selected industrial atmospheric contaminants in alphabetic order are listed in Appendices A and B. These TLVs were originally recommended by the American Conference of Governmental Industrial Hygienists (21) based on the best available information from industrial experience and experimental studies and various criteria of toxic effects or on marked discomfort. They should be used as guides in the control of health hazards by persons trained in this field. Technically, TLVs represent conditions under which it is believed that nearly all workers may be repeatedly exposed, day after day, without adverse effects. The values refer to time-weighted average concentrations for a normal workday. The TLVs of certain acutely acting (effects are felt immediately) substances may not provide a safety factor comparable to that of chronically acting (effects

Table 14
Heat Stress Standard Established by the Occupational Safety and Health Administration

Type of work	Max. WBGT[a] with low air movement[b]		Max. WBGT with high air movement	
	°F	°C	°F	°C
Light (400-800 Btu/h, or 28.56 g-cal/s)	86	30	90	32.2
Moderate (800–1200 Btu/h, or 56.84g-cal/s)	82	27.8	87	30.6
Heavy (above 1200 Btu/h, or 84 g-cal/s)	79	26.1	84	28.9

[a]WBGT is the wet-bulb globe temperature.
[b]Low air movement means that the air velocity is below 91.44 m/min.

are felt after a period of time) substances for which a time-weighted average applies. A "C" designation or "ceiling" has been affixed to such values, indicating that the TLV should not be exceeded.

Respirable dusts, such as silica, silicates, talc, or graphite, in the atmospheric environment are also hazardous. Appendix C indicates the TLV of respirable dusts recommended by the American Conference of Governmental industrial Hygienists. The unit of the threshold limits is "millions of particles per cubic foot of air" (mp/ft^3), based on impinger samples counted by light-field technics. The equivalent metric unit can be either "millions of particles per cubic meter" (mp/m^3), or "particles per cubic centimeter" (p/cm^3). The conversion factors to be used are

$$1 \ mp/m^3 = 1 \ p/cm^3 = 35.3 \ mp/ft^3$$

The percentage of crystalline silica in the formula is the amount determined from airborne samples.

Other suspended pollutants in air are plant spores (10–30 μm in diameter), plant pollens (20–60 μm in diameter), and bacteria (1–15 μm in diameter). Pollens cause hay fever in 2–4% of the population living east of the Rocky Mountains, and hay fever, it is reported, is increasing. Bacteria are not suspended in the air alone, but are carried by particles of dust, organic matter, or moisture. Winslow (22) has reported indoor dust up to 50 million particles per gram. One in 4000 of the indoor bacteria and 1 in 1000 of the street bacteria are of the intestinal type. The number of bacteria in indoor air may vary from 35 to 35,000 per cubic meter. According to the literature (21), infections causing respiratory diseases may occur by means of pathogenic bacteria attached to droplets discharged into the air during coughing, sneezing, and talking; infection of open wounds from air is also possible. Pathogenic bacteria may be carried to other persons or to food by air currents in a room.

Perspiration mingled with organic compounds from the skin and clothing of workers and decompositions taking place in workers' mouths can cause a serious odor problem in a room or building. Although long exposure to vitiated air and unpleasant odors causes no specific disease, certain illnesses may result (1).

Simple air conditioning, with filtration and cooling as the only air treatments, is generally all that is applied in most homes, stores, schools, business offices, restaurants, and hotels. Complete air conditioning with total control of atmospheric environment in a closed structure has been found to be important and profitable for use in certain industrial plants, which generate heat and small amounts of hazardous fumes, dusts, and odors. The operational cost can be justified by improving the health conditions of workers, in turn reducing medical expenses. Air conditioning also increases the output of workers by increasing efficiency and by reducing lost time and labor turnover.

For industrial plants producing a significant amount of hazardous fumes and dusts but a small amount of heat, only efficient ventilation is required. If both heat and hazardous gases are produced in significant amounts, a special cooling system and separate ventilation may be needed.

Apart from human considerations, there are other requirements in industry for control of the indoor environment, as in the case of flowering plants and animals in the scope of industrial processing (e.g., greenhouses and caged-rodent laboratories). Here, precise ventilation and control of temperature and humidity is often mandatory (23,24).

5.2. Typical Applications

5.2.1. Determination of Threshold Limit Values

Each year, the Threshold Limit Committee of the American Conference of Governmental Industrial Hygienists (ACGIH) reviews the list of TLVs published the preceding year. The most recent list appears in Appendices A–C. Knowing the TLV and the actual concentrations of atmospheric pollutants in the industrial plant, one can decide whether facilities of air conditioning or ventilation, or a combination of the two, should be installed. A brief discussion of basic considerations involved in developing TLV for mixtures is presented in this subsection (4).

When two or more hazardous substances are present, their combined effect, rather than that of each individually, should be given primary consideration. In the absence of information to the contrary, the effects of the different hazards should be considered as additive.

5.2.1.1. ADDITIVE EFFECTS (GENERAL CASE)

$$E_a = \sum_{i=1}^{i=n} \left(C_i / L_i \right) \tag{48}$$

where E_a is the combined effect of n different hazards (dimensionless), C_i is the observed concentration of the ith atmospheric pollutant (in ppm, mg/m^3, or mp/ft^3), and L_i is the corresponding TLVs (in ppm, mg/m^3, or mp/ft^3). The units of C_i and L_i must be identical. If E_a exceeds unity, then the threshold limit of the mixture should be considered as having been exceeded.

Sometimes, the chief effects of the different harmful substances are not, in fact, additive, but independent, such as when purely local effects on different organs of the body are produced by the various components of the mixture.

5.2.1.2. INDEPENDENT EFFECTS (GENERAL CASE)

$$E_i = C_i/L_i \tag{49}$$

where E_i is the independent effect of the ith hazard (dimensionless). In such exceptional cases, the TLV ordinarily is exceeded only when at least one member of the series (E_1, E_2, etc.) itself has a value exceeding unity.

When the contaminant source is a mixture and the atmospheric composition is similar to that of the original source material (i.e., the vapor pressure of each component is the same at the observed temperature), then a special case of additive effects arises.

5.2.1.3 ADDITIVE EFFECTS (SPECIAL CASE)

$$C_m = \sum_{i=1}^{i=n} C_i \tag{50}$$

$$C_m/L_m = \sum_{i=1}^{i=n} (C_i/L_i) \tag{51}$$

where C_m is the actual concentration of a mixture of atmospheric pollutants and L_m is the TLV of the corresponding mixture. The units of C_m and L_m must be identical. From Eqs. (50) and (51), L_m can be determined when C_i and L_i are known.

When a given operation or process characteristically emits a number of harmful dusts, fumes, vapors, or gases, it frequently will be only feasible to attempt to evaluate the hazard by measurement of a single key substance. In such cases, the TLV used for this key substance should be reduced by a suitable factor, depending on the number, toxicity, and relative quantity of the other pollutants ordinarily present. This is a special case of the "independent effects" consideration and will be illustrated further in Section 6.

5.2.2. Air Conditioning Operations

A spray-type central air conditioning system is shown in Fig. 23. Either the outdoor air or the recalculated air will pass through a primary treatment unit by which soots, dusts, and sonic fumes are removed from the air. The pretreatment unit can be a filter, a washer, or an electrical precipitator. Dry filters are made of polyurethane, polyester, urethane, or activated-carbon fiber. When cotton- or cloth-type activated-carbon fibers are used as the filter media, certain hazardous fumes and odor-causing substances can be adsorbed (25). Mats of crimped metal ribbon or glass wool that are coated with a viscous nondrying oil are used in viscous filters. For large-scale operations, automatic viscous filters moving over endless chains and having self-cleaning ability or a bag-house filter can be used. The residual dust particles in the filtered airstream can be carried down by water drops or caught on the eliminator plates, as shown in Fig. 23.

Electrostatic precipitators are used as primary treatment units of the central air conditioning systems in power plants, metallurgy, refining, heavy industries, and large hospitals for the removal of fumes, dusts, and acid mists. Particles, in passing through a high-voltage electrical field, are charged and then attracted to a plate of the opposite charge, where they collect.

For small-scale air conditioning operations, humidification may be obtained by passing air over a water surface or over cloth strips, which are kept wet by capillary action.

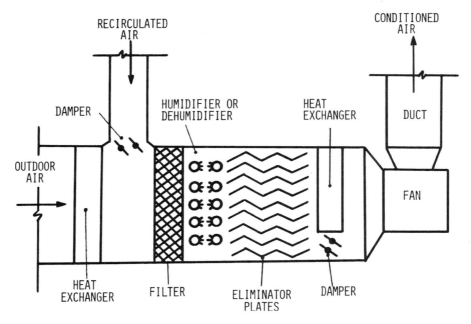

Fig. 23. Spray-type central air conditioning system. (Courtesy of MCGraw-Hill.)

When a large volume of air is to be moistened, spray humidifiers should be incorporated into the central air conditioning system. When the weather is cold and heaters are in use, humidification is generally necessary because the relative humidity in a heated room is very low. Humidification may reduce remove the drying problem.

When the weather is very hot, hot air evaporates and carries water. Because a high humidity may cause discomfort, dehumidification is necessary in conjunction with cooling of air. Dehumidification can be done by first lowering the room temperature below the dew point and condensing out the necessary amount of moisture and then reheating with dry heat to the desired temperature. Unusual examples in which the simultaneous control of temperature and humidity plays an important role are to be found in the life support systems utilized in spacecrafts and underground shelters (26–28).

Cooling of air is the major function of a complete air conditioning system and can be the sole function of a partial air conditioning unit. There are several methods for the cooling of the air in a building. The most common method is mechanical refrigeration of a type similar to that used in a household refrigerator. The power source can be either electricity or solar energy (29,30).

Heating of air in a central air conditioning system can be done in two general ways. In most industrial plants, houses, and public buildings, some of the heat loss from the building during cold weather is provided by the built-in heating facilities in the building, either by a warm-air furnace or by coils through which steam or hot water circulates. The outside cool air, introduced by the forced draft, or plenum system, must also be warmed by the heaters. When warm-air furnaces are used, the heating and cooling facilities use the same ductwork.

Another common central air conditioning system is the so-called "heat pump," which is used for both heating and cooling, particularly for buildings in areas with mild winters (30–32). Because of the growing cost and uncertainty of obtaining oil and supplies for building heating, plant owners and homeowners are looking increasingly to this most efficient type of electric heat. In the winter, the heat pump can produce heat at one-half the cost of electrical resistance systems, either baseboard or central furnace. The operation of a heat pump under two different weather conditions is shown in Fig. 24. The heart of any heat pump is its compressor (Fig. 24), which circulates a vapor refrigerant (i.e., a heat transfer fluid); its condensation, in which heat is released, is used for heating, and its evaporation after expansion in which heat is absorbed, is used for cooling. How the heat pump works on the heating cycle is demonstrated in Fig. 24a. Heat from the cold atmospheric air, what small amount there is, is absorbed by the expanded refrigerant in the evaporator. Thus, it increases the heat of the refrigerant after compression. In the condenser, heat is transferred to the coolant, which is the room air to be heated. The cooling cycle of a heat pump is illustrated in Fig. 24b. During hot weather, heat is extracted from the indoor hot air at the evaporator. At the condenser, either air or water can be used as the coolant to absorb heat and is then discharged. In summation, the cooling cycle of a heat pump is almost identical to those used in common air conditioners. The heating cycle, however, is exactly the reverse. The heat pump has an automatically controlled thermostat, which gives users exactly the temperature they want in winter, summer, and in between. Certain dirts, smokes, or odors can be removed by the filtration units in the heat pump.

Disinfection of air in a central air conditioned building should also be considered when possible. Hospitals, in particular, require exceptionally bacteria-free and dust-free air. Infection can be generally prevented by a high dilution of room air with filtered and conditioned new air. Filtration-washing can be effective for air disinfection because bacteria are not suspended in the air alone but are carried by particles of dust, organic matter, or moisture. Considerable air interchange provided by the fan of the air conditioner can also prevent the possibility of carbon monoxide (CO) poisoning resulting from gas heating appliances or other sources.

Various other methods of disinfecting air have been employed in central air conditioning systems. It has been demonstrated that quaternary ammonium compounds are effective germicides or bactericides (33–36). Washing air with a quaternary ammonium salt solution in an air conditioning system is a new approach to air disinfection. Ultraviolet light is also well known for its bactericidal effect and can be incorporated into the complete air conditioning system. The use of disinfectant vapors such as triethylene glycol in conjunction with air conditioning can be effective. The chemical has a high bactericidal effect and is free from odors, toxicity, and corrosiveness to metal surfaces.

6. DESIGN EXAMPLES

Example 1

The size of a laundry is 30 m long by 9 m wide with an average ceiling height of 5 m. Determine the necessary ventilation rate (fan capacity) using the rate of air change method and the rate of minimum air velocity method.

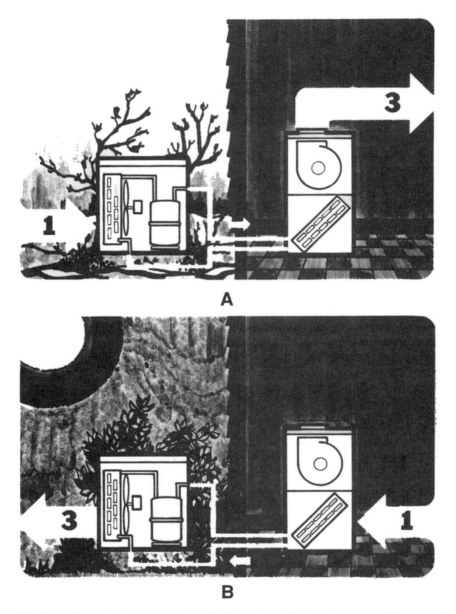

Fig. 24. Operations of a heat pump. **(a)** Cold weather: (1) heat is extracted from outdoor air; (2) heat transfer fluid carries heat to indoor unit; (3) circulating indoor air picks up heat and carries throughout home. **(b)** Hot weather: (1) heat is extracted from indoor air; (2) heat transfer fluid carries heat to outdoor unit; (3) heat is expelled to the outdoors.

Solution

1. *Rate of air change method.* From Table 1, it is seen that the laundry requires a complete air change every minute if adequate ventilation is desired. The necessary ventilation rate (Q) can then be calculated according to Eq. (1):

$$Q = C_V / R_a$$
$$= (30 \text{ m} \times 9 \text{ m} \times 5 \text{ m})/1 \text{ min} = 1350 \text{ m}^3/\text{min}$$

2. *Rate of minimum air velocity method.* Table 2 indicates that the recommended air velocity for the laundry is 45.7 m/min. Based on pulling air through 30 m, the required fan capacity is determined by multiplying the cross-sectional area through which the air is to move by the selected longitudinal air velocity:

$$Q = (5 \text{ m} \times 9 \text{ m}) \times 45.7 \text{ m/min} = 2056.5 \text{ m}^3/\text{min}$$

Example 2

The size of art auditorium is 25 m × 50 m × 6 m (i.e., $W \times L \times H$). Determine the required ventilation rate using the appropriate methods presented in Section 2.

Solution

1. *Rate of air change method.* This method is considered to be satisfactory for small buildings or rooms. It is not recommended for the auditorium because it is over 1416 m³ in volume and longer than 30.48 m in length.

Total volume of the auditorium = 25 m × 50 m × 6 m = 7500 m³

2. *Rate of minimum air velocity method.* The length of the auditorium is 50 m. The minimum air velocity recommended by American Coolair Corporation is 61 m/min (*see* Table 2).

Required ventilation rate 25 m × 6 m × (61 m/min) = 9150 m³/min

3. *Volumetric airflow rate per unit floor area method.* This method is more suitable for large assembly areas, such as auditoriums or gymnasiums. Assume that the local code is 1.22 m³/min/m² (4 ft³/min/ft²).

Required ventilation rate

= (Total floor area) × (Volumetric rate per unit floor area)
= 25 m × 50 m × (1.22 m³/min/m²) = 1525 m³/min

4. *Heat-removal method.* Because the ventilation of the auditorium does not involve a heat problem, this method is not applicable.
5. *Conclusion.* The selected ventilation rate (or exhaust rate) should be between 1525 and 9150 m³/min. The volumetric airflow rate per unit floor area method is the most economic method for the large assembly-type areas. However, air distribution and air velocity must be well controlled when using the method.

The rate of minimum air velocity method, although not always economical, is the most effective for comfort ventilation. When designing systems by this method, care must be taken to avoid air leakage or "short-circuiting" through unplanned openings.

Example 3

A dry-cleaning plant generates about 14,000 g-cal/s of heat. The average outside shade temperature and the maximum inside temperature are 26.7 and 29.4°C, respectively. Determine the ventilation rate using the heat removal method:

Solution

Use Eq. (3):

$$Q = \frac{0.208q}{(T_i - T_o)} = \frac{0.208 \times 14000}{(29.4 - 26.7)}$$
$$= 1078.52 \text{ m}^3/\text{min}$$

Example 4

A fan has a capacity of 22,000 ft³/min at 1-in. static pressure and 5-brake horsepower (hp). Determine the static efficiency of the fan.

The static pressure is the resistance, expressed in terms of the height of water column, against which the fan must operate. Resistance includes ductwork, dampers, louvers, and so forth, whether on the supply or exhaust side of the fan (or both).

Solution

The static efficiency (SE) of a fan can be calculated from the following formula:

$$SE = \frac{\text{Fan capacity} \times \text{Static pressure}}{6356 \times \text{Brake horsepower}}$$

$$= \frac{(22{,}000 \text{ ft}^3/\text{min}) \times (1 \text{ in.})}{(6356 \times 5 - \text{Brake hp})}$$

$$= 0.69 (\text{i.e., } 69\%)$$

Example 5

A 80-cm-diameter fan operating at 1060 rpm delivers 130.8 m³/min at 12.7 cm static pressure. What size fan of the same series would deliver 318 m³/min at the same static pressure?

Solution

There is a change in fan size, but no change in fan speed and gas density. Fan Law 2a can be applied: Q varies as the cube of wheel diameter.

$$Q'/Q = (D'/D)^3$$
$$D' = D(Q'/Q)^{0.333} = 80(318/130.8)^{0.333}$$
$$= 107.5 \text{ cm (i.e., 42.3 in.)}$$

Example 6

A fan is exhausting 345.5 m³/min of air at 316.6°C (density = 0.60 kg/m³) at 10.16 cm static pressure from a drier. The fan speed is 632 rpm, and 9.69 kW is required. What will be the required power if air at 21°C (density = 1.20 kg/m³) is pulled through the system?

Solution

Gas density has changed, but system, fan speed, and fan size are kept constant. Use Fan Law 4c: P varies as gas density.

$$P' = P(d'/d)$$
$$= 9.69 \text{ kW}(1.20/0.60)$$
$$= 19.38 \text{ kW (i.e., 26 hp)}$$

Example 7

A fan operating at 850 rpm delivers 230 m³/min at 15 cm static pressure and requires 8.95 kW (i.e., 12 hp). It is desired to increase the output to 340 m³/min in the same system. Determine the increased fan speed, the new static pressure, and the required power.

Solution

1. Fan Law 1a: Q varies as fan speed.

$$Q/Q' = N/N'$$
$$N' = N(Q'/Q) = 850(340/230)$$
$$= 1256 \text{ rpm (the increased fan speed)}$$

2. Fan Law 1b: H varies as fan speed squared.

$$H/H' = (N/N')^2$$
$$H' = H(N'/N)^2$$
$$= 15(1256/850)^2$$
$$= 32.75 \text{ cm water column (new static pressure)}$$

3. Fan Law 1c: P varies as fan speed cubed.

$$P/P' = (N/N')^3$$
$$P' = P(N'/N)^3$$
$$= 8.95(1256/850)^3$$
$$= 28.87 \text{ kW (the required new power, 38.7 hp)}$$

Example 8

Consider a paint spray booth 3 m wide by 2 m high. Work may be 1.5 m in front of the booth face at times. A nearly draftless area requires 30.5 m/min at the point of spraying. Determine the required exhaust rate and the face velocity.

Solution

Table 5 indicates that spray booths of the open-face type are generally designed to have a face indraft velocity of 30.5–61 m/min, which is usually adequate to ensure complete capture of all over spray if the spraying is done within the confines of the booth and the spray gun is always directed toward the interior. For this particular problem, a nearly draftless area requires 30.5 m/min at point of spraying. Use Eq. (10); the volumetric flow rate (i.e., the exhaust rate) required is

$$Q = V_x(2A \times 10X^2)/2$$
$$= 30.5(2 \times 3 \times 2 + 10 \times 1.5^2)/2$$
$$= 526.13 \text{ m}^3/\text{min}$$

The face velocity can be calculated using Eq. (8):

$$V_f = Q/A_f = 526.13/(3 \times 2)$$
$$= 87.68 \text{ m/min}$$

Example 9

Abrasive blasting booths are similar to spray booths except that a complete enclosure is always required. Determine the required exhaust rate for a small abrasive blasting enclosure 1.2 m wide by 0.9 m high by 0.9 m deep. The total open area is equal to 0.12 m².

Solution.

From Table 5, an air velocity of 152.4 m/min through all openings is required for an abrasive blast cabinet. A minimum of 20 air changes per minute also must be maintained. The volumetric airflow rate at 152.4 m/min through all openings [Eq. (8)] is

$$Q = A\,V = (0.12 \text{ m}^2) \times (152.4 \text{ m/min})$$
$$= 18.28 \text{ m}^3/\text{min}$$

The volumetric airflow rate required for 20 air changes per minute is

$$Q = (\text{volume of booth}) \times (20 \text{ volumes/min})$$
$$= (12 \text{ m} \times 0.9 \text{ m} \times 0.9 \text{ m}) (\ 20/\text{min})$$
$$= 19.44 \text{ m}^3/\text{min}$$

Example 10

A chrome-plating tank, 0.6 m wide by 1.0 m long, is to be controlled by parallel slot hoods along each of the 1.0-m-long sides. Determine the total ventilation rate required and the slot width.

Solution

This is an open-surface tank with two parallel slot hoods. From Fig. 18, the required minimum ventilation rate and the slot width are selected to be approx 36.2 m³/min/m and 2.9 cm, respectively, according to the width of the chrome-plating tank (0.6 m):

$$Q = \text{(minimum ventilation rate)} \times \text{(tank length)}$$
$$= (36.2 \text{ m}^3/\text{min/m}) \times (1.0 \text{ m})$$
$$= 36.2 \text{ m}^3/\text{min}$$

Example 11

Determine the total exhaust rate and slot width required for the same open-surface tank as in Example 10, except that a slot hood is to be installed along one side only. The other side is flush against a vertical wall.

Solution

The minimum required ventilation rate (in m³/min/m of tank length) is taken as one-half of the rate for a tank twice as wide from Fig. 18. Use a tank width of 1.2 m.

$$Q/L = (81.8 \text{ m}^3/\text{min/m})/2$$
$$= 40.9 \text{ m}^3/\text{min/m}$$

The total exhaust rate required is

$$Q = (40.9 \text{ m}^3/\text{min/m}) \times 1.0 \text{ m}$$
$$= 40.9 \text{ m}^3/\text{min}$$

The slot width is read directly from Fig. 18 for twice the tank width, and it is 6.35 cm.

Example 12

A zinc-melting pot is 1.2 m in diameter with a temperature of 475°C. A high-canopy hood is to be used to capture emissions. Because of interference, the hood must be located 3.2 m above the pot. The ambient air temperature is 28°C. Determine the size of hood and the required exhaust rate.

Solution

Use Eqs. (11b)–(15b) to solve the problem. The distance between the hypothetical point source and the hot source (*see* Fig. 19) can be calculated using Eq. (12b):

$$z = 2.5929 D_s^{1.138}$$
$$= 2.5929(1.2)^{1.138}$$
$$= 3.19 \text{ m} = X_f - y$$

Then, the distance from the hypothetical point source to hood face is

$$X_f = z + y = 3.19 \text{ m} + 3.2 \text{ m} = 6.39 \text{ m}$$

The diameter of rising airstream at the hood face from Eq. (11b) is

$$D_c = 0.4336 X_f^{0.88}$$

$$= 0.4336(6.39)^{0.88}$$
$$= 2.22 \text{ m}$$

The area of rising airstream at the hood face is

$$A_c = (3.14D_c^2)/4 = 3.14(2.22)^2/4 = 3.87 \text{ m}^2$$

The required hood diameter (D_f) including an increase to allow for waver of jet and effect drafts is

$$D_f = D_c + 0.8y = 2.22 \text{ m} + 0.8 \times (3.2 \text{ m})$$
$$= 4.78 \text{ m (use 4.8-m-diameter hood)}$$

The area of the hood face is

$$A_f = (3.14D_f^2)/4 = 3.14(4.8^2)/4 = 18.10 \text{ m}^2$$

The velocity of rising air jet at the hood face [Eq. (15b)] is

$$V_f = 3.9972A_s 0.333(1.8T_d + 32)^{0.417}(X_f)^{-0.25}$$
$$= 3.9972(0.7854 \times 1.2^2)^{0.333}(1.8 \times 447 + 32)^{0.417}(6.39)^{-0.25}$$
$$= 43.34 \text{ m/min}$$

The total volumetric airflow rate required for the hood is

$$Q = V_r(A_f - A_c) + V_r A_c$$
$$= 30.48(18.10 - 3.87) + 43.34(3.87)$$
$$= 601.46 \text{ m}^3/\text{min}$$

In the above equation, V_f is the required velocity through the remaining area of the hood, $A_f - A_c$ (m/min). The selection of V_r depends on the draftiness, height of the hood above the hot source, and the seriousness of permitting some of the contaminated air to escape capture. In general, V_r is chosen in the range 30.48–61.00 m/min. The low value of 30.48 m/min is selected for solving this problem.

Example 13

Determine the dimensions of a rectangular high-canopy hood and the exhaust rate required for a rectangular lead-melting furnace 0.76 m wide by 1.22 m long. The metal temperature is 382°C; the ambient air temperature is 28°C. The hood to be used is located 2.50 m above the furnace.

Solution

Initially, the distance between the hypothetical point source and the hot source (z), the distance from the hypothetical point source to the hood face (X_f), the dimensions of the rising-air jet at the hood (W_c, L_c, and A_c), the dimensions of the rectangular hood (W_f, L_f, and A_f), and the dimensions of the hot source (W_s, L_s, and A_s) should be determined using Eqs. (28)–(35):

$$z = 2.5929W_s^{1.138} = 2.5929(0.76)^{1.138} = 1.90 \text{ m}$$
$$X_f = z + y = 1.90 + 2.50 = 4.40 \text{ m}$$
$$W_c = 0.4336X_f^{0.88} = 0.4336(4.4)^{0.88} = 1.60 \text{ m}$$
$$L_c = L_s = (W_c - W_s) = 1.22 + (1.60 - 0.76) = 2.06 \text{ m}$$
$$A_c = W_c L_c = 1.60 \times 2.06 = 3.30 \text{ m}^2$$

$$W_f = W_c + 0.8y = 1.60 + 0.8 \times 2.50 = 3.60 \text{ m}$$

$$L_f = L_c + 0.8y = 2.06 + 0.8 \times 2.50 = 4.06 \text{ m}$$

$$A_f = W_f L_f = 3.60 \times 4.06 = 14.62 \text{ m}^2$$

$$A_s = W_s L_s = 0.76 \times 1.22 = 0.93 \text{ m}^2$$

Then, the velocity of the rising-air jet (V_f) and the total volumetric airflow rate required for the hood (Q) can be determined:

$$V_f = 3.9972 A_s^{0.333}(1.8\, T_d + 32)^{0.417}(X_f)^{-0.25}$$
$$= 3.9972(0.93)^{0.333}(1.8 \times 354 + 32)^{0.417}(4.4)^{-0.25}$$
$$= 40.61 \text{ m/min}$$

$$V_r = 61 \text{ m/min (selected value)}$$

$$Q = V_f A_c + V_r(A_f - A_c) = 40.61 \times 3.30 + 61(14.62 - 3.30)$$
$$= 824.53 \text{ m}^3/\text{min}$$

Example 14

Design a circular low-canopy hood for capturing tile emissions during fluxing arid slagging of brass in a 60-cm-diameter ladle. The metal temperature during this operation will not exceed 1315°C, the hood will be located 62 cm above the metal surface, and the ambient temperature will be about 28°C.

Solution

Select a hood diameter (D_f) 31 cm larger than the source:

$$D_f = 60 \text{ cm} + 31 \text{ cm} = 91 \text{ cm} = 0.91 \text{ m}$$

The temperature difference between the hot source and ambient air is

$$T_d = 1315 - 28 = 1287°C$$

The total exhaust rate required from Eq. (17b) is

$$Q = 2.1203(D_f)^{2.33}(1.8 T_d + 32)^{0.417}$$
$$= 2.1203(0.91)^{2.33}(1.8 \times 1287 + 32)^{0.417}$$
$$= 43.3 \text{ m}^3/\text{min}$$

Example 15

A zinc die-casting machine with a 0.62-m-wide by 0.95-m-long holding pot is used for the molten zinc. The metal temperature and the ambient air temperature are 455°C and 33°C, respectively. Design a rectangular low-canopy hood that will be provided 75 cm above the pot.

Solution

Equations (38b)–(41b) are used for designing the rectangular low-canopy hoods. Initially, the dimensions of the rectangular hood (W_r, L_r, and A_r) should be decided. Use a hood 31 cm wider and 31 cm longer; then

$$W_f = W_s + \Delta W = 0.62 + 0.31 = 0.93 \text{ m}$$

$$L_f = L_s + \Delta L = 0.95 + 0.31 = 1.26 \text{ m}$$

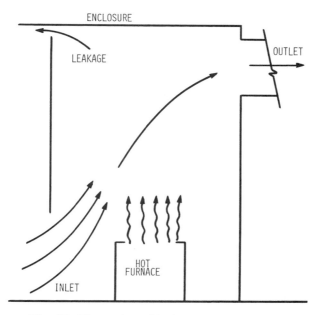

Fig. 25. Illustration of leakage from top of hood.

$$A_f = W_f L_f = 0.93 \times 1.26 = 1.17 \text{ m}^2$$

The temperature difference between the hot source and ambient air is

$$T_d = 455 - 33 = 422°C$$

The exhaust rate required can then be calculated using Eq. (41b):

$$Q = 2.8069(W_p)^{1.333}(1.8T_d + 32)^{0.417}L_f$$
$$= 2.8069(0.93)^{1.333}(1.8 \times 422 + 32)^{0.417}(1.26)$$
$$= 51.91 \text{ m}^3/\text{min}$$

Example 16

Determine the minimum face velocity and total exhaust rate required to prevent leakage of contaminated air through the upper openings of a hood assuming it is not possible to prevent the leakage at the top of the enclosure. Figure. 25 shows that several oil-fired crucible furnaces are hooded and vented. The enclosure is 6.5 m long. All openings are sharp-edge orifices. The total area of the leakage opening is 0.1 m². The fuel rate is 120 L/h and the heating value is 9320 kg-cal/L. The average temperature of gases in the hood and the ambient air temperature are 68°C and 27°C, respectively. The vertical distance above the hood face to the location of the orifice is 3.35 m.

Solution

The rate of heat generation is

$$q = (120 \text{ L/h}) \times (9320 \text{ kg-cal/L}) ((1 \text{ h/60 min})$$
$$= 18,640 \text{ kg-cal/mm}$$

The total open area of the orifice is

$$A_c = (6.50 \text{ m} \times 2.13 \text{ m}) + 0.1 \text{ m}^2 = 13.95 \text{ m}^2$$

The escape velocity through the leakage orifice [Eq. (42b)] is

$$V_e = 53.3951[Yq/A_o (273 + T)]^{0.333}$$
$$= 53.3951[3.35 \times 18640 / 13.95(273 + 68)]^{0.333}$$
$$= 125.85 \text{ m/min}$$

The required exhaust rate is

$$Q = A_o V_e = (13.95 \text{ m}^2) \times (125.85 \text{ m/min}) = 1755.59 \text{ m}^3/\text{min}$$

Check the mean hood air temperature with Eq. (43):

$$q = QdST_d$$
$$T_d = q/Qds$$
$$= \frac{(18640 \text{ kg-cal/min})}{(1755.59 \text{ m}^3/\text{min})(1.2015 \text{ kg/m}^3)(0.24 \text{ kg-cal/kg-°C})}$$
$$= 36.82°C$$

The average temperature of the air inside the hood is

$$T = 27°C + 36.82°C = 63.8°C$$

This adequately approximates the original assumption.

Example 17

What kind of comfort will be attained at an effective temperature of 22.2°C (72°F), relative humidity of 50%, a dry-bulb temperature of 25°C (77°F), and a wet-bulb temperature of 18.3°C (65°F).

Solution

According to the comfort chart established by the American Society of Heating, Refrigerating and Air-Conditioning Engineers, Inc. (*see* Fig. 22), a comfortable feeling will be attained. Note that relative humidities have only a slight effect on comfort, and then only at humidities above 50–60%, at which points the feeling curves in Fig. 22 turn slightly toward lower dry-bulb and effective temperatures.

Example 18

A new ventilation system has been installed in an industrial plant and is being tested. According to an environmental engineer's analysis, air still contains 5 ppm of carbon tetrachloride, 20 ppm of ethylene dichloride, and 10 ppm of ethylene dibromide when the new ventilation system is operated at full capacity (i.e., all mechanical fans are on). Is this new ventilation system acceptable? Is the TLV of the gaseous mixture exceeded?

Solution

From Appendix A, the TLV of the three atmospheric pollutants are obtained:

1. Carbon tetrachloride, TLV = 10 ppm
2. Ethylene dichloride (1,2-dichlorocthane), TLV = 50 ppm
3. Ethylene dibromide (1,2-dibromoethane), TLV = 25 ppm (tentative)

Use Eq. (48), assuming the general case of additive effects:

$$E_a = \sum_{i=1}^{i=n}(C_i/L_i)$$
$$= 5/10 + 20/50 + 10/25$$
$$= 65/50 = 1.3$$

The E_a value exceeds unity; therefore, the threshold limit of the mixture should be considered as being exceeded. The new ventilation is not acceptable. It should be improved by increasing the exhaust rate, changing the air distribution pattern, preventing the leakage, and so forth.

Example 19

The air in a small factory contains 0.5 mg/m³ of lead and 0.7 mg/m³ of sulfuric acid. Does this factory need any special air conditioning system in addition to temperature and humidity control?

Solution

Appendix A, TLVs are obtained:

1. Lead, TLV = 0.2 mg/m³
2. Sulfuric acid, TLV = 1 mg/m³.

It has been known that the effects of lead and sulfuric acid are not additive, but independent purely local effects. Use Eq. (49):

$$E_i \text{ (for lead)} = C_i/L_i = 0.15/0.20 = 0.75$$
$$E_i \text{ (for acid)} = C_i/L_i = 0.70/1.00 = 0.7$$

Both E_i values do not exceed unity; therefore, the threshold limit is not exceeded. The small factory needs only a conventional air conditioning system for temperature and humidity control.

Example 20

A special case arises when the contaminant source is a mixture and the atmospheric composition is assumed to be similar to that of the original source material; that is, the vapor pressure of each component is the same at the observed temperature. Determine the threshold limit of such a mixture containing equal parts of trichloroethylene and methylchloroform.

Solution

A solution is applicable to "spot" solvent mixture usage, where all or nearly all of the solvent evaporates. The TLVs of trichloroethylene and methylchloroform are 100 and 350 ppm, respectively, from Appendix A. Use Eqs. (50) and (51):

$$C_m = \sum_{i=1}^{i=n} C_i = C_1 + C_2 = 2C_1 = 2C_2 \ \left(\text{because } C_1 = C_2\right)$$

$$\frac{C_m}{L_m} = \sum_{i=1}^{i=n} \frac{C_i}{L_i} = \frac{C_1}{L_1} + \frac{C_2}{L_2} = \frac{C_1}{100} + \frac{C_2}{350}$$

$$\frac{2C_1}{L_m} = \frac{7C_1}{700} + \frac{2C_2}{700}$$

$$L_m = 700 \times (2/9) = 155 \text{ ppm}$$

Example 21

From naphtha (TLV = 500) containing 10 mol% percent benzene (TLV = 25), the narcotic effects can be considered as approximately the same as that of benzene-free naphtha. The blood effects can be considered a result of benzene alone. Consider an industrial plant that handles such chemicals. What thought needs to be given to the air quality before a ventilation system or a central air conditioning system is designed?

Solution

For intermittent exposure, a TLV of 500 ppm may be used as long as the average concentration does not exceed $25(100/10) = 250$ ppm; the TLV is based on the benzene content. This is the special case of "independent effects."

Example 22

Determine the size of a round duct and its friction loss. Assume that the required exhaust rate and air velocity are 280 m^3/min and 490 m/min, respectively.

Solution

The Exhaust rate is 280 m^3/min = 9888 ft^3/min. The air velocity is 490 m/min = 1607 ft/min. From Fig. 21, the duct size and the friction loss can then be determined. The 9888-ft^3/min horizontal line and the 1607-ft/min diagonal line intersect at 34-in. duct diameter. Dropping vertically from the same intersection, the duct friction loss is 0.10 in. of water for every 100 ft of the 34-in. diameter duct.

Example 23

Determine the size of a rectangular duct that can deliver 280 m^3/min air at 490 m/min air velocity.

Solution

From Example 22, it is known that a 34-in. duct can do the job. An equivalent rectangular duct size can be determined from Appendix D. For instance, 25 × 40-in. and 15 × 80-in. ducts are equivalent to the 34-in. round duct.

Example 24

Determine the pressure loss in a 12 × 24-in. rectangular 90° bend. The inside radius of the bend is 12 in.

Solution

From Table 10, the equivalent straight duct is 8 diameters, because the W/D (width/depth) ratio = 2 and $R' = 1.00D$. From Appendix D, 12 × 24 in. is equivalent to an 18-in.-diameter round duct. Therefore, the pressure loss in the bend is 8 × 18 in., or the same as 12 ft of a 18-in.-diameter round duct.

Example 25

The material in Section 3.3 was designed by the Air Moving and Conditioning Association (AMCA) and American Coolair Corporation (ACC) to assist an environmental engineer in the selection of mechanical fans suitable for use in an existing building or a similar facility where additional ventilation is needed. This example illustrates the step by step procedure for selecting the ventilation fans with the following assumptions:

1. Room size (width × length × height) = 20 × 40 × 15 ft.
2. Materials of room surfaces are floor, concrete; walls, concrete block, painted; ceiling, gypsum board.

3. All proposed wall fans are to be installed on the back wall near the ceiling and directly above individuals working on an assembly line. The walls are 15 ft high, and fans will be approx 12 in. above the ceiling opening. The distance from the proposed fans to the occupied space is measured to be about 10 ft.
4. The LwA of three proposed fans are 85, 88, and 83.
5. By use of a sound meter set on the "A" scale, the sound level in the occupied space is found to be 83 dBA.
6. The maximum sound level specified is 89 dBA.

Solution

1. Calculate the area for each surface and assign absorption coefficient for each from Appendix 1. This information is inserted in the chart below

Room surface	A	B	C_1	C_2	D_1	D_2	Σ
Area (ft^2)	800	800	300	300	600	600	3400
Absorption coefficient α	0.01	0.07	0.07	0.07	0.07	0.07	—

The average room absorption coefficient (β) is

$$\beta = \frac{\text{Each surface } \alpha \times [\text{each surface area(ft}^2)]}{\text{Total surface area (ft}^2)}$$
$$= \frac{(0.01 \times 800) + (0.07 \times 800) + (2 \times 0.07 \times 300) + (2 \times 0.07 \times 600)}{3400}$$
$$= 0.05$$

The room constant (R) is

$$R = \frac{[\text{Total surface area(ft}^2) \times \beta]}{1 - \beta}$$
$$= \frac{3400 \times 0.05}{1 - 0.05} = 179$$

2. The proposed location of the fan in relationship to wall surfaces determines the directivity factor Q' to be applied. There are four possibilities:
   (a) $Q' = 1$. Fan located in an open space away from room surfaces (on a pedestal or suspended from ceiling or column).
   (b) $Q' = 2$. Fan adjacent to a large flat surface (near center of ceiling or wall; no closer than 4 fan diameters to an adjacent wall and other surface).
   (c) $Q' = 4$. Fan adjacent to two surfaces (intersection of two walls or ceiling and wall).
   (d) $Q' = 8$. Fan adjacent to three surfaces (near floor and in corner of two walls; or near ceiling and in corner).

Accordingly, the directivity factor Q' should be equal to 4 because the proposed wall fans are to be installed on the back wall near the ceiling.

3. The distance from the fans to the occupied space (X) has been measured to be 10 ft.
4. It has been known that $R = 179$, $Q' = 4$, and $X = 10$. Read the chart in Appendix F and find the room attenuation effect: RAE = 5.
5. There are four possible locations for the fan in terms of fan installation factor (FIF):
   (a) Location type A: FIF = +3. Both inlet and exhaust within room (air circulator, fan cooler, etc.).
   (b) Location type B: FIF = 0. Either inlet or outlet within room—but only one (typical wall fan or roof ventilator).

(c) Location type C: FIF = −17. Duct connected fan with both inlet and outlet of duct outside the room.

(d) Location Type D: FIF = 0. Duct connected fan with one end of duct (or fan) open to room.

It has been assumed that all wall exhaust fans are to be installed on the back wall of the room. By reviewing the four possible locations in terms of the FIF, one finds that type B is applicable; thus, FIF = 0.

6.   Three fans are to be installed:

Fan no.	LwA	RAE	FIF	dBA
1	85	−5	0	80
2	85	−5	0	83
3	83	−5	0	78

From Appendix H, the total on-the-job fan sound level for the three fans is determined to be 86 dBA:

$$\left.\begin{array}{c} 83 \\ 80 \end{array}\right\} + 2 = 85 \; \left.\begin{array}{c} \\ \\ 78\ldots \end{array}\right\} \; \cdots + 1 = 86 \text{ dBA}$$

7.   The sound level in the occupied space is 83 dBA. The total on-the-job fan sound level of fans is 86 dBA. Then, the total sound level in the occupied space after the addition of the three proposed fans is calculated to be 88 dBA as below follows:

$$\left.\begin{array}{c} 86 \\ 83 \end{array}\right\} \; \cdots + 1 = 86 \text{ dBA}$$

The maximum sound level specified is 89 dBA. Therefore, the addition of three fans will not exceed the specified level.

7. HEALTH CONCERN AND INDOOR POLLUTION CONTROL

7.1. Health Effects and Standards

Because an average person spends approx 90% of his/her time indoors, an increasing concerning has been expressed over the quality of indoor air and those associated with personal activities. Studies have shown that the concentration of toxic organic and inorganic chemicals indoors can be several times greater than the concentration in outdoor air (37).

In the last two decades, the increase in the cost of energy has led to the improved construction and retrofitting of homes and commercial buildings for enhanced energy conservation. A reduction in the infiltration of fresh air is cost-effective and is widely practiced among the various energy-saving schemes that are presently being used. As a result, a large portion of people is living in tightly sealed structures in which most of the air is recirculated to reduce energy consumption, without full realization of the air quality problem that arises from the pollutants generated and retained indoors.

Inorganic gases including ozone, nitrogen oxides, carbon monoxide, sulfur dioxide, and hydrogen sulfide are the common indoor pollutants. Nitrogen dioxide and carbon monoxide are much more frequently studied because of their hazards. Carbon monoxide combines with hemoglobin and myoglobin to form carboxyhemoglobin and CO-myoglobin, resulting in reduced transport of oxygen to tissues.

Little is known about the acute and chronic effects of a mixture of volatile organic chemicals, especially for those lower concentrations in indoors on human health. Symptoms associated with this health are collectively referred to as sick building syndrome (SBS) or tight building syndrome can be related to (1) irritation of eyes, nose, and throat; (2) dry mucous membranes and skin; (3) erythema; (4) mental fatigue and headaches; (5) airway infections and coughing; (6) hoarseness and wheezing and (7) unspecific hypersensitivity reactions, nausea, dizziness, and so forth.

A number of sources provide further information about the human health effects posed by organic chemicals, but many of the findings remain controversial. The TLVs of hazardous substances suggested by the American Conference of Governmental Industrial Hygienists (ACGIH) are used as a basis. The Occupational Safety and Health Association (OSHA) has established one-tenth of the TLV as a guideline for indoor air contaminants. The American Society of Heating, Refrigerating and Air-Conditioning Engineers, Inc. (ASHRAE) Standard 62-1981R committee recommended as a preliminary guideline for residential, office, or retail spaces a concentration of one-tenth of the TLV.

Recommended threshold limit values of hazardous substances is listed in Appendix A.

7.2. Indoor Air Quality

A great variety of toxic materials has been identified in indoor air. These include aliphatic and aromatic hydrocarbons, chloroinated hydrocarbons, and various ketones and aldehydes. Some of these have been suggested as possible carcinogens. There are many sources for volatile organic compounds (VOCs), each with varying composition. These sources may be the result of the contamination of indoor air by emission of VOCs from a variety of sources, including construction materials, fabrics, furnishings, maintenance supplies, combustion byproducts, adhesives, paints, printed paper pastes, deodorizers, floor coverings, and cleaning products.

Pollutants are introduced indoors in several ways: (1) through normal biological processes, people and pets generate carbon dioxide, moisture, odors, and microbes; (2) by combustion appliances such as wood stoves, gas stoves, furnaces, fireplaces, and gas heaters; (3) from the use of consumer products such as spray cans, air fresheners, spray cleaners, construction materials, furnishings, and insulation; (4) from cigarett smoke; (5) from the soil under and around buildings; (6) from appliances such as humidifiers, air conditioners, and nebulizers.

A vast number of studies in the past have realized a direct relationship between the level of indoor air pollution and health problems such as headaches, nausea, allergies, humidifier fever, respiratory infections, Legionnaire's disease, and lung cancer. Particulate matters such as bioaerosols can have an immediate impact on public health. The presence of other indoor contaminants can have longer-term effects. Junker et al. presented an assessment of indoor air contaminants in buildings (38). In a typical office, there are over several dozens of volatile organic compounds detected using a diffusive sampling method; those included aromatic compounds and halogenated substances (39). Some common indoor air pollutants and their health effects are summarized follows (40,44):

1. VOCs
   - Source: VOCs are released from burning fuel, combustion gas, solvents, paints, glues, building materials, furnishing, aerosol cans, humans, and other products used at work or at home.
   - Health effects: Most of VOCs can cause serious health problems such as cancer and other effects.
   - Control strategies: Source control, increased ventilation, air cleaning.

2. Inorganic gases (SO_2, NO_2, NO, CO)
   - Source: Burning of gasoline, natural gas, coal, oil, etc.; industrial processes (paper, metals).
   - Health effects: Lung damage, illnesses of breathing passages and lungs (respiratory system); reduces ability of blood to bring oxygen to body cells and tissues; cells and tissues need oxygen to work. Carbon monoxide may be particularly hazardous to people who have heart or circulatory (blood vessel) problems and people who have damaged lungs or breathing passages.
   - Control strategies: Source control, increased ventilation, air cleaning.

3. Particulate matter (PM 2.5) (dust, smoke, soot)
   - Source: Burning of wood, diesel and other fuels, industrial plants, agriculture (plowing, burning off fields), unpaved roads.
   - Health effects: Nose and throat irritation, lung damage, bronchitis, early death.
   - Control strategies: Mechanical filtration, electronic air cleaning, and absorption.

4. Ozone
   - Source: Chemical reaction of pollutants, VOCs and NO_x, photocopy machine, laser machine.
   - Health effects: Breathing problems, reduced lung function, asthma, irritates eyes, stuffy nose, reduced resistance to colds and other infections, may speed up aging of lung tissue.
   - Control strategies: Source control, increased ventilation, air cleaning.

5. Heavy metal
   - Source: Leaded gasoline (being phased out), paint (houses, cars), smelters (metal refineries), manufacture of storage batteries, tobacco smoke, dust.
   - Health effects: Brain and other nervous system damage; children are at special risk. Some lead-containing chemicals cause cancer in animals. Lead causes digestive and other health problems.
   - Control strategies: Removal of source, regular cleaning.

6. Bioaerosols
   - Source: Various microorganisms, humidifiers, variety of fungi and bacteria, cooling towers, airborne antigens.
   - Health effects: Introduced into human body either through inhalation or deposition in wounds and can localize or migrate to other portion of body, bringing in: viruses, bacteria, hemophilius influenza, fungi, antigens.
   - Control strategies: Mechanical filtration, electronic air cleaning, and absorption.

7. Radon
   - Source: Soil and rock beneath or surrounding building structure, water supplies, building materials, and natural gas.
   - Health effects: Colorless and odorless inert gas, causes cancer.
   - Control strategies: Source removal, source control (sealing of entry path, subslab ventilation, basement pressurization), air cleaning (filtration, electronic cleaner, and absorption).

Indoor air quality has not been satisfied enough because of a substantial number of people in many buildings suffering from SBS symptoms. The use of source reduction, air handling systems, filtration, and sanitation methods to improve indoor air quality may prevent or ameliorate some of these effects, but the continued coordination of engineering and public health methodologies is required to design better buildings and better methods to maintain adequate interior air.

7.3. Pollution Control in Future Air Conditioned Environments

Today, air conditioning is used in many parts of the world, often in combination with heating and ventilation systems (HVAC systems). The image of such systems, however, is not always positive. The purpose of most systems is to provide thermal comfort and an acceptable indoor air quality for human occupants. However, numerous field studies have documented substantial rates of dissatisfaction with the indoor environment in many buildings. One of the main reasons is that the requirements of existing ventilation standards and guidelines are quite low (42).

The philosophy behind the design of HVAC systems has led in practice to quite a number of dissatisfied persons (as predicted); however, few seem to be ready to characterize the indoor environment as outstanding. At the same time, numerous negative effects on human health are reported: many persons suffer from SBS symptoms and a dramatic increase in cases of allergy and asthma have been related to poor indoor air quality (IAQ).

Our aim should be to provide indoor air that is perceived as fresh, pleasant, and stimulating, with no negative effects on health, and a thermal environment perceived as comfortable by almost all occupants. In achieving this aim, due consideration must be given to energy efficiency and sustainability.

7.3.1. A Good Indoor Environment Pays

A survey showed a dramatic increase in cases of allergy and asthma has been related to poor IAQ. It has been estimated the economic losses caused by poor IAQ resulting from illness, absenteeism, and lost production. The results from these blind studies show that improved air quality increases productivity significantly. This increase should be compared with the cost of conditioning the indoor environment, which for office buildings in the developed countries is typically less than 1% of the labor cost. There is, therefore, a strong economic incentive to improve the indoor air quality.

7.3.2. Pollution Source Control and Ventilation

Avoiding unnecessary indoor air pollution sources is the most obvious way to improve indoor air quality. Source control has also been used with great success outdoors and is the reason why the outdoor air quality in many cities in the developed world is much better today than it was 20 or 50 yr ago. Systematic selection of materials to avoid the well-known cases of SBS caused by polluting materials is common practice. Pollution sources in the HVAC system are a serious fault, degrading the quality of the air even before it is supplied to the conditioned space. The selection of materials, development of components and processes, as well as maintenance of the HVAC system should be given high priority in the future. Source control is the obvious way to provide good indoor air quality with a simultaneous decrease in the consumption of energy. However, increased ventilation also improves the indoor air quality and decreases SBS symptoms.

7.3.3. Serve the Air Cool and Dry

New comprehensive studies at the Technical University of Denmark have demonstrated that perceived air quality is strongly influenced by the humidity and temperature
of the air we inhale (42). People prefer rather dry and cool air. Recent studies showed
that people perceive the indoor air quality better at 20°C and 40% relative humidity
(RH) and a small ventilation rate of 3.5 L/s/person (liters per second per person) than at
23°C and 50% RH at a ventilation rate of 10 L/s/person. It is advantageous to maintain
a moderately low humidity and a temperature that is at the lower end of the range
required for thermal neutrality for the body as a whole. This will improve the perceived
air quality and decrease the required ventilation. It is surprising to note that even in
warm and humid climates, it may save energy to maintain a moderate indoor air temperature and humidity. Field studies show that moderate air temperatures and humidities also decrease SBS symptoms. There are, therefore, several good reasons to follow
this advice: Serve the air cool and dry for people.

7.3.4. Serve the Air Where It Is Consumed

In many ventilated rooms, the outdoor air supplied is of the order of magnitude of 10
L/s/person. Of this air, only 0.1 L/s/person, or 1%, is inhaled. The rest (i.e., 99%) of the
supplied air is not used. Also, the 1% of the ventilation air being inhaled by human occupants is not even clean. It is polluted in the space by bioeffluents, emissions from building
materials, and, sometimes, even by environmental tobacco smoke before it is inhaled.

The idea would be to serve to each occupant clean air that is unpolluted by the pollution sources in the space. Such "personalized air" (PA) should be provided so that the
person inhales clean, cool, and dry air from the core of the jet where the air is unmixed
with polluted room air. It is essential that the air is served "gently" (i.e., has a low velocity
and turbulence that do not cause draught).

8. HEATING, VENTILATING, AND AIR CONDITIONING

8.1. Energy and Ventilation

8.1.1. Energy Conservation

Ventilation is essential for the maintenance of good indoor air quality. However,
despite its essential need, there is much evidence to suggest that energy loss through
uncontrolled or unnecessary air change is excessive and that much can be done to minimize such loss. Considerable losses may also be associated with ventilating pollutants
that can be more effectively controlled by their elimination at the source. Air infiltration
exacerbated by poor building airtightness adds further to the lack of control and energy
waste. The relative importance of building energy consumption in developed countries
shows building energy demand to be of comparable significance more than twice that
of industrial demand (43). It is important to recognize the structure of airborne energy
loss and to identify the related methods of control.

In essence, distinct types of airborne energy losses would be as follows:

1. Venting of waste heat: Excess heat is often developed inside buildings. Examples include
   the generation of heat from office equipment and cooking appliances and gains from solar
   radiation. Minimizing waste heat loss is dependent on improving the energy efficiencies of
   appliances and processes.

2. Flue emissions: Much airborne heat is lost through combustion flues. Condensing appliances provide a means for capturing some of this waste. Flues and chimneys that are open to rooms provide a further uncontrolled route for air escape even when the appliance itself is not in use.
3. Loss of thermally conditioned air: This is associated with the loss of intentionally conditioned (heated or cooled) air from a space by ventilation or air infiltration.

It has been noted that air change heating losses (purely arising from the loss of enthalpy) are as important as other forms of energy loss. Heating air change and conduction losses are both associated with a proportion of the heating equipment losses. Therefore, a reduction in one of these will cause a proportional drop in heating equipment losses. Studies indicated the magnitude of energy consumption because of air change (e.g., air infiltration) alone accounts for roughly 18% of the heating load in all office buildings in United States and ventilation accounts for 20% of the energy losses in the UK service sector buildings (43).

Although it is difficult to estimate accurately by how much the energy consumption is associated with air change, because of the large uncertainties and deficiencies in essential data, it has still been possible to derive an approximation of the extent by which air change energy consumption may be reduced. If the ASHRAE Standard 62 guideline (44) were to be universally followed, it is conceivable that air change heat losses may be reduced to one-quarter of the current level (not allowing for reduced equipment losses). The assumption is that fresh outdoor air is only supplied on the basis of occupant requirements and ventilation air is supplied continuously throughout the year, for each of the service and residential sectors.

Sufficient ventilation must always be provided to satisfy the health and comfort needs of occupants. Such requirements are the subject of standards, regulations, and codes of practice (e.g., ASHRAE Standard 62). Minimum levels normally take into account metabolic needs and pollutant loads. In addition, considerably greater amounts of air change may be necessary to satisfy the cooling needs when mechanical cooling is not used. Practical measures for the reduction of air change energy consumption would be linked to the following:

1. Avoiding unnecessary air change (i.e., dealing with leaky buildings)
2. Introducing good control strategies (e.g., avoiding the use of window and door opening during periods of active cooling or heating)
3. Minimizing the heat load during cooling periods to restrict excessive heat gains
4. Optimizing fan efficiency
5. Optimizing other equipment efficiencies, such as heating equipment
6. Introducing guidelines indicating the expected best levels

8.1.2. Natural Ventilation and Night Ventilation

Natural ventilation occurs when doors and widows are open. This method is extremely effective in reducing pollutant levels when short-term activities, such as cooking or cleaning, are taking place. The airflow rate through the house is often increased by opening doors or windows for cross-ventilation. The main disadvantage of natural ventilation is the increased energy cost associated with extra heating or cooling load (45).

Alternatively, passive cooling techniques present a very important alternative to conventional air conditioning of buildings. The development of efficient passive cooling techniques is a first priority for building scientists. Recent research has shown that night ventilation techniques, when applied to massive buildings, can significantly reduce the cooling load of air conditioning buildings and increase the thermal comfort levels of non-air-conditioning buildings.

Night ventilation techniques are based on the use of the cool ambient air as a heat sink, to decrease the indoor air temperature as well as the temperature of the building's structure. The cooling efficiency of these techniques is mainly based on the airflow rate as well as on the thermal capacity of the building and the efficient coupling of airflow and thermal mass. Based on the flow regime, three main operational concepts can be mentioned, as follows:

> *Ventilation by natural means* (i.e., through the building's openings). In this case, the airflow is variable and random and depends on the temperature and wind-driven pressure differences between the indoor and outdoor environment. Thus, the efficiency of this technique is affected by the interdependence of the environmental parameters.
> *Ventilation by mechanical means* (i.e., by using supply and exhaust fans). The supply fans maintain a constant flow of the ambient air circulated to the building. Additionally, exhaust fans should be used in order to avoid the overpressurization of the building that causes a decrease in the efficiency of the fans. Thermostatic controllers can be installed to drive the fans and to shut them down when the outdoor temperature is higher than the indoor one.
> *Ventilation using both mechanical and natural means* (i.e., by using fans and the openings of the building). In this case, only supply or exhaust fans can be used while openings assist the outflow or the inflow of the air, respectively. In order to increase the efficiency of the technique, thermostatic controllers can be installed to turn the fans on and off as a function of the indoor and ambient temperature.

Recent research on the efficiency of night ventilation techniques is more concentrated on specific experiments as well as on the development of evaluation methodologies. Extended experiments using night ventilation have resulted in the development of an empirical formula to predict the indoor maximum temperature as well as the cold storage and the diurnal cooling capacity of the building. Other experiments carried out in an existing building have permitted the development of indices to characterize the energy gain resulting from night ventilation as well as the possible comfort improvements.

Night ventilation techniques can contribute to decrease significantly the cooling load of air conditioning and improve the comfort levels of free-floating buildings. The exact contribution of night ventilation for a specific building is a function of the building structural and design characteristics, the climatic conditions and the building's site layout, the applied airflow rate, the efficient coupling of airflow with the thermal mass of the building, and the assumed operational conditions. Appropriate design of night ventilation systems requires exact consideration of all the above parameters and optimization of the whole procedure by using exact thermal and airflow simulation codes.

8.1.3. Optimization of Operation

It has been estimated that air change accounts for approx 36% of total space conditioning energy and contributes to almost half of heating equipment losses (43). Computerization and simulation of optimal control HVAC processes in buildings can help to minimize energy losses and lower operation cost (46–48). The time-scheduled heating, ventilating, and air conditioning process is an example of the attempt (48).

8.2. HVAC Recent Approach

Because environmental requirements have banned the use of chlorofluocarbons (CFCs), further concerns about phasing out hydrochlorofluocarbons (HCFCs) are ongoing at the present (49). CFCs, ozone-depleting substances (ODSs), started out seemingly innocuous when they were invented in 1928. They began being used in refrigeration, air conditioners, solvents, fire extinguishers, aerosols, foams, and so forth. Unfortunately, scientists later realized that CFCs are long lived and that their emissions deplete the stratospheric ozone. This ozone depletion was dramatically confirmed through the discovery of the Antarctic "ozone hole" in 1985. Another concern is that increasing global climate change might interfere with the ozone layer's healing process. This issue is addressed by the Scientific Assessment Panel in their 2002 report on the scientific assessment of the ozone layer (http//www.unep-ch/ozone/unep wmo-sa2002.shtml).

Chlorofluocarbons also are subject to regulation under the Clean Air Act Amendments (CAAA) of 1990, which required the US Environmental Protection Agency (EPA) to phase out the production and importation of five ozone depleters: CFCs, halons, carbon tetrachloride, methyl chloroform (all called Class I substances), and hydrochlorofluorcarbons (HCFCs) (called Class II substances). The CAAA also gave the EPA the responsibility to add other ozone depleters to the phase-out list.

The response of the chemical industry to the ban has been to develop new synthetic refrigerants without the offending chlorine atoms. Table 15 lists some environmental effect of refrigerants and its global warming potential (GWP) effect (50). Although hydrofluocarbons (HFCs) were possible replacement of ozone depleting potential (ODP) refrigerant developed for commercial use, their long-term effects on the environment and human health are, however, uncertain. As a consequence, this has led to a total solution by the use of "natural" substances as working fluid.

8.2.1. CO_2 Desiccant Air Conditioning System

Carbon dioxide is regarded as one of the potential refrigerant candidates for air conditioning systems. CO_2 is not only a natural substance but also has a successful history as a refrigerant. Many investigators have discussed the well-known advantages of using CO_2 as s refrigerant. An excellent review of CO_2 as a refrigerant is reported by Kruse et al. (51). Rozhentsev and Wang had promoted design features of a CO_2 air conditioner (52). They studied the influence of a recuperative heat exchanger subject to practical constraints within semihermetically and hermetically sealed air conditioners.

The main shortcoming of CO_2 in air conditioning systems is high power consumption and loss of capacity at high ambient temperatures. According to recent studies, the coefficient of performance of a CO_2 air conditioning system can be improved. Compared to the effect of tube size on water chilling and tap water heating system between R-22 and CO_2, CO_2 shows an 18% lower water chilling capacity and 14% higher water heating coefficient of performance than that of R-22 if the same size and number of tubes are applied. When a smaller-diameter tube is applied, the related coefficient of performance (COP) of CO_2 system can be improved.

8.2.2. Ammonia Absorption Refrigerator Systems

Absorption refrigerator systems, ammonia and/or water–lithium bromide, have attracted increasing interests for the replacement of ozone-depleting refrigerant. Unlike

Table 15
Environmental Effect of Refrigerants

Substance	ODP	GWP
CFC-12	1.0	8500
HCFC-22	0.055	1700
HFC-134a	0	1300
HFC-143a	0	4400
Isobutane	0	3
Propane	0	3
Ammonia	0	0

mechanical vapor compression refrigerators, these systems could be superior to electricity-powered systems in that they harness inexpensive waste heat, solar, biomass, or geothermal energy sources for which the cost of supply is negligible in many cases. This makes heat-powered refrigeration a viable and economic option (53,54). The most common absorption systems are H_2O–LiBr and NH_3–H_2O cycles. Research has been performed for NH_3–H_2O systems theoretically and experimentally. The advantage for refrigerant NH_3 is that it can evaporate at lower temperatures (from −10° to 0°C) compared to H_2O (from 4°C to 10°C). Therefore, for refrigeration, the NH_3–H_2O cycle is most often used.

The ammonia–water absorption cycle is mainly used for refrigeration temperatures below 0°C. Absorption pairs of ammonia–lithium nitrate and ammonia–sodium thiocyanate cycles is an alternative refrigerant. Ammonia–lithium nitrate and ammonia–sodium thiocyanate cycles give better performance than the ammonia–water cycle, with higher COP (54).

8.2.3. Latent Heat Storage and Ventilation Cooling System

Turnpenny et al. introduced a latent heat storage (LHS) unit incorporating heat pipes embedded in phase change material (PCM) for use with night ventilation (55,56). The system can eliminate or reduce the use of air conditioning in commercial/industrial buildings and thereby reduce energy costs. The system makes use of nighttime cooling in order to store "coolth," which is then used during the daytime to absorb internal and external heat gains and maintain a comfortable internal environment.

The newly introduced system offers important benefits, which offer enhancement by providing storage capacity in a more convenient form and by increasing the rate of heat transfer into storage medium. The novelty arises from the use of PCM to store the "coolth" at a nominally constant temperature and heat pipes to obtain enhanced heat transfer between the air and the PCM. The demonstration was that an extractor fan delivering about eight air changes per hour was sealed over the inside of vent to draw cool air through the room from vent at night. At night, cool air was drawn inward from upper-level vents, guided above the units by a movable board and passed over the heat pipes using the ceiling fan blowing downward. The extract fan switched off, and the ceiling fan still blew downward to cool the room through convention as well as LHS.

8.2.4. Desiccant Cooling Systems

Desiccant cooling systems employ either solid materials, such as silica gel, zeolites, and carbon, or liquid absorbents, such as calcium chloride, lithium chloride, and lithium bromide. Air conditioning using desiccants is based on the principle that air is dehumidified using an absorbent, cooled through a sensible heat exchanger, and finally cooled to the desired space condition using an evaporative process.

Solid desiccant systems commonly employ a honeycomb wheel impregnated with a solid material to adsorb water vapor from moist air passing through it. Liquid desiccants may employ packed beds, packed towers, or spray chambers and, therefore, have several advantages over solid desiccants, including lower pressure drop of air across the desiccant material, suitability for dust removal by filtration, ease of manipulation, and greater mobility. Liquid desiccant systems are also potentially more efficient than solid desiccant ones: a set point coefficient of performance (COP) of 2 compared to 1 (57).

8.2.5. Geothermal Heating and Cooling Technology

Geothermal heating and cooling technology is a cleaner technology for saving energy and protecting the environment (58). The direct use of groundwater in a conventional geothermal heating and cooling process system has been proven to be the most efficient methodology for providing air heating, air cooling, and hot water supply for buildings. When large quantities of water must be responsibly returned to the water table and sufficient groundwater is not available from a simple well, the vertical standing water column (VSWC) technology can then be applied to an innovative geothermal heating and cooling process system.

8.2.5.1. INNOVATIVE GEOTHERMAL HEATING AND COOLING TECHNOLOGY

The basic ground source heating and cooling technology is derived from the use of a water source heat pump. The heat pump is simply another name for a refrigeration machine, which can be reversed for heating purposes. A refrigerator works by moving heat from a "box" and depositing that heat to another location outside the "box." The heat pump in the cooling mode is, likewise, a refrigeration machine removing heat from one source and depositing it in another. In the winter heating mode, the heat is absorbed from an outside source and delivered to the "box" (room). In the process of removing or adding heat, the heat pump makes use of a compressor and a liquid refrigerant. The liquid refrigerant flows through the heat pump, absorbing heat energy from the groundwater. The liquid refrigerant evaporates to a gas at well water temperatures and below. After it evaporates, it is passed through a compressor, which efficiently raises its temperature as the pressure increases. A hot liquid refrigerant produced in this manner, with the use of a heat exchanger, enables the heat pump to give off heat into the building. In the summer cooling mode, the opposite of this process occurs. As the refrigerant evaporates, it absorbs heat from the inside of the building and transfers this heat to the groundwater through the heat pump while lowering the temperature of the building side liquid and passing it through a fan coil that efficiently cools the building (58).

There are two sides to a groundwater heat pump. The incoming side is the groundwater source side. The opposite side is the building conditioning side. The actual heat pump is located in between and consists of a heat exchanger between the groundwater side and a compressor/refrigerant and a second heat exchanger on the building conditioning side. The use of the two separate heat exchangers effectively prevents contact between refrigerant and water on either side of the mechanism. A heat exchanger is primarily two pipes coiled together, with one pipe containing the refrigerant and the other pipe containing the water. Heat is transferred by conduction between the two pipes.

The VSWC is essentially a long pipe placed in the ground (or a long hole drilled in bedrock) kept filled with water to make sufficient transfer of heat energy possible. Briefly, VSWC technology is employed to assist and, in certain instances, replace the use of natural groundwater when sufficient groundwater is unavailable from a simple well. The basic concept in any earth source heat pump system is to use the relatively constant temperatures of the earth (50°F) in the temperate zone as a source from which heat can be obtained during the winter or to which heat may be deposited during the summer.

8.2.5.2. WINTER HEATING OPERATION

During the winter heating season, heat is extracted from the ground by pumping the groundwater through the heat pump, thereby reducing its temperature a few degrees (about 10°F, or 6.8°C), after which, that water is returned to the top of the VSWC. Because this water is now cooler than the surrounding ground, the natural conduction process reheats the water in the column back to ground temperature as it travels from the pipe to the water pump located at the bottom of the VSWC.

8.2.5.3. SUMMER COOLING OPERATION

In the cooling season, the returning water is a few degrees warmer and is recooled by the natural conduction process allowing continuous 50°F water to pass through the heat pump (now working as a refrigeration machine). Any natural groundwater available from a well is actually a bonus, attributable only to a vertical standing water column system. All other closed-loop systems (horizontal or vertical) make no use of available groundwater, thereby losing the opportunity of offering the most efficient means for providing geothermal heating, cooling, and hot water supply. In most wells drilled, some groundwater can be found, especially if the geology of the area has been properly researched and studied, and the information used by the well driller. Even when sufficient water is not available from the well, potable water can be augmented from a separate source. The quantity of augmented water can be calculated and computer-controlled by the use of thermostats, becuase the incoming source water and the augmented return water are at different temperatures. This quantity of augmented water depends on the natural flow of the well, the amount of the draw from the well for heat pump use, and the adequacy of contact between the VSWC and the earth.

When using the groundwater heat pump system, groundwater is pumped by a mechanical pumping device called a water pump, which pushes water from the ground for delivery above ground. There is nothing new in the process of moving water here. This water is now passed through another mechanical device called a heat pump. With the two heat exchangers inside the heat pump, there is no practical way that the refrigerant can come in contact with the heat pump water on its way back to the water column.

8.2.5.4. ENERGY CONSERVATION

Geothermal energy is smart energy because it uses the free energy of the earth to provide the least expensive solutions for providing air heating, air cooling, and hot water supply for residential, commercial, and industrial uses. The US Department of Energy, the US Environmental Protection Agency, state energy and environmental agencies, and electric utility companies throughout the nation are "swinging in behind" the use of geothermal energy simply because it is smart energy. Many electric power companies across the nation are offering financial assistance to building owners to take part in the energy savings and demand-side management savings that are available through geothermal energy systems.

8.2.5.5. ENVIRONMENTAL PROTECTION

Federal and state environmental agencies are extremely supportive because the use of geothermal energy will reduce the need for future power-generating plants whether nuclear, hydroelectric, gas fired, oil fired, or coal fired. More important, perhaps, is the reductions in emissions of global warming gases of the compounds of carbon dioxide, methane, sulfur dioxide, and nitrogen oxides and the reduction of airborne heavy metals. Fire insurance companies are likewise looking favorably at the new geothermal heating and cooling technology, as there are no open flames or pilot lights and, therefore, no possible chance of explosions or other fires caused by leaking gas or oil. Likewise, fires from coal- and/or wood-burning stoves are also effectively eliminated.

More than 400 systems throughout New York's Capital District, the Adirondack Region, and western Vermont have been installed (58). This new technology has been proven to be not only the least expensive closed-loop installation for any ground source heat pump system, but also the least expensive operation for any type of air heating, air cooling, and hot water supply installation.

8.3. HVAC and Indoor Air Quality Control

8.3.1. HVAC Shutdown Case History

When it comes to commonly found indoor air quality (IAQ) problems, HVAC airflow and contaminants issues top the list. For economic savings, HVAC is frequently shut down and, subsequently, many IAQ problems arise because of its shutdown. This shutdown action can lead to prolific growth of fungus in the working areas. Cook and McDaniel (59) have suggested one major rule to follow: There is no place for energy savings in vital document storage areas and special collections. These areas require 24 h a day environmental controls, and independent control of temperature and relative humidity is desirable.

8.3.2. HVAC Tampering Case History

Tampering with HVAC systems can also lead to IAQ problems. During a hot summer month, for instance, unhappy apartment tenants may complain about the room temperature. Rather than addressing the system's cooling capacity, facilities personnel close the fresh air returns to get maximum cooling. This poor action exacerbates an existing moisture problem and concentrates various fumes in the apartment building.

8.3.3. HVAC Conversion Case History

Indoor air quality problems arise when the function of an existing space is converted to another use. One case occurred when an area that had been an automotive garage was converted to a school gymnasium. After multiple complaints about skin and eye irritation from the students, a certified industrial hygienist was retained. After exhaustive testing and examination under the gym flooring and declarations that the area was free of any IAQ problems, the complaints continued. Then, after examination and discussions with BMS Catastrophe's Special Technologies Department, the IAQ problem was finally defined. In the automotive garage, mercury vapor lighting had been provided for economy. With the former garage completely enclosed and inadequately ventilated, the heavy ultraviolet spectral output of these lamps was now generating ozone gas, the source of the irritation. All symptoms disappeared after the lighting was removed and converted (59).

8.3.4. HVAC Excessive Moisture Case History

The HVAC system can be the cause or the cure of IAQ when it comes to indoor mold growth (60,61). Improperly maintained, the HVAC can become the mold-amplification site resulting from condensation and the buildup of debris. Excessive moisture-generating activities can contribute to the problem and raise indoor relative humidity above 70%, which is ideal for mold growth. Cook and McDaniel (59) have offered some strategies for operating the HVAC system aiming at mold and fungus control: (1) dilute humid indoor air with dry air via the HVAC system's air handling unit, (2) control the source of the water by repairing water leaks, (3) properly ventilate the air space that is humidified by activities that increase indoor moisture levels, (4) control sources of water or leaks outside of the building envelope by improving drainage, thus moving water away from the building, (5) waterproof windows and other potential entry points when a HVAC system is in operation, (6) increase ventilation because stagnant air tends to increase in relative humidity quicker than moving air, increasing the potential for migration of indoor humidity into hidden wall cavities, and (7) employ dehumidification to remove moisture from the air space and building materials. The US EPA website on mold remediation in schools and commercial buildings using the HVAC systems is available and useful (62). The Center for Disease Control National Center for Environmental Health also has a web site on molds control and HVAC proper maintenance (63).

8.3.5. Municipal Utility District Odor Control Case History

Increased development in the Oakland, California area made controlling odor emissions a priority for the East Bay Municipal Utility district's main wastewater treatment plant. In response to this need, the district created an odor control master project providing both short-term and long-term solutions to mitigate neighbors' concerns and improve both indoor and outdoor air quality. The project team planned and pursued the following ventilation improvements: (1) routing the foul-air exhaust from the coarse-screen room to the odor control system; (2) covering all open gratings in the coarse-screen room with bolted-down, checkered plate steel covers to keep foul air below the floor; (3) extending exhaust-air intake ducts below the floor elevation into channels and closer to the odor source; (4) designing supply-air and exhaust-air fans to create negative air pressure in the coarse-screen and fine-screen areas; (5) complying with the recommendations of National Fire Protection Association, Quincy, MA; and providing at least 12 air changes per hour for the ventilation of influent pump station's foul air to the odor control system (64)

NOMENCLATURE

a	A constant (dimensionless)
A	Area of the opening (L^2)
A_c	Area of the rising column of contaminated air at the hood face (L^2)
A_f	Total area of the hood face (L^2)
A_o	Area of the orifice (L^2)
A_n	Area of the hot source (L^2)
b	A constant (dimensionless)
c	A constant (dimensionless)
C_i	Observed concentration of the ith atmospheric pollutant (ppm, mg/m^3, or mp/ft^3)
C_m	Actual concentration of a mixture of atmospheric pollutants (ppm, mg/rn^3, or mp/ft^3)
C_v	Volume of building to be ventilated (L^3)
d	Gas density (ML^{-3})
D	diameter (L)
D_c	Diameter of the hot column of air at the level of the hood face (L)
D_f	Diameter of the round hood (L)
D_s	Diameter of hot source (L)
E_a	Combined effects of different hazards (dimensionless)
E_i	Independent effect of the ith hazard (dimensionless)
f	A coefficient (dimensionless)
F_r	Loss fraction of velocity pressure difference (dimensionless)
h	Hour
h	Natural convection heat loss coefficient (Btu/h-ft^2-°F or kg-cal/h-m^2-°C)
H	Fundamental force per unit area
H_o	Orifice pressure loss, height of water column (L)
H_{s1}	Static pressure at point 1
H_{s2}	Static pressure at point 2
H_v	Velocity pressure or head, height of water column (L)
H_{v1}	Velocity pressure at point 1
H_{v2}	Velocity pressure at point 2
K	A coefficient (dimensionless)
L	Liter
L_c	Length of the rising jet at the hood (L)
L_f	Length of the rectangular hood (L)
L_i	Threshold limit value of the ith atmospheric pollutant (ppm, mg/m^3, mp/ft^3)
L_m	Threshold limit value of a mixture of atmospheric pollutants (ppm, mg/m^3, or mp/ft^3)
L_s	Length of the hot source (L)
ΔL	Increased length (L)
m	A constant (dimensionless)
N	Fan speed(t^{-1} [i.e., rpm])
p	Percentage of the velocity at the opening found at a point X on the axis of the duct

P	power (ML^2/t^3)
q	Rate at which the heat is transferred to the rising column of air (Btu/min or kg-cal/min); rate of heat generation (Btu/h or g- cal/s)
Q	Measurement of airflow through a fan or system (L^3/t)
Q'	Directivity factor (dimensionless)
r	A constant (dimensionless)
R	Room constant (dimensionless)
R'	Throat radius (L)
R_f	Regain factor expressed as a fraction of the velocity pressure difference between two points (dimensionless)
R_a	Recommended rate of air change (t)
s	A constant (dimensionless)
S	Average specific heat of the mixture (Btu/lb-°F or kg-cal/kg-°C)
T	Average temperature of the air inside the hood (°F or °C)
T_d	Temperature difference between the hot source and the ambient air, (°F or °C)
T_i	Inside temperature (°F or °C)
T_o	Outside temperature (°F or °C)
u	A constant (dimensionless)
v	A constant (dimensionless)
V	velocity (L/t)
V_a	Velocity of air (L/t)
V_e	Velocity of escape through orifices in the upper portions of a hood (L/t)
V_f	Velocity of the hot air jet at the level of the hood face (L/t)
V_r	Required velocity through the remaining area of the hood ($A_f - A_e$, L/t)
V_X	Velocity of air at point X (L/t)
w	A constant (dimensionless)
W_c	Width of the rising-air jet at the hood (L)
W_f	Width of the rectangular hood (L)
W_s	Width of the hot source (L)
ΔW	Increased width (L)
X	Distance (L)
X_f	Distance from the hypothetical point source to the hood face (L)
y	Distance between the hot source and the hood face (L)
Y	Vertical distance above the hood face to the location of the orifice (L)
z	Distance between the hypothetical point source and the hot source (L)
α	Absorption coefficient of materials (dimensionless)
β	Average absorption coefficient of a room (dimensionless)
θ	angle (deg)

ACKNOWLEDGMENTS

This chapter is updated based on the first edition (4), which was originally written by Dr. Mu Hao Sung Wang and Professor Lawrence K. Wang. Dr. Mu Hao Sung Wang granted Dr. Zucheng Wu permission to update her chapter for this new book. The authors would like to thank Minghua Zhou, Jingying Ma, and Yunrui Zhou who are graduate students at Zhejiang University and have provided valuable assistance in reference search and electronic manuscript conversion. The current authors have updated this chapter in honor of Professor Tan'en Tan at Zhejiang University for his lifelong accomplishment and dedication in the field of environmental engineering. Professor Tan's concern and encouragement to younger professional generations are forever appreciated.

REFERENCES

1. V. M. Ehlers and B. W. Steel, *Municipal and Rural Sanitation*, 6th ed., McGraw-Hill, New York, 1965, pp. 353–371.
2. ASHRAE, *ASHRAE Guide and Data Books*, American Society of Heating, Refrigerating and Air-Conditioning Engineers, New York, 1963.
3. US DOL, *The Occupational Safely and Health Act, Federal Register*, US Department of Labor Government Printing Office, Washington, DC, 1971, Part 2.
4. M. H. S. Wang and L. K. Wang, in *Handbook of Environmental Engineering, Volume 1* (L. K. Wang and N. C. Pereria eds.), Humana P, Totowa, NJ, 1979.
5. ILG Industries, Inc., *Industrial Ventilation Guide*, Chicago ILG Industries, Inc., IL, 1972.
6. American Coolair Corporation, *Handbook on Ventilation and Cooling of Commercial and Industrial Buildings*. Jacksonville, FL, 1975.
7. G. Murphy, *Similitude in Engineering*, Ronald Press, New York, 1950.
8. ACC, *Procedure for Fan Selection to Meet a Specified Sound Level Limit*, American Coolair Corporation, Jacksonville FL, 1973.
9. Editor, *Chem. Eng.* **109**(9) (2002).
10. Editor, *Pollut. Eng.* **34**(12) (2002).
11. Editor, *Environ. Protect.* **14**(2) (2003).
12. Editor, *Environ. Tech.* **11**(6) (2002).
13. J. Bouley, *Pollut. Eng.* **25**(8), 510 (1993).
14. J. M. Dalla Valle, *Exhaust Hoods, 2nd ed.*, Industrial Press, New York, 1952.
15. J. A. Danielson, *Air Pollution Engineering Manual*, 2nd ed., US Environmental Protection Agency, Research Triangle Park, NC, 1973.
16. W. C. L. Hemeon, *Plant and Process Ventilation*, 2nd ed., Industrial Press, New York, 1963.
17. O. G. Sutton, *J. Meteorol.* **7**, 307 (1950).
18. ACGIH, *Industrial Ventilation*, 7th ed., American Conference of Governmental Industrial Hygienists, Lansing, MI, 1962.
19. L. K. Wang, *Environmental Engineering Glossary*, Calspan Corp., Buffalo, NY, 1974.
20. World Health Organization, *Air Pollution*, Columbia University Press, New York, 1961.
21. Governmental Industrial Hygienist, *Threshold Limit Values of Industrial Atmosphere Contaminants, 25th Annual Meeting of the American Conference of Governmental Industrial Hygienist*, 1966.
22. C. E. A. Winslow, *Fresh Air and Ventilation* Dutton, New York, 1958.
23. E. L. Besch, *Environmental Requirements for Laboratory Animals*, Report to NIH, Institute for Environmental Research, Kansas State University, 1970.
24. L. T. Fan and N. C. Pereira *Can. J. Chem. Eng.* **52**, 251 (1974).
25. L. K. Wang, *Filter Package*, Technical Report No. NT.5255-M-2, Calspan Corp., Buffalo, NY, 1974.
26. E. Nakanishi, N. C. Pereira, L. T. Fan, et al., *Build. Sci.* **8**, 39 (1973).
27. E. Nakanishi, N. C. Pereira, L. T. Fan, et al., *Build. Sci.* **8**, 51 (1973).
28. E. Nakanishi, N. C. Pereira, L. T. Fan, et al., *Build. Sci.* **8**, 65 (1973).
29. P. Urycak, *Proc. Inst. Environ. Sci.* **23**, 157 (1977).
30. E. A. Farber, C. A. Morrison, H. A. Ingley, et al., *Proc. Inst. Environ. Sci.* **22**, 319 (1976).
31. M. Meckler, *Proc. Inst. Environ. Sci.* **23**, 118 (1977).

32. Carrier Corporation, *A Guide for Residential Heat Pumps*, Carrier Corporation, Syracuse, NY, 1974.
33. L. K. Wang and G. G. Peery, *Water Resources Bull.* **11**, 919 (1975).
34. L. K. Wang and S. C. Pek, *Ind. Eng. Chem.* **24**, 308 (1975).
35. L. K. Wang, D. B. Aulenbach, and D. F. Langley, *Ind. Eng. Chem.* **15**, 68 (1976).
36. H. Schwartz, *Ind. Wastes* **22**, 31 (1976).
37. Z.C. Wu, *Ph. D thesis*, The University of Hong Kong, 1995.
38. M. Junker, T. Koller and C. Monn, *Sci. Total Environ.* **246**, 139–152 (2000).
39. Y. S. Fung and Z. C. Wu, *Analyst* **121**, 1959 (1996).
40. A. L. Hines, T. K. Ghosh, S. K. Loyalka, et al., *Indoor Air Quality and Control*, PTR Printice-Hall, Englewood Cliffs, NJ, 1993.
41. M. Boss, *Environ. Protect.* **12**(5), 16 (2001); 12(9), 51(2001).
42. P. Ole Fanger, *Int. J. Refrig.* **24**, 148–153 (2001).
43. M.W. Liddament and M. Orme, *Appl. Thermal Eng.* **18**, 1101–1109 (1998).
44. ASHRAE, *Ventilation for Acceptable Indoor Air Quality*, Report No. ASHRAE 62-1989R, American Society of Heating, Refrigerating, and Air-Conditioning Engineers, New York, 1996.
45. V. Geros, M. Santamouris, A. Tsangrasoulis, et al., *Energy Build.* **29**, 141–154 (1999).
46. G. Singh, M. Zaheer-uddin, and R.V. Patel, *Energy Convers. Manag.* **41**, 1671–1685 (2000).
47. M. Kolokotroni, M. D. A. E. S. Perera, D. Azzi, et al., *Appl. Thermal Eng.* **21**, 183–199 (2001).
48. M. Zaheer-uddin and G. R. Zheng, *Energy Convers. Manag.* **41**, 49–60 (2000).
49. A. Neville, *Environ. Protect.* **12**(11), 6 (2001).
50. E. K. Aisbett and Q. Tuan Pham, *Int. J. Refrig.* **21**(1), 18–28 (1998).
51. H. Kruse, R. Heideld and J. Suss, *IIR Bull.* **99**, 1 (1999).
52. A. Rozhentsev and C. C. Wang, *Appl. Thermal Eng.* **21**, 871–880 (2001).
53. D.-W. Sun, *Energy Convers. Manag.* **38**(5), 479–491 (1997).
54. D.-W. Sun, *Energy Convers. Manag.* **39**(5/6), 357–368 (1998).
55. J. R. Turnpenny, D. W. Etheridge, and D. A. Reay, *Appl. Thermal Eng.* **20**, 1019–1038 (2000).
56. J. R. Turnpenny, D. W. Etheridge, and D. A. Reay, *Appl. Thermal Eng.* **20**, 1203–1217 (2001).
57. A. C. Oliveira, C. F. Afonso, S. B. Riffat, et al., Doherty, *Appl. Thermal Eng.* **20**, 1213–1223 (2000).
58. L. K. Wang, H. E. Rist, and M. H. Wang, *OCEESAs J.* **11**(3), 21–25 (1994).
59. C. Cook and McDaniel, *Environ. Protect.* **13**(5), 24 (2002).
60. US EPA, www.epa.gov/iaq/pubs (accessed 2003).
61. A. Neville, *Environ. Protect.* **13**(5), 6 (2002).
62. Environmental Support Solution Inc., *Environ. Protect.* **13**(10), 50 (2002).
63. Center for Disease Control National Center for Environmental Health, www.cdc.gov/nceh/airpollution/molds/stachy.htm (accessed 2003).
64. D. Kiang, J. Yoloye, J. Clark, et al., *Water Environ. Tech.*, **14**(4), 39–44 (2002).
65. ASHRACE (1963) *Handbook of Fundamentals,* American Society of Heating, Refrigerating, and Air-Conditioning Engineers, New York, NY.

APPENDICES

Appendix A: Recommended Threshold Limit Values of Hazardous Substances

Substance	ppm[a]	mg/m³[b]	Substance	ppm[a]	mg/m³[b]
Acetaldehyde	200	360	2-Butanone	200	590
Acetic acid	10	25	2-Butoxy ethanol (Butyl		
Acetic anhydride	5	20	Cellosolve)—skin	50	240
Acetone	1000	2400	Butyl acetate[f]		
Acetonitrile	40	70	(*n*-butyl acetate)	—	—
Acetylene dichloride; *see*			Butyl alcohol	100	300
1,2 dichloroethylene			*tert*-Butyl alcohol	100	300
Acetylene tetrabromide	1	14	Butylamine[c]—skin	5	15
Acrolein	0.1	0.25	*tert*-Butyl chromate[c]		
Acrylonitrile—skin	20	45	(as CrO_3)—skin	—	0.1
Aldrin—skin	—	0.25	*n*-Butyl glycidyl ether (BCE)	50	270
Allyl alcohol—skin	2	5	Butyl mercaptan	10	35
Allyl chloride	1	3	*p-tert*-Butyltoluene	10	60
Allyl glycidyl ether[c] (AGE)	10	45	Cadmium oxide fume	—	0.1
Allyl propyl disulfide	2	12	Calcium arsenate	—	1
2 Aminoethanol; *see*			Calcium oxide	—	5
ethanolamine			Camphor	—	2
Ammonia	50	35	Carbaryl[d] (Sevin) (R)	—	5
Ammoniurn sulfamate			Carbon dioxide	5000	9000
(Ammate)	—	15	Carbon disulfide—skin	20	60
n-Amyl acetate	100	525	Carbon monoxide[f]	—	—
Aniline—skin	5	19	Carbon tetrachloride—skin	10	65
Anisidine[d]			Chlordane—skin	—	0.5
(*o, p*-isomers)—skin	—	0.5	Chlorinated		
Antimony and compounds			camphene—skin	—	0.5
(as Sb)	—	0.5	Chlorinated diphenyl oxide	—	0.5
ANTU (alpha naphthyl			Chlorine[f]	—	—
thiourea)	—	0.3	Chlorine dioxide	0.1	0.3
Arsenic and compounds			Chlorine trifluoride[c]	0.1	0.4
(as As)	—	0.5	Chloroacetaldehyde[c]	1	3
Arsine	0.05	0.2	Chlorobenzene		
Barium (soluble compounds)	—	0.5	(monochlorobenzene)	75	350
Benzene[c] (benzol)—skin	25	80	Chlorobromomethane	200	1050
Benzidine[e]—skin	—	A¹	2-Chloro-l,3 butadiene;		
p-Benzoquinone; see quinone			*see* chloroprene		
Benzoyl peroxide[d]	—	5	Chlorodiphenyl (42%		
Benzyl chloride	1	5	chlorine)—skin	—	1
Beryllium	—	0.002	Chlorodiphenyl (54%		
Biphenyl[f]; see diphenyl			chlorine)—skin	—	0.5
Boron oxide	—	15	1-Chloro, 2,3 epoxypro-		
Boron trifluoride[c]	1	3	pane; *see* epichlorhydrin		
Bromine	0.1	0.7	2, Chloroethanol; *see*		
Butadiene (1,3-butadiene)	1000	2200	ethylene chlorohydrin		
Butanethiol; see butyl			Chloroethylene; *see* vinyl		
mercaptan			chloride		

(Continued)

Appendix A: *(Continued)*

Substance	ppm[a]	mg/m³ [b]	Substance	ppm[a]	mg/m³ [b]
Chloroform[c]			Dichlorotetrafluoroethane	1000	7000
(trichloromethane)	50	240	Dieldrin—skin	—	0.25
1-Chloro-1-nitropropane	20	100	Diethylamine	25	75
Chloropicrin	0.1	0.7	Diethylether; see ethyl ether		
Chloroprene (2-chloro-1,3-			Difluorodibromomethane	100	860
butadiene)—skin	25	90	Diglycidyl ether[c] (DGE)	0.5	2.8
Chromic acid and chromates			Dihydroxylbenzene; see		
(as CrO₃)	—	0.1	hydroquinone		
Cobalt[c]	—	—	Diisobutyl ketone	50	290
Copper fume	—	0.1	Dimethoxymethane; see		
Dusts and mists	—	1.0	methylal		
Cotton dust[d] (raw)	—	1	Dimethyl acetamide—skin	10	35
Crag (R) herbicide	—	15	Dimethylamine[d]	10	18
Cresol (all isomers)—skin	5	22	Dimethylaminobenzene,		
Cyanide (as CN)—skin	—	5	see xylidene		
Cyclohexane[f]	—	—	Dimethylaniline (*N*-		
Cyclohexanol	50	200	dimethylaniline)—skin	5	25
Cyclohexanone	50	200	Dimethylbenzene,		
Cyclohexene[c]	—	—	see Xylene		
Cyclopentadiene[d]	75	200	Dimethyl 1,2-dibro-2,2-		
2,4-D	—	10	dichloroethyl phosphate[d]		
DDT—skin	—	1	(Dibrom) (R)	—	3
DDVP—skin	—	1	Dimethylformanide[d]—skin	10	30
Decaborane—skin	0.05	0.3	2,6-Dimethylheptanone;		
Demeton (R)—skin	—	0.1	see diisobutyl ketone		
Diacetone alcohol (4-			1, 1-Dimethylhydrazine—		
hydroxy-4-methyl-2-			skin	0.5	1
pentanone)	50	240	Dimethylsulfate—skin	1	5
1,2 Diaminoethane; see			Dinitrobenzene		
ethylenediamine diborane			(all isomers)—skin	—	1
1,2-Dibrornoethane[f]			Dinitro-*o*-cresol—skin	—	0.2
(ethylene dibromide)—			Dinitrotoluene—skin	—	1.5
skin	—	—	Dioxane(diethylene		
o-Dichlorobenzene[c]	50	300	dioxide)—skin	100	360
p-Dichlorobenzene	75	450	Dipropylene glycol methyl		
Dichlorodifluoromethane	1000	4950	ether—skin	100	600
1,3-Dichloro-5-dimethyl			Di-*sec*-octyl phthalate[d]-		
hydantoin[d]	—	0.2	(di-2-ethylhexylphthalate)	—	5
1,1-Dichloroethane	100	400	Endrin—skin	—	0.1
1,2-Dichloroethane	50	200	Epichlorhydrin—skin	5	19
1,2-Dichloroethylene	200	790	EPN—skin	—	0.5
Dichloroethyl ether[c]—skin	15	90	1, 2-Epoxypropane; see		
Dichloromethane; see			propyleneoxide		
methylenechloride			2,3-Epoxy-1-propanol; see		
Dichloromonofluoro-	1000	4200	glycidol		
methane			Ethanethiol; see ethyl		
1,1-Dichloro-1-nitroethane[c]	10	60	mercaptan		
1,2-Dichloropropane, see			Ethanolamine	3	6
propylenedichloride			2-Ethoxyethanol—skin	200	740

Appendix A: *(Continued)*

Substance	ppm[a]	mg/m³[b]	Substance	ppm[a]	mg/m³[b]
2-Ethoxyethylacetate (cellosolve acetate)—skin	100	540	Freon 113; see 1,1,2-tri-Chloro-1,2,2-trifluoroethane		
Ethyl acetate	400	1400	Freon 114; see dichlorotetrafluoroethane		
Ethyl acrylate—skin	25	100			
Ethyl alcohol (ethanol)	1000	1900	Furfural—skin	5	20
Ethylamine[f]	—	—	Furfunyl alcohol	50	200
Ethylbenzene[c,f]	—	—	Gasoline[e]	—	A[6]
Ethyl bromide	200	890	Glycidol (2,3-epoxy-1-propanol)	50	150
Ethyl chloride	1000	2600			
Ethyl ether	400	1200	Glycol monoethyl ether; see 2-ethoxyethanol		
Ethyl formate	100	300			
Ethyl mercaptan[c,f]	—	—	Guthion[s]; see azinphosmethyl		
Ethyl silicate	100	850			
Ethylene chlorohydrin—skin	5	16	Hafnium	—	0.5
Ethylenediamine	10	25	Heptachlor—skin	—	0.5
Ethylene dibromide; see 1,2-dibromoethane			Heptane (*n*-heptane)	500	2000
			Hexachloroethane[d]—skin	1	10
Ethylene dichloride; see 1,2-dichloroethane			Hexane (*n*-hexane)	500	1800
			2-Hexanone	100	410
Ethylene glycol dinitrate[c]—skin	0.2	1.2	Hexone	100	410
			sec-Hexyl acetate	50	295
Ethylene glycol mono-methyl ether acetate; see methyl cellosolve acetate			Hydnazine—skin	1	1.3
			Hydrogenbromide	3	10
			Hydrogen chloride[c]	5	7
Ethylene imine[f]—skin	—	—	Hydnogencyanide—skin	10	11
Ethylene oxide	50	90	Hydrogen fluoride	3	2
Ethylidine chloride; see 1,1-dichloroethane			Hydrogen peroxide, 90%	1	1.4
			Hydrogen selenide	0.05	0.2
Ferbam	—	15	Hydrogen sulfide[d]	10	15
Ferrovanadium dust	—	1	Hydroquinone	—	2
Fluoride (as F)	—	2.5	Iodine[c]	0.1	1
Fluorine	0.1	0.2	Iron oxide fume[f]	—	—
Fluorotrichloromethane	1000	5600	Isoamyl alcohol	100	360
Formaldehyde[c]	5	6	Isophorone	25	140
Freon 11; see fluorotrichloromethane			Isopropyl alcohol	400	980
			Isopropylamine	5	12
Freon 12; see dichlorodifluoromethane			Isopropylether	500	2100
			Isopropyl glycidyl ether (IGE)	50	240
Freon 13B1; see trifluoromonobromethane			Ketene	0.5	0.9
			Lead	—	0.2
Freon 21; see dichloro-monofluoromethane			Lead arsenate	—	0.15
			Lindane—skin	—	0.5
Freon 112; see 1,1,2,2-tetrachloro-1,2-difluoroethane			Lithium hydride	—	0.025
			LPG[d] (liquified petroleum gas)	1000	1800

(Continued)

Appendix A: *(Continued)*

Substance	ppm[a]	mg/m³[b]	Substance	ppm[a]	mg/m³[b]
Magnesium oxide fume	—	15	Methylene bisphenyl		
Malathion—skin	—	15	isocyanate[c] (MDI)	0.02	0.2
Manganese[c]	—	5	Methylene chloride		
Mercury—skin	—	0.1	(dichloromethane)	500	1740
Mercury (organic			Molybdenum		
compounds)—skin	—	0.01	Soluble compounds	—	5
Mesityl oxide	25	100	Insoluble compounds	—	15
Methanethiol; see methyl			Monomethyl aniline—skin	2	9
mercaptan			Morpholine[d]—skin	20	70
Methoxychlor	—	15	Naphtha (coal tar)	200	800
2-Methoxyethanol; see			Naphtha (petroleum)	500	2000
methyl cellosolve			Naphthalene[d]	10	50
Methyl acetate	200	610	β-Naphthylamine	—	A²
Methyl acetylene			Nickel carbonyl	0.001	0.007
(propyne)	1000	1650	Nickel,[d] metal and soluble		
Methyl acetylene–			compounds	—	1
propadiene mixture			Nicotine—skin	—	0.5
(MAPP)[d]	1000	1800	Nitric acid[d]	2	5
Methyl acrylate—skin	10	35	p-Nitroaniline—skin	1	6
Methylal			Nitrobenzene—skin	1	5
(dimethoxymethane)	1000	3100	p-Nitrochlorobenzene[d]—	—	1
Methyl alcohol (methanol)	200	260	skin		
Methyl amyl alcohol; see			Nitroethane	100	310
methyl isobutyl carbinol			Nitrogen dioxide[c]	5	9
Methyl bromide[c]—skin	20	80	Nitrogen trifluoride[d]	10	29
Methyl butyl ketone; see			Nitroglycerin + EGDN—	0.2	2
2-hexanone			skin		
Methyl cellosolve—skin	25	80	Nitromethane	100	250
Methyl cellosolve			1-Nitropropane	25	90
acetate—skin	25	120	2-Nitropropane	25	90
Methyl chloride[c]	100	210	N-Nitrosodimethyl-		
Methyl chloroform	350	1900	amine (dimethyl-		
Methylcyclohexane	500	2000	nitrosoamine)—skin	—	A³
Methylcyclohexanol	100	470	Nitrotoluene—skin	5	30
o-Methylcyclohexanone—			Nitrotrichloromethane; see		
skin	100	460	chloropicrin		
Methyl ethyl ketone			Octane	500	2350
(MEK); see 2-butanone			Oil mist (mineral)	—	5
Methyl formate	100	250	Osmium tetroxide	—	0.002
Methyl isobutyl carbinol—			Oxygen difluoride[d]	0.05	0.1
skin	25	100	Ozone	0.1	0.2
Methyl isobutyl ketone;			Parathion—skin	—	0.1
see hexone			Pentaborane	0.005	0.01
Methyl mercaptan[c,d]	10	20	Pentachloronaphthalene—		
Methyl methacrylate[d]	100	410	skin	—	0.5
Methyl propyl ketone; see			Pentachlorophenol—skin	—	0.5
2-pentanone			Pentane	1000	2950
α-Methyl styrene[c]	100	480	2-Pentanone	200	700

Appendix A: *(Continued)*

Substance	ppm[a]	mg/m³[b]	Substance	ppm[a]	mg/m³[b]
Perchloroethylene	100	670	Strychnine	—	0.15
Perchloromethyl			Styrene monomer[c]		
mercaptan	0.1	0.8	(phenyl ethylene)	100	420
Perchloryl fluoride	3	13.5	Sulfur dioxide	5	13
Phenol—skin	5	19	Sulfur hexafluoride	1000	6000
p-Phenylene diamine[d]—skin	—	0.1	Sulfuric acid	—	1
Phenylethylene; see styrene			Sulfur monochloride	1	6
Phenyl glycidyl ether (PGE)	50	310	Sulfur pentafluoride	0.025	0.25
Phenyihydrazine—skin	5	22	Sulfuryl fluoride	5	20
Phosdrin (Mevinphos)			Systox; see demeton 2,4,5 T	—	10
(R)—skin	—	0.1	Tantalum	—	5
Phosgene[d] (carbonyl			TEDP—skin	—	0.2
chloride)	0.1	0.4	Teflon (R) decomposition		
Phosphine	0.3	0.4	products	—	A[4]
Phosphoric acid	—	1	Tellurium	—	0.1
Phosphorus (yellow)	—	0.1	TEPP—skin	—	0.05
Phosphorus pentachloride	—	1	1,1,1,2-Tetrachloro-2,2-		
Phosphorus pentasulfide	—	1	difluoroethane[d]	500	4170
Phosphorus trichloride	0.5	3	1,1,2,2-Tetrachloro-1,2-		
Phthalic anhydride[d]	2	12	difluoroethane	500	4170
Picric acid—skin	—	0.1	1,1,2,2-Tetrachloroethane—		
Platinum (soluble salts)	—	0.002	skin	5	35
Polytetrafluoroethylene			Tetrachloroethylene; see		
decomposition products	—	A4	perchloroethylene		
Propane[d]	1000	1800	Tetrachloromethane; see		
Propyne; see methyl			carbon tetrachloride		
acetylene			Tetraethyl lead	—	0.075
β-Propiolactone	—	A[5]	(as Pb)—skin		
n-Propyl acetate	200	840	Tetrahydrofuran	200	590
n-Propyl nitrate	25	110	Tetranitromethane	1	8
Propylene dichloride	75	350	Tetryl (2,4,6-trinitrophenyl-		
Propylene imine[f]—skin	—	—	methylnitramine)—skin	—	1.5
Propylene oxide	100	240	Thallium (soluble		
Pyrethrum	—	5	compounds)—skin	—	0.1
Pyridine	5	15	Thiram	—	5
Quinone	0.1	0.4	Tin (inorganic compounds		
Rotenone (commercial)	—	5	except oxide)	—	2
Selenium compounds[d]			Tin (organic compounds)	—	0.1
(as Se)	—	0.2	Titanium dioxide	—	15
Silver, [d]metal and soluble			Toluene (toluol)	200	750
compounds	—	0.01	Toluene-2,4-diisocyanate[c]	0.02	0.14
Sodium fluoroacetate			o-Toluidine—skin	5	22
(1080)—skin	—	0.05	Toxaphene; see chlorinated		
Sodium hydroxide	—	2	camphene		
Stibine	0.1	0.5	1,1,1-Trichloroethane; see		
Stoddard solvent	500	2900	methyl chloroform		

(Continued)

Appendix A: *(Continued)*

Substance	ppm[a]	mg/m³[b]	Substance	ppm[a]	mg/m³[b]
Trichloroethylene	100	535	Uranium		
Trichloromethane; see chloroform			Soluble compounds	—	0.05
			Insoluble compounds	—	0.25
Trichloronaphthalene—skin	—	5	Vanadium[c] (V_2O_5 dust)	—	0.5
			(V_2O_5 fume)	—	0.1
1,2,3-Trichloropropane	50	300	Vinyl benzene; see styrene		
1,1,2-Trichloro 1,2,2-tri fluoroethane	1000	7600	Vinyl chloride[c]	500	1300
			Vinylcyanide; see acrylonitrile		
Triethylamine	25	100			
Trifluoromonobromo-ethane	1000	6100	Vinyl toluene	100	480
			Warfarin	—	0.1
2,4,6-Trinitrophenl; see picric acid			Xylene[c] (xylol)	—	—
			Xylidine—skin	5	25
2,4,6-Trinitrophenylmethyl-nitramine; see tetryl			Yttrium[d]	—	1
			Zinc oxide fume	—	5
Trinitrotoluene—skin	—	1.5	Zirconium compounds	—	5
Triorthocresyl phosphate	—	0.1	(as Zr)		
Triphenyl phosphate	—	3			
Turpentine	100	560			

Radioactivity: For permissible concentrations of radioisotopes in air, see US Department of Commerce, National Bureau of Standards, Handbook 69, *Maximum Permissible Body Burdens and Maximum Permissible Concentrations of Radionuclides in Air and in Water for Occupational Exposure*, June 5, 1959. Also, see US Department of Commerce National Bureau of Standards, Handbook 59, *Permissible Dose from External Sources of Ionizing Radiation*.

"A" values: A[1]   Benzidine. Because of the high incidence of bladder tumors in man, any exposure, including skin, is extremely hazardous.

A[2]   β-Naphthylamine. Because of the extremely high incidence of bladder tumors in workers handling this compound and the inability to control exposures, β-naphthylamine has been prohibited from manufacture, use, and other activities that involve human contact by the state of Pennsylvania.

A[3]   *N*-Nitrosodimethylamine. Because of extremely high toxicity and presumed carcinogenic potential of this compound, contact by any route should not be permitted.

A[4]   Polytetrafluoroethylene decomposition products. Thermal decomposition if the fluorocarbon chain in air leads to the formation of oxidized products containing carbon, fluorine, and oxygen. Because these products decompose by hydrolysis in alkaline solution, they can be quantitatively determined in air as fluoride to provide an index of exposure. No TLV is recommended pending determination of the toxicity of the products, but air concentrations should be minimal.

A[5]   β-Propiolactone. Because of high acute toxicity and demonstrated skin tumor production in animals, contact by any route should be avoided.

A[6]   Gasoline. The composition if gasoline varies greatly and thus a single TLV for all types of gasoline is no longer applicable. In general, the aromatic hydrocarbon content will determine what TLV applies. Consequently the content of benzene, other aromatics, and additives should be determined to arrive at the appropriate TLV.

Trade Names: Algoflon, Fluon, Halon, Teflon, Tetran.

[a]Parts of vapor or gas per million parts of air plus vapor by volume at 25°C and 760 mm Hg pressure.

[b]Approximate milligrams of particulate per cubic meter of air.

[c]Indicates a value that should not be exceeded.

[d]1966 addition.

[e]See A values in Appendix B.

[f]See tentative limits, Appendix B.

Appendix B: Tentative Threshold Limit Values of Hazardous Substances

Substance	ppm[a]	mg/m³[b]	Substance	ppm[a]	mg/m³[b]
Acrylamide—skin	—	0.3	Ethyl butyl ketone		
2-Aminopyridine	0.5	2	(3-heptanone)	50	230
sec-Amyl acetate	125	650	Ethylene glycol		
Azinphos-methyl—skin	—	0.2	dinitrate and/or		
Bromoform—skin	0.5	5	nitroglycerin[d,e]—skin	0.02[f]	0.1[f]
n-Butyl acetate	150	710	Ethylene imine—skin	0.5	1
sec-Butyl alcohol	150	950	Ethyl mercaptan[d]	10	25
tert-Butyl acetate	200	950	*N*-Ethylmorpholine—skin	20	94
sec-Butyl alcohol[c]	150	450	Fibrous glass	—	5
Cadmium (metal dust			Formic acid	5	9
and soluble salts)	—	0.2	Gasoline[c]	A[6]	
Carbon black	—	3.5	*sec*-Hexyl acetate	50	300
Carbon monoxide	50	55	Hexachloronaphthalene[c]—		
α-Chloroacetophenone[c]			skin	—	0.2
(phenacychloride)	0.05	0.3	Iron oxide fume	—	10
o-Chlorobenzylidene			Isoamyl acetate	100	525
malononitrile (OCBM)	0.05	0.4	Isobutyl acetate	150	700
Chlorine[d]	1	3	Isobutyl alcohol[c]	100	300
Chromium,[c]			Isopropyl acetate	250	950
Sol. chromic, chromous			Maleic anhydride[e]	0.25	1
salts, as Cr	—	0.5	Methylamine	10	12
Metallic and insoluble salts	—	1.0	Methyl (*n*-amyl)		
Coal tar pitch volatiles			Ketone (2-heptanone)	100	465
(benzene soluble fraction)			Methyl iodide—skin	5	28
(anthracene, BaP,			Methyl isocyanate—skin	0.02	0.05
phenanthrene, acridine,			Monomethyl		
chrysene, pyrene)	—	0.2	hydrazine[d]—skin	0.2	0.35
Cobalt,[c] metal fume			Naphtha[e] (coal tar)	100	400
and dust	—	0.1	Nitric oxide[c]	25	30
Crotonaldehyde	2.0	6.0	Octachloronaphthalene[c]—		
Cumene—skin	50.0	245.0	skin	—	0.1
Cyclohexane	300.0	1050	Oxalic acid	—	1
Cyclohexene	300.0	1015	Parquat[c]—skin	—	0.5
Diazomethane	0.2	0.4	Phenyl ether (vapor)	1	7
1,2f-Dibromo-			Phenyl ether–biphenyl		
ethane[d]—skin	25	190	mixture (vapor)	1	7
Dibutyl phosphate[e]	1	5	Phenyl glycidyl ether[e]		
Dibutdylphosphate[e]	—	5	(PGE)	10	62
Diethylamino ethanol—			Pival (2-pivalyl-1,3-		
skin	10	50	indandione)	—	0.1
Diisopropylamine[c]—skin	5	20	Propyl alcohol[c]	200	450
Dimethylphthalate[c]	—	5	Propylene imine—skin	2	5
Diphenyl[c]	0.2	1	Rhodium, metal fume		
Ethylamine	10	18	and dusts	—	0.1
Ethyl-*sec*-amyl ketone			Soluble salts	—	0.001
(5-methyl-3-heptanone)	25	130	Ronnel[c]	—	15
Ethyl benzene	100	435	Selenirum hexafluofride	0.05	0.4

(Continued)

Appendix B: *(Continued)*

Substance	ppm[a]	mg/m³[b]	Substance	ppm[a]	mg/m³[b]
Tellurium hexafluoride	0.02	0.2	Tremolite	5 mppcf	—
Terphenyls[c,d]	1	9.4	Tributyl phosphate[c]	—	5
Tetrachloronaphthalene[c]—			1,1,2-Trichloroethane—		
skin	—	2	skin	10	45
Tetramethyl lead (TML)			Xylene	100	435
(as lead)—skin	—	0.075	Zinc chloride[c]	—	1
Tetramethyl					
succinonrtrile—skin	0.5	3			

Notes: These substances, with their corresponding tentative limits, comprise those for which a limit has been assigned for the first time or for which a change in the "Recommended" listing has been made. In both cases, the assigned limits should be considered trial values that will remain in the tentative listing for a period of at least 2 yr, during which time, definitive evidence and experience is sought. If acceptable at the end of 2 yr, these substances and values will be moved to the "Recommended" list. Documentation for tentative values are available for each of these substances.

[a]Parts of vapor or gas per million parts of air plus vapor by volume at 25°C and 760 mm Hg pressure

[b]Approximate milligrames of particulate per cubic meter of air.

[c]1966 additions.

[d]Indicates a value that should not be exceeded.

[e]1966 revision.

[f]For intermittent exposures only.

Appendix C: Respirable Dusts Evaluated by Count

Substance	mp/ft³
Silica	
Crystalline	
Quartz, threshold limit calculated from the formula	$\dfrac{250}{\%S_iO_2 + 5}$
Cristobalite calculated from the formula	$\dfrac{250}{\%S_iO_2 + 5}$
Amorphous, including natural diatomaceous earth	20
Silicates(less than 1% crystalline silica)	
Asbestos	5
Mica	20
Soapstone	20
Talc	20
Portland cement	50
Miscellaneous (less than 1% crystalline silica)	50
Graphite (natural)	
"Inert" or nuisance particulates	50 (or 15 mg/m³ whichever is the smaller)

Source: American Conference of Governmental Industrial Hygienists.

Appendix D: Converting from Round to Rectangular Ductwork (Source: ILG Industries)

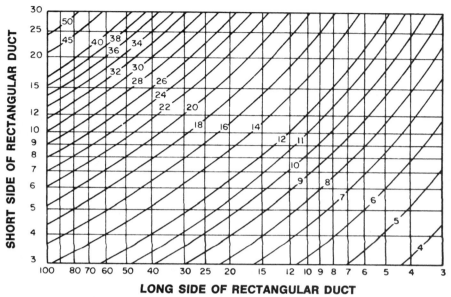

Appendix E: Procedure for Fan Selection to Meet a Specific Sound Level Limit (Source: American Coolair Corporation)

A.   Summary of Selection Procedure

The first step is to determine what effect the acoustical characteristics of the room will have on the sound that comes from the fan. The next step is to consider the effect of the fan's location with regard to adjacent room surfaces. The third step is to measure the distance from the fan to the point where it will be heard by employees. With these factors, one can determine from a chart indicating the amount of the reduction (or addition) in the sound made by the fan that will occur when a specific mechanical fan model is installed in the selected location. This amount is deducted from a sound rating assigned to the fan by the fan manufacturer. This net figure is further adjusted by a factor that relates to the method by which the fan is installed. (Information necessary for making all of these calculations is outlined below.) From the above procedure, one can obtain an approximation of the amount of sound that will actually be heard when the selected fan is put in use. If more than one fan is to be used, the procedure includes a method to total these sounds. With this information on fan sound, the existing sound level in the room or enclosure (after measurement with a sound level meter) is added. With this total (in dBA), a comparison with the authorized maximum sound level can be made.

B.   Explanation of Step-by-Step Calculations

Step 1: Calculation of Room Constant (R)
The acoustical characteristics of the wall, floor, and ceiling surfaces in the room are usually the most important influences on the fan sound. The room constant R is a number that will be calculated and used in the determination of room attenuation effect.

Step 2: Determination of Directivity Factor (Q').

The proposed location for the fan in relation to adjacent room surfaces must be determined. There are four general locations that affect to varying degrees the amount of sound reaching the occupied space in the room. Knowledge of the proposed location will allow one to select the correct directivity factor.

Step 3: Measurement of Distance from Fan to Occupied Space (X)

The actual distance from the fan location to the point at which it is heard by a worker in the occupied space is also important. This distance X is utilized along with the factors determined in Steps 1 and 2 to find the total room attenuation effect on the fan sound.

Step 4: Determination of Room Attenuation Effect (RAE)

A chart (Appendix F) is provided to determine the amount of reduction (or increase) in fan sound that will occur from the factors calculated in Steps 1–3 These factors are read into the chart and the RAE (in dB) determined.

Step 5: Determination of Fan Installation Factor (FIF)

This is the final step necessary to arrive at the net fan sound in the occupied space. An FIF has been assigned by the manufacturer to each of the conventional methods of installing mechanical fans. Selection of the method proposed for the specific fan will provide the correct FIF.

Step 6: Determination of Fan Sound Actually Heard in Occupied Space (dBA)

(This is exclusive of existing background noise resulting from other sound sources.) Each fan model has been assigned a single number sound rating (LwA). This number is a logarithmic summation of the sound level for eight octave bands adjusted to represent the effect of the "A" weighted network. If it is necessary to calculate the LwA for a sound source, see Appendix G for the procedure to be used. This single-number sound-power level (LwA) provides a convenient bridge for the simplified sound-level calculations used in this fan selection procedure. The formula for arriving at the on-the-job dBA sound level of the fan is as follows:

$$\text{LwA} - \text{RAE} \pm \text{FIF} = \text{dBA of the fan}$$

in which the LwA used is from the manufacturer, the RAE is derived from Step 4, and the HF is calculated from Step 5.

Step 6a: Determination of Sound Level for More Than One Fan (Optional)

If more than one fan is to be installed in the room, the procedure outlined in Steps 1–6 should be followed for each fan. Because no variation will occur in the room constant R, substitution of the other factors is relatively simple. When the answer in Step 6 for each fan is obtained, these are added together logarithmically to obtain a total dBA for fan sounds. Appendix H provides the necessary procedure for this calculation.

Step 7: Determination of Total Sound Level Including Existing Background Noise in Occupied Space

On the assumption that fans are to be installed in an existing facility, the total sound level after the fans are in use is critical. To obtain this information, it is necessary to measure the noise level in the occupied space before the addition of the fan sound. This is easily done with a conventional sound-level meter using the "A" scale on the meter. With this measurement in dBA, the amount of fan sound determined in Step 6 (or Step 6a) is added to it. The same procedure as described in Step 6a is used (see Appendix H). The total dBA that results is then the basis for a final decision regarding the acceptability of the sound level resulting from the addition of the selected fan(s).

Step 8: Suggested Alternatives to Reduce Sound Level

(It is assumed that fans selected will add to background noise and total sound is then excessive.)

1. If the noise level has become a critical problem, it is recommended that the customer employ professional help, trained in acoustical engineering, to make recommendations.

2. OSHA Bulletin 334 contains excellent information and suggestions for reduction of noise from machinery and similar sources.
3. Fan sound can be reduced by selecting fans with a lower sound rating. It may be possible to relocate some or all of the fans to reduce the amount of sound reaching occupied space.
4. Acoustical treatment of duct work or use of baffles to trap some of the fan sound is another possible solution.

Appendix F: Method for Determination of Room Attenuation Effect (RAE) (Source: ASHRAE)

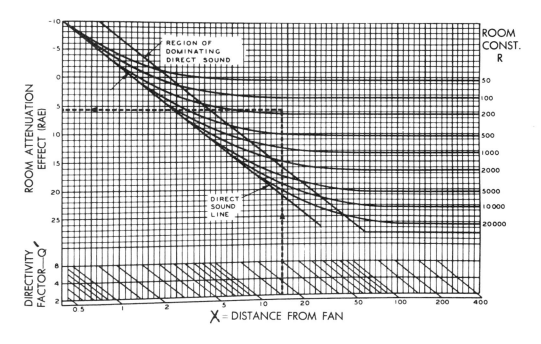

A. Procedure for Use of Chart

1. To determine the room attenuation effect (RAE) on a fan or other sound source, it is first necessary to calculate the following factors: R, room constant; Q', directivity factor; X, distance from fan to occupied space.
2. Locate the point that equals the R factor on right side of chart. Follow the curved line over to intersection with vertical line from step 3 below.
3. Locate the point at which the Q' factor on horizontal line intersects with the R factor on diagonal line and read up to the intersection of the R factor line.
4. From the intersection found in step 3 above, read the left-hand abscissa to obtain the RAE. This value is in dBA.

B. Example

Assume the following values: $R = 174$, $Q' = 4$, and $X = 5$. Read these values in chart to find RAE = 5.7.

Appendix G: Calculation of a Single-Number Sound-Power Level Adjusted to "A" Weighted Network (LwA)

Octave band	1	2	3	4	5	6	7	8
Octave band Center frequency LW[a]	63	125	250	500	1000	2000	4000	8000
Correction dB[b] LwA	−25	−15	−8	−3	0	+1	+1	−1

[a]LW, octave band sound power furnished by manufacturer of air moving devices.
[b]Correction dB. These are standard factors to weight dB to "A" weighted sound power.
Source: American Coolair Corporation.

Procedure
1. Construct a table similar to the one illustrated. Fill in all of the numbers precisely as shown in the illustration.
2. Obtain the correct sound power in octave bands for the fan or other air-moving devices and insert these numbers in line 3.
3. Apply Correction dB (line 4) and final LwA for each octave band.
4. Logarithmically add LwA values for octave bands. Use Appendix H for this purpose.

Appendix H: Determination of Composite Sound Level

Adding unequal sounds		Adding equal sounds	
(1) DBA difference between two levels	(2) dBA to add to higher level	(3) Number of equal sources	(4) Increase in dBA level
0	3.0	2	3
1	2.5	3	5
2	2.0	4	6
3	2.0	5	7
4	1.5	6	8
5	1.5	7	8.5
6	1.0	8	9
7	1.0	9	9.5
8	0.5	10	10
9	0.5		
10	0.0		

Procedure
1. When two or more sound sources are present, the composite sound level is obtained by adding the values logarithmically. Necessary values for use in this procedure are shown.
2. For unequal sounds, list the one with highest value first. List the others in order of descending value.
3. Subtract the sound source ranked second from the one ranked highest; take the dBA difference to column 1 to find correct dBA value (column 2) to add to the higher sound level. This becomes the composite dBA level for the two. Take the next two values in the working column (ranked 3 and 4) and make the same calculations. Continue this until a composite value for each pair is obtained. This procedure is then repeated for the composite values until a single total is obtained.
4. If several sound sources of equal value are to be added, the procedure can be simplified. Find the correct number of units in column 3; look opposite this number in column 4 for the increase in dBA level this number represents. The figure shown is then added to the sound level of 1.

Appendix I: Noise Absorption Coefficients of General Building Materials (Source: American Coolair Corporation)

The following list of building materials includes the more common materials used. The absorption coefficient indicated for each one is an average for the range of sound frequencies from 250 Hz to 2000 Hz.

Materials	Average absorption coefficient
Brick. unglazed	0.04
Brick, unglazed painted	0.02
Carpet, heavy, on concrete	0.30
Same on 40-oz hairfelt or foam rubber	0.35
Same, with impermeable latex backing on 40-oz hairfelt or foam in rubber	0.37
Concrete block, coarse	0.36
Concrete block, painted	0.07
Fabrics	
Light velour, 10 oz/yd^2, hung straight, in contact with wall	0.14
Medium velour, 14 oz/yd^2, draped to half-area	0.56
Heavy velour, 18/oz per sq yd^2, draped to half-area	0.58
Floors	
Concrete or terrazzo	0.01
Linoleum, asphalt, rubber, or cork tile on concrete	0.03
Wood	0.08
Wood parquet in asphalt on concrete	0.06
Glass	
Large panes of heavy plate glass	0.04
Ordinary window glass	0.15
Gypsum board, $\frac{1}{2}$ in. nailed to 2×4's 16 in. o.c.	0.07
Marble or glazed tile	0.01
Plaster gypsum or lime, smooth finish on tile or brick	0.03
Plaster gypsum or lime, rough finish on lath	0.04
Same with Smooth finish	0.03
Plywood paneling, $\frac{3}{8}$ in. thick	0.14

7

Indoor Air Pollution Control

Nguyen Thi Kim Oanh and Yung-Tse Hung

1. INDOOR AIR QUALITY: INCREASING PUBLIC HEALTH CONCERN

Indoor air pollution has occurred since prehistoric times when people moved to live indoors and fire was brought into closed shelters for cooking and space heating (1). Today, indoor air pollution caused by burning of traditional solid fuels such as woodfuel, agricultural residues, and dried animal dung in unvented cookstoves in rural areas of developing countries is not much different from that of the past. Problems associated with indoor air pollution, however, have developed a new dimension, because of energy-efficient measures (tightly constructed buildings, increased insulation, and reduced ventilation) implemented since the early 1970s in response to the oil crisis. Tightly constructed buildings, for instance, reduce the amount of fresh air for dilution and for purging out of pollutants, which, in turn, builds up high levels of toxic substances indoors. The fact that indoors air quality is not an exact reflection of the ambient air quality was recognized only recently. Scientific evidence has indicated that the air within homes and other public and office buildings can be more seriously polluted than the outdoor air. Public concern about indoor air pollution effects on health has thus attracted expanded research on the topic.

Numerous sources emit air pollution indoors, including unvented cookstoves, space heaters, tobacco smoking, and building materials. Accumulation of indoor air pollutants such as friable asbestos, formaldehyde (HCHO), radon, and so forth is increasingly recognized as causing a number of fatal diseases, including cancer. Toxic organic compounds such as formaldehyde, pesticides, polycyclic aromatic hydrocarbons (PAHs) are often found indoors at levels much higher than outdoors (2,3). In the absence of indoor sources, the concentrations of NO_x, CO, and respirable suspended particles (RSPs) indoors may be approximately the same as outdoors. With cooking and smoking in

From: *Handbook of Environmental Engineering, Volume 2: Advanced Air and Noise Pollution Control*
Edited by: L. K. Wang, N. C. Pereira and Y.-T. Hung © The Humana Press, Inc., Totowa, NJ

homes, the pollutant levels indoor may be much higher than outdoors. Only during air pollution episodes are concentrations of ambient air pollutants such as SO_2, O_3, and other oxidants higher than indoors (1,4). Because many people spend 80–90% of their lives indoors, the high level of indoor air pollutants increases the personal exposure and subsequent health effects, which probably has more impact than exposure to ambient air. In addition, those who are exposed to indoor air pollutants for longer periods of time are often the ones most susceptible, such as children, the elderly, and the chronically ill, especially those suffering from respiratory or cardiovascular disease.

In the past, ambient air quality was the only basis for assessing human exposure to air pollution outside the workplace. In spite of increasing awareness about indoor air pollution today, most air quality management measures still heavily address outdoor air issues, which may be different from those of indoor air quality. For example, most ambient air quality standards in the world do not include biological contaminants, which is an important category of indoor air pollutants. Until recently, the definition of air pollution did not include the indoor air environment. Presently, textbooks have modified the definition stating that "air pollution may be defined as the presence in the outdoor and/or indoor atmosphere of one or more contaminants or combinations thereof in such quantities and of such duration as may be or may tend to be injurious to human, plant, or animal life, or property or which unreasonably interferes with the comfortable enjoyment of life or property or the conduct of business" (5).

Solving indoor air quality problems requires a continued coordination of engineering and public health methodologies (6). Synchronized efforts of engineers, architects, public health experts, and regulatory bodies are required for controlling and mitigating indoor air pollution. Engineering aspects of indoor air pollution include the design and construction of "healthy" buildings, the design and manufacture of low-emission household materials, investigation and mitigation of potential indoor air pollution during the building operation. To achieve these goals, an integrated approach would be necessary, which requires applications of both regulatory measures, such as product emission standards or bans, and non-regulatory educative measures. These aspects will be further discussed in this chapter, with a focus on mitigation of potential health effects from indoor air pollution.

2. INDOOR AIR POLLUTION AND HEALTH EFFECTS

2.1. Sources of Indoor Air Pollution

Principal categories of indoor air pollutants consist of combustion products, chemical products, radon, and biological agents. The main issues in indoor air pollution are related to the buildup of these pollutants at high concentration resulting from the presence of emission sources and inadequate ventilation. Indoor air pollutants are mainly emitted directly from numerous indoor sources such as cookstoves, heaters, tobacco smoking, building materials, and pest control chemicals. They may also originate outdoors (e.g., soil gas from the basement, outdoor pollutants entering through improperly designed fresh air intakes [of ventilation systems], or during ambient air pollution episodes). Common indoor air pollutants and possible indoor sources are given in Table 1. Most of the listed air pollutants are also commonly found outdoors but originate from different sources such as outdoor VOCs (volatile organic compounds) from petrochemical industries or vehicles versas indoor VOCs from building materials. Not listed in Table 1 are those pollutants that are

Table 1
Indoor Air Pollutants and Emission Sources

Pollutants	Indoor sources	Average levels	Remarks
Radon	Earth, soil beneath building, building materials, well water	Indoor: 1.3 pCi/L[a] Out-door: 0.4 pCi/L	Predominantly indoor problem
Asbestos, mineral fiber, synthetic fiber	Deteriorating, damaged or disturbed ACM: fire retardant, acoustic, thermal or electric insulation	A few ng/m^3 to hundreds of ng/m^3 [a] indoor level higher than outdoor	Predominantly indoor problems
Organic pollutants: VOC, HCHO	Building materials, furnishings, and interior finishes (adhesives, solvent, paint cosmetics, wood preservatives, insulation); tobacco smoke; disinfectants, air fresheners; dry-cleaned clothing; fuel combustion; metabolic activity	Indoor VOC: a few hundreds of μg/m^3 Indoor levels two to five times that of outdoor levels[b] Indoor HCHO: a dozen to hundreds ppb, higher than outdoors	VOC: indoors and outdoors HCHO: predominantly indoor problems
Pesticides	Products for household pest control; products for lawn and gardens	Measurable levels of dozen pesticides indoors, <5 μg/m^3 [a]	Indoors and outdoors
Respirable particles	Fuel combustion in cookstoves, space heaters; tobacco smoking	Commonly 50 μg/m^3, rarely exceeds 200 μg/m^3 [a]; Smoking indoors: RSP level higher than outdoors	Indoors and outdoors
Biogenic particles	Wet or moist places indoors; carpets; poorly maintained humidifiers, dehumidifiers, and air conditioners; bedding; household pests	Outdoor levels are higher than indoor levels House dust mites: indoor level is higher than outdoor level	Indoors and outdoors
NO$_x$	Unvented kerosene and gas cookstoves and space heaters; tobacco smoking	In the absence of indoor sources, levels normally lower than outdoor levels	Indoors and outdoors
CO	Unvented cookstoves; space heaters; tobacco smoking; automobile exhaust from attached garages	0.5–30 ppm,[a] commonly indoor levels equal the outdoor levels With sources, indoor levels higher	Indoors and outdoors
CO$_2$	Unvented cookstoves; space heaters; tobacco smoking; metabolic activity	With indoor sources: up to 0.45% or higher[a]	Health effects predominantly indoors
Ammonia	Cleaning products; metabolic activities	8.1 ppb[c]	Indoors and outdoors
Lead	Lead-based paints; contaminated soil; dust from outdoor	No data available	Indoors and outdoors

Note: ACM = asbestos-containing material.
[a]data from ref. 4.
[b]data from ref. 7.
[c]data from ref. 8.

predominantly found outdoors, although they may also be emitted from indoor sources, such as SO_2 from burning of sulfur-containing fuels, soil particles from unpaved parts of floor, and O_3 from electrostatic air cleaners and photocopying machines.

The relative importance of any single source depends on source strength and the hazardous nature of pollutants emitted. Some sources such as building materials and furnishings release pollutants more or less continuously. Other sources related to activities carried out in homes release pollutants intermittently. These include tobacco smoking, operating unvented or malfunctioning stoves, furnaces, or space heaters, the use of solvents in cleaning and hobby activities, the use of paint strippers in redecorating activities, and the use of cleaning products and pesticides in housekeeping. High pollutant concentrations can remain in the indoor air long after these activities (2).

Typical indoor pollution sources are similar worldwide, the relative importance of each source category may differ from one part of the world to another. Although tobacco smoking is the common important source both in developing and developed countries, other residential combustion sources may be different. Cooking and space heating using low-grade and noncommercial solid fuels are the foremost sources of indoor air pollution in developing countries. Nearly half of the world's population uses solid fuels for cooking. Approximately 60–90% of households in developing countries, or around 3.5 billion people, still rely on biomass as the primary source of energy (9). Common lack of flues or chimneys to vent out combustion smoke and lack of ventilation devices in kitchens normally lead to high concentration of combustion-related air pollutants indoors. The indoor levels were reported to be higher than those in the ambient air and much higher than the corresponding levels in developed countries (3). Millions of people living in the rural areas of developing countries and primarily the housebound women and children, thus have a higher risk of being adversely affected by indoor air pollution.

Residential combustion using clean space heating and cooking devices such as vented gas and oil space heaters, and electric cookstoves and heaters can help to maintain "smoke-free" homes in developed countries. Chimneys installed in the wood stoves, furnaces, and fireplaces vent emission into the ambient air and hence reduce indoor air pollution levels. Nevertheless, the total emission into the environment may still be substantial. It was estimated that residential wood combustion alone accounted for more than 30% of anthropogenic PAH emission in the eastern North America (10).

Building materials, furnishings, and interior finishes are among the most important sources of indoor air pollution in developed countries. Pollutants of most concern are radon, VOCs, and biogenic particles. Buildings with energy-efficient measures that have reduced ventilation recirculate these pollutants indefinitely inside.

2.2. Health Effects of Indoor Air Pollutants

2.2.1. General Features

Indoor air pollution may have create far-reaching impact involving not only human health but also subsequent societal and economical conditions. Multimillion dollar lawsuits have been reported for compensation and building renovation (11). This chapter focuses on health effects, a driving force for indoor air quality improvement. In general, the study of environmental health effect involves many uncertainties owing the numerous variables involved. In addition to many unknown air pollutants and their synergistic and/or antagonistic effects, the indoor environment also has many other factors such as humidity, temperature, artificial lighting, and noise that may have a great effect on illness or perception of illness (11).

Strong evidence links long-term exposure to indoor air pollution with the increased risks of many respiratory illnesses. Indoor air pollution accounts for more than 34% of the global burden of disease measured by disability-adjusted life-years (DALYs) lost (9). Health effects from indoor air pollutants may be experienced soon after exposure or, possibly, years later. Immediate and short-term effects, including irritation of the eyes, nose, and throat, headaches, dizziness, and fatigue, may be observed after a single exposure or repeated exposures. These effects are normally treatable and sometimes the treatment is simply elimination of the exposure to the pollution sources. Symptoms of some diseases, including asthma, hypersensitivity pneumonitis, and humidifier fever, may also show up soon after exposure to certain indoor air pollutants. The longer-term effects can be severely weakening or fatal; they include some respiratory diseases, heart disease, and cancer. These effects may be observed years after exposure or only after long or repeated periods of exposure (2).

The seriousness of effects depends on the levels of indoor pollutants, the exposure period, and the nature of the pollutants. Sensitivity to indoor air pollutants also differs significantly from person to person. Individual reactions to indoor pollutants depend on several factors such as age, pre-existing medical conditions, and individual sensitivity to the pollutants. Children and immune-compromised people are more sensitive to pollutants and even low pollutant levels may cause irreparable effects. During the immune cycle, a normal person may also become temporarily immune compromised and suffer from abnormal toxicological effects (6).

Considerable uncertainty exists about what pollutant levels or exposure periods are necessary to produce specific health problems. Data on health effects resulting from the exposure to average normal levels of indoor air pollution is not readily available. However, health effects associated with high indoor air pollution have been documented; a few prominent cases are presented below.

2.2.1.1. LEGIONNAIRE'S DISEASE

This disease was first identified after an investigation of the death and illness of attendees of an American Legion Convention in a Philadelphia hotel in 1976. It was named after the legion members who became fatally ill (11). The disease had an indoor route of exposure and was transferred through the central air handling systems. It was characterized by pneumonialike symptoms (high fever, headache, lung consolidation, respiratory failure, and death) and found to be caused by rod-shaped bacteria, which were then named *Legionnella pneumonophila* (lung loving). The outbreak of the disease in the 1976 Legion convention was linked to aerosols drifting from a cooling tower (of the central air conditioning system) of the hotel. The bacteria thrived in the cooling tower, where the temperature was in the bacteria's favorable range. They then became aerosolized and drifted through mist eliminators (of the cooling tower) and were carried into the hotel lobby air handling system.

Any aerosol-producing devices such as cooling towers, whirlpool baths, showers, and respiratory equipment can spread the disease. In the past, a number of outbreaks, mainly in hotels or hospitals, was recorded in the United States or other parts of the developed world (4). Legionnaire's disease has a low attack rate but a high mortality rate among those affected. The bacteria also causes another nonpneumonic and nonfatal disease known as Pontiac fever. Because *Legionella pneumonophila* is commonly found in stream waters and soil, it can be widely distributed. Control measures for the disease focus on treating cooling tower and evaporative condenser water with biocidal chemicals (1).

2.2.1.2. SICK BUILDING SYNDROME

"Sick building syndrome" refers to a set of symptoms that affects a large portion of building occupants and may be larger than 30% (1), occur during the time they spend in the building, and diminish or go away during the periods when they leave the building. The symptoms are of a nonspecific nature, such as headaches, eye, and nose irritation, respiratory diseases (throat irritation, shortness of breath), neurotoxic disease (dizziness), and general fatigue and malaise. These symptoms have no identifiable causes (i.e., cannot be traced to specific pollutants or sources within the building).

"Building-related illness" refers to health complaints when a discrete, identifiable disease or illness can be traced to a specific pollutant or source in a building. The main cause of both sick building syndrome and building-related illness is energy-conserving designs compounded by poor management and ventilation. Attempts made to identify the cause–effect relationship in the "sick buildings" revealed three majors causes (listed in decreasing frequency): inadequate ventilation, chemical contamination, and microbial contamination. A study conducted on public access buildings in the United States for which illness complaints have been reported showed that 50% had ventilation problems leading to a buildup of CO_2 exceeding 1000 ppm, 23% resulting from chemicals from indoor sources (building fabric, furniture, product), 5% resulting from biological contamination, 11% resulting from outdoor pollution, and only 11% was not attributed causes (1). Because of the crucial emphasis placed on ventilation, the term "tight building syndrome" is also used interchangeably with "sick building syndrome." During the 1990s, a number of sick building syndrome cases were reported in the United States (e.g., county courthouses and school buildings) (12).

2.2.1.3. SMOKY COAL AND CARCINOGENIC MORTALITY

The high lung cancer mortality in Xuan Wei, a rural county of Yuannan province of China, has been linked to indoor smoky coal used for cooking and space heating in poorly vented houses. The unadjusted rate of annual lung cancer mortality per 100,000 in a commune where 89.7% of the population used low-quality smoky coal was 109.3, which is much higher than the rate of the reference commune of 2.3, where no smoky coal was used (13). In homes where smoky coal was used, high levels were reported for respirable particulate matter (PM_{10} of 38–39 mg/m^3 at a distance of 0.9 m from the fire), PAHs [e.g., benzo(*a*)pyrene of 10.3 µg/m^3], and mutagenicity associated with PM_{10} (14).

2.2.2. *Specific Indoor Air Pollutants and Health Effects*

2.2.2.1. RADON AND RADON PROGENY

Radon is a noble gas that is colorless, odorless, and radioactive. It is produced in the radioactive decay process of radium, which, itself, is a product of uranium decay and is found in almost all soils and rocks. Consequently, the most common source of indoor radon is uranium in the soil or rock under homes. Radon gas enters homes through dirt floors, cracks in concrete walls and floors, floor drains, and sumps. Sometimes, radon enters homes through well water. Building materials can give off radon, but they rarely cause radon problems by themselves. Thus, radon is a natural indoor contaminant, not a product of technology. Any home may have a radon problem, but the gas itself poses no hazard. However, radon undergoes radioactive decay, producing a series of short-lived progeny, which are solid particles. The decay process of radon produces α-particles and γ-rays. Some progeny (i.e., polonium-218 and polonium-214) also emit α-particles in the decay process. The major exposure route to the pollutants is through breathing air containing radon. Inhaled progeny give off the α energy particles when they decay, which can penetrate into the lung and cause damage to genetic materials and tissue injury with a high risk of lung cancer tumor development.

Radon is ubiquitous in the ambient air, with average concentrations in the range 0.20–0.25 pCi/L (4). Indoor levels may be several times or several orders of magnitude higher than the ambient. The US Environmental Protection Agency (EPA) guideline for radon is 4 pCi/L for remedial action. A radon level of 8 pCi/L is considered unacceptably high by the National Council on Radiation Protection (NCRP). A US national residential radon survey completed in the early 1990s showed the average indoor radon level to be 1.3 pCi/L, whereas the average outdoor level was about 0.4 pCi/L (2). Radon levels in the United States vary by geographical location. The highest level was reported in the northeast (4). Many studies have indicated a high frequency of indoor radon levels exceeding 4 pCi/L in US homes (1).

Although there are no immediate symptoms of exposure to radon, the predominant health effect associated with exposure to high levels of radon is lung cancer. A causal link exists between lung cancer and exposure to high radon and progeny levels among uranium and other underground miners. Laboratory studies on animals have confirmed this link. It is estimated that radon causes thousands of preventable lung cancer deaths each year. In the United States, the average annual death rate resulting from radon is 14,000 in a range from 7000 to 30,000 (2). High rates of lung cancer not related to tobacco smoking could be associated with radon progeny. There is also some evidence linking radon exposure to acute myeloid leukemia and other cancers (other than lung). It is estimated that around 6–12% of acute myeloid leukemia in Great Britain is associated with radon. The average radon level of 1.3 pCi/L in the world could be linked to 13–25% of acute myeloid leukemia cases (4). The US EPA (15) estimated risks resulting from exposure to radon at various levels are given in Table 2.

Tobacco smoking and radon produce synergistic effects resulting in a higher risk of radon-induced lung cancers. Radon progeny are electrically charged and tend to attach to suspended particles indoors rather than to walls or surrounding surfaces. In smoking areas, radon progeny are thus transported primarily by tobacco smoke particles to the lung. Upon inhalation, these radioactive particles are deposited in the respiratory airways. Cigarette smoke particles persist in the air with low mobility and are highly resistant to dissolution in the lung once deposited and hence increase the risks (1).

2.2.2.2. Pollutants from Tobacco Smoking

Tobacco smoking indoors is one of the major sources of combustion-related indoor air pollutants. Tobacco smoke is a complex mixture of both suspended particulate matter (PM) and gaseous pollutants. Some gases may also be adsorbed on PM. Tobacco

Table 2
Radon Risk Evaluation Chart

pCi/L	WL	Estimated number of lung cancer deaths as a result of radon exposure (out of 1000)	Comparable exposure levels	Comparable risk
200	1	440–700	1000 times average outdoor level	More than 60 times nonsmoker risk
100	0.5	270–630		4 pack-a-day-smoker
40	0.2	120–380	100 times average indoor level	20,000 chest X-rays per year
20	0.1	60–210	100 times average outdoor level	2 packs-a-day smoker
10	0.05	30–120	10 times average indoor level	1 packs-a-day smoker
4	0.02	13–50		5 times nonsmoker risk
2	0.01	7–30	10 times average outdoor level	200 chest X-rays per year
1	0.005	3–13	Average indoor level	Nonsmoker risk of dying from lung cancer
0.2	0.001	1–3	Average outdoor level	20 chest X-ray per year

WL: working level
Source: ref. 15.

smoke contains over 4000 compounds, more than 40 of which are known to cause cancer in humans or animals and many of which are strong irritants (2). Significant pollutants of concern are respirable suspended particulates (RSP), nicotine, nitrosamines, PAH, CO, CO_2, NO_x, acrolein, and hydrogen cyanide (4).

Pollutants from tobacco smoking are generated from the side-stream smoke (SS) (i.e., smoke given off between puffs with idling cigarets) and from the mainstream smoke (MS) (i.e., the exhaled smoke by smokers after inhalation). Quantitatively, SS smoking comprises about 55% of the mass of the cigaret column consumed. The SS is produced at low temperature and under strongly reducing conditions with more incomplete combustion products. Normally, this smoke contains higher levels of pollutants than the MS smoke. The combination of SS and MS is usually called "environmental tobacco smoke" (ETS). ETS is often referred to as "secondhand smoke" and exposure to ETS is as referred to as "passive smoking." ETS appears to be the dominant source of RSP, mostly submicron, in residential indoor environments (16). A comparison between air quality in smoking and nonsmoking indoors showed that RSP in smoking areas such as a bingo hall or bar and grill ranged from several hundreds to 1000 $\mu g/m^3$, whereas in nonsmoking areas, the levels were a few dozen micrograms per cubic meter (1).

The health effects of smoking on smokers are well documented, including lung cancer and other cancers, respiratory diseases, and cardiovascular diseases. Health effects of ETS include the combined effects of toxic pollutants present in the smoke as well as their synergistic effects with other indoor air pollutants. The World Health Organization (WHO) (17) reported that annually in developed countries, the disease burden attributable to tobacco (smoking, oral use, and ETS) is estimated at above 12% (of a total 214 million DALYs) and ranked first out of 10 selected leading risk factors. The burden is 2% (of a total 833 million DALYs) in developing countries with high mortality rates, which ranked ninth, and it is 4% (of a total 408 million DALYs) in the developing countries with low mortality rates, which ranked third out of the 10 leading risks factors. Worldwide tobacco is estimated to cause 8.8% of deaths (4.9 million) and 4.1% of DALYs (59.1 million) in a year.

Both carcinogenic and noncarcinogenic effects have been strongly linked with ETS. The health effects of ETS are more serious because of its synergistic effects with other pollutants indoors. The noncancer effects of ETS belong to two categories: cardiovascular and broncho-pulmonary diseases (16). It has been reported that exposure to ETS is responsible for approx 3000 lung cancer deaths each year in nonsmoking adults and impairs the respiratory health of hundreds of thousands of children in the United States (2). The adverse health effects are for both intrautero and extrautero children (18). Infants and young children (under 1.5 yr) whose parents smoke in their presence are at increased risk of lower-respiratory-tract infections (pneumonia and bronchitis) and are more likely to have symptoms of respiratory irritation like cough, excess phlegm, and wheeze. Asthmatic children are especially at risk. Exposure to ETS increases the number of episodes and severity of symptoms in hundreds of thousands (200,000 to 1,000,000) of asthmatic children and may cause thousands of nonasthmatic children in the country to develop the disease each year.

2.2.2.3. ASBESTOS

Asbestos is the commercial name of six naturally occurring fibrous minerals belonging to two groups: serpentine and amphibole. Chrysotile (*white asbestos*), a member of the serpentine group with a layered silicate structure, was the most common asbestos (90% of all worldwide commercial asbestos). Amphibole asbestos has a crystalline structure and comprises actinolite, amosite, anthophyllite, crocidolite, and tremolite. They differ in chemical composition, but all are hydrated silicates. Crocidolite, which is also known as blue asbestos, is commonly identified as the most dangerous mineral, but amosite (brown asbestos) may be equally harmful (19).

Asbestos possesses properties suitable for wide applications (i.e., excellent resistance to heat and fire for insulation and fire protection, friction and wear characteristics for brakes, and good mechanical strength with chemical durability to be used as reinforcing agent for various composites). In the past, asbestos was applied widely, in over 3000 applications, and most commonly for insulation and fire-retardant purposes (1). In the construction industry, it was used mostly in bound form, such as floor tiles, asbestos cements, roofing, and so forth. These asbestos-containing materials (ACMs) are unlikely to release fibers unless disturbed. The use of unbound or friable asbestos, on the other hand, has a high potential for fiber release. In the United States, the friable asbestos materials were widely used during a boom in the construction of school buildings during 1950–1973, but has been banned since 1973 both for school and large public buildings (1). Regulation of building demolition and renovation also began in 1973.

Several asbestos products have been banned in the United States and manufacturers have also voluntarily limited uses of asbestos (2). The use of asbestos other than white asbestos was banned or discouraged in most western countries for some time (19).

Because of the regulatory ban, it is unlikely that asbestos will be an indoor air pollution concern for new buildings with no ACM applications. Today, asbestos is most commonly found in older homes, pipe and furnace insulation materials, asbestos shingles, millboard, textured paints and other coating materials, and floor tiles. The presence of ACMs in homes does not itself lead to exposure to asbestos fibers. Improper attempts to remove ACMs can release asbestos fibers and increase airborne asbestos levels, hence endangering people living in those homes. Elevated concentrations of airborne asbestos fibers occur after ACMs are disturbed by cutting, sanding, or other remodeling activities. Also, over time, the friable ACMs may be damaged and sprayed-on asbestos can lose its adhesion to the surface and become airborne.

Asbestos fibers are ubiquitous both indoors and outdoors. Levels of indoor asbestos in US schools with asbestos-containing surfaces were one to two orders of magnitude higher than the ambient air level. The highest levels were found in schools with damaged surfacing materials, with a median value of over 120 ng/m^3 (1).

Asbestos fibers are thin, and long (several microns), with large length-to-width ratios. Health effects of asbestos fibers depend on chemical and physical characteristics such as size, number, and surface charge. First, they must be in the respirable size range. Once inhaled, the fibers are positioned parallel to respiratory airways. Fibers found in human lungs typically have diameters ≤ 3.5 μm. The longer the fibers, the more dangerous they are. Cancer-causing fibers (identified by Stanton and are named Stanton's fibers) have diameters $d < 1$ μm and lengths larger than a certain threshold between 5 and 20 μm (19).

Four types of disease are associated with exposure to asbestos: lung cancer, mesothelioma (cancer of the chest and abdominal linings), asbestosis (irreversible lung scarring that can be fatal), and nonmalignant pleural disease (formation of fibrous and calcified plaques). Symptoms of these diseases do not show up until many years after the initial exposure. Most people with asbestos-related diseases had occupational exposures to high concentrations; some developed disease from exposure to clothing and equipment brought home from job sites. People with occupational exposure to asbestos have five times the risk of lung cancer than nonsmokers with no professional exposure to the pollutant. Exposure to asbestos fibers and smoking has synergistic effects, with a combined risk of 50 times greater than nonexposed nonsmokers. The lifetime risk for lung cancer of school children exposed to chrysotile asbestos of 0.001 fiber/cm^3, for an average period of 6 yr, is 0.6×10^{-6} (1).

2.2.2.4. Organic Indoor Air Pollutants

Organic pollutants are one of the main causes of indoor air pollution problems. Indoor levels of about a dozen common organic pollutants were found to be two to five times higher than outdoor levels (2). VOCs form a subgroup of organic pollutants. VOCs are carbon-based organic chemicals that are present as vapors at room temperature. Hundreds of VOCs present indoors comprise a wide variety of hydrocarbons and hydrocarbon derivatives, including aliphatics, aromatics, alkylbenzenes, ketones, and chlorinated and polycyclic hydrocarbons. Based on sampling methods, VOCs are defined as organic compounds that have a lower boiling point limit between 50°C and 100°C and upper boiling limit between 240°C and 260°C (7). Organic compounds with boiling points above 400°C are solids and compounds with the boiling points in the intermediate range are semi-VOCs (4). The semi-VOCs are present indoors both in particles and in the gaseous phase.

Many household products that contain organic chemicals can release vapors indoors while they are in use or stored. Indoor VOC sources include various building materials, furnishings, paints, vanishes, solvents, adhesives, office equipment, household products, bioeffluents, and pesticides. Combustion sources such as cooking, space heating, and tobacco smoking also emit VOCs; they will be discussed separately. The human body releases bioeffluents containing various VOCs and inorganic compounds. Both the number and levels of VOCs increase in the presence of humans, with significant increases observed for acetone and ethanol (1).

Volatile Organic Compounds from building materials are released at higher rates when buildings are new and gradually decrease with time. Alkyd paint, which is frequently applied indoors, was found to release VOCs at a high rate, with over 90% emitted during the first 10 h in a small dynamic chamber emission test (20). A long-term test found an emission rate of formaldehyde from a conversion of varnishes of 0.17 mg/m²/h vanishes after 115 d (21). Some building materials can release VOCs slowly for years. In addition, many indoor materials such as carpets and gypsum boards can first adsorb/absorb VOCs, (i.e., serve as temporary sinks) and off-gas the sorbed VOCs slowly, which results in chronic and low-level exposures to VOCs (22).

Reported levels of individual VOCs (other than formaldehyde) indoors are at microgram per cubic meter levels, which are typically lower than permissible exposure limits. Total VOC (TVOC) concentrations may reach above 1 mg/m³. The extent and nature of health effects depend on chemical compounds, levels, and length of exposure periods. Some common health effect symptoms from exposure to VOCs include headaches, nausea, eyestrain, dizziness, diarrhea, rashes, persistent coughing, and insomnia and can include ailments as severe as respiratory disorders, nervous system disorders, heart problems, and cancers (12). Some VOCs may be adsorbed/absorbed to aerosols indoors. Fine particles, which are inhalable, possess large surface areas for adsorption and, hence, pose more health hazards when inhaled. Among indoor organic pollutants, formaldehyde and pesticides are of special health concern:

Formaldehyde (HCHO). Formaldehyde, a colorless and pungent-smelling gas, is one of the most important and ubiquitous organic air pollutants indoors. Formaldehyde is used widely by industries to manufacture building materials and numerous household products. In the past, synthetic urea-formaldehyde (UF) resins and phenol-formaldehyde (PF) resins were widely applied as binding materials in insulation, building materials, and numerous household products: furniture, textile containing adhesives, floor covering, and carpet backing. HCHO is also used as a preservative in some paints and coating products. In addition, combustion for cooking, space heating, and tobacco smoking also release HCHO indoors.

The most significant indoor sources of HCHO are likely to be pressed-wood products using adhesives that contain UF such as particleboard, plywood paneling, and medium-density fiberboard. The latter (used for drawer fronts, cabinets, and furniture tops) contains a higher resin-to-wood ratio than any other UF pressed-wood products and is generally recognized as being the highest formaldehyde-emitting pressed-wood product (2). Pressed-wood products contain the dark or red/black-colored PF resin, which generally release formaldehyde at considerably lower rates than those containing UF resin.

Pressed-wood products nowadays emit HCHO at levels less than 10% of those in early 1980s (4). Additionally, HCHO emissions generally decrease as products age. In homes where UF foam insulation (UFFI) was installed in the 1970s and 1980s as an energy conservation measure to cope with the energy crisis, initial indoor concentrations of HCHO could be higher than 0.4–0.5 ppm but decreased to an average of 0.06 ppm 1–2 yr. after installation. Thus, these homes are unlikely to have high levels of formaldehyde now. Homes with UFFI are not common in the United States but still may exist in other countries (4). Mobile homes and manufactured homes have been reported to have the highest levels of HCHO, varying from 0.05 to above 1 ppm with an average from 0.09 to 0.18 ppm. New mobile homes today have lower emission, with the initial indoor levels mostly below 0.20 ppm. For comparison, outdoor peak levels of HCHO observed during photochemical smog episodes are around 0.10 ppm and common levels in the ambient air are less than 0.01 ppm (4).

Formaldehyde can cause both acute and chronic health effects. At 0.1–5 ppm, it irritates the mucous membranes of the eyes and upper respiratory tract. Immediate symptoms are watery eyes, burning sensations in the eyes and throat, nausea, and difficulty in breathing for some people exposed at elevated levels. High concentrations may trigger attacks in asthma sufferers. There is evidence that some people can develop sensitivity to HCHO. The cause–effect link between HCHO and nasal cancer has been confirmed in laboratory studies using rats. Epidemiological studies of exposed workers and mobile home residents indicate that HCHO may cause cancer in humans (4).

The Occupational Safety and Health Administration (OSHA) regulated professional exposure of 8-h time-weighted average of 0.75 ppm as permissible levels with an action level of 0.5 ppm, whereas the WHO guideline value for 30 min exposure is 0.082 ppm. A number of countries have promulgated guideline values for HCHO of 0.1–0.12 ppm for homes. Some studies show that health effects may be observed even at lower levels than these guideline values (4).

Pesticides. Pesticides are widely used in and around homes with the intention to control insects (insecticides), termites (termiticides), rodents (rodenticides), fungi (fungicides), and microbes (antimicrobials/disinfectants). Insecticides and antimicrobials are most commonly used indoors. Pesticides are sold as sprays, liquids, sticks, powders, crystals, balls, and foggers. Indoor pesticides may come from contaminated soil or dust entering from outside, stored pesticide containers, wood preservatives, and household surfaces that sorb and then release pesticides. A dozen pesticides have been found indoors at measurable levels. It has been estimated that 75% of US households used at least one pesticide product indoors and that 80% of most people's exposure to pesticides occurs indoors. In 1990, the American Association of Poison Control Centers reported that some 79,000 children were involved in common household pesticide poisonings or exposures (2).

Most pesticides are semivolatile (i.e., indoors they can be present both in the gas phase and attached on aerosol particles). Pesticides are made up of both active agents and carrier agents, which are called "inerts" because they are not toxic to the targeted pests; nevertheless, some inerts are capable of causing health problems. The levels of pesticides indoors are normally less than 5 µg/m^3 (4), provided there are no misapplications.

There is insufficient understanding regarding the level at which pesticides can adversely affect health. It is important to note that pesticides are by nature toxic and should be used properly. Health concerns associated with pesticides indoors include both acute toxic effects reported immediately after applications and potential carcinogenic effects of some pesticides. Gradually, many pesticides have been banned in the world. In the United States, for example, there is no further sale or commercial use permitted for the following cyclodiene or related pesticides: chlordane, aldrin, dieldrin, and heptachlor. The only exception is the use of heptachlor by utility companies to control fire ants in underground cable boxes (2).

2.2.2.5. Biological Contaminants

These contaminants are of biological origin and are also called biogenic particles or viable particles. Numerous sources of indoor biological contaminants have been identified, including pollens originating from plants; viruses transmitted by people and animals; bacteria carried by people, animals, and soil and plant debris; and saliva and animal dander from household pets. Contaminated central air handling systems can also become breeding grounds for mold, mildew, and other biological contaminants. The systems can then distribute these contaminants throughout the home (2). Legionnaires' disease, discussed above, is a pronounced example of the effects of biological indoor air pollution.

The initial colonization of variable particles originated from both indoor and outdoor environment. In homes, these sources are located at wet or moist places, such as bathrooms, flooded basements, furniture, and poorly maintained humidifiers, dehumidifiers, and air conditioners. In the absence of indoor fungi sources, the indoor levels of pollen and fungi are lower than outdoor levels.

Biogenic particles are well-known allergens. Their allergic reactions comprise hypersensitivity pneumonitis, allergic rhinitis, and some types of asthma. These particles are also manifested as infectious illnesses, such as influenza, measles, tuberculosis, and chicken pox transmitted through the air. Also, molds and mildews release disease-causing toxins. Symptoms of health problems caused by biological pollutants include sneezing, watery eyes, coughing, shortness of breath, dizziness, tiredness, fever, and digestive problems. Children, elderly people, and people with breathing problems, allergies, and lung diseases are particularly susceptible to disease-causing biological agents in the indoor air.

House dust mites, the source of one of the most powerful biological allergens, grow in damp and warm environments. The species are of small size (250–300 µm) and translucent, hence not visible by unaided eyes. They consume skin scales of humans or animals and are commonly found in humans' dwellings. Therefore, indoor levels of house dust mites are higher than outdoor. House dust, which is heavily contaminated with the fecal pellets of these mites, is one of the most strongly allergenic materials found indoors. It is probably the most important cause of asthma in North America, as well as the major cause of common allergies (1).

2.2.2.6. Pollutants from Combustion Sources

Principally, complete combustion of carbonaceous fuels in pure oxygen releases only carbon dioxide and water vapor. Combustion of noncarbonaceous contaminants in fuels, such as sulfur and heavy metals, produces sulfur oxides, heavy metal vapors, and various oxidation products. In all practical applications, however, the combustion is incomplete, which emits various products of incomplete combustion (PIC) such as CO, soot particles (unburned carbon element), a variety of organic compounds, and PM. Nitrogen oxides, on the other hand, are the product of the combustion of nitrogen contained in fuels and nitrogen in the combustion air and are formed at high combustion temperature.

Combustion sources of indoor air pollution include open cooking fires, tobacco smoking, unvented cookstoves and space heaters, and vented appliances with improperly installed flue pipes. Emission from fireplaces and woodstoves with no adequate outdoor air supply can be "back-drafted" from the chimney into the living space.

Burning of solid fuels such as wood, agricultural waste, cattle dung, and low-quality smoky coal for cooking and space heating is the foremost source of indoor air pollution in developing countries. In rural areas of developing countries, these fuels are commonly burned in low-efficiency cookstoves such as tripods or three-stone stoves, which result in high fuel consumptions and emissions of large amount of pollutants. Moreover, the fuels are often used in poorly ventilated kitchens. As a result, the levels of many air pollutants in homes in developing countries are higher than the outdoor levels. They are also higher than indoor levels in developed countries. This is clearly demonstrated for products of incomplete combustion such as CO, PAHs, and formaldehyde (3). In homes of developing countries, PAH levels were reported in the range 100–10,000 ng/m^3 as compared to 20 ng/m^3 in traffic areas or 20–100 ng/m^3 in cigaret-smoking areas. Worldwide, burning of solid fuels indoors is estimated to cause 2.7% of DALYs. A strong link exists between burning solid fuels indoors and a number of diseases. It has been reported that solid fuel smoke causes around 35.7% of lower respiratory infections, 22% of chronic obstructive pulmonary disease, and 1.5% of trachea, bronchus, and lung cancer (17).

Polycyclic organic hydrocarbons. Among the organic compounds emitted from combustion, of special interest are polycyclic organic matters (POMs) (16). POMs is a chemical group that contains two or more benzene rings. One particular set of POMs that are known to be toxic and carcinogens or mutagens are the PAHs, which constitute a major group of carcinogens and mutagens in our environment.

Emissions of PAHs vary with combustion systems. Large-scale combustion, (e.g., in industry with a burning rate of hundreds of kilograms per hour) is normally better controlled, more complete, and results in lower formation of PAHs than small-scale combustion such as domestic cookstoves with a burning rate of a few hundred grams to a few kilograms per hour. The Emission factor of benzo(*a*)pyrene (B*a*P), a well-established carcinogenic PAH, from small-scale woodstoves can exceed that from coal, on an energy equivalent basis with a factor of 100 (16). Most PAHs are found in the gas phase of cooking smoke. The PAHs adsorbed on PM in smoke contributes only less than 10% of total PAHs emitted (23–25). It has been reported that burning coal and wood increased B*a*P and PM levels by three to five times in kitchens. Organic extracts of smoke samples from domestic cooking have demonstrated genotoxic effects in experiments (16,24). There is a general paucity of PAH emission data

Table 3
PAH and Toxicity Emission From Domestic Combustion

Fuel stove systems	PAH[a] (mg/kg)	Genotoxic PAH[b] (mg/kg)	Microtox Toxicity (10^3 TU/kg)	Mutagenicity (10^6 rev/kg fuel) (TA98+S9)
Eucalyptus—open fire[c]	110	13	—	—
Pterocarpus ind.—ceramic cookstove[d]	65	21	440	2.8
Metal heating woodstoves	73[e]–270[f]	9[e]–84[f]	—	2.9–15.8
Metal heating stove compressed wood[f]	96	28	—	—
Coal briquette—ceramic cookstove[c]	101	6.5	—	—
Peat—metal heating stove[f]	123	27	—	—
Bituminous—metal heating stove[f]	87	14	—	—
Charcoal—ceramic cookstove[c]	25	2.2	—	—
Sawdust briquette—ceramic cookstove[d]	258	20	490	1.8
Kerosene—wick cookstove[d]	67	28	460	NR
Residential oil furnaces[g]	—	—	—	0.04–0.36

Note: [a]Total 16 US EPA priority PAHs (1: naphthalene; 2: acenaphthylene; 3: acenaphthene; 4: fluorene; 5: phenanthrene; 6: anthracene; 7: fluoranthene; 8: pyrene; 9: benzo[a]anthracene; 10: chrysene; 11: benzo[b]fluoranthene; 12: benzo[k]fluoranthene; 13: benzo[a]pyrene; 14: dibenzo[a,h]anthracene; 15: benzo[g,h,i]perylene; 16: indeno[1,2,3-c,d]pyrene).

[b] Including compounds from 7–16. NR-not responsive.
[c]16 PAHs (data from ref. 23).
[d]16 PAHs (data from ref. 24).
[e]13 PAHs: 16 listed PAHs minus 1, 6, and 12 (data from ref. 16.).
[f]11 PAHs: 16 listed minus 2, 11, 12, 14, and 16 (data from ref. 16).
[g]Data from ref. 16.

TA98+S9: Ames test using the tester strain, *Salmonella typhymurium* TA98 with metabollic activation S9 (24).

for domestic combustion. Available data from literature varies greatly mainly as a result of the variations of the PAH emission on combustion conditions, fuels and stoves, and as a result of the lack of standard monitoring procedures. Table 3 presents PAH emission factors and toxicity, and genotoxicity from different domestic combustion systems.

Carbon monoxide. Carbon monoxide levels in kitchens with operating gas stoves have been reported to be in the range 10–40 ppm (4). CO levels in a residence of 236 m^3 volume with operating airtight woodstoves, at 0.36–0.48 air changes per hour (ACH), was 0.8–2.8 ppm on average and peaked at 3.8 ppm. For operating nonairtight stoves, at 0.56–0.67 ACH, the average and peak values were 1.8–14 ppm and 43 ppm, respectively (26). Carbon monoxide interferes with the oxygen-carrying capacity of blood because of its high affinity for red blood cells, which is 200–250 times higher than that of oxygen. A range of symptoms can be expected from CO exposure, such as headaches, dizziness, weakness, nausea, confusion, disorientation, and fatigue in healthy people and episodes of increased chest pain in people with chronic heart diseases. Exposure to CO concentrations of 10–50 ppm can impair a person's ability to estimate time intervals and affect visual acuity. A high CO concentration causing a number of poisoning cases in Canada from 1973 to 1982 was related to spillage and back-drafting of flue gas (1). The symptoms of CO poisoning are sometimes mistaken for those of the flu or food poisoning. Fetuses, infants, elderly people, and people with anemia or with a history of heart or respiratory diseases can be especially sensitive to CO. Exposure to CO of 50 ppm for 8 h is defined as "significant harm" by the US EPA (5).

Carbon dioxide. CO_2 is emitted from combustion and human metabolism. At a ventilation rate of 1 ACH, the steady-state CO_2 concentration in an experimental chamber with kerosene heaters was reported to be around 1% (4). Depending on occupant density and ventilation rate, levels of CO_2 indoors varies widely from the outdoor level of around 370 ppm to above 4500 ppm (27). At high concentrations, CO_2 is toxic. Exposure of healthy individuals for a prolonged period to 1.5% CO_2 apparently causes mild metabolic stress. At 7–10%, CO_2 can cause unconsciousness in a few minutes (28). The amount of CO_2 normally exhaled by an adult with an activity level representative of office workers is about 200 mL/min. Because CO_2 is an indicator of bioeffluent and is easy to measure, it is used as the basis for estimating ventilation requirements. Ventilation standards are normally set to maintain the CO_2 indoor concentration at a level that appears not to adversely affect persons with normal health (e.g., 0.25%) (28).

Nitrogen dioxide. Gas stoves and ovens are listed as the major sources of nitrogen oxides (NO_x) in residential indoor air. In homes with gas cooking, the NO_2 level may reach 18–35 ppb, whereas in homes without gas cooking, it is around 5–10 ppb (4). Without indoor sources, the ratio of indoor to outdoor NO_2 is normally less than 1. NO_2 irritates the mucous membranes in the eyes, nose, and throat and causes shortness of breath after exposure to high concentrations. There is evidence that high concentrations or continued exposure to low levels of nitrogen dioxide increase the risk of respiratory infection. Animal studies showed that repeated exposures to elevated nitrogen dioxide levels could lead or contribute to the development of lung disease such as emphysema. People at particular risk to exposure of nitrogen dioxide include children and individuals with asthma and other respiratory diseases. Exposure to a concentration of NO_2 above 150 ppm (282 mg/m^3) results in lethal effects. At 50–150 ppm (94–282 mg/m^3), NO_2 can cause lung disease such as bronchial pneumonia and bronchitis.

Particles. Wood-burning appliances emit fine particles and a range of gaseous air pollutants including CO, NO_x, SO_2, aldehydes, and VOCs. Levels of submicron particles (<0.6 μm) in a residence operating airtight woodstoves, with a ventilation rate of 0.36–0.48 ACH were reported to average 11–36 μg/m^3 and peak at 290 μg/m^3. When nonairtight stoves were operating at 0.58–0.67 ACH, the average and peak values in the residence were 210–1900 μg/m^3 and 10,000 μg/m^3, respectively (26). Tobacco smoking is also a significant source of fine particles indoors. Levels of RSP indoors with smokers was quoted to be 4.4 times of that of outdoors (4). Once inhaled, fine particles deposit in the lungs and irritate or damage lung tissues. A number of carcinogenic pollutants, including radon and PAHs, attach to fine particles and can be carried deep into the lungs.

Sulfur dioxide. SO_2 is released when fuels containing sulfur, such as coal and oil, are burned. Kerosene-burning devices indoors such as heaters and cookstoves may be significant sources of SO_2. At a ventilation rate of 1 ACH, the steady-state SO_2 concentration in an experimental chamber with kerosene heater operation was reported to be above 1 ppm. In normal conditions, indoor levels of SO_2 are normally only 30–50% of the outdoor levels (4). At levels above 1 ppm, the gas can be suffocating and irritating to the upper respiratory tract. At levels above 3 ppm, it has a pungent odor. In combination with particulate matter and moisture, SO_2 in the form of sulfate particles produces the most damaging effects.

2.2.2.7. OTHER INDOOR AIR POLLUTANTS

Ozone and lead are among other pollutants of concern. Ozone indoors may be emitted from office and domestic equipment, including photocopying machines, domestic and commercial electrostatic air cleaners, and ion generators, which will be discussed in Section 3.3.3. Health effects of ozone include damage to lung tissues and respiratory functions. Exposure to O_3 at concentration of 0.3 ppm for continuous working hours causes nose and throat irritation and chest constriction. At concentrations as high as 2 ppm, it causes severe coughing (5). Ozone can enter homes from outside through ventilated air. With severe photochemical smog (outdoors), the ozone level indoors can reach 100 ppb (8). The level of ozone indoors is normally 10–30% of that outdoors (4).

Before its toxic effects were known, lead (Pb) was used in paint, gasoline, water pipes, and many other products. Before leaded gasoline was banned, gasoline-powered vehicles were the major source of lead outdoors. Lead indoors mainly comes from old lead-based paints. Lead may also originate from certain indoor activities such as soldering and stained-glass making. There are many ways in which humans are exposed to lead. Airborne lead enters the body when an individual inhales or swallows lead particles. Harmful exposures to indoor lead can be created when lead-based paint is improperly removed from surfaces by dry scraping, sanding, or open-fire burning. High concentrations of airborne lead particles in homes can also result from lead dust from outdoor sources, including contaminated soil.

Lead affects practically all systems within the body. At high levels, it can cause convulsions, coma, and even death. Lower levels of lead can adversely affect the brain, central nervous system, blood cells, and kidneys. The effects of lead exposure on fetuses and young children can be severe. These include delays in physical and mental development, lower IQ levels, shortened attention spans, and increased behavioral problems. Children are more vulnerable to lead exposure than adults because lead is more easily absorbed into growing bodies, and the tissues of small children are more sensitive to the damaging effects of lead. Children may have higher exposure because they are more likely to get lead dust on their hands and then put their fingers or other lead-contaminated objects into their mouths. Lead in house dust is considered to be the major risk factor for high lead levels in blood in children.

3. INDOOR AIR POLLUTION CONTROL

This section focuses on the measures to mitigate indoor air pollution in existing homes and not on building design aspects. Important steps for mitigation of indoor air pollution in existing building are highlighted in Fig. 1.

3.1. Identifying Indoor Air Pollution Problems

Symptoms of illness and general health effects can be useful indicators of indoor air quality problems. However, because of the similarity between these symptoms and some common ailments such as cold or viral diseases, it is important to note the time and place the symptoms occur. If the symptoms occur when a person comes home or moves to a new residence, remodels or refurnishes a home, or treats a home with pesticides, an effort should be made to identify possible causes among indoor air pollution sources (2).

The health-effects-based approach, is a reactive approch, but the most common approach, for identifying indoor air quality problems. However, except for serious cases, most of the problems go unrecognized. A more proactive method is to identify potential sources of indoor air pollution. Although the presence of such sources does not necessarily introduce an indoor air quality problem, awareness of the type and number of potential sources is an important step toward assessing the indoor air quality. The

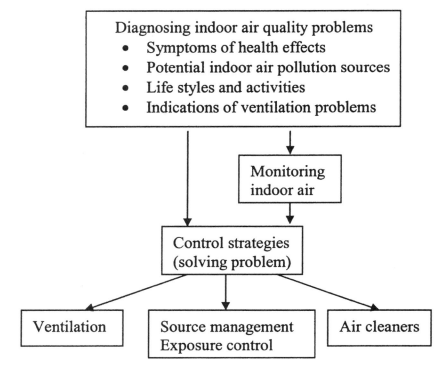

Fig. 1. Principle tasks for indoor air pollurtion management.

lifestyle and human activities of residents/occupants could also be significant sources of indoor air pollution (2), which should be examined. A third approach is to look for signs of poor ventilation in homes such as smelly or stuffy air, dirty central heating and air cooling equipment, and the presence of moisture condensation and moldy items.

3.2. Monitoring Indoor Air Quality

Monitoring indoor air quality involves taking samples and performing subsequent analyses, or conducting direct real time measurements. Monitoring is necessary to establish the potential cause–effect relationship between indoor air pollutant levels and the illness or symptoms of illness of occupants. Many indoor air pollutants such as radon or CO (the "silent killer") are colorless, odorless gases that can be identified only by monitoring. Simple measurements for radon using "do-it-yourself" radon test kits, in particular, are inexpensive; hence, the US EPA recommends that residents/occupants check the radon levels indoors (2).

Monitoring for many indoor pollutants may be expensive and requires special equipment and trained personnel. It is also normally difficult because of low levels of the pollutants. The indoor air pollution investigative protocol developed by US EPA/NIOSH (29) does not devote much attention to air monitoring. A limited number of parameters is recommended for monitoring, including CO_2 to determine ventilation adequacy, and temperature, humidity, and air movement for thermal comfort. In practice, for most pollutants, measurements are conducted when there are either health

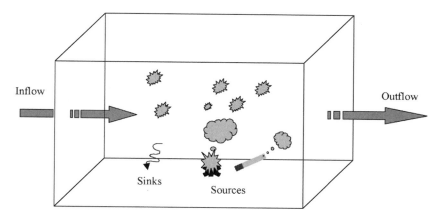

Fig. 2. Processes affecting indoor air pollutant levels.

symptoms or signs of poor ventilation, and specific sources or pollutants have been identified as possible causes of indoor air quality problems. Apart from the technical and economical difficulties, monitoring for indoor air quality may pose other social problems because of its intrusive nature, especially if monitoring is conducted in residences. Therefore, monitoring is conducted only in problematic homes with the full cooperation from owners/occupants. Passive sampling methods would be less intrusive, but the reliability of the monitoring results depends largely on the care and integrity of the occupants (1).

Routine monitoring, in principle, could be conducted for CO, total hydrocarbons, and formaldehyde. Additional pollutant categories may be included but are situation dependent, such as allergens, mycotoxin, endotoxins, pathogenic microbes, fine particles, carcinogens/mutagens, and others. To identify potential sources of allergens, comparative sampling may be conducted by simultaneously taking samples at different sites (both problem areas indoors and reference area outdoors). Various active and passive air sampling techniques and devices are commercially available (30). Examination of settled dust on some indoor surfaces or dust from a vacuum cleaner bag may be useful in many circumstances.

3.3. Mitigation Measures

Conceptual model for indoor air quality control is presented in Fig. 2. If the indoor environment is considered a box, the concentrations of indoor air pollutant C is a function of "accumulation," which is estimated as follows:

$$Accumulation = Sources + Inflow - Outflow - Sink$$

Thus, reduction of indoor air pollution level can be achieved through (1) source emission reduction, (2) reduction of pollutants inflow from outdoors and increase outflow of pollutants from indoors, and (3) increase of air pollutant sinks. Source reduction is the most effective measure and should be the first choice for indoor air pollution control (31,32). However, in many cases, the total elimination of sources may not be practical. Ventilation is the second most effective measure, but it may not always be possible to use, for example, for all weather conditions. In addition, the energy cost for heating and cooling the incoming air may be high, and in some cases, the outdoor air may be highly polluted itself (32). If the first two measures do not work, a third measure should be considered, (i.e., to increase sinks or removal of airborne pollutants through indoor air cleaning devices). However, indoor air cleaning is helpful only in combination with efforts to remove the sources.

3.3.1. Source Emission Management

For most indoor air quality problems in homes, source management is usually the most effective solution. This includes source exclusion, source removal, and source treatment (4). Source management measures are pollutant-specific. Some sources, such as those of asbestos fibers, can be sealed or enclosed; others, such as gas stoves, can be adjusted to decrease the amount of emissions. Use of increasingly available, low-emitting building materials can substantially reduce VOC levels indoors.

3.3.1.1. RADON

Proper house design during the construction phase can prevent radon from entering residences. Techniques to minimize radon entry applicable during the construction of new houses include, among others, (1) selection of construction sites with low permeable soil, (2) modification of sites by removing high-radium-containing subsoil, or cover the ground surface with low-radium-containing soil, which has high resistance to soil gas flow, and (3) designing houses with minimum radon exposure. The third technique may be realized by increasing substructure ventilation or using freestanding supporting columns such as those homes in some tropical countries, by use of high-quality mortar and concrete, impermeable plastic sealants, and waterproof basement walls.

For existing houses, modifications can also be made to minimize radon entry. Retrofit measures comprise (1) use of diffusion barrier/sealants made of nonpermeable materials to seal cracks to prevent pressure-driven inflow of radon, (2) use of suction fans to exhaust soil gas from beneath building and substructure to prevent radon entry into the basement, and (3) house pressurization to create outward airflow direction. Additionally, well water is a possible source of radon; hence, treatment of radon-contaminated well water can also eliminate this radon source (1).

3.3.1.2. ENVIRONMENTAL TOBACCO SMOKE

The most effective measure to reduce exposure to ETS is to ban smoking in homes or public buildings. The physical separation of smokers and nonsmokers in a common air space, such as different rooms within the same house, may reduce the exposure of nonsmokers to ETS. Increased ventilation in the area where smoking takes place will reduce but not eliminate exposure to ETS because of the large amounts of pollutants produced. Prohibiting smoking indoors has additional health benefit, as it also excludes the synergistic effects of smoking and other indoor air pollutants (e.g. radon).

3.3.1.3. COMBUSTION PRODUCTS IN HOMES

Source control measures for combustion appliances include (1) source modifications, (e.g., use cleaner fuels and cleaner-burning devices), (2) optimization of operation and maintenance of appliances, and (3) proper flue gas system installation to vent out smoke. Use of electrical appliances for cooking and space heating produces a much cleaner indoor environment than oil and gas stoves or woodstoves. In some cases, proper operation and maintenance of appliances can reduce emission significantly. For example, an improperly adjusted gas stove can emit significantly more carbon monoxide than a properly adjusted stove. Blocked, leaking, or damaged chimneys can release harmful combustion gases and particles and even fatal concentrations of carbon monoxide.

3.3.1.4. ASBESTOS

Source control appears to be the only applicable measure for asbestos (1). If ACMs are present in a building and in good conditions, they will not release asbestos fiber. Thus, it is usually the best to leave undamaged ACMs alone and, to the extent possible, prevent them from being damaged, disturbed, or touched. Overtime, the ACM (containing friable asbestos) ages, becomes damaged, or may loosen the adhesion to the substrate to which it was applied. Periodical inspection for damage or deterioration of ACM should be conducted. If asbestos material is damaged, or if changes are to be made in home that might disturb ACMs, professional handling is required. The US EPA website provides valuable contacts and links to the available information on-line (www.epa.gov/iaq/). Three common interim source control measures for asbestos are as follows:

> *Building operation and maintenance.* Measures related to the building operation and maintenance practices are applied when the asbestos sources have been identified but more permanent measures have not yet been used. The presence and location of ACMs in the buildings should be first identified and any damage or disturbance to it should be avoided. Precaution labels should be placed to alert workers/occupants to ACM locations. Dust removal from indoor surfaces should be done with a damp cloth, and the floor should be wet mopped instead of dry swept to minimize asbestos fiber suspension. Handling and disposal of damaged ACM items should be done properly. In some cases, the damaged ACM can be repaired easily. For example, damaged ACM thermal insulation can be fixed by patching the wrapping cloth of the insulation.
>
> *Enclosure.* This involves the construction of airtight walls and ceiling around ACMs to minimize air movement across the enclosure boundary, hence reducing the entrainment of asbestos fibers into indoor air. This method is appropriate where the ACM is located in a small area and there is an unlikelihood of entry or disturbance of the enclosed area. This low-cost method requires periodic inspections for damage.
>
> *Encapsulation.* The encapsulation method is to spray encapsulants to bind asbestos fibers in the ACM and prevent fiber release. This method should be applied when the ACM is in good condition and will not be dilapidated and fall off afterward. Encapsulation is recommended when the ACM is very difficult to remove or is troweled-on or cemented in nature, which do not pose as much exposure problem as spongy/fluffy ACMs.

The above-mentioned measures are defined as interim measures only. They do not eliminate asbestos sources on a permanent basis because they do not involve ACMs removal. Removal of ACMs is dangerous, expensive, and tedious and should be done strictly following guidelines and by experienced professionals. Site inspection and monitoring for asbestos in ambient air must be followed after ACMs removal is completed (1).

3.3.1.5. ORGANIC POLLUTANTS FROM HOUSEHOLD MATERIALS

Common measures to reduce emission from household products are of both preventive and mitigative in nature. The preventive measures include the efforts of manufacturers to produce safe products and the consumers' know-how to avoid using high-pollutant-emitting products indoors.

Development and production of low-emitting products. This should be the goal set for manufacturers. Reduction of formaldehyde emission, for example, can be achieved through a number of approaches, such as changes in resin formulation (low-emitting or nonemitting resin adhesives such as phenol formaldehyde, polyvinyl acetate, or lignosulfonates) and production variables (proper curing and production conditions to reduce free formaldehyde), addition of scavenging compounds, quality control of products, and postproduction measures (1). The addition of scavenging compounds such as urea, which chemically binds free formaldehyde and forms stable complexes, can reduce emission. A wide range of chemicals can be used for postcuring treatment. For example, urea, sodium sulfite, and sodium bisulfite applied to a hardwood plywood surface can reduce formaldehyde emission by 99%, 98%, and 100%, respectively (1). Lamination or coating of particleboard reduces formaldehyde emission by 20% to 75%, depending on coating materials and on the product types. Formaldehyde-free particleboard can be manufactured using isocyanate resins (iso-board). However, this may be a potential health hazard because of the release of isocyanate vapor.

Prevention measures by consumers. Five important steps for the reduction of source emission have been identified (7), that are relevant for the house owners who plan to install new house materials:

1. Evaluate and select low-VOC-impact building materials and products
2. Precondition materials such as carpets and furniture, to minimize emissions before installation.
3. Install building materials and products based on their VOC decay rates (aging)
4. Ventilate buildings during and after installation of new materials and products, (i.e., "flush out" for at least 7 d)
5. Delay occupancy until VOC concentrations have been reduced adequately

The first step intends to exclude/avoid products producing high VOCs indoors. In conjunction with excluding VOC-producing activities indoors, this is the best way of minimizing exposures to the pollutants in theory. To reduce indoor formaldehyde levels, products with high emission potential such as hardwood plywood, particleboard bonded with urea–formaldehyde resins, urea–formaldehyde foam insulation, and urea–formaldehyde wood finishes, should be avoided (1,4). Alternative low-HCHO-emission products are available and should be used. It is important to identify alternatives that have a lower VOC emission and do not create any other health hazards. Emission testing data should be used as references in selection of alternative products, but with caution because they may be incomplete and inaccurate because of the lack of standard VOC emission testing and reporting methods (7). The second step involves preconditioning of house materials such as carpets and furniture to minimize emissions before installation. This means storing materials in well-ventilated areas until emission is reduced. VOCs emission is normally high for new materials and decreases with time when the materials age. A proper schedule of occupancy allowing a delay period after building completion while providing good ventilation would further reduce exposure levels to VOCs. Aging new carpets and installing new carpets only when there is good natural ventilation (to accelerate aging) is recommended.

"Bake-out" is a process designed to "artificially age" house materials by increasing the temperature (35–39°C) of unoccupied newly built or renovated buildings at a normal ventilation rate for a period of time. The buildings are "flushed out" with maximum ventilation after the "bake-out" period to reduce indoor VOC levels. However, the effectiveness of "bake-out" for VOC reduction is not firmly proven by the available field trial research. In some cases, VOC was reduced immediately after "bake-out" but increased after some time; in other cases, the VOC either remained the same or even increased (7). Other problems associated with "bake-out" include (1) technical difficulties to raise temperature while providing sufficient ventilation, (2) potential damage of building materials because of the high temperature, and (3) possible adsorption of VOC on porous materials and re-emission later. In general, "bake-out" was reported to be most effective when initial VOC levels were high and the optimal temperature and "bake-out" period were attained (27).

Mitigation measures. These measures are related to the existing installations of VOC-emitting materials indoors. They consist of source removal/elimination, source treatment, and climate control. When possible, source elimination is the most effective way to control VOC. Identification of such sources is the first step.

Source treatment comprises two approaches. The first approach is an interim solution, which involves applying coating on the source surface or to place physical barriers between the sources emitting VOC and surrounding airspace. Such barriers can be alkyd resin paint, vinyl wallpaper, and paper/plastic laminate. The first two barriers were reported to reduce formaldehyde concentration indoors by more than 85% and the third barrier by 67%. Applying formaldehyde-scavenging paint on particleboard is reported to reduce emission substantially. Another method applicable to formaldehyde reduction is to apply gaseous ammonia or ammonia fumigation, notably in mobile homes. The efficiency of this method is in the range of 70–75% measured over 32 wk, and 60% measured after 3.5 yr of the treatment. The use of ammonia, however, raises concern of corrosion and cracking of brass connections on various appliances in the presence of water on the connectors. Another concern is residual ammonia accompanied by an irritating odor after treatment in mobile homes (1).

Climate control aims to manipulate environmental conditions such as temperature and humidity to reduce emission and hence concentration of VOC indoors. The VOC emission rate from house materials is a function of temperature and relative humidity (RH). Thus, theoretically, control of these environmental parameters can achieve low emission of VOCs and hence reduce the exposure level. A reduction in temperature from 30°C to 20°C (at a fixed RH of 70%) in a mobile home caused a reduction in HCHO concentration from 0.36 ppm to 0.12 ppm, (i.e., by 67%). At a fixed temperature of 20°C, a reduction in RH from 70% to 30% caused a reduction in HCHO concentration from 0.12 to 0.07 ppm (i.e., by 42%) (27). The cost of energy necessary for manipulation of temperature and humidity should also be considered.

3.3.1.6. POLLUTANTS FROM HOUSEHOLD CHEMICALS

Use of household chemicals such as pesticides, wood preservatives, cosmetics, disinfectants, air fresheners, and so forth is mostly intentional by homeowners. The applications are often discontinuous, such as for controlling termites or cockroaches. The proper choice of chemical formulations, appropriate application method, application rate, and practices can greatly reduce the hazards associated with these chemicals (1). Chemicals with high volatility will result in a high short-term concentration indoors. Spray applications would result in higher airborne concentration than controlled-release (paint-on) formulations. Pesticide problems, however, are not limited to the gas-phase pollutants. Because of their semivolatile nature, they may be associated with particulate matters.

It is essential that users of such household chemicals closely follow directions on labels. Preparation of chemicals such as mixing or dilution should be done outdoors. Indoor areas should be ventilated after chemical application and prior to entry. Integrated pest management practices for pest control should be used when possible.

Dry cleaning introduces another common exposure to chemicals indoors, notably to perchloroethylene, the most widely used chemical for dry cleaning. Perchloroethylene has been shown to cause cancer in laboratory animals. The chemical is recaptured during the dry-cleaning process for reuse and also further removed during the pressing and finishing processes. Proper handling of dry cleaning will remove the chemical from cleaned goods, leaving no strong chemical odor on the items.

3.3.2. Ventilation

Ventilation allows outdoor air to dilute indoor air pollutants and supplies sufficient oxygen for normal respiration. It also removes pollutants emitted from indoor sources by venting them out. The rate at which outdoor air replaces indoor air is described as the air exchange rate, measured as air changes per hour (ACH). Theoretically, the indoor pollutant concentration in a house is inversely proportional to the air exchange rate. Ventilation of houses occurs due to natural forces driving air through unintentional openings of houses (infiltration/exfiltration) or through opened doors and windows, and due to mechanical devices. Ventilation dissipates thermal loads in occupied spaces. Incoming outdoor air for ventilation also pressurizes houses, hence reducing the infiltration of soil gas. When there is little ventilation, the air exchange rate is low and pollutant levels increase. It is particularly important to increase ventilation during short-term activities generating high levels of pollutants, such as painting, paint stripping, heating with kerosene heaters, cooking, and so forth.

3.3.2.1. NATURAL VENTILATION

Natural ventilation is significant when windows and doors are opened and air can move through houses. This natural ventilation is not only desirable and affordable, but it is the primary means to increase air exchange and to reduce contaminant levels indoors during warm weather conditions. Home designs, which stimulate natural ventilation, should be used in tropical climates.

In the infiltration/exfiltration process, outdoor air enters a house by natural flow of infiltration through unintentional openings, joints, and cracks in walls, floors, and ceilings, and around windows and doors. Indoor air is exfiltrated through the openings. The infiltration/exfiltration rate varies depending on the tightness of the building construction and on environmental conditions, namely diurnal and seasonal ambient temperature and wind. The average ACH resulting from infiltration/exfiltration ranges from 0.1–0.2 in tight buildings to 3.0 in leaky houses, with a median value of 0.5. The larger the differences in indoor–outdoor temperature and wind speed, the higher the rate. Ventilation rates can exceed 0.5 ACH if the temperature difference is above 20°C. In heating seasons, the large difference of indoor–outdoor temperature results in a pressure difference that draws outdoor air and soil gas into the basement and forces air out of the top (exfiltration). This effect is most pronounced in tall buildings (4).

3.3.2.2. MECHANICAL VENTILATION

In mechanical ventilation, a number of devices are employed to force air through a building. These range from local exhaust fans that intermittently remove air from bathrooms or kitchens, to air handling systems that use fans and ductwork to continuously remove indoor air and distribute filtered and conditioned outdoor air throughout the house. Mechanical ventilation is most widely used in large public buildings. Mechanical ventilation is, however, more effective for episodic pollution problems such as tobacco smoking, indoor combustion activities, or constant emission such as human bioeffluents; it is less effective for pollutants such as VOCs and HCHO that are released from indoor sources through diffusion (4). When sources of high emissions are known, local exhaust ventilations can be used (e.g., to control odor from kitchen, laboratories, or lavatories). Outdoor air for ventilation should be clean so as not to introduce more pollutants indoors.

A mechanical ventilation system is designed to provide sufficient air to maintain a healthy and comfortable environment. Ventilation guidelines specify the quantity of outdoor air required per unit of time (CFM [cubic feet per minute] or L/s [liters/second]) per person to provide acceptable air quality related to human odor and comfort. The guidelines, which are provided by ASHRAE, WHO, and governmental agencies or regional collaborations (4), aim to maintain steady-state CO_2 levels at maximum design occupancy of buildings.

Energy conservation should also be taken into account when maintaining a reasonable cost for ventilation. It was reported that increasing ventilation rate up to10 L/s per person would result in a decrease in occupant health symptoms occurrence. Further increase of ventilation rate is relatively ineffective (4). A ventilation rate of 10 L/s (around 20 CFM) per person would maintain the CO_2 level at 800 ppm (27). ASHARE (33) recommends a minimum rate of 15 CFM per person in education buildings such as classroom or libraries, and a rate of 20 CFM per person for nonsmoking general office environment, dining rooms, or conference rooms. Higher ventilation rates are recommended for other environments, such as ballrooms and patient rooms in hospitals (25 CFM) or bars/cocktail lounges (30 CFM). In hotel smoking lounges, a rate of 60 CFM per person is recommended, which is also required to control high PM emission associated with smoking.

3.3.3. Indoor Air Cleaning

Two major categories of indoor air pollutants are typically considered in air-cleaning discussions: gaseous pollutants and particulate matters (PM). Air cleaners are normally designed to control PM or a group of gaseous pollutants. Certain air cleaners can remove particles and some gaseous pollutants. They cannot be applied to control all indoor airborne pollutants. Control techniques of gaseous and PM follow the general principles. Physical, chemical, and physical–chemical processes are commonly used. Some air cleaners are designed to generate charged ions or oxidizing agent (O_3) to control target indoor air pollutants.

Radon is a special group, as radon itself is a gas but its short-lived progeny are particles. A radon control principle is to either remove the radon gas itself or remove the short-lived progeny particles. Radon gas can be removed by adsorption on activated carbon (AC), although this requires a large quantity of AC and frequent regeneration of adsorption beds. Because radon progeny are electrically charged, they can become attached to aerosol particles and removed by PM control devices. Unattached radon progeny may deposit or plateout on building surfaces. Air-mixing fans can reduce radon progeny indoors by forcing them to deposit on fan blades and other surfaces. Removal of radon progeny alone will not be effective because the radon gas itself is not eliminated and will continuously produce progeny. Source control to reduce radon entry into homes is the most effective control measure, for this group of pollutants.

3.3.3.1. AIR CLEANER PERFORMANCE ASSESSMENT: EFFICIENCY VERSUS EFFECTIVENESS

Unlike the case of stack gas emission control in industry or utility, in indoor air pollution cleaning there is constant mixing between treated outlets from cleaners and untreated air present indoors. The mixture is then circulated back to the cleaners. The common term "efficiency," referring to the percentage of pollutants removed by a cleaner, is not the measure of the effectiveness of the device in reducing indoor pollution level. An important parameter determining the effectiveness of the device is the clean air delivery rate, which is the volumetric flow rate of treated air, normally in cubic feet per minute. An air cleaner of 50 CFM, which removes PM at 80% efficiency, means that the device provides 50 cubic feet of treated air per minute (with 20% PM remained) to dilute the untreated air indoors. It is essential to have enough volume of treated air for reasonable dilution of available volume of indoor air. The common parameter incorporating both the volumetric flow rate of air cleaners and the room size is the circulation rate or the number of volume air changes per unit of time, (i.e., ACH mentioned earlier). The American Lung Association (34) quoted that at the recirculation rate of 5–6 ACH, any filter with a single-pass efficiency greater than 50% can provide more than 80% reduction of the room indoor particle concentration.

Other important parameters for determining the effectiveness of cleaners include the degradation rate of the efficiency caused by loading, the placement of inlet and outlet of the devices, air discharge patterns that enhance or suppress the mixing of treated air with air indoors, and the strength of existing pollutant sources (32,34).

In the case of PM control, the weight-based removal efficiency for an overall complex mixture of particles may be misleading because it emphasizes large particles, which are not normally of health concern. There is a wide range of particle sizes found indoors, from large particles with a size of 100–150 μm in diameter to viruses (<0.1 μm). The efficiency of air cleaners for PM varies with the particle size and there is normally a submicron size for which the removal efficiency is minimum. For example, a filter, that has an overall PM removal efficiency of 95% (ASHRAE dust spot rating) may actually remove only 50–60% of particles of common size range of 0.1–1 μm found indoors. The size-specific removal efficiency (grade efficiency) for particles is thus of special interest. High-efficiency particulate air (HEPA) filters, for example, are reported to have a minimum removal efficiency of 99.97%, which is the removal efficiency for the most difficult to remove particle size, and in this case, it is 0.3 μm (27). As a matter of fact, there is a PM size at which a control device has minimum grade efficiency. This is because of differences in optimum removal mechanisms for large as compared to fine particles. The minimum and maximum removal efficiency of various air cleaners for particle sizes less than 10 μm are summarized in Table 4.

Large particles, that are suspended in the air for a short time are commonly found settled on the indoor surfaces and therefore not removed by air cleaners unless disturbed and become resuspended. Residence time of particles commonly found in air varies significantly. Overly large particles such as human hair suspend in the air for a few seconds, whereas fine particles such as tobacco smoke (0.1–1 μm) can remain suspended in the air for a few hours. It takes common pollens (15–25 μm) and mite allergens (10–20 μm) a few minutes to settle through 1 m of air (34).

Table 4
Approximate Minimum and Maximum Grade Collection Efficiency of Air Cleaners for Particles Size Range 0.03–10 μm

Air cleaners	Minimum efficiency		Maximum efficiency	
	Efficiency (%)	Size (μm)	Efficiency (%)	Size (μm)
Two-stage ESP	80	0.2	>95	>1
Charged-media filter	8	0.2	60	2
Single-stage ESP	0	0.1–0.4	35	3
Furnace-filter	0	0.03–0.4	12–13	2–4
HEPA	99.97	0.3		

Source: Adapted from ref. 34.

3.3.3.2. TYPES OF AIR CLEANERS

Many air cleaners are commercially available. These devices can be classified based on size, their removal mechanisms, or the pollutants they are designed to control. They can be incorporated into the central heating/air conditioning systems, i.e., in-duct units or as portable units. The latter ranges from small low-efficiency tabletop units to larger room consoles.

Classification by Unit Sizes

Tabletop units. A typical tabletop unit contains a fan and an electronic or other low-efficient panel filter. It normally has a small airflow and can be effective only for a small space. Usually, these devices are insufficient to clean air even for a single closed room.

Room units. These portable units have higher efficiency and larger airflow and are considerably more effective than tabletop units for cleaning air in a room. They employ various air-cleaning mechanisms with variable efficiency. In general, the units containing electrostatic precipitators (ESP), negative-ion generators, or pleated filters and hybrid systems are more effective than flat-filter units, for example, in removing cigarett smoke (32). Portable high-efficiency devices can reduce PM levels in a single room when operated continuously. One study reported the PM level reduction of 70% by a HEPA filter placed in a bedroom (27).

In-duct units. They are installed in the central air handling systems and are also referred to as the "central filtration systems." By continuously recirculating air, they provide air cleaning for an entire building. Most in-duct air cleaners have a dust spot efficiency of 90% or higher. Table 5 presents typical applications and limitations of filters according to ASHRAE's atmospheric dust spot efficiency.

Classification by Cleaning Technologies and Target Pollutants

Control devices for PM. Most air cleaners are designed to remove PM. Simple physical principles adapted from industrial air pollution control are applied to collect and remove particles from ventilation air. Fibrous filtration and electrostatic precipitation are the two primary particle collection principles employed for PM control in nonindustrial indoor air. Hybrid mechanisms may also be employed. The concentration of indoor particles is less than 50 μg/m^3 on average and rarely exceeds 200 μg/m^3 (27). Hence, the dust load is much lower than the case of industrial flue gas. The airflow rate through the cleaners is also low. Mechanical fibrous filters are by far more commonly used than electrostatic precipitators.

Control devices for gaseous pollutants. The control technologies for indoor gaseous pollutants are relatively new and complex compared to PM control. Gaseous pollutants found

Table 5
Typical Filter Applications for In-Duct Systems Based on ASHRAE Atmospheric Dust Spot Efficiency

Air cleaner efficiency rating					
10%	20%	40%	60%	80%	90%
Used in window air conditioners and heating systems	Used in air conditioners, domestic heating, and central air systems	Used in heating and air conditioning systems and as prefilters for high-efficiency cleaners	Use same as 40%, but better protection	Generally used in hospitals and controlled areas	Use same as 80%, but better protection
Useful on lint	Fairly useful on ragweed pollen	Useful on finer airborne dust and pollen	Useful on all pollens, the majority of particles causing smudge and stain, and coal and oil smoke particles	Very useful on particles causing smudge and stain, and coal and oil smoke particles	Excellent protection against all smoke particles
Somewhat useful on ragweed pollen	Not very useful on smoke and staining particles	Reduce smudge and stain materially		Quite useful on tobacco smoke particles	
Not very useful on smoke and staining particles		Slightly useful on nontobacco smoke particles	Partially useful on tobacco smoke particles		
		Not very useful on tobacco smoke particles			

Note: The atmospheric dust spot test, also described in ASHRAE Standard 52–76, is usually used to rate medium-efficiency air cleaners (both filters and electronic air cleaners). It is designed to measure the devices ability to remove relatively fine particles (<10 μm). It, however, addresses the overall efficiency of removal of a complex mixture of dust. Removal efficiencies for particles in the size range of 0.1–1 μm are much lower than the ASHRAE dust spot rating (32).
Source: ref. 32.

indoors possess different chemical properties and can be selectively removed by a particular device. Techniques employed for removal of gases from indoor or outdoor air drawn into ventilation systems include adsorption (most common), catalytic oxidation/reduction, absorption, botanical air cleaning, and ozonation.

3.3.3.3. CONTROL DEVICES FOR PARTICULATE MATTER

Mechanical Filtration. Filters range in size from small portable units to in-duct units used in air handling equipment in large buildings and homes. Fibrous media are commonly made of glass or cellulose fibers, nonwoven textile cloths, and synthesis fibers. Several physical mechanisms for PM removal are employed in filteration, including impaction, interception, and diffusion of particles toward collecting fibers. Large particles are removed by impaction, interception and gravitational settling with the efficiency increasing with particle size. Small particles are predominantly removed by diffusion due to the Brownian motion. For the intermediate particle size range, commonly from 0.05 to 0.5 μm, both interception, and diffusion mechanisms function but neither provides high collection efficiency. As the result, for each filter there is a particle size in this range at which the grade efficiency is minimum. For HEPA filters, this size is 0.3 μm, as mentioned earlier. Filter performance depends on particle size, packing density, packing depth, fiber size, and filtration velocity. Collection efficiency increases with the increase of filter media thickness, but the pressure drop also increases. Consequently, the airflow rate through filters and volume of treated air are also reduced.

Flat or panel in-duct filters. These cleaners contain low-density packing fibrous media, which may be dry or coated with viscous oil to increase particle adhesion. The units commonly have low removal efficiency, especially for respirable particle size. They collect large particles and therefore are used as dust-stop filters to protect heating, ventilation, and air conditioning (HVAC) systems, and to protect furnishings or decor rather than to reduce health effects of indoor air pollution. The pressure drop in this type of filter is small. Replacement of filter media for dry-type panel filters, for example, is done when pressure drop reaches 0.5–0.75 in. H_2O (27).

Pleated filters. These air cleaners have extended filter surface areas for particle collection, which is done by pleating the filter medium. Pleated filters can increase collection efficiency substantially while avoiding the high resistance to airflow and pressure drop typically encountered in thick, high-density filter media. The pleat media depth may be a dozen inches. The collection efficiency of pleated filters ranges from medium, to high, to very high. The dust-spot efficiency ranges from 40% to 60% for the medium-efficiency filters and 80% to above 90% for high-efficiency filters (27).

High-efficiency particulate air filters. HEPA is a special type of pleated media filter with very high efficiency. HEPA traditionally uses submicronic glass fiber, of dry-type and high-packing density, which yields the minimum grade collection efficiency of 99.97% (for particles of 0.3 μm), as discussed above. Another advantage of the traditional HEPA filter is their long maintenance-free life cycle (e.g., up to 5 yr when used with a prefilter) (34). Some new types of HEPA, using less efficient filter media are also referred to as HEPA but the collection efficiency for particles of 0.3 μm may be only 55% or less. These new HEPA filters, however, have higher airflow and lower costs than the traditional HEPA. In general, HEPA filters have a higher efficiency than other conventional filters, but their operation and maintenance cost is also higher. The latter is related to the high energy costs for operating a powerful fan and the high cost for the replacement filters.

Renewable-media filters. These filters have a filtering surface that advances in response to pressure drop or a timer, and they have found applications in HVAC systems. The media may be both dry and viscous. The replacement is made when the filter roll is used up. In the viscous type, the media can be recoated when passing through a reservoir of the viscous substance (27).

Electronic Air Cleaners. Devices such as ESPs, charged-media filters, and ion generators use an electrical field to trap particles. They may be used as portable units with fans and may be free standing or hung on ceilings. They can also be installed in central air handling systems as in-duct units.

Electrostatic precipitators. ESPs for indoor applications are the most common type of electronic air cleaners. They are widely used in residential, commercial, and office buildings. The design and operation principles are similar to ESPs used in industries. Indoor ESPs may be of single-stage or double-stage design. In the double-stage ESP, particles in the incoming airstream are charged in the first stage by ionizing wires and are collected on a series of oppositely charged metal plates in the second stage. In the single-stage ESP, the charging wires are placed between the collecting plates. The efficiency of the ESP depends on particle migration velocity, surface area of collecting plates, particle migration path length through the collection field, and volumetric airflow rate. The particle migration path length for ESP used indoors is typically 6–12 in., which is about half of that in ESPs used for industrial flue gas. The collection efficiency of ESP for indoor applications is normally less than 95%, compared to 99% of ESP used in industry. Because of the shorter path length of the particle migration, the single-stage type is less efficient than the double-stage type, but is also less expensive and requires less space (1,27). The collecting plates must be cleaned regularly to remove the collected particles. The cleaning frequency of collecting plates depends on the dust load and should be at least once every few months (34) to maintain the proper electric field strength between the collectors and the discharge electrode. A prefilter can be used to reduce the dust load to ESP and consequently the plate-cleaning frequency.

Charged-media filter. This device employs a hybrid removal mechanism (i.e., the filtration and the electrostatic forces). Hybrid systems are reputed to have better performance than mechanical filters, but they also require much higher initial equipment cost (32,34). Two variants exist: nonionizing and ionizing charged-media filters. The nonionizing type consists of a dielectric medium made from fibrous materials, which is inherently charged to form a strong electrostatic field through the medium. Particles are not ionized, but as they pass through the charged medium, they are polarized, and drawn to and get deposited on it (1). The media has to be replaced periodically.

In the ionizing charged-media filters, particles are first charged when passing through a corona discharge ionizer and then collected on a charged filter media. Thus, the sole ESP principle is employed, but the collecting plate is a charged filter medium. Several versions of the ionizing charged-media filter are available in which particles are charged negatively or alternating positively and negatively to enhance particle deposition (34). The devices require high initial equipment cost.

Negative-ion generators. These are the simplest and inexpensive type of electronic air cleaners and are available only as portable units. Several variants of ion generators are available. The simplest variant generates ions that diffuse into air indoors and attach to particles, which are then charged, attached, and deposited on the indoor surfaces. The deposited particles, however, may cause soiling problems. More advanced versions are designed to reduce surface soiling by using a collector to attract the charged particles back to the unit (34). In one type of ionizers, a suction fan, is used to draw charged particles into the air cleaner and deposit them on a charged filter panel. In another type of ionizers, the negative ions are generated in pulses and the charged particles move back to the positively charged cover of the ionizers.

Electronic air cleaners may unintentionally produce ozone. The ESP, for example, uses high voltages and can produce a significant quantity of ozone if continuously operated in closed rooms. Because of toxic effects of ozone, the US Food and Drug Administration (FDA) limits ozone emission from medical service products so that indoor ozone levels do not exceed 0.05 ppm (27).

3.3.3.4. Control Devices for Gaseous Pollutants

An in-depth discussion on the common techniques for gaseous pollutants can be found in the respective text books (e.g. ref. 5). This subsection presents the applications of the techniques for indoor air cleaning only.

Physical Adsorption. Various adsorbents are used to retain selectively gaseous pollutants on their surface. Activated carbons (AC) are the most widely used solid adsorbent for indoor gaseous pollutants control. Other absorbents such as silica gel, activated alumina, zeolites, and porous clay minerals are also frequently used.

Activated carbon has high porosity and is a nonpolar adsorbent, which permits preferential removal of organic gases from air with high moisture content. AC is effective for organic vapors with molecular weight higher than 45. In general, VOCs containing four or more atoms (exclusive of hydrogen) are considered to have reasonable carbon affinity. Organic compounds with smaller molecules are highly volatile and have lower adsorbability on AC. Thin-bed AC panels of 1–1.25 in. are normally used indoors. The beds can be made of V-shaped configurations to reduce the pressure drop. In thin-bed units used indoors, AC cannot efficiently remove formaldehyde and other light organics. However, with its catalytic abilities, AC can destroy ozone and other oxidants such as peroxides. It can also catalytically oxidize H_2S, an odorous gas, to elemental sulfur in the presence of oxygen (27), thus providing removal of these gases from indoor air.

In adsorption, the mass transfer zone or adsorption wave moves through the beds. When the zone moves to the end of the bed, the breakthrough occurs and its useful life ends. After the breakthrough, the adsorbent becomes saturated and cannot remove pollutants anymore. Instead, they may re-emit adsorbed pollutants back to indoor air. Thus, replacement of adsorbent should be done in time, (i.e., before breakthrough occurs). Used beds are discarded or may be sent back to the manufacturers for regeneration. The service time of an adsorption bed depends on pollutant properties and concentration, the adsorbent properties and quantity, and volumetric airflow rate. It can be determined using empirical equations. For real-world conditions, the determination of adsorption bed service time is more complicated. Several practical approaches may be employed for this purpose (27).

Chemisorption. In this adsorption process, chemical reactions form strong chemical bonds between the sorbed contaminants and adsorbent substrate. These reactions transform pollutants into other compounds; hence, chemisorption is not reversible. Adsorbents such as AC, activated alumina, and silica gel may be impregnated with active agents, which selectively react with gaseous pollutants. Such active agents may be catalysts, including bromine, metal oxides, element sulfur, iodine, potassium iodide, and sodium sulfide. Activated alumina has the ability to remove low-molecular-weight-organic gases. Activated alumina impregnated with $KMnO_4$ is widely applied in industrial and commercial air cleaning but not in homes. In the application, organic vapors are first retained on the surface of activated alumina and then oxidized by the impregnated $KMnO_4$. If complete oxidation occurs, only water vapor and carbon dioxide are produced, which are released back into the gas stream. However, there is the possibility that the oxidation is not complete and a number of other toxic compounds may be formed (27).

Absorption. In absorption, a scrubbing liquid, normally water or a chemically reactive liquid is sprayed over a gas stream to remove target gaseous pollutants. This technology is widely used in industrial flue gas control but not in homes.

Filsorption. This is a recent development that incorporates an adsorbent such as AC, permanganate/alumina, or zeolite into mechanical filters. A mixture of adsorbents may be used in the filter bed to remove a broader range of gaseous pollutants. The device can remove both PMs and gases and can be employed as in-duct devices in HVAC or as portable air cleaners (34).

Low Temperature Catalyst. Catalytic treatment of indoor pollutants at room temperature is an emerging technology (34). In some catalytic chemical air filters, a mixture of copper and palladium chlorides was mixed with AC to serve as the medium for a thin-bed filter. Performance studies show that the device is effective for removal of O_3, H_2S, SO_2, CO, and NH_3, but relatively ineffective for NO_x and benzene (27).

Another version of catalytic treatment, called photocatalysis, employs a catalyst (commonly titanium dioxide), which is illuminated by an ultraviolate (UV) source, and a fan that passes air over the catalyst surface where oxidation/reduction occurs. When used for VOCs the pollutants are catalytically oxidized to carbon dioxide and water vapor. Initial tests show that the technology can rapidly destroy toxic components of tobacco smoke such as formaldehyde, acrolein and benzene (34). It has been reported that at low concentrations normally found indoors, a few parts per billion, photocatalyst oxidation could remove nearly 100% of the formaldehyde and acetone (35).

Treatment of indoor air by catalytic processes at room temperature is theoretically possible but may involve a high cost. A study showed that a photocatalytic oxidation unit would have an installation cost 10 times higher and an operating annual cost 7 times higher than a granular AC unit when treating 1 m^3/s indoor air with an inlet VOC concentration of 0.27 mg/m^3, assuming that the VOCs have reasonable carbon affinity. The high cost of the photocatalyst unit is attributed to the large amount of UV energy and catalyst surface required. The technology, however, can be a good alternative for treatment of light organics when activated carbon is less efficient (36).

Ozone Generators. Theoretically, ozone as a strong oxidant can oxidize gaseous pollutants found indoors (e.g., VOCs can be oxidized to CO_2 and H_2O). Ozone generators intentionally produce ozone indoors for this purpose. The resulting level of indoor ozone depends on the device types/sizes, the number of devices placed in a room, as well as on the volume of air available for dilution and ventilation conditions. In certain circumstances, the ozone level may exceed public health standards (e.g., 8-h standards of OSHA at 100 ppb, US EPA at 80 ppb outdoors, WHO guideline at 60 ppb). Reportedly, a single unit of certain types of ozone generator operating at a high setting with interior doors closed would frequently produce O_3 levels of 200–300 ppb (32). The cited values do not include the ozone contribution from outdoor air, which is typically 20–30 ppb and in some cases may reach 30–50 ppb or higher. Ozone generators thus increase the risk of occupants being exposed to excessive ozone levels.

On the other hand, high levels of ozone are required for effective control of indoor air pollutants. This is potentially harmful to human beings. To effectively prevent survival and regeneration of biological organisms indoors, for example, the O_3 level acquired should be 5–10 times higher than the public health standards. The low levels of ozone (to be within the standards) can inhibit the growth of some biological agents but is unlikely to fully decontaminate the air (32). At low levels, ozone does not effectively remove CO (38) and formaldehyde (39). Several scientific studies showed that ozone reacts with odorous chemicals and effectively removes odors (e.g., acrolein in tobacco smoke) (32), but not all odorous chemicals, including formaldehyde, can be removed. For many indoor gaseous pollutants, the oxidation reaction by ozone may take a long time (i.e., months or years) hence practically providing no removal (an ozone generator does not remove particles unless it is combined with an ion generator in the same unit). Additionally, ozone can produce adverse effects on indoor materials such as fabrics, rubber, or plastic (5).

Little is known about byproducts formed in complex reactions of the mixture of indoor pollutants with ozone. Harmful byproducts are possibly introduced, which, in turn, may be active and capable of producing other harmful byproducts. Consequently, concentration of organics indoors may be increased, not reduced. A research study showed that the level of aldehydes increased when ozone was mixed with chemicals from new carpets, although many of the chemicals, including the odor causing, were reduced. Because of the potential harmful effects of ozone generators and their ineffectiveness in the removal of many pollutants indoors, the US EPA has not recommended ozone generators as air-cleaning devices (32). Ozone generators may only be used to generate high O_3 levels sufficient for decontamination of unoccupied spaces to control biological pollutants, odor, and some other pollutants.

Botanical Air Cleaning. This method was investigated in chamber studies. In principle, green plants uptake and hence remove various VOCs and other gases present indoors. However, this process is passive while pollutant generation is dynamic, which makes it unlikely that effective cleaning by plants can be achieved (27). To many people, green plants indoors mostly serve to enhance the decor; cleaning the air would be an added value. However, care must be taken so that the plants do not generate additional biogenic pollutants indoors.

4. REGULATORY AND NONREGULATORY MEASURES FOR INDOOR AIR QUALITY MANAGEMENT

Unlike ambient air, which as common property can be controlled by public policies and regulations, the air in buildings basically relates to property rights and privacy. This is especially true for individual residences in which exposure to indoor air pollution, to various degrees is subject to homeowner control. Hence, both voluntary and involuntary potential risks are involved. For public access buildings, in principle, the regulatory approach for indoor air quality control may be considered for the public interest and may be practiced. The regulatory measures that may be considered include setting standards for indoor air quality, product emission standards, application standards, prohibitive bans and use restrictions, warnings, and others. Each measure has its advantages and disadvantages or difficulties.

Indoor air quality standards. Setting indoor air quality standards requires scientific database for threshold values for health effects (i.e., the dose–effect relationship), which needs intensive research. Political and economic conditions should also be considered. Subsequently, the standards may not sufficiently protect the population from health hazards. Compliance assessment for indoor air quality standards would be troublesome because monitoring, with the inherent intrusive nature as discussed earlier, should be conducted. Passive monitoring is possible but not reliable. In addition, monitoring in every residence and public building would not be practically implementable. As the result, the indoor air quality standards would not be easily enforceable. Several attempts were made in the past to set indoor air quality standards for HCHO in the United States but they did not prove to be successful (27).

Product emission standard. This standard limits emission of toxic substances from products to be used indoors. Such standards can be imposed on emission of formaldehyde and other

VOCs from particleboards, paints, carpets, varnishes, and lacquers, which are produced for indoor applications. This is a relatively simple approach and manufacturers have to meet the emission standards before they can sell the products. The standards have potential for improving indoor air quality. However, the measure will not be effective for existing in-use products unless they are replaced by new products. The standard measurement methods for pollutants emissions should be developed for the enforcement. Lack of such methods for VOC emission from building materials (7), for example, means that compliance can only be encouraged on voluntary basis.

Source emission standards are available to protect ambient air, and if sources are placed indoors, the indoor air also benefits from these standards. For instance, wood-burning stoves, known to emit a high quantity of PM10 and CO, must comply with the US EPA new source standards (27), which will also help to improve indoor air quality.

Application standard. Misapplication of products results in high indoor air pollution. Misapplications of pesticides are common, resulting in adverse health effects. Thus, performance standards are necessary. Application standards are available for termiticides and other pesticides (27).

Prohibitive bans and use restrictions. Bans can be placed on friable asbestos, urea–formalde-hyde foam insulations, tobacco smoking in public assessed places, or paints with high lead content. Bans or use restrictions are also imposed on various chemicals used indoors such as chlordane for termite control and pentachlorophenol for wood preservation. Both regulatory actions and voluntary industry agreements are used to implement these measures.

Warning. Manufacturers have a legal duty to place warning labels on specific products that have hazardous or potentially hazardous emission. The label usually describes health risks associated with the product and conditions under which the product can be used safely. Typically, the label would be placed on paint strippers or lacquers, advising consumers to apply in well-ventilated places. Yet, despite warning labels placed on cigaret packages, millions of people still smoke. Thus, other measures should be used in conjunction with the warnings, such as bans and nonregulatory education measures to effectively reduce exposure risks.

Nonregulatory measures are also important for improving indoor air quality. These include health guidelines, ventilation guidelines and public information/education. Health guidelines are specified by governmental agencies or professional groups, and carry the power of scientific consensus (1), but do not have regulatory standing. They are considered to more likely to reflect true health risks and public health needs than air quality standards. Guidelines are not enforceable and compliance with them is voluntary. The most common indoor air quality health guidelines are those for indoor radon by the United States, Canada, and some European countries (27). The US EPA recommended a guideline level of indoor radon at the annual average concentration of 4 pCi/L. The Department of Housing and Urban Development (HUD) guideline values for lead are of 1 mg/cm^2 in paints, 100 µg/ft^2 in floor dust, 400 µg/ft^2 in window-sill dust, and 800 µg/ft^3 in soil. If the values are known to homeowners, they have the regulatory duty to inform potential purchasers. Ventilation guidelines are used by building design engineers as the recommended values to determine airflow rates required for new buildings at design occupancy. ASHRAE guidelines have the force of law when incorporated in building codes.

Development and operation of public information and education programs to increase awareness are important nonregulatory measures. As the responsibility of identifying and migrating indoor air quality rests mainly with homeowners and tenants, having the appropriate information and education would help them with the necessary knowledge to improve indoor air quality in their homes.

REFERENCES

1. T. Godish, *Indoor Air Pollution Control*, 3rd ed., 1991. Lewis, Chelsea, Michigan
2. US EPA, *The Inside Story: A Guide to Indoor Air Quality*, US Environmental Protection Agency and the United States Consumer Product Safety Commission, Office of Radiation and Indoor Air (6604J), EPA Document #402-K-93-007, 1995; National Center for Environmental Publications, Cincinati, OH. available from www.epa.gov/iaq/pubs/insidest.html.
3. UNEP, *Urban Air Pollution*, UNEP/GEMS Environment Library No. 4, 1991.
4. T. Godish, *Air Quality,* 3rd ed., Lewis, Boca Raton, FL, 1997.
5. K. Wark, C. F. Warner, and W. T. Davis, *Air Pollution: Its Origin and Control*, Addison-Wesley longman, Reading, MA, 1998.
6. M. Boss, D. Day, and B. Wight, *Environ. Protect.* **12** (2001).
7. L. E. Alevantis, *Reducing Occupant Exposure to Volatile Organic Compounds (VOC) from Office Building Construction Materials: Non-binding Guidelines*, California Department of Health Services. Berkeley, CA, 1999.
8. C. J. Wescheler, A. T. Hodgon , and J. D. Wooley, *Environ. Sci. Technol.* **26**, 2371–2377 (1992).
9. World Bank, *At A Glance: Indoor Air Pollution*; available from wbln0018.worldbank.org/HDNet/hddocs.nsf/.
10. US EPA, *Deposition of Air Pollutants to the Great Lakes*. Report No. EPA453/R93-055, US Environmental protection Agency, Washington, DC, 1994.
11. D. D. Buchman, *The Hidden Time Bomb in Medical Mysteries: Six Deadly Cases*, Scholastic, New York, 1992.
12. R. R. von Oppenfeld, M. E. Freeze, and S. M. Sabo, *J. Air Waste Manage. Assoc.* **48**, 995–1006 (1998).
13. J. L. Mumford, R. S. Chapman, D. B. Harris, et al., *Environ. Int.* **15**, 315–320 (1989).
14. J. L. Mumford, D. B. Harris, K. Williams, *Eniron. Sci. Technol.* **21**, 308–311 (1971).
15. US EPA, *A Citizen's Guide to Radon*, Report No. EPA-86-004, US Environmental protection Agency, Washington, DC, 1986.
16. K. R. Smith, *Biofuels, Air Pollution, and Health–A Global Review*, Plenum, New York, 1987.
17. WHO, *World Health Report 2002: Reducing Risks, Promoting Life, WHO, Geneva, 2002;* available from www.who.int/whr/en/.
18. M. D. Lebowitz, *Environ. Int.* **15**, 11–18, 1989.
19. R. C. Brown, J. A. Hoskins, and J. Young, *Chem. Brit.* October 910–915 (1992).
20. R. Fortmann, N. Roache, J. C. S. Chang, et al., *J. Air Waste Manage. Assoc.* **48**, 931–940 (1998).
21. E. M. Howard, R. C. McCrillis, K. A. Krebs, et al., *J. Air Waste Manage. Assoc.* **48**, 924–930 (1998).
22. C. S. J. Chang, L. E. Spark, Z. Guo, et al., *J. Air Waste Manage. Assoc.* **48**, 953–958 (1998).
23. N. T. Kim Oanh, L. Bætz Reutergårdh, and N. T. Dung, *Environ. Sci. Technol.* **33**, 2703–2709 (1999).
24. N. T. Kim Oanh, L. H. Nghiem, and L. P. Yin, *Environ. Sci. Technol.* **36**, 833–839 (2002).
25. B. M. Jenkins, A. D. Jones, S. Q. Turn, et al., *Atmos. Environ.* **30**, 3825–3835 (1996).
26. G. W. Traynor, M. G. Apte, A. R. Carruthers, et al., *Environ. Sci. Technol.* **21**, 691–697 (1987).
27. T. Godish, *Indoor Environmental Quality*, Lewis, Boca Raton, FL, 2000.
28. ASHRAE Standard 62-1981, *Ventilation for Acceptable Indoor Air Quality*. American Society of Heating, Refrigerating and Air-Conditioning Engineers. Atlanta, GA, 1981.
29. USEPA/NIOSH, *USEPA/NIOSH Building Investigation Protocol*, Report No. EPA/400/1-91/003, US Environmental Protection Agency, Washington, DC, 1991.

30. K. Hess, *Environmental Sampling for Unknowns*, Lewis, Boca Raton, FL, 1996.
31. US EPA, *Indoor Air Facts No. 7—Residential Air Cleaners, Office of Air and Radiation*, Washington, DC, 1990.
32. US EPA, *Residential Air Cleaning Devices: A Summary of Available Information*, Office of Air and Radiation Washington, DC, 1990.
33. ASHRAE Standard 62–1989, *Ventilation for Acceptable Indoor Air Quality*, American Society of Heating, Refrigerating and Air-Conditioning Engineers, Atlanta, GA, 1989.
34. American Lung Association, *Residential Air Cleaning Devices: Types, Effectiveness, and Health Impact.*, American Lung Association Washington, DC, www.lungusa.org/air/air00_aicleaners.html.
35. L. Stevens, J. A. Lanning, L. G. Anderson, et al., *J. Air Waste Manage. Assoc.* **48**, 979–984. (1998).
36. D. B. Henschel, *J. Air Waste Manage. Assoc.* **48**, 985–994 (1998).
37. US EPA, *Ozone Generators That Are Sold as Air Cleaners: An Assessment of Effectiveness and Health Consequences,* US Environmental Protection Agency, Washington DC, 1995; available from www.epa.gov/iaq/pubs/ozonegen.html.
38. R. J. Shaughnessy, E. Levetin, J. Blocker, et al., *Indoor Air* **4**, 179–188 (1994).
39. E. J. Esswein and M. F. Boeniger, *Appl. Occup. Environ. Hygiene* **12**, 535–542 (1994).

8

Odor Pollution Control

Toshiaki Yamamoto, Masaaki Okubo, Yung-Tse Hung, and Ruihong Zhang

CONTENTS

1. INTRODUCTION

Humans perceive odors by chemical stimulation of the chemoreceptors in the olfactory epitheliurn located in the nose. Odorants are the chemicals that stimulate the olfactory sense. This interaction between sensory cells and volatile molecules, which produces a nerve impulse, enables humans to detect and differentiate between different odors and to detect odor intensity.

Odor emissions may create serious annoyance in the neighborhood of the emission source especially in densely populated areas. A variety of common ailments may be related to exposure to odors. In most cases, the concentrations of odor-causing chemicals were well below the threshold for toxicity. The exposure to unpleasant odors may result in adverse physiological and neurogenic responses, including stress and nausea.

1.1. Sources of Odors

Many industrial, agricultural, and domestic activities may create odor nuisance. Sources of offensive odors may include the following:

- Industrial plants
- Incinerators for solid-waste processing
- Industrial and municipal wastewater-treatment plants and sewer systems
- Livestock farms

1.2. Odor Classification

Odors can be classified according to descriptors. However, to date, no unique or wholly satisfactory scheme has emerged (1). In a study of 600 organic compounds,

From: *Handbook of Environmental Engineering, Volume 2: Advanced Air and Noise Pollution Control*
Edited by:L. K. Wang, N. C. Pereira, and Y.-T. Hung © The Humana Press, Inc., Totowa, NJ

including camphor, musk, floral, peppermint, ether, pungent, and putrid (2,3), Amoore
has initially proposed seven primary odors. In subsequent studies, it was found that this
number may vary and may depend on the previous exposure of the assessor to the odor
and on the application or the product sector (1,4). *The Atlas of Odor Character Profiles*
(5) includes a list of 146 odor descriptors, whereas the American Society of Testing and
Materials (ASTM) (6) includes a comprehensive list of 830 odor descriptors.

1.3. Regulations

1.3.1. United States

In the United States, odors are regulated by states and by local governments. There
are no federal regulations for odors. For example, in Massachusetts, there was a draft
odor policy for composting facilities, which requires that new or expanding facilities not
exceed 5 dilutions/threshold (D/T) at the property line. Some states have hydrogen sulfide standards and other states have D/T limits. Many states have general nuisance lan-
guage that stipulates that odors do not create a nuisance. Nuisance is usually defined as
interfering unreasonably with the comfortable enjoyment of life and property or the
conduct of business.

1.3.2. United Kingdom

The Environment Protection Act of the United Kingdom covers laws regarding
control over odor annoyance. A comprehensive overview of the legal context of odor
annoyance in the United Kingdom was described by Salter (7). There are no general
valid emission standards concerning odor. The regulations do not address impact odor
concentrations or percentages of time, but they include the more general statements
concerning odor nuisance.

1.3.3. Germany

The Directive on Odor in Ambient Air (8) sets an impact odor concentration of 1 odor
unit (ou)/m^3, which is the limit impact concentration, and then limits the percentage of
time during which a higher impact concentration may be tolerable, which is insubstantial
annoyance. The time percentages are 15%, and 10% for industrial area and residential
areas, respectively. Frechen (9) provided more current information in Germany.

1.3.4. The Netherlands

The odor regulations in the Netherlands have the objective to keep the population free
from annoyance. The aim in 2000 was that not more than 12% of the population would
be annoyed by industrial odors. The percentage of people strongly annoyed by industrial
odors should drop below 3% in the future. The percentage results are renewed every year.

1.3.5. Belgium

There is no general law regarding odor in Belgium except a general statement that it
is forbidden to cause unacceptable annoyance by dust, smoke, odors, fumes, and so
forth. There are no specific odor regulations for various sectors. There was no specific
regulation for wastewater-treatment plant odors. However, there are minimum distances
between intensive animal farming houses and residential areas. There are general emission limits for organic and inorganic compounds. In general, the odor policy is usually
a case-to-case policy in Belgium.

1.3.6. Switzerland

The laws in Switzerland state that too high of an impact from odor is not permitted. A too high impact exists when a relevant portion of the population is significantly annoyed. A questionnaire was used to determine the annoyance levels on a scale of 1 to 10. There are no emission standards concerning odor concentration, but there are emission standards for 150 odor-causing substances. It is normally expected that no serious annoyance will occur if these standards are met.

1.3.7. Japan

National laws regarding odor prevention in Japan were first established in 1972 and have been modified several times up to the present. There are some emission standards concerning both odor concentration and odor-causing substances. As main cause substances of odor pollution, 22 substances are the targets for regulation at the present. The head of a local self-governing body or a prefecture has the authority of issuing improvement advices or commands for correcting the environment around factories and offices that have discharged odor substances.

1.4. Odor Control Methods

Before odor control methods are used, it is essential to look at the sources of the generation of odors. Prevention or reduction of odor generation at the source must be explored. Change of processes and use of different raw materials with less odor potential must be investigated before considering odor control methods.

In this chapter, various odor control methods are discussed. Two main categories of odor control consist of biological methods and nonbiological methods. Emphasis will be placed on the nonbiological methods, as biological treatment processes for air pollutant removal are covered in other chapters.

The odor control methods covered in this chapter are as follows:

- Emission control
- Air dilution
- Odor modification
- Adsorption
- Wet scrubbing or gas washing oxidization
- Incineration
- Nonthermal plasma method
- Ozone or radical injection
- Electrochemical method
- Biological methods

The advantages and disadvantages of various methods are described in this chapter. Principles of various methods are presented and several design examples are provided to illustrate the application of principles in actual conditions.

2. NONBIOLOGICAL METHOD

In order to prevent offensive odor, it is first important to choose raw materials that generate the least amount of odorous substances, to improve manufacturing process, and to decrease the emission of odorous substances from a generating source. Many generation sources of offensive odors, such as a factory, an incinerator, a sewage disposal plant, a livestock farm, a refuse incinerator plant, and a human waste-treatment plant, exist near the human environment (10–22). It is important to find the generation source, to improve the system, and to prevent the generation of odors. Air dilution, diffusion of odors, or washing with water are simple and effective methods. On the other hand, there is a method of odor modification (10,20). In this method, good smelling substances cover the odor. However, it is difficult to consider a drastic method because the application is restricted to small-scale equipment and there is also individual differences in the standard of a good smell.

When it is difficult to satisfy the public standard of offensive odor only using these policies for the generation sources, the postprocessing methods such as a diffusion method (into the atmosphere) or odor control equipment are effective in order to protect the human habitation environment. Various kinds of books and reports have reported on odors and its control method (10–28). Typical offensive odor substances and its main effective postprocessing methods are summarized in Table 1. Some of these methods have already been commercialized and extended as large-sized equipment and some methods are still being research. Hereafter, these methods are explained with some commercialized examples.

In this chapter, the principles of odor control technologies are discussed and some examples of the operation are shown.

2.1. Emission Control

Because many kinds of odor control technology for emission control or direct treatment of a generation source exist, it is difficult to explore all methods effectively. In this subsection, several common examples are discussed.

2.1.1. Local Ventilation

Ventilation is one of the most important techniques available for controlling odor levels in the workplace. Ventilation systems can be designed for commercial/residential or industrial applications. When an odorous source exists in an indoor environment, it is essential to discharge it to the outside effectively so that the odor will not remain trapped indoors. For this purpose, a local ventilation apparatus equipped with an effective hood cover (*see* Fig. 1) or air curtain (*see* Fig. 2) should be installed and operated near the source of odor.

2.1.2. Control of Direction and Pressure of Odor Stream

In factories where odor exists, zoning prevents the odor from streaming into work areas. Techniques for isolating the source of odor from the working zone, flow control between different areas, pressure control, and odor interception are employed simultaneously for air purification and prevention of dust particulates, volatile organic compounds (VOCs), and odor.

Figure 3 presents an example of zoning in an automobile painting factory (29). The paint shop is separated into zones having different levels of air pressure. This configuration allows the airflow to be regulated in order to prevent the entry of fumes, dust, and dirt into sensitive systems. Clean conditions for the entire paint shop are crucial to ensuring the glossiness and smoothness of paint films.

Table 1
Odor Control Systems and Comparison of Removal of Major Odor Components

Odor control methods	Ammonia (NH_3)	Trimethyl amine [$(CH_3)_3N$]	Hydrogen sulfide (H_2S)	Methyl mercaptan (CH_3SH)	Dimethyl sulfide [$(CH_3)_2S$]	Dimethyl disulfide [$(CH_3)_2S_2$]	Acetaldehyde (CH_3CHO)
Adsorption	A	A	A	A	A	A	A
Acid and alkali scrubbing	A	A	A	B	C	C	A
NaClO scrubbing	A	A	A	B	A	A	A
Direct incineration	A	A	A	A	A	A	A
Catalytic incineration	A	A	A	A	A	A	A
Ozone or plasma method	A	A	A	B	B	B	A
Electrochemical method	A	B	B	B	B	B	B
Biological method	A	A	A	A	A	A	A

Note: A = highly effective; B = somewhat effective; C = ineffective.

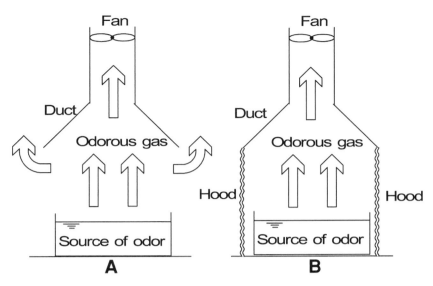

Fig. 1. Local ventilation apparatus equipped with an effective hood cover: **(a)** without hood cover; **(b)** with hood cover.

Achieving cleanliness of this high degree requires not only that the direct entry of outdoor air be prevented but also that movement of air between each sector of the paint shop be restricted. This is done through an air-pressure zoning system in which air pressure increases together with the required levels of air purity. For instance, air from the painting area may flow to the cleaning area, but not vice versa. Painting activities and car body storage are divided into separate areas. In each area, the required cleanliness is maintained through carefully balanced air pressure. Figure 4 illustrates the air-pressure ratio (29).

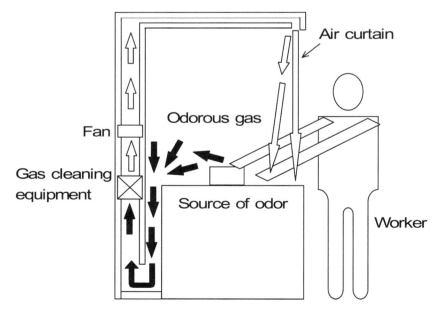

Fig. 2. Local ventilation apparatus equipped with an air curtain.

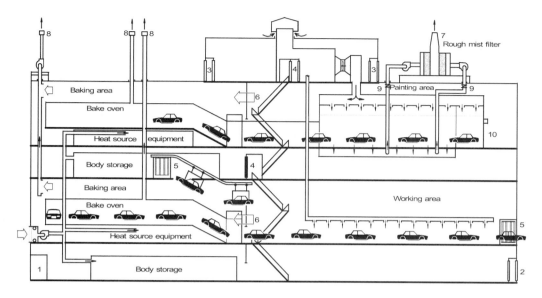

Prevention of dust and dirt during operation
1. Entrance/exit for workers: air shower
2. Passage to other factories: double door
3. Entrance/exit for outside maintenance staff: double door
4. Staircase room: double door
5. Opening for conveyor: air curtain
6. Area separation: partition wall
7. Spray booth exhaust: high-speed discharge
Prevention of dust and dirt during downtime
8. Draft prevention: air pressure damper
9. Draft prevention: interlocked operation of fire prevention damper with booth
10. Protection of equipment against dust and dirt: shutter at entrance and exit of paint booths and bake ovens

Fig. 3. Automobile painting factory with zoning. (Courtesy of Taikisha Corp.)

2.1.3. *Control of Pressure and Minimization of the Times of Ventilation*

When odor ventilation is performed in an isolated space containing gas and liquid phases, minimization of the ventilation times effectively prevents odor diffusion. Furthermore, higher gas-phase pressure in the isolated space causes the amount of generated odor substances to decrease. The principle is explained below.

The degree to which odorous gas with partial pressure p is dissolved in the solution in contact with the gas is expressed as Henry's law as follows:

$$p = C_H X \tag{1}$$

where p is a partial pressure of odorous substance in gas phase (Pa), C_H is Henry's constant (Pa / molar fraction), and X is the concentration of odor substance dissolved in the liquid phase or solution (molar fraction). This law is valid for water-insoluble gas in the range of low partial pressure. However, it does not apply for organic steam (acetone, methanol) and gases that can be easily converted to ions in the water (SO_2 and chlorine).

Using the total pressure of the gas phase p_{total} (Pa), the volumetric concentration of odorous gas Y (dimensionless) is expressed as

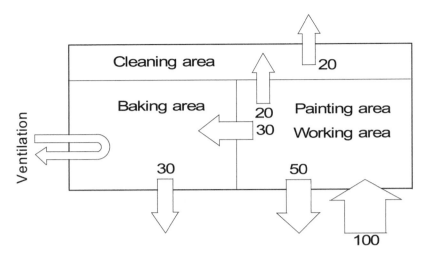

Fig. 4. Diagram of air-pressure balance and airflow direction in an automobile painting factory with zoning. (Courtesy of Taikisha Corp.)

$$Y = \frac{p}{p_{\text{total}}} = \frac{C_H}{p_{\text{total}}} X \tag{2}$$

It is known from Eq. (2) that the volumetric concentration of odorous gas Y decreases with increase in the total pressure p_{total} or increases with increase in the Henry constant C_H when the concentration of odor substance dissolved in the solution X is constant. Therefore, the generation of odor is prevented by increasing the pressure p_{total}. Henry's constants for representative odorous substances are shown in Table 2 (21).

Because the Henry constants of methyl mercaptan and hydrogen sulfide are large, they are easily released to the atmosphere from the liquid phase. A calculation example is presented in the next subsection.

2.1.4. Design Procedure Example of Pressure Control Method
Problem

Hydrogen sulfide is dissolved in water with a mass concentration of 1 ppm. Calculate the equilibrium volumetric concentration in air contacted in water. Henry's constant is $C_H = 47$ MPa/molar fraction.

Table 2
Henry's Constant of Various Odor Gas Substances in Water

Odor gas substances	Henry's constant, (MPa / molar fraction)
Methyl mercaptan	12
Hydrogen sulfide	47
Dimethyl sulfide	6.1
Dimethyl disulfide	5.1
Phenol	0.91
Trichloroethylene	0.18
Ammonia	0.061

Table 3
Influence of Hydrogen Sulfide on the Human Body

Concentration (ppm)	Influence on the human body
1000–3000	Instantaneous death
600	Fatal poisoning in 1 h
200–300	Acute poisoning in 1 h
100–200	Smell paralysis
50–100	Respiratory tract stimulus, conjunctivitis
10	Tolerate concentration
3	Unpleasant odor
0.03	Lower limit for odor

Solution

Because the molecular weights of hydrogen sulfide and water are 18 and 34, respectively, the molar fraction of 1 ppm hydrogen sulfide in water is expressed as

$$\frac{\dfrac{1\times10^{-6}}{34}}{\dfrac{1}{18}} = 5.29\times10^{-7}\,\text{molar fraction} \qquad (3)$$

The partial pressure of hydrogen sulfide in the gas phase is

$$p = (47)\times(5.29\times10^{-7}) = 2.5\times10^{-5}\ \text{MPa} \qquad (4)$$

Therefore, the gas-phase volumetric concentration at 1 atm is

$$\frac{2.5\times10^{-5}}{0.1} = 250\ \text{ppm} \qquad (5)$$

2.1.5. Chemical Treatment of an Odor Source

The odor and corrosion originating from human and industrial waste-treatment plants are often caused by the generation of hydrogen sulfide below the ground. Sulfate-reducing bacteria contained in moisture or the ground decomposes organic matter (supplied from sludge/paper/wood in the ground) from a sulfuric acid ion (supplied from gypsum board/groundwater) as an oxidization chemical, and the induced energy is consumed as self-multiplication. As a result, sulfuric acid is reduced and hydrogen sulfide is generated as follows:

$$SO_4^{2-} + \text{organic compound} \rightarrow S^{2-} + H_2O + CO_2\,(\text{by anerobic bacteria}) \qquad (6)$$

$$S^{2-} + 2H^+ \rightarrow H_2S \qquad (7)$$

Hydrogen sulfide (H_2S) is harmful to the human body and it occasionally results in death. Furthermore, it becomes sulfuric acid by oxidation and corrodes concrete. The effects of hydrogen sulfide on the human body are severe, as Table 3 shows.

To prevent chemical reactions (6) and (7), the following methods have been proposed.

1. The oxidation chemicals such as iron chloride ($FeCl_3$), iron sulfate ($FeSO_4$), and hydrogen peroxide (H_2O_2) to a liquid odor generation source are investigated. The hydrogen sulfide reacts with the chemicals to form a water-soluble salt, such as ferric chloride, iron sulfate,

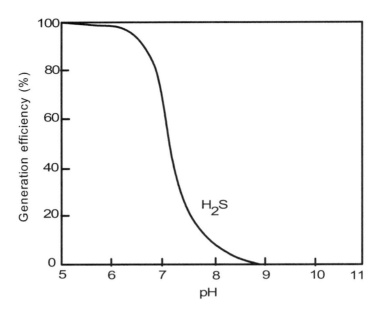

Fig. 5. Generation efficiency of hydrogen sulfide versus pH in wastewater tank.

or sulfuric acid sodium. Odor generation can be prevented, based on the following chemical reactions:

Iron chloride:	$3H_2S + 2FeCl_3 \rightarrow Fe_2S_3 + 6HCl$	(8)
Iron sulfate:	$H_2S + FeSO_4 \rightarrow FeS + H_2SO_4$	(9)
Hydrogen peroxide:	$H_2S + H_2O_2 \rightarrow S + 2H_2O$	(10)

2. The generation of hydrogen sulfide can be prevented by maintaining the pH level above 8 in the liquid phase using alkali chemicals (*see* Fig. 5).
3. Because sulfate reduction bacteria are anaerobic, their activity is weakened and the advance of a chemical reaction is prevented by air injection or aeration to the source.
4. An additive agent may be used to prevent the activity of sulfate-reducing bacteria.

The method of using an additive agent (method 4) is discussed in the following. This method has the ability to prevent the generation of hydrogen sulfide in both sewage disposal plants and industrial waste disposal facilities. In addition, because sulfate-reducing bacteria can absorb sulfuric acid and decompose organic matter simultaneously, they are especially useful for global purification. Use of strong chemicals such as $FeCl_3$, $FeSO_4$, and H_2O_2 often leads to the removal of bacteria; subsequently, extensive investment in these are not desirable.

One additive agent is marketed under the brand name Sulfur Control and has as the main ingredient anthraquinone ($C_{14}H_8O_2$) (30–32) first developed by Dupont Corporation. Sulfur Control is a water-based sludge that is transported using an ordinary pump. This additive agent acts only on the metabolism process of sulfate-reducing bacteria and controls the generation of hydrogen sulfide.

Figure 6 shows one way to decrease the hydrogen sulfide concentration generated from the sludge storage tank by adding the additive agent to the upper part of the liquid.

(Near the surface of the sludge storage tank)

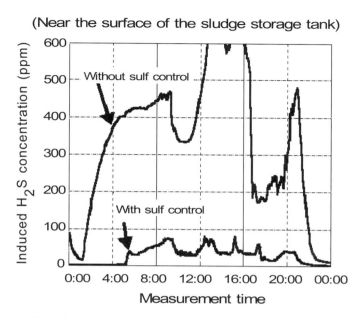

Fig. 6. Concentration of the induced hydrogen sulfide with and without Sulfur Control in the sludge storage tank. (Courtesy of Kawasaki Kasei Chemicals LTD.)

Figure 7 illustrates the result of the Sulfur Control additives in the industrial waste disposal site. In this example, there were five generating spots of hydrogen sulfide in a 10,000-m² area. At each generating point, the water-diluted additive chemical (Sulfur Control SC-381) was sprinkled. The concentration of hydrogen sulfide could be reduced by more than 95%. The effectiveness lasted for more than 6 mo.

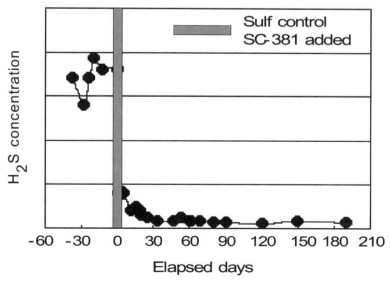

Fig. 7. Effects of Sulfur Control (SC-381) additive in the industrial waste disposal site. (Courtesy of Kawasaki Kasei Chemicals LTD.)

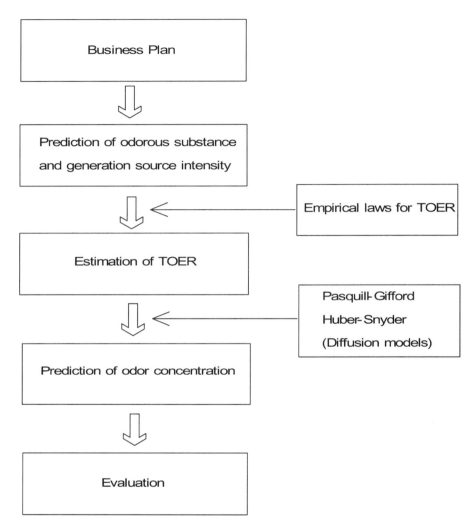

Fig. 8. Typical procedure for environmental assessment and evaluation of odor pollution.

2.2. Air Dilution

2.2.1. Procedure for Prediction of Odor Air Diffusion

Odor exhausted from factories may be removed by diffusion to the atmosphere. However, a sufficient environmental assessment is first required, especially in a residential area. Environmental assessment and evaluation are often conducted based on the procedure shown in Fig. 8.

Generally, it is difficult to strictly define odor concentration. The "triangle bag" method is used for odor sensory measurement and was first developed in Japan. In this method, three polyester gasbags containing no-odor air (odor bags) are prepared. Odor is then sampled in one of the gasbags with a certain volume. Several panelists are asked to smell that odor and judge the existence of odor. The magnification of the dilution is increased gradually, and a similar judgment is performed. The odor

Table 4
Relation Bbetween the Total Odor Emission Rate and the Range of Odor Pollution

TOER (m^3/min)	Odor pollution	Odor attainable distance (km)	Main complaint range
6 to <10	None	—	—
10^5–10^6	Small-scale influence or possibility of immanency	1–3	<500 m
10^7–10^8	Small- or middle-scale influence	3–5	<1 km
10^9–10^{10}	Large-scale influence	<10	<2–3 km
10^{11}–10^{12}	Greatest generation source	Several 10 km	<4–6 km

threshold is determined by all panelists who agree on odor sensory evolution. The final magnification times of the dilution is defined as odor concentration. In Europe or the United States, the olfactory meter is usually implemented as the measuring method of odor.

The odor emission rate (OER) is defined as the products of exhaust gas volumetric flow rate and dimensionless odor concentration at the exit of the source. The total odor emission rate (TOER) is defined by the sum of OER for two or more exits. The value of TOER is calculated for the target of the source. In a rough estimation, odor complaints usually occur in many cases when the value is greater than 10^5 m^3/min, as shown in Table 4.

In order to precisely predict odor concentration at a specified location, atmospheric diffusion should first be determined. According to the results, the limitation value of TOER should be determined so that the concentration does not exceed the threshold of the human sense of odor. This prediction method is described next.

2.2.2. Pasquille and Gifford Atmospheric Diffusion Model

Since the 1970s, semiempirical methods of predicting environmental odor concentration have been developed instead of strict analysis based on the numerical simulation of the equations of fluid mechanics. Environment assessment on odor diffusion is mainly based on these results. The method presented by Pasquille and Gifford is a prototype of this kind of prediction technique. Because this procedure is convenient, easy to understand, and can be performed easily accompanied with the investigation of weather conditions, it is often adopted as a first approximation irrespective of the size of the enterprise scale. In this subsection, this method is explained according to refs. 33–35.

When odor from a stack emission is diffused into the atmosphere, the distribution of odor concentration is assumed to be express by a normal (Gaussian) plume formula. In this case, the concentration of an arbitrary point C is expressed as

$$C = \frac{\text{TOER}}{2\pi\sigma_y\sigma_z u} \exp\left(-\frac{y^2}{2\sigma_y^2}\right)\left[\exp\left\{-\frac{(z-H_e)^2}{2\sigma_z^2}\right\} + \exp\left\{-\frac{(z+H_e)^2}{2\sigma_z^2}\right\}\right] \quad (11)$$

where TOER is the total odor emission rate (m^3/s), u is the wind velocity (m/s), H_e is the effective stack height (m), σ_y is the diffusion width in the y direction (m), and σ_z is

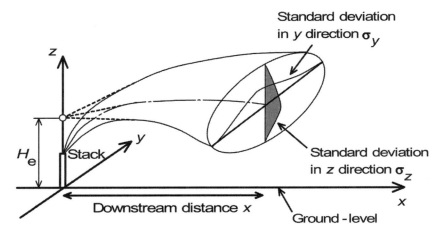

Fig. 9. The x, y, z coordinates in the odor diffusion model.

the diffusion width in the z direction (m). The x, y, z coordinates are selected as shown in Fig. 9. The concentration reaches the maximum in the downstream direction along the x-axis. The concentration on the ground or ground-level concentration can be expressed by Eq. (11) at $z = 0$.

$$C_g = \frac{\text{TOER}}{\pi \sigma_y \sigma_z u} \exp\left(-\frac{y^2}{2\sigma_y^2}\right) \exp\left(-\frac{H_e^2}{2\sigma_z^2}\right) \tag{12}$$

In order to calculate the odor concentration using Eq. (12), TOER, u, sy, sz, and He are required for a given y. The procedure of determining the ground-level concentration Cg is as follows:

1. Measure or estimate TOER, u, and H_e.
2. From Pasquille's table for atmospheric stability categories (*see* Table 5) (34), the stability category is determined for the specified date. This table has been revised according to the atmospheric conditions of each country.
3. From the Pasquille–Gifford diagram (*see* Fig. 10) or Table 6, in which the diagrams are summarized as numeric equations, the diffusion widths or the standard deviations of the concentration distribution in the downstream of the source are calculated as (36)

$$\sigma_y = \gamma_y x^{\alpha_y} \tag{13}$$

$$\sigma_z = \gamma_z x^{\alpha_z} \tag{14}$$

4. The ground-level concentration is obtained from Eq. (12).

2.2.3. Design Example for Pasquille and Gifford Model

Problem

On a day with thin cloudy cover, odor gas is exhausted at an emission rate of 1.0 m³/min from a stack with an effective height of 60 m. The odor concentration in the source is 2×10^6. When the wind velocity is 7 m/s, calculate the maximum ground-level concentration at 500 m and 4 km downstream of the stack. It is assumed that the influence of the wake behind the building can be neglected.

Table 5
Pasquille's Table for Atmospheric Stability Categories

Ground wind speed (m/s)	Insolation			Night	
	Strong	Moderate	Slight	Thinly overcast or ≥ 4/8 low cloud	≤ 3/8 Cloud
< 2	A	A–B[a]	B	–(G)	–(G)
2–3	A–B	B	C	E	F
3–5	B	B–C	C	D	E
5–6	C	C–D	D	D	D
> 6	C	D	D	D	D

Notes
1. Strong insolation corresponds to sunny midday in midsummer in England; slight insolation corresponds to similar conditions in midwinter.
2. Night refers to the period from 1 h before sunset to 1 h after dawn.
3. The neutral category D should also be used, regardless of wind speed, for overcast conditions during day or night, and for any sky conditions during the hour preceding or following night as defined in Note 2.

[a]For A − B, take average of values for A and B, and so forth.
Source: ref. 36.

Solution

It is known from Table 5 that the stability category of atmosphere is D for daytime of thin clouds. From Table 6, $\alpha_y = 0.929$, $\gamma_y = 0.1107$, $\alpha_z = 0.826$, and $\gamma_z = 0.1046$ for category D and $x = 500$ m. $\alpha_y = 0.889$, $\gamma_y = 0.1467$, $\alpha_z = 0.632$, and $\gamma_z = 0.4$ for category D and $x = 4000$ m. Therefore,

For $x = 500$ m point:

$$\sigma_y = 0.1107 \times 500^{0.929} = 36 \text{ m} \tag{15}$$

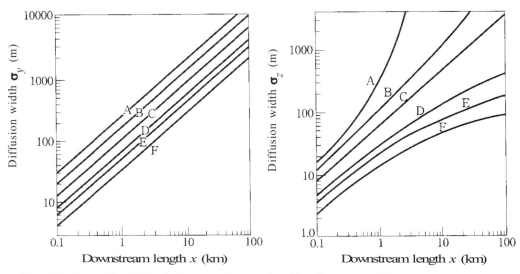

Fig. 10. Pasquille–Gifford diagram for relationships between diffusion width and downstream distance. (From ref. 36.)

Table 6
Coefficients in Approximation Equations for Diffusion Widths (σy and σz)

Stability category	α_y	γ_y	x (m)
(a) $\sigma_y\,(x) = \gamma_y x^{\alpha_y}$			
A	0.901	0.426	0–1000
	0.851	0.602	1000–
B	0.914	0.282	0–1000
	0.865	0.396	1000–
C	0.924	0.177	0–1000
	0.885	0.232	1000–
D	0.929	0.1107	0–1000
	0.889	0.1467	1000–
E	0.921	0.0864	0–1000
	0.897	0.1019	1000–
F	0.929	0.0554	0–1000
	0.889	0.0733	1000–
G	0.921	0.038	0–1000
	0.896	0.0452	1000–
(b) $\sigma_z\,(x) = \gamma_z x^{\alpha_z}$			
A	1.122	0.08	0–300
	1.514	0.00855	300–500
	2.109	0.000212	500–
B	0.964	0.127	0–500
	1.094	0.057	500–
C	0.918	0.1068	0–
D	0.826	0.1046	0–1000
	0.632	0.4	1000–10,000
	0.555	0.811	10,000–
E	0.788	0.0928	0–1,000
	0.565	0.433	1000–10,000
	0.415	1.732	10,000–
F	0.784	0.0621	0–1000
	0.526	0.37	1000–10,000
	0.323	2.41	10,000–
G	0.794	0.0373	0–1000
	0.637	0.1105	1000–2000
	0.431	0.529	2000–10,000
	0.222	3.62	10,000–

Source: ref. 36.

$$\sigma_z = 0.1046 \times 500^{0.826} = 18 \text{ m} \tag{16}$$

For $x = 4$ km point:

$$\sigma_y = 0.1467 \times 4000^{0.889} = 233 \text{ m} \tag{17}$$

$$\sigma_z = 0.4 \times 4000^{0.632} = 76 \text{ m} \tag{18}$$

The TOER is $1.0 \times 2 \times 10^6 = 2 \times 10^6$ m³/min $= 3.3 \times 10^4$ m³/s. The maximum concentration occurred at $y = 0$. Therefore, $y = 0$ and other numerical data are substituted into Eq. (12); we obtain the ground-level concentrations as

For $x = 500$ m point:

$$C_g = \frac{3.3 \times 10^4}{\pi \times 36 \times 18 \times 7} \exp(0) \times \exp\left(-\frac{60^2}{2 \times 18^2}\right) = 0.0090 \tag{19}$$

For $x = 4$ km point:

$$C_g = \frac{3.3 \times 10^4}{\pi \times 233 \times 76 \times 7} \exp(0) \times \exp\left(-\frac{60^2}{2 \times 76^2}\right) = 0.062 \tag{20}$$

In order to prevent odor pollution, the ground-level concentration should be much lower than 1 at the specified point. Other examples of calculations using these equations are presented in ref. 37.

2.2.4. Huber and Snyder Atmospheric Diffusion Model

When odor diffusion exhaust from a stack near a building is investigated in detail, the effects of obstacles, such as the wake of the building and the stack, must be considered. When a high building is near the stack, a turbulent flow will become intense and smoke will be drawn into this turbulent flow region. Therefore, the ground-level concentration of the smoke becomes remarkably high. This phenomenon has been well known for many years.

In order to develop the model of this phenomenon, many field tests and wind tunnel experiments have been conducted since the 1970s in order to understand the behavior of smoke streams behind buildings, or the width of smoke diffusion. The empirical law that the height of the stack should be generally 2.5 times higher than that of the building was driven by the collection of such experimental results. However, there are many cases in which smoke is directly emitted from the building because various restrictions do not allow installing a sufficiently high stack. In these cases, it is important to predict the concentration of odor near the source, and a semiempirical Huber and Snyder diffusion model considering the building wake effect is often used.

Huber and Snyder (38,39) performed a series of experiments on flow measurements using many types of rectangular building model at a US EPA wind tunnel, and they proposed a diffusion model with the building wake effect. The results are summarized as an empirical formula by modifying the standard deviation in the method shown in the Section 2.2.2. The result is described below according to the original articles (38,39) and a review (40). In the range shown in Fig. 11, modification of the diffusion width is needed. The procedure is as follows:

1. Calculation of plume principal axis height: The plume principal axis height H is calculated as

$$H = H_0 + \Delta H - \Delta H' - \Delta H'' \tag{21}$$

where H_0 is the height of the stack, ΔH is the height of rising plume without downwash, $\Delta H'$ is the decrease in plume principal axis height resulting from turbulence behind the stack, and $\Delta H''$ is the decrease in plume principal axis height resulting from turbulence behind the building. The involvement phenomenon of the smoke in the peak of the stack is called the downwash.

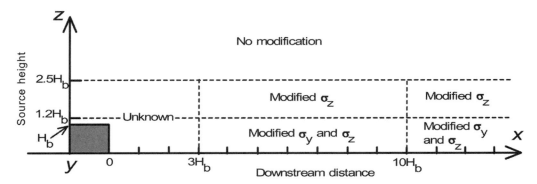

Fig. 11. Suggested use for modified dispersion equations. (From ref. 38.)

(i) The height of rising exhaust gas without downwash, ΔH, can be calculated from the following Briggs's equation. Here, the equation without thermal buoyancy effect is shown.

$$\Delta H = 3\frac{V_s}{u}D \tag{22}$$

where V_s is the velocity of the odorous exhaust, u is the wind velocity, and D is the diameter of stack.

(ii) The decrease in plume principal axis height by the stack, $\Delta H'$, is calculated from

$$\begin{aligned} V_s \geq 1.5u \qquad & \Delta H' = 0 \\ V_s < 1.5u \qquad & \Delta H' = 2\left(1.5 - \frac{V_s}{u}\right)D \end{aligned} \tag{23}$$

(iii) The decrease in the plume principal axis height as a result of turbulence behind the building, $\Delta H''$, is calculated from

$$\begin{aligned} H_0/H_b \leq 1.2 \qquad & \Delta H'' = 0.333\Delta H \\ 1.2 < H_0/H_b \leq 2.5 \qquad & \Delta H'' = 0.333\Delta H - \left[\left(\frac{H_0}{H_b} - 1.2\right)(0.2563\Delta H)\right] \\ H_0/H_b > 2.5 \qquad & \Delta H'' = 0 \end{aligned} \tag{24}$$

where H_b is the height of the building.

2. The width of the diffusion in the vertical direction, σ_z, is calculated from

$$\begin{aligned} x < 3L_b \qquad & \sigma_z = 0.7L_b \\ 3L_b \leq x \leq 10L_b \qquad & \sigma_z = 0.7L_b + 0.067(x - 3L_b) \\ x > 10L_b \qquad & \sigma_z = \gamma_z(x + x_0)^{\alpha_z} \end{aligned} \tag{25}$$

where L_b is the shorter dimension of either the height of building or the width of building, x is distance in the downstream, x_0 is the distance to the virtual smoke source (in the Pasquille and Gifford diagram as shown in Fig. 10, the difference between $10L_b$ and x for $\sigma_z = 1.2L_b$)

3. The width of the diffusion in the horizontal direction, σ_y, is calculated from

$$H_0/H_b < 1.2; \qquad \sigma_y = \gamma_y x^{\alpha_y} \tag{26}$$

$$H_0/H_b \geq 1.2$$

$$x < 3L_b \qquad \sigma_y = 0.35L_b'$$

$$3L_b \leq x \leq 10L_b \qquad \sigma_y = 0.35L_b' + 0.067(x - 3L_b) \qquad (27)$$

$$x > 10L_b \qquad \sigma_y = \gamma_y \left(x + x_0'\right)^{\alpha_y}$$

where $L_b' = W_b$ for $W_b < 10H_b$ and $L_b' = H_b$ for $W_b \geq 10H_b$, x_0' is the distance to the virtual smoke source (in the Pasquille and Gifford diagram, the difference between $10L_b$ and x for $\sigma_y = 0.35L_b' + 0.5L_b$)

4. The ground-level odorous concentration C_g is calculated from an equation similar to Eq. (12):

$$C_g = \frac{\text{TOER}}{\pi \sigma_y \sigma_z u} \exp\left(-\frac{y^2}{2\sigma_y^2}\right) \exp\left(-\frac{H^2}{2\sigma_z^2}\right) \qquad (28)$$

The prediction method of odor diffusion in the atmosphere is effective as a simple calculation technique for the environmental assessment of odor. In a more detailed analysis, it is necessary to evaluate the thermal convection diffusion by building heat and so forth. Moreover, in a prediction for broader area, the effect of the geographical feature cannot be ignored. For various kinds of odor generation source, only a single stack can be treated as a point generation source. It is necessary to consider a waste disposal place as a plane generation source, and some industrial factories as a spatial generation source. Although a simple prediction method for these kinds of generation source does not exist presently, the method of integrating the results with a point source is usually used. Numerical simulations of the fluid mechanics equations are effective, but they require a significant amount of labor to perform an exact and valid prediction of atmosphere (41,42).

2.2.5. Design Example for Huber and Snyder Model

Problem

During days with thin clouds, odor gas is exhausted with the emission rate of 1.0 m³/min from a stack. The plume principal axis height is 60 m. The stack on the building has a $H_b = 50$ m height and $W_b = 100$ m width. The odor concentration in the source is 2×10^6 If the wind velocity is 7 m/s, calculate the maximum ground-level concentration in 500 m and 4 km points downstream of the stack. It is assumed that the influence of the wake behind the building cannot be neglected.

Solution

From Fig. 10, Table 6, and Eqs. (25)–(27), we obtain the length $L_b = 50$ m, $x_0 = 1500$ m, $L_b' = 100$ m, and

For $x = 500$ m point:

$$\sigma_y = 0.35 \times 100 + 0.067 \times (500 - 150) = 58 \text{ m} \qquad (29)$$

$$\sigma_z = 0.7 \times 50 + 0.067 \times (500 - 150) = 58 \text{ m} \qquad (30)$$

For $x = 4$ km point:

$$\sigma_y = 0.1467 \times (4000 + 500)^{0.889} = 259 \text{ m} \qquad (31)$$

$$\sigma_z = 0.4 \times (4000 + 1500)^{0.632} = 92 \text{ m} \qquad (32)$$

The ground-level concentrations are

For $x = 500$ m point:

$$C_g = \frac{3.3 \times 10^4}{\pi \times 58 \times 58 \times 7} \exp(0) \times \exp\left(-\frac{60^2}{2 \times 58^2}\right) = 0.26 \tag{33}$$

For $x = 4$ km point:

$$C_g = \frac{3.3 \times 10^4}{\pi \times 259 \times 92 \times 7} \exp(0) \times \exp\left(-\frac{60^2}{2 \times 92^2}\right) = 0.050 \tag{34}$$

In the $x = 500$ m point, the concentrations become high compared with the results obtained in Subsection 2.2.3 because of the involvement of smoke.

2.3. Odor Modification

Odor modification involves the addition of an odorous substance to a given odor emission in order to reduce the odor intensity or to change the odor character to one that is less objectionable. Three types of odor modification are described in the following subsections.

2.3.1. Deodorization Chemicals

Masking agents change the characteristics of an odor but also increase its resistant intensity. Counteractants are chemicals that may change the character of an odor and also reduce the intensity of an odor emission. Some deodorization chemicals are often used in offices filled with cigaret smoke, toilets, and so forth. In these places, low-concentration odor substances are broadly spread throughout. It is well known that Phytoncide (refined vegetables oil) easily reacts with odor substances and decomposes them. It is often used as a deodorization chemical in automobiles and air conditioning units.

2.3.2. Chemical Fragrance

Deodorization or odor removal is an important odor control. The method of odor modification using fragrances or flavorings is also effective if used properly. Figure 12 shows a classification of fragrance generally used in a variety of field applications (43). They are classified into natural flavorings or synthetic flavorings.

Solid air fresheners are often used at home or in automobiles. "Aromatization airconditioning" has often been performed in large spaces such as an atrium and the lobby of the hotel and in limited spaces such as a conference room or a restroom. However, because it is aimed for unspecified persons, the concentrations should be limited to levels tolerable by most individuals. In the following example, aromatization airconditioning is utilized in a Japanese corporate office.

2.3.3. Aromatization AirConditioning Example

Before installation of commercial equipment, the office workers were asked to answer a questionnaire on the effect of aromatization airconditioning (44). The experimental conditions are as follows:

- Object space: an office building in Japan (second through fourth floor, area of each floor was 610 m³).
- Number of respondents: 122 persons (without aromatization), 117 persons (with aromatization).
- Flavoring: rose fragrance (penetrated in ceramics pellets).

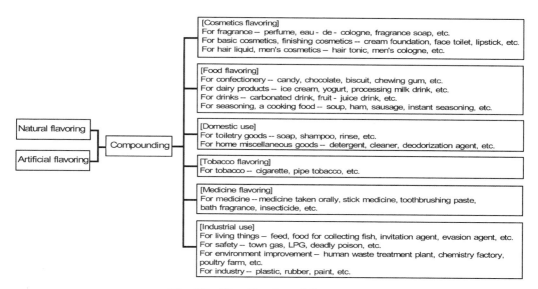

Fig. 12. Classification of fragrances.

- Methods of aromatization: a container containing the rose-fragranced ceramics pellets placed in the duct of an ordinary air-conditioning machine. The fragrance naturally vaporizes from the pellets and is released to the air.

Figure 13 shows partial results of the investigation indicating that aromatization has a considerable effect on the masking of odor, especially the odor of cigaret smoke. In this graph, the air conditioner emits an offensive odor when aromatization is not performed. On the other hand, it smells of the fragrance when air-conditioning aromatization is performed. It is known from this result that the offensive odor from the air conditioner can be masked with this method.

In ordinary aromatization airconditioning, liquid fragrances are sprayed into the air in the duct of the air conditioning unit using nebulizers. However, in small- or medium-scale offices, it is necessary to control the rate of generation. Low concentrations are difficult to maintain in ordinary nebulizers because it is difficult to keep the diameters of the induced mists small. Furthermore, unvaporized liquid fragrances often remain in the duct as liquid droplets. In order to address these difficulties, new aromatization air-conditioning equipment was proposed. The schematic diagram of the equipment is shown in Fig. 14. In this equipment, many small pellets (porous spheres of silicic acid calcium), in which fragranced liquid was infused, were used as a source of aromatization. Using a layer of pellets, a precise and well-controlled generation is possible. Because only volatile liquid flavoring is used in this method, liquid residual substances do not attach to the duct. However, because the intensity of fragrance coming from the pellets decreases with time, proper operating conditions should be determined to maintain consistency of the fragrance intensity.

Pellets of three kinds of fragrance (A, B, C) were used according to individual preference, and the system is operated intermittently according to a set schedule. The flow rate

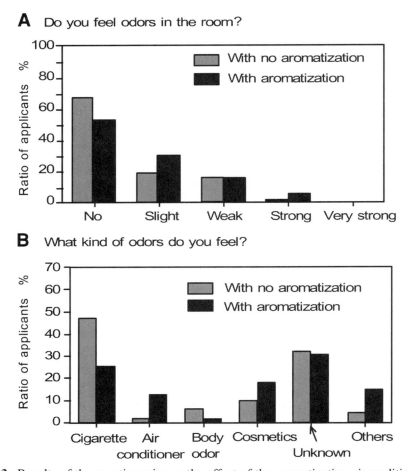

Fig. 13. Results of the questionnaire on the effect of the aromatization airconditioning.

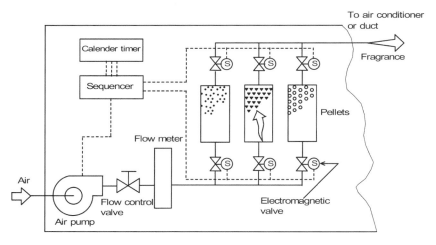

Fig. 14. Schematic diagram of the aromatization airconditioning equipment. (Courtesy of Asahi Kogyosha Co. LTD.)

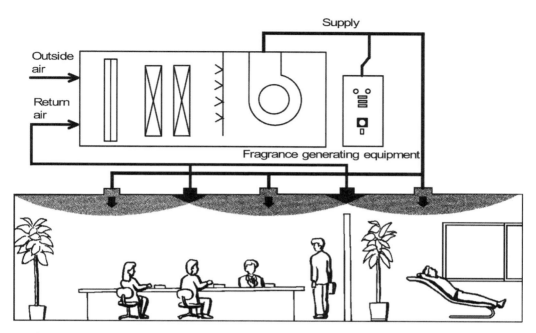

Fig. 15. Example of application of aromatization airconditioning system. (Courtesy of Asahi Kogyosha Co. LTD.)

of air was adjusted to 5 L/min, forcing the air to pass through the pellets layer and adding high-concentration fragrances before entering the duct of the air-conditioning unit.

Figure 15 shows another example of an aromatization airconditioning system (45). In this system, the fragrance source is placed behind a fan at the exit of the duct. An experiment was carried out to find the optimum operation condition for aromatization air-conditioning of 8 h/d using 1-kg pellets. The experiment was conducted at the entrance hall and the lobby of a resort hotel in Japan. The volume of the space was around 2060 m^3. This field test was carried out using two sets of aromatization induction equipment. The experimental results indicate that intermittent operation as shown in Fig. 16 and Table 7 was demonstrated to be effective for sustaining comfort levels for a month.

2.4. Adsorption Method

In this method, offensive odor components are passed through the adsorption tower containing various adsorbent media, such as activated carbon, a filter, and a molecular sieve. The absorbent absorbs odors or chemical compounds physically or chemically (46–52). The odor control is effective using this method, even when other methods are not effective. Therefore, this method is widely used as an independent technology or a final approach combined with other methods in order to adhere to public odor regulations. A small package is often used for deodorization of the refrigerator. An activated carbon filter is often used for an indoor air cleaner. The durability of the adsorbents is more important than the efficiency in odor removal. The adsorption performance is evaluated using the amount of adsorption equivalent to various odor ingredients

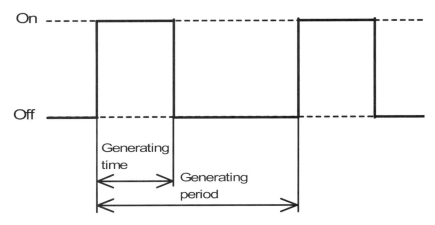

Fig. 16. Aromatization generating pattern in operation of equipment.

(37,38,48,49). In more advanced systems, the reuse or regeneration of adsorbent media is achieved by desorption with steam, heat additions, or plasma (53–55). Moreover, a functional filter described in Section 2.4.2. has been developed (51,52) which can be used as a bag filter or a filter for an indoor air cleaner as a adsorbent media. The details of the adsorption method are explained in the following subsection.

2.4.1. Activated Carbon or Zeolite Method

In the adsorption method using activated carbon or zeolite particles (48,49), odor components are adsorbed physically into small holes of the absorbent materials. Then, they are removed or decomposed chemically by the acid or alkali chemicals coated on the adsorption materials. This method is an effective odor-removal technology known for many years, and an example of the practical application is shown.

Figure 17 shows deodorization equipment using activated carbon adsorption (50). In the adsorption tower, alkalized carbon, acidized carbon, and neutral activated carbon are fills as layers according to the types and the strength of odorous components. The layer of alkalized carbon removes hydrogen sulfide (H_2S), methyl mercaptan [$(CH_3)_2SH$], and so forth. The layer of acidized carbon removes ammonia (NH_3), trimethyl amine [$(CH_3)_3N$], and so forth. The layer of neutral activated carbon removes dimethyl sulfide [$(CH_3)_2S$], dimethyl disulfide [$(CH_3)_2S_2$], hydrocarbons, and so forth.

When the concentration of odorous components is high, the process is combined with the biological deodorization system, acid and alkali scrubbing method, hypochlorous

Table 7
Operating Condition for a Month

Working period (d)	Generating time (s)	Operating time
1–10	90	8 h 00 min
11–20	150	13 h 20 min
21–30	300	26 h 40 min
Total	—	48 h 00 min

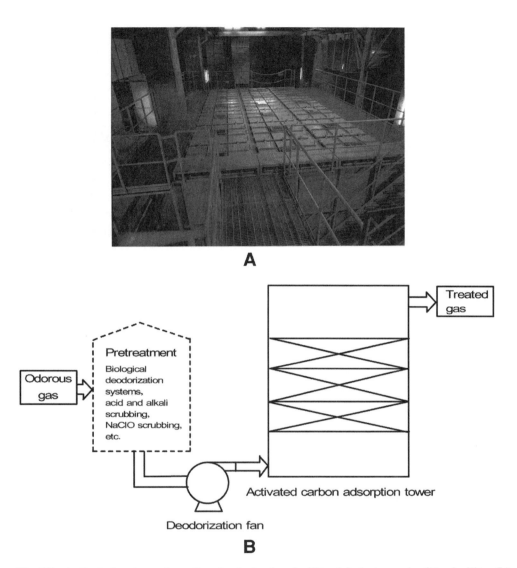

Fig. 17. Activated carbon adsorption deodorization facility: **(a)** photograph of the facility; **(b)** schematic diagram. (Courtesy of NGK Insulators, LTD.)

acid soda (NaClO) scrubbing method, and so forth and becomes effective in increasing the life of activated carbon.

2.4.2. Functional Cloth Filter

Highly functional air filters for cleaning exhaust gas from factory and indoor environments have been reported to improve the odor removal properties using the plasma –chemical treatment. In this subsection, the manufacturing process, measurement result of the functional properties, and applications are described (51,52).

The nonthermal plasma is first applied and hydrophilic monomers are then graft-polymerized to the cloth surface. The morphology of the cloth has been dramatically

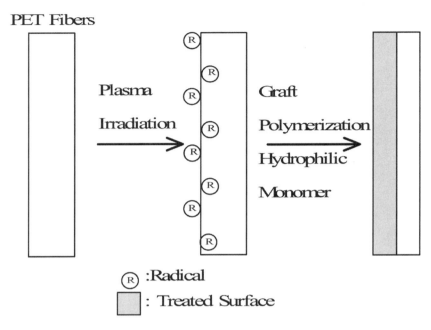

Fig. 18. Plasma graft-polymerization process.

changed to breathe moisture and to absorb offensive odor simultaneously. The process
of a new plasma graft-polymerization is shown in Fig. 18. As a plasma source, the
atmospheric pressure radio frequency (RF, 13.56 MHz) glow discharge using argon and
helium mixed gas is applied uniformly on the surface of the high-polymer cloth (e.g.,
PET [polyethylene terephthalate] fiber). The surface of the cloth is modified to generate many radicals. Next, aerosols of the hydrophilic monomer (acrylic acid) are sprayed
on the surface to produce a graft-polymerization. The surface morphology of the cloth
is modified to become porous (*see* Fig. 19) and the cloth adsorbs moisture and offensive
odors simultaneously.

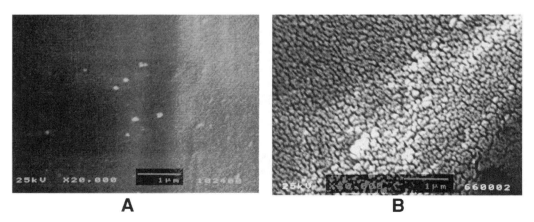

Fig. 19. Scanning electron micrograph (**a**) photograph of the PET cloth surface before and
after (**b**) the plasma-graft polymerization process.

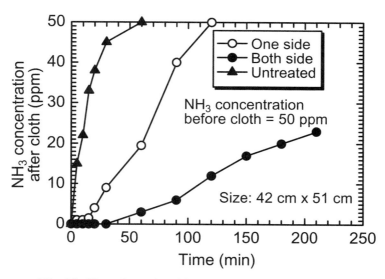

Fig. 20. Time-dependent NH_3 adsorption characteristics.

In order to evaluate odor control properties of this functional cloth, experiments were conducted. A 50-ppm and 1.0-L/min NH_3 gas balanced with dry air (relative humidity = 4%) was prepared and passed through the test section, where the tested cloth (51 cm × 42 cm) wrapped a porous polyethylene tube. The gas penetrates through the holes in the tube and then the cloth to be tested. The adsorption characteristics of NH_3 were evaluated by measuring the concentrations of NH_3 before and after the test section using gas-detection tubes.

The adsorption characteristics of three different functional cloths are shown in Fig. 20. For the untreated cloth, the concentration after the cloth was 15 ppm in 5 min. It reached 50 ppm in 60 min and the adsorption to the cloth was saturated. A total volume of adsorption became 0.68 mL. On the other hand, for the cloth treated on both sides, 100% NH_3 was adsorbed for the first 30 min and the concentration reached 22 ppm after 210 min. A total volume of adsorption was about 10 mL. For the cloth, treated on one side, over 90% NH_3 was adsorbed for 20 min and was saturated in 120 min, resulting in a total adsorption amount of 3.4 mL.

Various potential applications of this functional cloth have been proposed (51,52): underwear with moisture breath and odor control, seat covers of automobile, partitions in hospital, bag filters for dust and odor control, and filters of indoor air cleaner system (56).

2.5. Wet Scrubbing or Gas Washing Oxidation

Wet scrubbing and gas washing oxidation for odor control were employed extensively (10–13,15,20,21) because more than 90% of odorous compounds are water soluble. Offensive odor substances were decomposed, oxidized, or converted into nonodorous substances with chemical reactions using chemical solutions. This method is considered one of the most fundamental and universal deodorization technologies. Sulfuric acid (H_2SO_4), hydrochloric acid (HCl), sodium hydroxide (NaOH), sodium hypochlorite (NaClO), hydrogen peroxide (H_2O_2), sodium thiosulfate ($Na_2S_2O_3$), sodium sulfite (Na_2SO_3), and

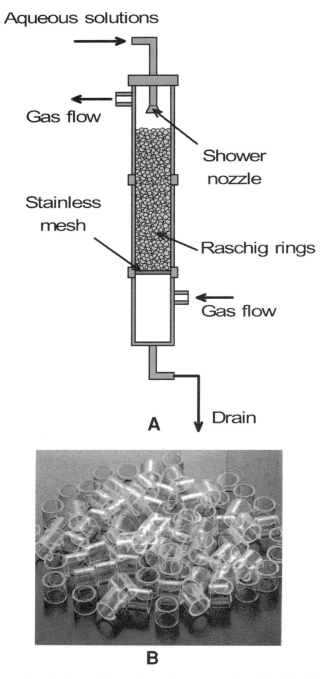

Fig. 21. Two-phase chemical scrubber using a shower nozzle with a Raschig rings layer: **(a)** chemical scrubber; **(b)** typical Raschig rings.

so forth are used as chemicals in washing solutions. Using various forms of chemical scrubber (10–13,15,20,21), odor components are deodorized by oxidization, decomposition, separation, or fixation in the absorption liquids in the form of the salts. Either a spray-type or Venturi scrubber is often used. Figure 21 shows a typical chemical scrubber called a spraytype in which Raschig rings are packed in order to promote gas–liquid contact and chemical reactions. Odor components consist of many components and are often in low concentration. A gas–liquid phase reaction does not always take place according to the chemical reaction formula even in simple substances and the required chemicals often exceed the theoretical values. The odor-removal efficiency greatly depends on the contact efficiency of the gas–liquid, the gas/liquid ratio, the concentration of the chemicals, pH, oxidization and reduction potential, and generated salt concentration. In the case of human waste odor, it is processed in many cases using a mixture of calcined soda and hypochlorous acid soda solutions.

The features of this method are as follows: Equipment is relatively simple, the maintenance is easy, and running cost and energy consumption are low. The efficiency is not largely affected by the concentration of gas components and the variation of compositions, inlet concentration, and flow rate. Additional experimental and theoretical studies for controlling low-concentration gases are essential. The commercialized examples of this kind of deodorization system and the method of design are explained in the following subsections.

2.5.1. Acid and Alkali Scrubbing Method

In the method of acid and alkali scrubbing, odor components are neutralized and absorbed in alkali water solutions. Figure 22 shows an example of the commercial acid and alkali scrubbing system (50). Usually, acid scrubbing is performed in the first stage and alkali scrubbing is done in the second stage. In the acid scrubbing process, dilute HCl or H_2SO_4 water solution (pH = 3–4) is used and the acid solution is recirculated. Primarily, ammonia and trimethyl amine can be removed in this process. The chemical reactions are as follows:

$$\text{Ammonia:} \qquad 2NH_3 + H_2SO_4 \rightarrow (NH_4)_2 SO_4 \qquad (35)$$

$$NH_3 + HCl \rightarrow NH_4Cl \qquad (36)$$

$$\text{Trimethyl amine:} \qquad (CH_3)_3 N + H_2SO_4 \rightarrow (CH_3)_3 NH_2SO_4 \qquad (37)$$

$$(CH_3)_3 N + HCl \rightarrow (CH_3)_3 NHCl \qquad (38)$$

In the alkali scrubbing process, dilute NaOH water solution (pH = 10–12) is recirculated. Hydrogen sulfide, methyl mercaptan, and so forth can be removed in the following process:

$$\text{Hydrogen sulfide:} \qquad H_2S + 2NaOH \rightarrow Na_2S + H_2O \qquad (39)$$

$$\text{Methyl mercaptan:} \qquad CH_3SH + NaOH \rightarrow CH_3SNa + H_2O \qquad (40)$$

2.5.2. NaClO Scrubbing Oxidation Method

Figure 23 shows the schematic diagram of the odor control system using the hypochlorous acid soda (NaClO) solution scrubbing method (50). Odor components are

A

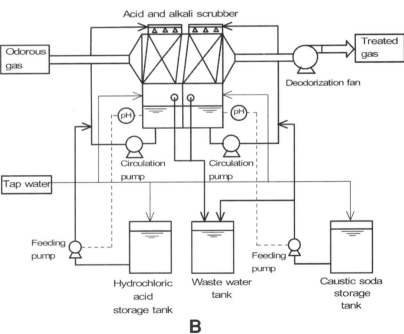

B

Fig. 22. Acid and alkali scrubbing system: (**a**) photograph of the facility; (**b**) schematic diagram. (Courtesy of NGK Insulators, LTD.)

oxidized and decomposed to nontoxic compounds by the gas washing method using a NaClO water solution. NaClO is inexpensive, safe, easily available, and a strong oxidization chemical. It can remove chemically-neutral odor components that acid and alkali scrubbing methods cannot remove. The operation is usually performed for con-

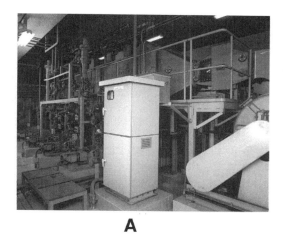

A

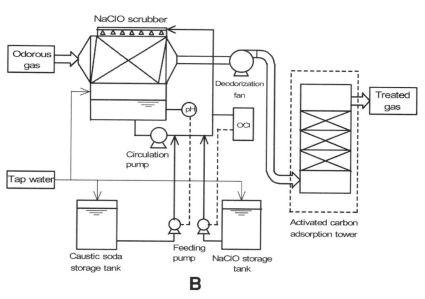

B

Fig. 23. NaClO scrubbing system: (**a**) photograph of the facility; (**b**) schematic diagram. (Courtesy of NGK Insulators, LTD.)

trolling pH = 10. The amount of NaClO addition is monitored and controlled economically using a chlorine concentration monitor. The operating cost can be reduced with this method. The odor components can be removed primarily according to the following chemical reactions:

Alkali gas

$$\text{Ammonia:} \qquad 2NH_3 + 3NaClO \rightarrow N_2 + 3NaCl + 3H_2O \qquad (41)$$

$$\text{Trimethyl amine:} \qquad (CH_3)_3 N + NaClO \rightarrow (CH_3)_3 NO + NaCl \qquad (42)$$

Table 8
Design Criteria for the Chemical Scrubbing System

Gas flow rate:	200 m³/min
Gas temperature:	25°C
Gas pressure:	1 atm

Odor concentration (ppm)	Inlet (ppm)	Outlet (ppm)	N_{OG}	Needed tower height Z (m) (H_{OG} =0.5 is assumed)
Ammonia	5	0.7	1.97	0.98
Trimethyl amine	0.1	0.002	3.91	1.96
Hydrogen sulfide	0.3	0.005	4.09	2.05
Methyl mercaptan	0.02	0.0008	3.21	1.61
Dimethyl sulfide	0.3	0.003	4.61	2.30
Dimethyl disulfide	0.3	0.003	4.61	2.30
Acetaldehyde	2	0.5	1.39	0.69

Note: H_{OG} is height of an overall gas transfer unit, N_{OG} is number of gas transfer units

Acid gas

Hydrogen sulfide: $H_2S + 2NaOH \rightarrow Na_2S + H_2O$ (43)

$$Na_2S + 4NaClO \rightarrow Na_2SO_4 + 4NaCl \quad (44)$$

$$Na_2S + NaClO + H_2O \rightarrow S + NaCl + 2NaOH \quad (45)$$

Methyl mercaptan: $CH_3SH + 3NaClO \rightarrow CH_3SO_3H + 3NaCl$ (46)

Neutral gas

DiMethyl sulfide: $(CH_3)_2S + 3NaClO \rightarrow (CH_3)_2SO_3 + NaCl$ (47)

DiMethyl disulfide: $(CH_3)_2S_2 + 5NaClO \rightarrow 2CH_3SO_3H + 5NaCl$ (48)

Acetaldehyde: $CH_3CHO + NaClO + NaOH \rightarrow CH_3COONa + NaCl + H_2O$ (49)

A higher level of odor control can be achieved by combining this method with the activated carbon adsorption method.

2.6. Design Example of Wet Scrubbing or Gas Washing Oxidation

The design procedure of odor control equipment with a flow rate of 200 m³/min by acid and alkali washing (20,21) is described based on the design criteria listed in Table 8. The density of the gas or air at the standard condition (0°C, 1 atm) is

$$\rho_{gas} = \frac{29\left(\frac{kg}{kmol}\right)}{22.4\left(\frac{Nm^3}{kmol}\right)} = 1.295 \frac{kg}{Nm^3} \quad (50)$$

The volume flow rate of the gas at the standard condition (0°C, 1 atm) becomes

$$Q_{gas} = 200\left(\frac{m^3}{min}\right) \times 60\left(\frac{min}{h}\right) \times \frac{273}{273+25}\left(\frac{Nm^3}{m^3}\right) = 10{,}993\frac{Nm^3}{h} \qquad (51)$$

2.6.1. Determination of the Diameter of the Chemical Scrubber

The mass velocity of gas per unit cross-sectional area is usually taken as $G = 5000\text{--}15{,}000$ kg/(m² h). In the present calculation, $G = 9000$ kg/(m² h) is assumed. The area and the diameter of the chemical scrubber are determined as follows:

$$A = \frac{\rho_{gas}Q_{gas}}{G} = \frac{1.295\left(\dfrac{kg}{Nm^3}\right) \times 10{,}993\left(\dfrac{Nm^3}{h}\right)}{9{,}000\left(\dfrac{kg}{m^2 h}\right)} = 1.582 \text{ m}^2, \text{ therefore, } d = 1.5 \text{ m} \qquad (52)$$

2.6.2. Determination of Mass Flow Rate of the Solutions

Considering the effective use of washing chemicals, the volumetric ratio of gas to liquid is usually selected in the range of around $r_{g/l} = 500$. The flow rate of the liquid, Q_{liq}, and the mass flow rate per unit cross-sectional area, L, can be determined as follows:

$$Q_{liq} = \frac{Q_{gas}}{r_{g/l}} = \frac{10{,}993}{500}\left(\frac{Nm^3}{h}\right) \times \frac{273+25}{273}\left(\frac{m^3}{Nm^3}\right) = 24.0\frac{m^3}{h} = 24{,}000\frac{kg}{h} \qquad (53)$$

$$L = \frac{Q_{liq}}{\pi(d/2)^2} = \frac{24{,}000}{3.14 \times 0.75^2} = 13{,}588\frac{kg}{m^3 h} \qquad (54)$$

2.6.3. Determination of the Height of the Washing Scrubber

Acid washing and alkali washing towers as shown in Fig. 21 are installed separately. For the acid washing tower, ammonia and trimethyl amine are removed by H_2SO_4. For the alkali washing scrubber, hydrogen sulfide, methyl mercaptan, dimethyl sulfide, dimethyl disulfide, and acetaldehyde are removed with NaOH and NaClO mixed solutions. The height of the washing tower, Z, is determined by the products of N_{OG} value for various odor components and the H_{OG} value for the packed Raschig rings:

$$Z = N_{OG} \times H_{OG} \qquad (55)$$

Where H_{OG} is height of an overall gas transfer unit, N_{OG} is number of gas transfer units. N_{OG} values for various odor components are calculated from the inlet and outlet concentrations of odor components:

$$N_{OG} = \ln\frac{\ln\left(1 - C_{inlet}\right)}{\ln\left(1 - C_{outlet}\right)} \qquad (56)$$

where C_{inlet} and C_{outlet} are inlet and outlet odor concentrations, respectively. The values of N_{OG} for various odor components are listed in Table 8. On the other hand, the value of H_{OG} depends on the types of packed pellet, odor component, L, G, C_{inlet}, and pH and should be determined experimentally. Some experimental results are summarized in ref. 21. From the results, we can assume $H_{OG} = 0.5$ m for the odor components. The required height of the washing tower for each component is shown in Table 8. From the maximum value of these heights, the heights of the acid washing tower and the alkali washing tower are determined considering 20% other parts length of the tower:

$$\text{Height of the acid scrubbing tower:} \quad Z_t = 1.96 \times 1.2 = 2.35 \, \text{m} \tag{57}$$

$$\text{Height of the alkali scrubbing tower:} \quad Z_t = 2.30 \times 1.2 = 2.76 \, \text{m} \tag{58}$$

2.6.4. Determination of the Required Chemicals

The amount of required chemicals are calculated based on the chemical reactions (35), (37), (39), (40), and (43)–(49). $Q_{gas} = 10{,}993$ m³/h is used in the following calculation.

1. The amounts of H_2SO_4 (molecular weight = 98) needed are

$$\text{for ammonia:} \quad M_1 = 10{,}993 \left(\frac{N \, m^3}{h} \right) \times (5 - 0.7) \times 10^{-6} \times \frac{98 \left(\frac{kg}{kmol} \right) \times 0.5}{22.4 \left(\frac{N \, m^3}{kmol} \right)} = 0.103 \frac{kg}{h} \tag{59}$$

$$\text{for trimethyl amine:} \quad M_2 = 10{,}993 \times (0.1 - 0.002) \times 10^{-6} \times \frac{98}{22.4} = 0.005 \frac{kg}{h} \tag{60}$$

$$\text{A total amount of needed } H_2SO_4 : M_t = M_1 + M_2 = 0.11 \left(\frac{kg}{h} \right) \tag{61}$$

2. The amounts of NaClO (molecular weight = 74.4) needed are

$$\text{for hydrogen sulfide:} \quad M_3 = 10{,}993 \times (0.3 - 0.005) \times 10^{-6} \times \frac{74.4 \times 2.5}{22.4} = 0.0269 \frac{kg}{h} \tag{62}$$

$$\text{for methyl mercaptan:} \quad M_4 = 10{,}993 \times (0.02 - 0.0008) \times 10^{-6} \times \frac{74.4 \times 1.5}{22.4}$$

$$= 0.0011 \frac{kg}{h} \tag{63}$$

$$\text{for dimethyl sulfide:} \quad M_5 = 10{,}993 \times (0.3 - 0.003) \times 10^{-6} \times \frac{74.4 \times 3}{22.4} = 0.0325 \frac{kg}{h} \tag{64}$$

$$\text{for dimethyl disulfide:} \quad M_6 = 10{,}993 \times (0.3 - 0.003) \times 10^{-6} \times \frac{74.4 \times 5}{22.4}$$

$$= 0.0542 \frac{kg}{h} \tag{65}$$

$$\text{for acetaldehyde:} \quad M_7 = 10{,}993 \times (2 - 0.5) \times 10^{-6} \times \frac{74.4}{22.4} = 0.0548 \frac{kg}{h} \tag{66}$$

$$\text{A total amount of needed NaClO:} \quad M_t = M_3 + M_4 + M_5 + M_6 + M_7 = 0.17 \frac{kg}{h} \tag{67}$$

3. The amounts of NaOH (molecular weight = 40) needed are

for hydrogen sulfide: $M_8 = 10,993 \times (0.3 - 0.005) \times 10^{-6} \times \dfrac{40 \times 2}{22.4}$

$$= 0.0116 \frac{kg}{h} \tag{68}$$

for methyl mercaptan: $M_9 = 10,993 \times (0.02 - 0.0008) \times 10^{-6} \times \dfrac{40 \times 0.5}{22.4}$

$$= 0.00019 \frac{kg}{h} \tag{69}$$

for acetaldehyde: $M_{10} = 10,993 \times (2 - 0.5) \times 10^{-6} \times \dfrac{40}{22.4} = 0.0294 \dfrac{kg}{h} \tag{70}$

In the alkali scrubber, NaOH strongly reacts with CO_2 in the gas. Therefore, an additional large amount of NaOH is required. The reaction greatly depends on the pH of the chemical solutions and the needed NaOH can be calculated from the inlet and oulet CO_2 concentrations. When the inlet and oulet CO_2 concentrations. are 1000 ppm and 980 ppm respectively, the NaOH needed is

for carbon dioxide: $M_{11} = 10,993 \times (1,000 - 980) \times 10^{-6} \times \dfrac{40 \times 2}{22.4} \tag{71}$

$$= 0.785 \frac{kg}{h}$$

A total amount of NaOH needed: $M_t = M_8 + M_9 + M_{10} + M_{11}$

$$= 0.83 \frac{kg}{h} \tag{72}$$

2.6.5. Volume of the Chemical Tanks

The chemical supply cycle is assumed to be rescheduled every month. The minimum volume of the chemical tanks needed for the month are calculated as follows:

The volume of 70 % H_2SO_4 solution (density=1.61 kg/L) tank)

$$= \frac{0.11 \left(\dfrac{kg}{h}\right) \times 24 \left(\dfrac{h}{d}\right) \times 30(d)}{0.75 \times 1.61 \left(\dfrac{kg}{L}\right)} = 66 \text{ L} \tag{73}$$

The volume of 12% NaClO solution (density $= 1.15$ kg / L) tank $= \dfrac{0.18 \times 24 \times 30}{0.12 \times 1.15} \tag{74}$

$$= 940 \text{ L}$$

The volume of 20% NaOH solution (density $= 1.22$ kg / L) tank $= \dfrac{0.83 \times 24 \times 30}{0.20 \times 1.22} \tag{75}$

$$= 2450 \text{ L}$$

2.7. Incineration

Typical flowcharts of this method are shown in Fig. 24. This method is classified as follows: direct incineration and catalyst incineration. The heat recovery, heat exchange,

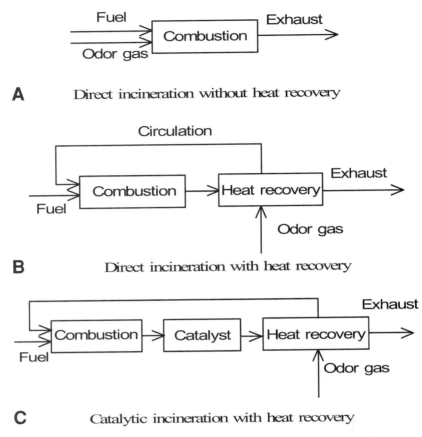

Fig. 24. Various types of incineration deodorization method.

or the existence of catalyst has been an important point in the design. In the following subsections these methods (14,15,20,21) are explained.

2.7.1. Direct Incineration

Direct incineration is a method in which flammable odor components are burned and decomposed by a combustion furnace or separate afterburners. If the performance of a burner, the mixing of a flame and odor components, combustion temperature, and resident time are proper, this method is very reliable and has wide applications. Although this method required a high energy consumption and an odor of combustion exhaust gas is emitted, it is a reliable method for many substances. In some industries, incineration is the only available method, as other methods are unavailable for odor control.

The important parameters for direct incineration are as follows. The combustion temperature in the resident zone is designed as 650–800 °C and the residence time is 0.3–1 s. The amount of dust, mist, fume, moisture, and oxygen concentration in odorous gas should be accounted for properly in the design. The blowing angle of odor gas to a flame should be proper. The temperature inside the furnace should be uniform. A high-performance heat exchanger and high-quality thermal insulation material should be used. Corrosion and choking should be prevented. The inside temperature in the heat exchanger should be kept at higher than 150°C and the outlet temperature should be kept higher than 250–300 °C.

2.7.2. Catalytic Incineration

The catalytic incineration method is classified as an incineration method. In this method, flammable odor components such as hydrocarbons are burned, oxidized, or decomposed into water and carbon dioxide under a lower-temperature condition with an activated catalyst. Although the equipment looks very simple that odorous gas passes through, it can achieve stable deodorization effects under a relatively lower temperature of 250–300°C. When cold odorous gas is treated, the temperature of the catalysis needs to be elevated using a furnace or electric heater. As an example of a great success, the three-way catalyst that was used to clean up the exhaust gas of a passenger car using gasoline is mentioned. This method is often used for odor control in a painting factory (15), a sewage disposal plant (15,16), and diesel engine exhaust gas (57,58).

The main problem in this method is how to prevent the poisoning and fouling of the catalysts. Oxidization catalysts such as Pt, V_2O_5, Co, Mn, Fe, Ni, and others such as Co_3, MnO_2, Fe_2O_3, NiO are used, held in ceramic pellets or ceramic honeycombs. It is important to choose the proper catalyst suitable for the kind of processed odor from viewpoints of gas temperature, performance, and durability.

Generally, odorous gas consists of many components, including not only gases but particulates, mist, and dusts. Some of these components are difficult to oxidize. Especially, mist removal is very important to achieve high performance. Pretreatment of the gas for the removal of mists is desired using the activated carbon or zeolite adsorption method (10,11,15,48,49).

Figure 25 shows a commercial plant of the exhaust gas processing system of the sludge incinerator from a sewer using the catalyst incineration (50). This system also works as a NO_x removal system. By the action of the catalyst, odor ingredients are oxidized at relatively low temperature (approx 350°C). The temperature of odorous gas is first elevated by the heat exchange with exhaust gas (approx 250°C). In the next stage, the temperature is elevated up to 350°C by the furnace and enters into the reaction tower in which the catalysts is packed. The treated gas is emitted into the atmosphere via a heat exchanger.

As an example of other catalysis odor control, the so-called photocatalyst (TiO_2) (59,60) is widely used for commercial equipment (61,62). The principle is shown in Fig. 26. If an ultraviolet ray is irradiated with an inverter lamp on the photocatalyst (TiO_2), an electron will jump from the surface. At this time, the portion from which the electron escaped is called a hole and has the positive charge. The hole has strong oxidization power and an electron is taken from OH^- (hydroxide ion) contained in the moisture in air. As a result, OH^- becomes an OH radical that is very unstable but chemically reactive. Because the OH radical has powerful oxidization power, it tends to be stable by taking electrons from organic matters, which are the main odorous compounds in the gas. As a result, organic matters are decomposed or converted into CO_2 and H_2O and dispersed into the atmosphere. The dust-collecting filter with the photocatalyst is frequently used for the commercial indoor electric air cleaner (62).

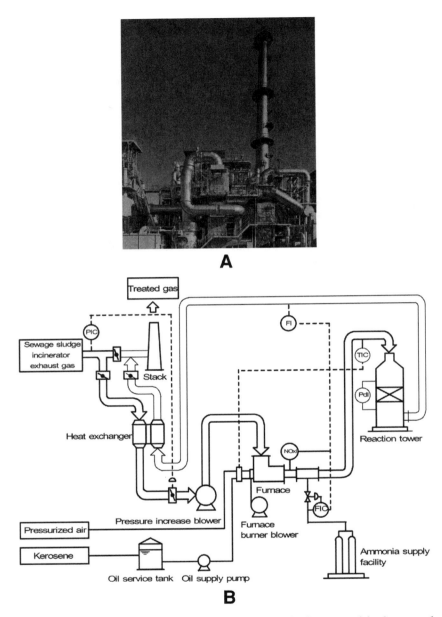

Fig. 25. Catalytic incineration plant for sewage sludge incinerator: **(a)** photographs of the facility; **(b)** schematic diagram. (Courtesy of NGK Insulators, LTD.)

2.8. Nonthermal Plasma Method

Nonthermal plasma or nonequilibrium plasma means an electrically neutral and chemically activated ionization state in which the electron temperature is siginificantly higher than the gas or ion temperature. Using an AC high voltage or a pulse high voltage with sharp rising time of several to several hundreds of nanoseconds, nonthermal plasma can be induced at atmospheric pressure and temperature. The hazardous air pollutants can be cleaned with nonthermal plasma (63–70). A barrier electric discharge,

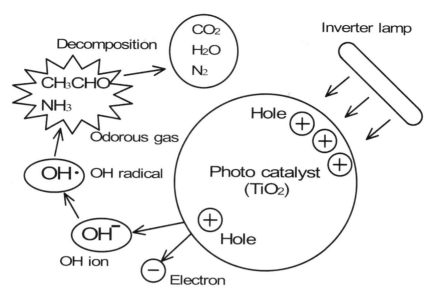

Fig. 26. Principle of odor decomposition using a photocatalyst.

glow discharge, or surface discharge is mainly used for nonthermal plasma treatment. The nonthermal plasma can be generated with lower energy than the thermal plasma. The plasma is able to promote chemical reactions such as oxidization or decomposition when exposed to odorous gases. A nonthermal plasma in which the average electron temperature is in the range 4–5 eV and an electron density of $10^{10}/m^3$ is often used. This technology was commercialized as the pulse corona-induced plasma chemical process (PPCP) system for treating hazardous air pollutants (66). The applications of the non-thermal plasma to odor control have been reported for the deodorization from a sewage disposal plant, a refuse disposal plant (69), and a livestock farm (70). In the following subsections, research on indoor air purification or odor control technology using the plasma are explained.

2.8.1. Indoor Air Cleaner Using Nonthermal Plasmas

Odor control from a living environment is of increased concern. One of the main odor sources in the living environment is cigaret smoke. More than 4000 chemical components are identified in the cigaret smoke (28). These are roughly classified into two groups: gaseous compounds, such as acetaldehyde (CH_3CHO) and ammonia (NH_3), and particulate matter, such as tar. Recently, various kinds of indoor electric air cleaner using electrostatic precipitators (ESPs) (71–76) have been manufactured to improve the living environment. Although a wide size range of airborne particles can be removed effectively with an ordinary electrostatic air cleaner, it is impossible to remove gaseous compounds from cigaret smoke using the ESP. Using the nonthermal plasma, experimental studies of new electric air cleaners have been reported concerning the removal of the gaseous compounds (77–83). One example of a new type of indoor electric air cleaner that realizes simultaneous removal of airborne particles and odors was developed (83). It is composed of a nonthermal (nonequilibrium) plasma reactor followed by a two-stage ESP as illustrated in Fig. 27.

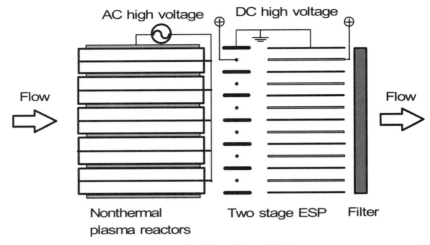

Fig. 27. Electric air cleaner composed of nonthermal plasma reactor, two-stage ESP, and functional filter. (© 2001 IEEE.). (*Source:* Reference 83)

The experimental results on CH_3CHO, NH_3, and cigaret smoke removal are reported in detail (56). Two kinds of AC nonequilibrium plasma reactor are employed: one is the packed-bed plasma reactor, and the other is the film-type plasma reactor consisting of laminated parallel steel plate electrodes. In this reactor, the electrically grounded plates are wrapped with a thin film of high dielectric material. The results are shown in Section 2.8.2.

2.8.2. Odor Removal Using the Packed-Bed Plasma Reactor

A barrier-type packed-bed plasma reactor as shown in Fig. 28 is used for the removal of gaseous CH_3CHO and NH_3. It consists of a 1.6-mm-diameter wire electrode and a Pyrex glass tube (20 mm inner diameter and 24 mm outer diameter) as an dielectric barrier around which the copper screen is wrapped as the other electrode. The 1.7 to 2.0-mm-diameter $BaTiO_3$ pellets (the relative dielectric constant is 10,000 at room temperature) are packed inside the tube. The width of the copper mesh electrode H is set at 260 mm. The AC high voltage (max 20 kV) of 60 Hz is applied to the reactor using a neon transformer. Nonequilibrium plasma is induced between the pellets. The experimental setup is shown in Fig. 29. Then, 1020 ppm CH_3CHO and 950 ppm NH_3 balanced with N_2 are prepared in the cylinders and the dry air (relative humidity = 4%) is supplied by the compressor through the dryer. The desired CH_3CHO and NH_3 concentrations are obtained with the mass flow controller on each line. The flow rates are varied at 1.0, 2.0, and 4.0 L/min. The removal efficiency of CH_3CHO and NH_3 are evaluated using an FID (flame ionization detector) gas chromatograph and gas-detection tubes, respectively. The concentrations of byproducts such as CO, CO_2, NO, NO_2, and N_2O are measured using a set of gas analyzers. The concentration of ozone was measured using the gas-detection tubes.

Figure 30 shows the removal efficiency of 100 ppm acetaldehyde balanced with dry air when the flow rate is at 1.0, 2.0, and 4.0 L/min in the packed-bed plasma reactor. The specific energy density, SED, based on the discharge power in the plasma reactor (SED = discharge power/flow rate) is used for the horizontal axis. At a fixed SED, the

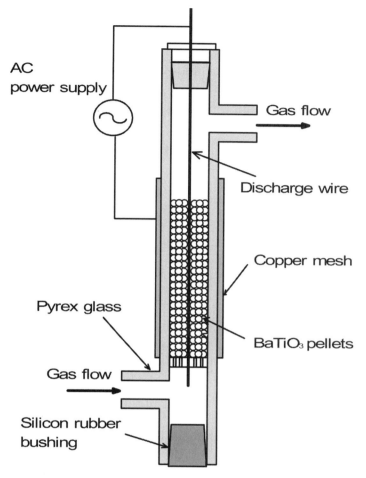

AC power supply

Gas flow

Discharge wire

Copper mesh

Pyrex glass

BaTiO₃ pellets

Gas flow

Silicon rubber bushing

Fig. 28. Barrier-type packed-bed plasma reactor. (© 2001 IEEE.). (*Source:* Reference 83)

removal efficiency increases with increasing flow rate. When the flow rate is 1.0 L/min, the concentration reaches minimum at a SED of 146 J/L (V_{p-p} = 16 kV). This optimum value of SED decreases with increased flow rate and becomes 94.5 J/L at a flow rate of 4.0 L/min. More than 95 % of removal efficiency is obtained with a flow rate of 4.0 L/min. The concentration of byproducts, such as CO and CO_2, increase with increase in SED as CH_3CHO is removed. However, the concentrations of NO, NO_2, and N_2O are relatively low. Considering the carbon balance, acetaldehyde is effectively converted to CO, CO_2, and other hydrocarbons by nonthermal plasma. It is clear that radical species such as active oxygen (O), nitrogen (N), and OH play a key role compared to O_3 in CH_3CHO removal.

Figure 31 shows the removal efficiency of 20, 60, and 100 ppm NH_3 balanced with dry air when the flow rate is set at 1.0 L/min. One hundred percent NH_3 removal is achieved at SED = 265 J/L. The concentration of N_2O is about 10 ppm at SED =219 J/L (V_{p-p}= 16 kV), but the concentrations of NO_x and CO are low. A large amount of white

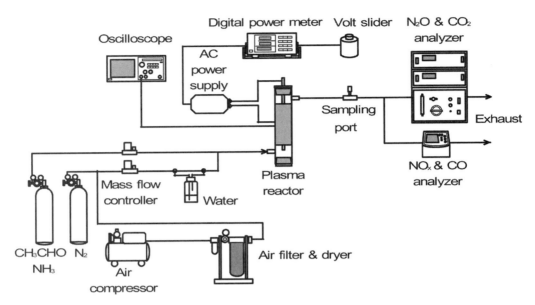

Fig. 29. Experimental setup for acetaldehyde and ammonia removal using the packed-bed plasma reactor. (© 2001 IEEE.). (*Source:* Reference 83)

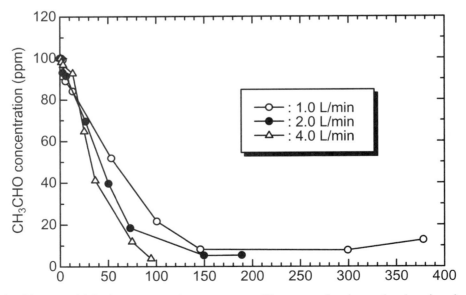

Fig. 30. Acetaldehyde concentration versus specific energy density under dry air using the packed-bed plasma reactor (initial concentration of CH_3CHO = 100 ppm). (©2001 IEEE.). (*Source:* Reference 83)

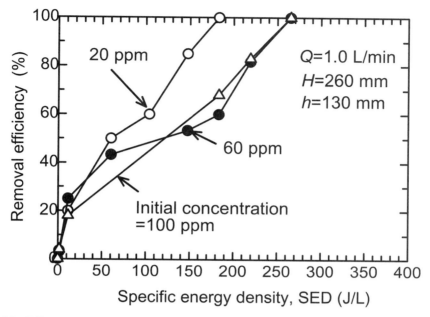

Fig. 31. Effect of initial concentration on ammonia removal. (© 2001 IEEE.), (*Source:* Reference 83)

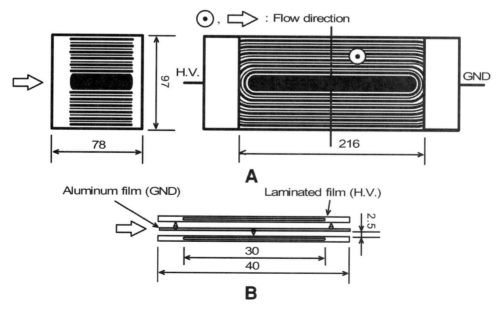

Fig. 32. Film-type plasma reactor: **(a)** frontal and cross-sectional views; **(b)** details of the film electrode section.

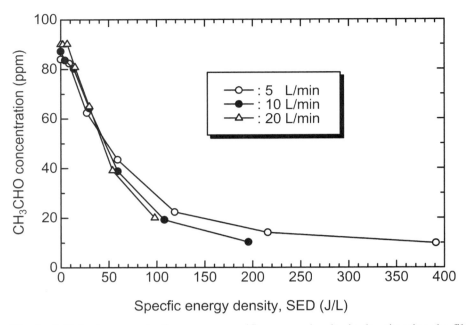

Fig. 33. Acetaldehyde concentration versus specific energy density in dry air using the film-type plasma reactor. (© 2001 IEEE.). (*Source:* Reference 83)

powder observed inside the wall of the plasma reactor is identified as NH_4NO_3 aerosol produced as the reaction of induced HNO_3 and NH_3 (66).

2.8.3. Odor and Cigaret Smoke Removal Using the Dielectric-Barrier Plasma Reactor

The plate-type, dielectric-barrier discharge plasma reactors are widely used for ozone generation and odor and VOC removal. A film-type plasma reactor shown in Fig. 32 is used for the removal of gaseous CH_3CHO, NH_3, and cigaret smoke. It consists of aluminum electrodes with many sharp projections and other aluminum electrodes wrapped with polyester film. The 60-Hz AC high voltage (max 8 kV) is applied to the wrapped electrode and the other electrode is grounded. The effective reactor length is 40 mm and the gap between the two electrodes is 2.5 mm. The 90 ppm CH_3CHO and 100 ppm NH_3 are balanced with dry air (relative humidity = 4% at 25°C). The flow rate is set at 5, 10, and 20 L/min and the corresponding residence times are 7.5, 3.8, and 1.9 s, respectively. The removal efficiency of CH_3CHO is evaluated with the sample using a gas chromatograph. The removal efficiency of NH_3 is measured using the gas-detection tubes and an FTIR (Fourier transform infrared) analyzer. The concentrations of byproducts such as O_3, HNO_3, $HCOOH$, CO, and N_2O were also measured using the FTIR analyzer.

Experiments were carried out using cigaret smoke. A Japanese cigaret (Mild Seven) is burned using dry air with a flow rate of 1.0 L/min. The particulates in the cigaret smoke were first removed with a glass fiber filter. After the plasma treatment, the concentrations of CH_3CHO and NH_3 in the smoke were measured using the gas chromatograph and the gas-detection tube.

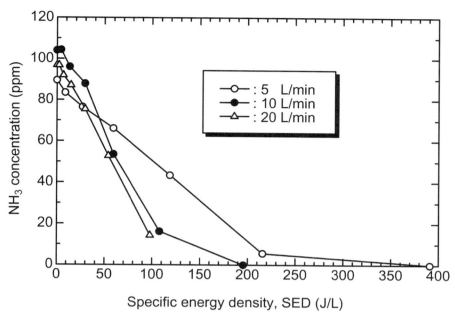

Fig. 34. Ammonia concentration versus specific energy density under dry air using the film-type plasma reactor. (© 2001 IEEE.). (*Source:* Reference 83)

Figure 33 shows the removal efficiencies of acetaldehyde in dry air when the flow rate was set at 5, 10, and 20 L/min, respectively. At a fixed SED, the removal efficiency increases with the increase in flow rate. The maximum removal efficiencies are 88% for 5.0 L/min, 87% for 10 L/min, and 78% for 20 L/min. It is known from the byproducts measurement using a FTIR analyzer that O_3, HNO_3, HCOOH, CO, CO_2, and N_2O are induced when CH_3CHO is removed.

Figure 34 shows the removal efficiency of 100 ppm NH_3 in dry air when the flow rate is set at 5, 10, and 20 L/min. In this figure, 100% removal is accomplished at the applied voltage of 8 kV for flow rates of 5.0 and 10 L/min (SED=196 and 391 J/L, respectively). The removal efficiency is 85% with a flow rate of 20 L/min. The reaction byproducts such as gas components of O_3, CO, CO_2, N_2O, HNO_3, and HCOOH and aerosol of NH_4NO_3 are identified when NH_3 is removed. A large amount of NH_4NO_3 white powder is observed on the grounded electrodes after the experiment. This aerosol is induced by the reaction between the NH_3 and the induced HNO_3. Those aerosols can be easily removed by the ESPs or filters.

The experiments are further carried out using real cigaret smoke (56). In the result, one of the components, NH_3, is completely removed at a SED of 150 J/L with an efficiency of over 90%. However, the removal efficiency of CH_3CHO is reduced to 50% because of its low concentration.

2.8.4. Design Example of Indoor Air Cleaner Using Nonthermal Plasma

The system as shown in Fig. 27 is designed so that odorous gases including NH_3 and CH_3CHO are decomposed or converted into aerosol by the plasma reactor in the first stage of the system, and in the second stage, the induced aerosols are completely collected by the electrostatic force. The volumetric flow rate is assumed to be 10 m³/min (25°C, 1 atm). In the first stage or the plasma reactor, the distance between the discharge wire and the outer grounded electrode is usually taken as $r_0 = 1$–3 cm in order to produce the strong plasma. Here, $r_0 = 3$ cm is assumed. The residence time should be taken around $T_r = 0.5$ s for over 90% removal efficiency. When the length of the plasma reactor is assumed to be equal to 50 cm, the following equations hold:

$$T_r = \frac{L}{V_{ave}}, \quad V_{ave} = \frac{q}{\pi r_0^2} \tag{76}$$

where V_{ave} is the mean velocity inside the reactor and q is the volumetric flow rate for a single reactor. The flow rate of a single reactor is

$$q = \frac{\pi r_0^2 L}{T_r} = \frac{3.14 \times 0.03^2 \times 0.5 \times 60}{0.5} = 0.170 \, \text{m}^3 \, / \, \text{min} \tag{77}$$

Therefore, the number of annular-type plasma reactors required for parallel connection is $10/0.170 = 59$. For a single reactor, a square cross-section of $6 \times 6 = 36$ cm² is needed. Moreover, the power consumption is calculated from the best result using an AC neon transformer with SED = 15 J/L = 15 kJ/m³:

$$P = \frac{15\left(\dfrac{\text{kJ}}{\text{m}^3}\right) \times 10\left(\dfrac{\text{m}^3}{\text{min}}\right)}{60\left(\dfrac{\text{s}}{\text{min}}\right)} = 2.5 \, \text{kW} \tag{78}$$

This is quite a large value. It is necessary to attain further optimization of the energy efficiency in the nonthermal plasma reactor. In addition, the design of the second stage of an electric precipitator system is beyond the scope of this chapter. The details of the design procedure are discussed in other books on the ESP (71) and aerosols (84,85).

2.9. Indirect Plasma Method (Ozone or Radicals Injection)

2.9.1. Ozonizer Using Surface Discharge

The indirect plasma method is based on the concept of remote plasma processing. Odor components are oxidized or decomposed by injecting ozone (86–88) or chemically activated radicals such as O, N, OH, and H_2O_2, which are produced by the discharge reactor. Figure 35 shows photographs of a small ozonizer for indoor and automobile odor control (89). Very small amounts of ozone lower than the safety standards concentration (<ppm) and nitrogen radicals are continuously supplied by the ozonizer only with several watts to several tens of watts power consumption. Offensive odor and uncomfortable gas components suspended inside the room or automobile are decomposed or removed to achieve a comfortable environment. The target odors are cigaret smoke, a sickroom, pet, toilet, exhaust gas, mold, and so forth. Moreover, it will be useful for sterilization of air.

Figure 36 shows an example of high concentration ozone generating equipment by adapting the principle of the surface-discharge-induced plasma chemical process

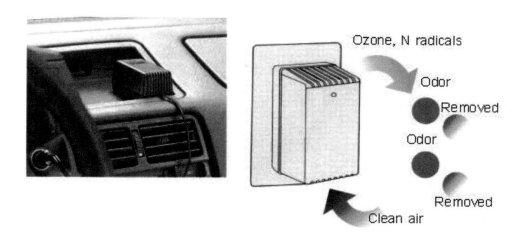

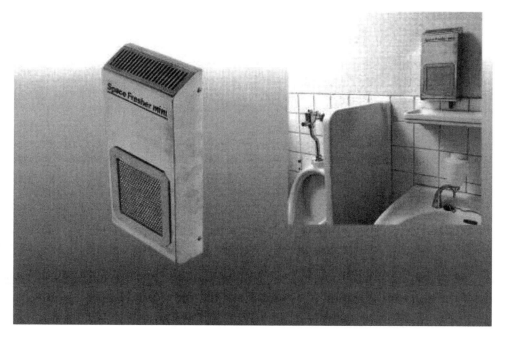

Fig. 35. Small ozonizers and principle of odor removal by ozone or radical injections. (Courtesy of Masuda Research Inc.)

(SPCP) (78,89) and the cross-section of the surface discharge electrode. Because the discharge electrode is protected with the glass coating, it has good reliability and durability. It is useful for ozone water processing. In addition, there are many other applications, such as semiconductor manufacturing, surface treatment, medical application, and food processing. Table 9 shows the main specification of these ozonizers.

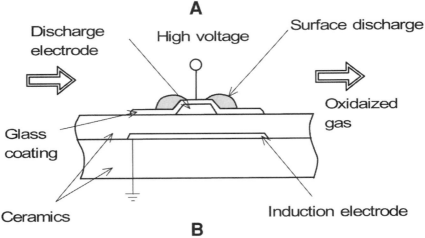

Fig. 36. Commercially used ozonizers and the shapes of a surface discharge electrode for high voltage: **(a)** photos of ozonizers; **(b)** details of the discharge electrode. (Courtesy of Masuda Research Inc.)

2.9.2. Commercial Plant for Odor Removal Using Ozone and Radical Injection

A schematic of an ozone and radical injection commercial plant is shown in Fig. 37 (50). Because ozone generally becomes more reactive under humidified conditions, a water washing scrubber is additionally built and water-soluble components are absorbed simultaneously (90,91). In addition, ozone is easily decomposed at the outlet of the system to a level that does not affect the human body by installing a decomposition equipment with activated carbon or a heating furnace ($T = 300°C$) after the chemical scrubber.

Odor control using ozone is an excellent method from the viewpoint of energy efficiency. This method is effective especially for ammonia, trimethyl amine, dimethyl

Table 9
Specification of High-Performance Ozonizers

	Model	
	OZS-MC702D-4WJ	OZS-FC-2/10-AC
Ozone concentration	140 g/Nm³ (10 NL/min)	90 g/Nm³ (1 NL/min)
Ozone production	150 g/h (25 NL/min)	15 g/Nm³ (5 NL/min)
Gas flow rate	10–25 NL/min	< 5 NL/min
Type of gas	Oxygen	Oxygen
Operation pressure	0.12 MPa (12 kgf/cm²)	0–0.05 MPa
Cooling water	6 L/min (at 15°C)	Air cooling
Power supply and power	AC 200V, 50/60 Hz, 2.7 KVA	AC 100V, 50/60 Hz, 0.8 KVA
Environmental condition	Indoor, 5–35°C	Indoor, 20°C (5–35°C)
Size	W600 × D700 × H1100 mm	W300 × D500 × H700 mm
Mass	About 130 kg	About 45 kg

Source: Courtesy of Masuda Research Inc.

disulfide, and so forth. The following chemical reactions occur when odor components are removed by ozone:

$$\text{Ammonia:} \qquad 2NH_3 + O_3 \rightarrow N_2 + 3H_2O \tag{79}$$

$$\text{Trimethyl amine:} \qquad 3(CH_3)_3 N + O_3 \rightarrow 3(CH_3)_3 NO \tag{80}$$

$$(CH_3)_3 N + 3O_3 \rightarrow CH_3NO_2 + 2CO_2 + 3H_2O \tag{81}$$

$$\text{Dimethyl sulfide:} \qquad 3(CH_3)_2 S + O_3 \rightarrow (CH_3)_2 SO \tag{82}$$

$$(CH_3)_2 S + O_3 \rightarrow (CH_3)_2 SO_3 \tag{83}$$

$$\text{Dimethyl disulfide:} \quad 2(CH_3)_2 S_2 + H_2O + O_3 \rightarrow 2CH_3SO_3H \tag{84}$$

$$3(CH_3)_2 S_2 + H_2O + 5O_3 \rightarrow 3(CH_3)_2 S_2O_5 \tag{85}$$

There are many unknown mechanisms in the reaction of radicals and odor components. It is known that radicals are used the decomposition of odor ingredients. When a high level of odor control cannot be obtained, the process combined with the NaClO scrubbing method is recommended.

2.9.3. Design Example of Ozone Oxidization System for Odor Removal

Ozone water is formed by a ozonizer as shown in Table 9, and odorous gas passes through the chemical scrubber containing Raschig rings and ozone water flows. Design conditions such as flow rate, temperature, and odor-removal efficiency are the same as the values shown in Table 8. Objective odorous gases are ammonia, trimethyl amine, dimethyl sulfide, and dimethyl disulfide.

The diameter of a washing tower, gas speed, the amount of water, and the height of the tower are determined based on the similar calculation to the example in Section 2.6. The details are omitted here. The most important point in designing is the estimation of the

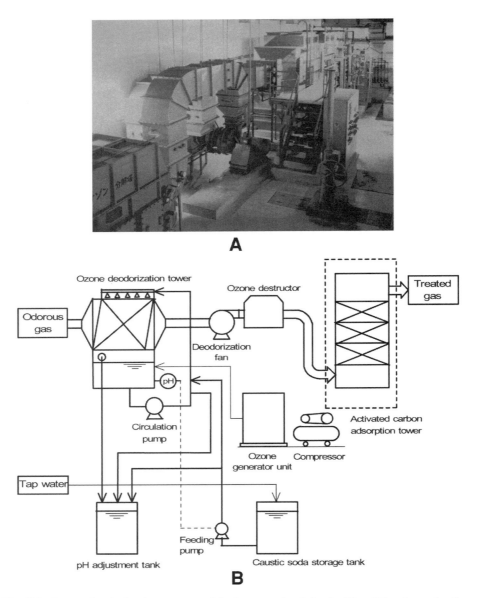

Fig. 37. Ozone deodorization system: **(a)** photograph of the facility; **(b)** schematic diagram. (Courtesy of NGK Insulators, LTD.)

ozonizer capacity. From the main chemical reactions between ozone and the odor ingredients as shown in Eqs. (79)–(85), the amount of required ozone can be calculated as follows:

$$\text{For ammonia: } M_{12} = 10{,}993\left(\frac{\text{Nm}^3}{\text{h}}\right) \times (5 - 0.7) \times 10^{-6} \times \frac{48\left(\frac{\text{kg}}{\text{kmol}}\right) \times 0.5}{22.4\left(\frac{\text{Nm}^3}{\text{kmol}}\right)} = 0.0506\frac{\text{kg}}{\text{h}} \quad (86)$$

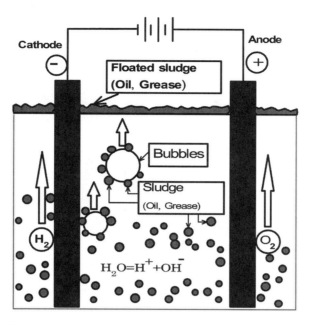

Fig. 38. Principle of electrochemical wastewater treatment.

For trimethyl amine: $M_{13} = 10,993 \times (0.1 - 0.002) \times 10^{-6} \times \dfrac{48}{22.4} = 0.0023 \dfrac{kg}{h}$ (87)

For dimethyl sulfide: $M_{14} = 10,993 \times (0.3 - 0.003) \times 10^{-6} \times \dfrac{48 \times 0.5}{22.4}$
$$= 0.0035 \dfrac{kg}{h}$$ (88)

For dimethyl disulfide: $M_{15} = 10,993 \times (0.3 - 0.003) \times 10^{-6} \times \dfrac{48 \times 1.2}{22.4}$
$$= 0.0083 \dfrac{kg}{h}$$ (89)

A total amount of needed O_3: $M_t = M_{12} + M_{13} + M_{14} + M_{15} = 65 \, g/h$ (90)

Therefore, the OZS-MC702D-4WJ type ozonizer (capacity = 150 g/h) shown in Table 9 has enough capacity.

2.10. Electrochemical Method

Many studies have been reported on industrial wastewater (12–14,16,18,21,92–94) treatment with various methods. Recently, an electrochemical treatment based on the nonbiological method was investigated and some studies were reported (94,95). The principle is based on the electrolysis of water, as shown in Fig. 38. A weak DC voltage is applied between the electrodes arranged inside the wastewater. The electrons that come out of the negative electrode (cathode) decompose a water molecule into oxygen

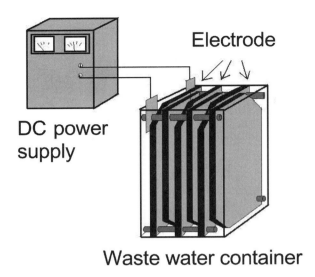

Fig. 39. Schematic of the reactor and power supply.

and hydrogen atoms and generate fine air bubbles. In general, the wastewater dissolves various ions and molecules and ionization, oxidization, and reduction take place under the activated chemical condition. Because the surfactants and chemical and oxidation agents are added depending on the case, the condensation of molecules and particle are promoted. As a result, the impurities become a large cluster that sticks to air bubbles, rises to the surface, and are removed mechanically. Odor components in the wastewater are removed chemically and mechanically with this process. Moreover, reduction of the degree of wastewater color (decolorization) can be achieved. Liquid, which contains little solute, has a low electric conductivity and requires high electric power or running cost. When the electric conductivity of wastewater is too low, the separation of contaminations and odor removal cannot be performed efficiently. The features of the electrolysis method are as follows (94,95):

- Continuous operation under atmospheric temperature and pressure
- Easy control and high-selectivity chemical reactions expected
- Small amount of residual sludge
- Simultaneous oxidation and reduction without chemicals addition

However, the details of chemical reactions in wastewater, including many kinds of chemical component, are not clarified at present and further investigation is needed.

An example of the application of the electrolysis method follows. The dust-collecting electrodes in the electric air cleaner for cigaret smoke removal installed in an airport are periodically washed using alkaline solvents. The main components of the wastewater is tar from cigaret smoke. Because the wastewater is muddy and black and has a bad smell, also it cannot be drained as it is. Then, the electrolysis method is studied as preprocessing before drainage. The schematic of the experimental apparatus is shown in Fig. 39. This reactor consists of acrylic plates 5 mm thick, 100 mm deep, 112 mm wide, and 165 mm

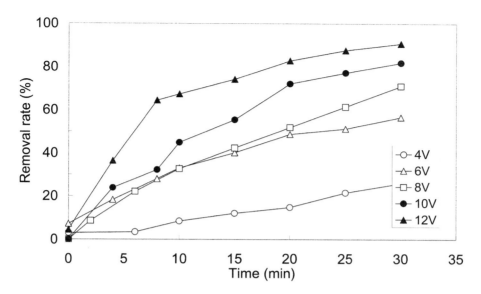

Fig. 40. Time dependent removal efficiencies of impurity components.

high with a tested waste solution of 800 mL. The electrodes are placed in parallel at an interval of 6–9 mm. As electrodes, three kinds of material (stainless steel, aluminum, carbon) are tested. The carbon electrode has excellent durability but is expensive. The other electrodes are degraded a certain degree. After the DC voltage application, the turbidity of waste solution is obtained using a spectrophotometer.

The time-dependent removal efficiency of organic matter in the sample solution based on the turbidity is shown in Fig. 40 when the carbon electrode is used. The removal efficiency increases with increase in time. For each voltage, higher efficiency is achieved with higher voltage. The power consumption is 180 W on average for a voltage of 12 V. The removal efficiency becomes maximum at 12 V and reaches 91% after 30 min. Because the turbidity of the sample waste solution does not change only with heat addition of the same energy using an electric heater, decolorization occurs with electrochemical reactions. Moreover, odor removal is also remarkable after the treatment.

Furthermore, in recent years, the method of discharge in the Aquarius solution has been attracted attention as a new water processing method. The electrodes are placed inside or outside the liquid phase and pulse high voltage is applied between the electrodes. The chemical reactions are promoted by the high-speed electron and induced radicals such as O and OH. Many experiments have been reported on the treatment of industrial wastewater, especially the removal of odor and included phenol (96,97).

3. BIOLOGICAL METHOD

3.1. Introduction

Odor abatement systems typically handle large volumes of air containing rather low concentrations of odorants and need to achieve high removal efficiencies (98). Odor

intensities and odorant concentrations are related to the so-called Weber–Fechner psy-
chophysical powerlaw, which can be mathematically expressed as

$$I = kC^n \qquad (91)$$

where I is odor intensity, C is odorant concentration, and k and n are odor-dependent
coefficients. High removal efficiencies of odorants are needed in order to achieve effective
odor control.

3.2. Biological Control

Biological odor control methods include biofiltration and bioscrubbing. The two methods are similar in fundamental physical, chemical, and biological processes involved for
odor control, which are adsorption, absortion, and biological oxidation. They are different
in physical structures and engineering designs. To remove odorous compounds from the
contaminated airstream, biofiltration uses a moist solid media, whereas bioscrubbing uses
a liquid media. Once the odorous compounds are sorped in the media, they are degraded
and oxidized by microorganims into odorless inorganic products, such as carbon dioxide,
water, and minerals. Biofilters use biofiltration and biotrickling filters and bioscrubbers
use bioscrubbing as the working mechanism for odor removal. All three biological systems have over 30 yr of history in their applications for odor control (99). Biofilters are
more commonly used than biotrickling filters and bioscrubbers, especially for composting
facilities and industrial and agricultural operations.

The driving forces for the removal of a given compound in the biological treatment system is its solubility in water and biodegradability. High treatment performance of biofiltration or bioscrubbing is normally associated with compounds of low
molecular weight and high water solubility. Most odorous compounds encountered
at wastewater-treatment plants, composting facilities, animal operations, and some
industrial operations are easily degradable compounds, having over 90% removal
efficiencies by biofilter treatment. They include aldehydes, alcohols, mercaptans,
sulfides, amines, ammonia, ketones, volatile fatty acids, and monocyclic aromatics.
The relative biodegradability of these compounds and other VOCs are described by
Michelsen (100). Hydrogen sulfide and ammonia are rapidly degradable inorganic
compounds; alcohols, aldehydes, amines, and volatile fatty acids are rapidly degradable organic compounds; esters, ketones, mercaptans are medium degradable organic
compounds; and hydrocarbons are slowly and very slowly degradable organic compounds.

Organic compounds are degraded and oxidized by heterotropic bacteria, fungi,
and other organisms, with end products being carbon dioxide, water, ammonia, phosphate, and sulfate. Ammonia is further oxidized into nitrite and nitrate by two
autotrophic bacteria, called nitrosomonas and nitrobacteria. Hydrogen sulfide is
oxidized by sulfur-oxidizing bacteria into sulfate. Production of nitric acids and sulfuric acids tend to decrease the pH of the filter media. The filter media needs to have
certain level of buffering capacity to mediate the pH change. Sometimes, limestone
is mixed with the packing materials to provide sufficient initial alkalinity in the filter media. If the acid accumulation is excessive, washing of filter media by using
water may be needed.

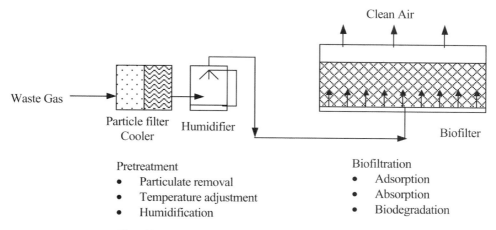

Fig. 41. Biofilter system and working principles.

3.3. Working Principles of Biological Treatment Processes

Biofilters are made of moist porous materials, which are used as a filter media and support the growth of microorganisms, including bacteria, actinomycetes, and fungi. When the waste gas stream passes through the filter media, odorous compounds are first adsorbed onto solid particles and then absorbed into the moist biofilm, which is formed by the microorganims on the solid particles, and finally degraded by the microorganisms. The filter media has three primary functions: providing large surface areas for microbial attachment, providing energy source and nutrients to the microorganisms, and providing a stable structural matrix to allow the gas stream to pass through.

The biofilters packed with natural organic materials normally rely on indigenous microorganisms present in the filter media to carry out biological degradation. For some applications, inoculation of special organisms may be necessary to achieve high removal efficiencies. For example, some studies show that inoculation of sulfur-oxidizing bacteria, such as *Thiobacillus thioparus* (101,102) improved the performance of biofilters for removing sulfides and mercaptants. The biofilters packed with synthetic materials need to be seeded with aclaimated bacterial culture, normally using activated sludge (*see* Fig. 41).

Biotrickling filters are similar to biofilters, but contain inert scrubbing packing materials instead of organic materials and operate with a recirculating liquid flowing over the packing (103). The recirculating liquid is initially inoculated with microorganisms, and over time, a biofilm layer establishes itself on the packing shortly after start-up. The bioscrubber, also called an aeration basin, uses a water tank or basin to remove contaminants from the airstream as it passes through a diffuser system. The schematics of the biotrickling filter and bioscrubber systems are shown in Figs. 42 and 43, respectively. Both systems generate liquid wastes that need to be managed properly. Nutrients and water are regularly added to replenish those lost via volatilization and wastewater removal. The pH of the liquid media is easily monitored and controlled by the automatic addition of acid or base.

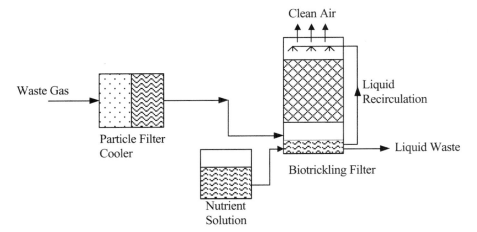

Fig. 42. Biotrickling filter system.

3.4. Design of Biofilters

Because biofilters are the mostly commonly used biological treatment system for odor control, its design considerations are outlined in this subsection. The designs and performance of biotrickling filters and bioscrubbers can be referred to in other publications (104–106). Creating and maintaining a healthy environment for microbial growth is critical for the success of biofiltration. A healthy environment is characterized as sufficient oxygen in the biofilm, absence of toxic compounds, adequate nutrients and moisture, and suitable levels of temperature and pH. Materials of high porosity and high water-retention capability are desirable for use as the filter media. Soil, compost, and peat are commonly used. Compost can be produced from wood chips, bark, leaves, sewage sludge, or other organic materials (98,107). Synthetic materials, such as activated carbon, ceramics, and plastics, have also been used in some applications. The

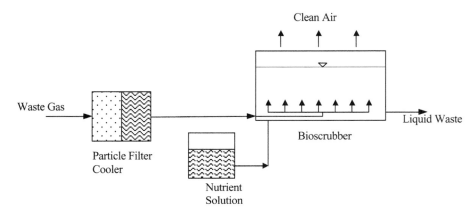

Fig. 43. Bioscruber system.

replacement of the filter media, if it is biodegradable, may be required after 3–5 yr because of degradation and compaction of the filter media.

A typical biofilter system includes the following components: gas collection and transport, pretreatment, biofilter, moisture control and waste removal, and posttreatment, as shown in Fig. 41. Pretreatment includes particulate removal and temperature and moisture adjustment. Particulate removal prevents clogging of the biofilter. Most biofilters operate in the mesophilic temperature range (25–40°C). Therefore, hot off-gas requires cooling to temperatures below 40°C to sustain the predominantly mesophilic microbial populations. Humidification of the waste gas stream is often necessary to achieve near 100% relative humidity prior to the biofilter. An appropriate moisture content of packing media is important to provide enough water to support the growth of microorganisms. A range of 40–70% has been recommended for the optimum moisture content. The optimum moisture content depends on the packing media used and should be as high as possible as long as no deficiency of oxygen and excessive pressure drop in the media are experienced. In addition to humidification of the waste gas stream, water may be added directly into the biofilter periodically using drip irrigation pipes to maintain appropriate moisture contents. If the waste gas contains high amounts of nitrogenous and sulfurous compounds, the biofilters may need to be washed regularly to remove nitrate and sulfate from the packing media to prevent pH drop and chemical toxicity. The wastewater from biofilter washing needs to be collected and managed properly. Generally, oxygen is not considered a limiting factor. However, filter media compaction may create local anaerobic zones. Periodic mixing of the filter media is a common and effective means of maintaining a desired porosity.

Based on structural configurations, biofilters are classified into two types: open bed and close bed. With each type, the biofilters can be designed to have a single bed or multiple beds. Both types have been applied to odor control, but open-bed biofilters are more common. Design features of different types of biofilter have been described in much detail by Skadany et al. (99).

The major design parameters of biofilters include contaminant loading rate, type and volume of packing media, moisture content, pH, temperature, residence time (also called contact time), and pressure drop across the media. The residence time is the single most important design parameter used to size the biofilter. It is defined as the volume of the filter media divided by the volumetric gas flow rate. The residence time required for odor control in most applications is less than 30 s. The overall performance of biofilters is normally evaluated by removal efficiency, which is defined as the difference in influent and effluent contaminant concentrations divided by the influence concentration. Other parameters, such as instantaneous and overall elimination capacities, are also used to describe the performance of biofilters (99). The engineering design of biofilters consists of two major tasks: (1) sizing the biofilters and (2) engineering the physical system. Biofilter sizing determines the type and volume of the packing media required to attain the desired odor-removal efficiency. Engineering of the physical system involves design and equipment selection for fans and pipes for gas collection and transport, particle filter, gas cooling and humidification units, biofilters, irrigation and wastewater drainage system. There are several methods available to size the biofilters, including empirical and mathematical modeling methods. To size the biofilter, the volume

(V) of the filter media is calculated from the required residence time (t_r) and gas flow rate (Q) by using

$$V = T_r Q \qquad (92)$$

where the units for V, T_r, and Q are m^3, s, and m^3/s, respectively. The residence time is related to the odor concentration in the gas stream, the properties of packing media, and the desired odor-removal efficiency. The residence time required for odor control in most applications is less than 30 s. The actual value for the residence time may be obtained from literature on similar applications, mathematical models, or data collected from laboratory and field pilot tests (99). Once the volume of the packing media is determined, the cross-sectional area of the biofilter can be calculated from the volume divided by the depth. The depth of each bed is usually designed to be 1–1.5 m in order to keep the pressure drop below 125 Pa initially.

The cost of biofiltration varies substantially for different applications. The biofilters for odor control are generally designed to be simple and use inexpensive materials and therefore have lower capital costs compared to the biofilters for VOC control. Reported capital costs range from $10–40 ft^3/min of air treated at flow rates less than 100,000 ft^3/min., to $3–5/ft^3/min of air treated at flow rates higher than 10,000 ft^3/min.; the operating costs have been estimated at $2–14/ft^3/min of air treated (108).

NOMENCLATURE

A	Area of chemical scrubber
C	Odorant concentration
C_g	Ground-level concentration
C_H	Henry's constant
C_{inlet}	Inlet odor concentration
C_{outlet}	Outlet odor concentration
d	Diameter of chemical scrubber
D	Diameter of the stack
G	Mass velocity of gas per unit cross-sectional area
h	Height of pellets layer
H	Width of copper mesh electrode
H_b	Height of the building
H_e	Effective stack height
H_{OG}	Height of an overall gas transfer unit
I	Odor intensity
k	Odor-dependent coefficients
L	Mass flow rate per unit cross-sectional area
L_b	Shorter dimension in either height of building or width of building
M	Mass of needed chemicals per unit time
n	Odor-dependent coefficients
N_{OG}	number of gas transfer units
OER	Odor emission rate
p	Partial pressure of gas-phase odorous substances

p_{total}	Gas-phase total pressure
P	Power consumption
q	Flow rate of single reactor
Q	Flow rate
Q_{gas}	Volumetric flow rate of gas at the standard condition
$r_{\text{g/l}}$	Volumetric ratio of gas to liquid
r_0	Distance between discharge wire and outer grounded electrode
SED	Specific energy density
TOER	Total odor emission rate
T_r	Residence time
u	Wind velocity
V	Volume of filter media
V_{ave}	Mean velocity inside reactor
W_b	Width of the building
x, y, z	Cartesian coordinates in Fig. 9
X	Concentration of odor substances resolved in the liquid phase
Y	Volumetric concentration of odorous substances
Z	Height of washing tower
ρ_{gas}	Density of gas or air at standard conditions
σ_y	Diffusion width in the y-direction
σ_z	Diffusion width in the z-direction

REFERENCES

1. J. W. Gardner and P. N. Barlett, *Electronic Noise: Principles and Applications,* Oxford University Press, New York, 1999.
2. J. E. Amoore, *Nature* **198**, 271–272 (1963).
3. J. E. Amoore, *Nature* **199**, 912–913 (1963).
4. R. H. Wright, *The Sense of Smell*, CRC, Boca Raton, FL, 1982.
5. A. Dravnieks, *The Atlas of Odor Character Profiles*, ASTM Data Series DS6I, American Society for Testing and Materials, Philadelphia, 1985.
6. G. Ohloff, *Scent and Fragrances*, Springer-Verlag, Berlin, 1994.
7. J. Salter, *Proc. International Meeting on Odor Measurement and Modelling*, 2000.
8. R. Both, *Proc. International Specialty Conference Air Waste Management Association, Odors: Indoor and Environmental Air*, 1995.
9. F. B. Frechen, *Water Sci. Technol.* **41**(6), 17–24 (2000).
10. B. Howe, *Water Wastes Digest* **43**(1), 16–17 (2003).
11. H. J. Rafson (ed.) *Odor and VOC Control Handbook*. McGraw-Hill, New York, 1998.
12. G. Martin and P. Laffort (eds.), *Odors and Deodorization in the Environment*, Wiley, New York, 1994.
13. Water Environment Federation Staff and American Society of Civil Engineers Staff, *Odor Control in Wastewater Treatment Plants*, Water Environment Federation, Alexandria, Virginia, 1995.
14. Task Force on Air Toxics Staff (ed.), *Odor and Volatile Organic Compound Emission Control for Municipal and Industrial Wastewater Treatment Facilities*, Water Environment Federation, Alexandria, Virginia,1994.
15. N. Hirayama, (ed.), In *JSME Mechanical Engineers' Handbook—Engineering C8: Environmental Equipment* Japan Society of Mechanical Engineers, Tokyo, 1997, pp. 234–238 (in Japanese).

16. J. Hermia, J. Chaouki, and S. Vigneron (eds.), *Proceedings of 2nd International Symposium on Characterization and Control of Odours and VOC in the Process Industries*, Elsevier Science, Amsterdam, 1994.

17. R. P. G. Bowker and J. M. Smith, *Odor and Corrosion Control in Sanitary Sewerage Systems and Treatment Plants,* Noyes, 2000.

18. V. C. Nielsen, P. L. L'Hermite, and J. H. Voorburg (eds.), *Volatile Emissions from Livestock Farming and Sewage Operations*, Elsevier Science, Amsterdam, 1988.

19. F. B. Frechen and R. Stuetz, *Odours in Wastewater Treatment: Measurement, Modelling and Control*, IWA Publishing, 2001.

20. A. J. Buonicore and W. T. Davis (eds.), *Air Pollution Engineering Manual.* Air & Waste Management Association Van Nostrand Reinhold, New York, 1992, pp. 147–154.

21. K. Hinokiyama, *Jiturei ni Miru Datsusyu Gijyutu* [*Odor Control Technologies with Industrial Applications*], Kogyo Chosa Kai, et al. Tokyo, 1999 (in Japanese).

22. T. Ando, M. Izumo, S. Komatsu, et al. (eds.), *Saisin Boudassyu Gijyutsu Syusei* [*New Odor Control Technologies*], Enu Tee Esu, Tokyo, 1997 (in Japanese).

23. I. Suffet, *Advances in Taste-and-Odor Treatment and Control*, American Water Works Association, Denver, Colorado, 2001.

24. American Chemical Society (ed.), *Odor Quality and Chemical Structures*, American Chemical Society, Washington, D.C., 1987.

25. L. Louden and J. Weiner, *Odors and Odor Control*, Institute of Paper Science & Technology, Atlanta, Georgia, 1976.

26. H. E. Hesketh and F. L. Cross, Jr., *Odor Control Including Hazardous-Toxic Odors*, CRC, Boca Raton, FL, 1988.

27. National Research Council, Board on Environmental Studies and Toxicology, *Odors from Stationary and Mobile Sources*, National Academy Press, Washington, DC, 1979.

28. S. Ishiguro and S. Sugawara, *Koryo* **130**, 31–39 (1981) (in Japanese).

29. Taikisha Corp., *Products Catalog*, Taikisha Corp., Tokyo, 2002.

30. I. Namiki and H. Fujimura, *Yosui to haisui* **44**(5), 388–393 (2002) (in Japanese).

31. F. B. Cooling III, C. L. Maloney, E. Nagel, et al., *Appl. Environ. Microb.* **62**(8), 2999–3004 (1996).

32. E. D. Burger, J. M. Odom, and L. B. Gosser, *Proc. 3rd Annual Int. Petroleum Environmental Conference*, 1996.

33. F. Pasquill, *Meteorol. Mag.* **90**, 33–49, Great Britain, (1961).

34. F. Pasquill, and F. B. Smith, *Atmospheric Diffusion*, 3rd ed., Ellis Horwood, London, 1983.

35. S. Okamoto, *Taiki Kankyo Yosoku Kougi* [*Lectures on Atmospheric Environment Prediction*], Gyosei, Tokyo, 2001 (in Japanese).

36. Environment Agency in Japan (ed.), *Chituso Sankabutu Souryo Kisei Manual* [*Manual on Regulation for Total Emission of Nitrogen Oxide*], Kogai Kenkyu Taisaku Center, Tokyo, 1982 (in Japanese).

37. D. B. Turner, *Workbook of Atmospheric Dispersion Estimates*, 2nd ed., Lewis, New York, 1994.

38. A. H. Huber and W. H. Snyder, *Atmos. Environ.* **16**(12), 2837–2848 (1982).

39. A. H. Huber and W. H. Snyder, *3rd Symposium on Atmospheric Turbulence, Diffusion, and Air Quality*, 1976, pp. 235–242.

40. S. Okamoto, *J. Japan Soc. Air Pollut.* **27**(2), A25–A35 (1992).

41. G. R. Carmichael, D. M. Cohen, S. Y. Cho, et al., *Comput. Chem. Eng.* **13**(9), 1065–1073 (1989).

42. R. D. Sylor and R. I. Fernandes, *Atmos. Environ.* 27A, 625–631, New York, New York, (1993).

43. K. Kogami, *Kenchiku Setsubi To Haikan Koji* 31–38 (1997) (in Japanese).

44. H. Kohno and S. Koganei, *Preprints of Society of Heating, Air-Conditioning and Sanitary Engineers Japan*, Society of Heating, Air-Conditioning and Sanitary Engineers Japan Koriyama-City, 1992, pp. 765–768 (in Japanese).

45. Asahi Kogyosha Co., *Aromatic Air-Conditioning System "Honoka,"* Products Catalog and Manual, Asahi Kogyosha Co, LTD, Tokyo, 1992 (in Japanease).

46. P. C. Wankat, *Separ. Sci.* **9**(2), 85–116 (1974).

47. D. Diagne, M. Goto, and T. Hirose, *J. Chem. Eng. Jpn.* **27**, 85–89, Tokyo, Japan, (1994).

48. H. Tominaga, *Zeolite no Kagaku to Ouyo* [*Science and Applications of Zeolite*], Kodansha LTD., 1987 (in Japanese).

49. M. Suzuki, *Adsorption Engineering*, Kodansha LTD Elsevier Science, Amsterdam, 1990.

50. NGK Insulators, *NGK Deodorization Systems*, Products Catalog, NGK Insulators, LTD, Nagoya, 2002.

51. M. Okubo, J. Mine, T. Kuroki, et al., *Proc. 2nd Asia Aerosol Conference*, 2001, pp. 361–362.

52. M. Okubo, T. Yamamoto, T. Kuroki, et al., *J. Inst. Electrostat. Jpn.* **25**(6), 328–329 (2001) (in Japanese).

53. T. Yamamoto and C. L. Yang, *Proc. IEEE/IAS Annual Meeting*, 1998, pp. 1877–1883.

54. A. Ogata, D. Ito, K. Mizuno, et al. *Proc. IEEE/IAS Annual Meeting*, 1999, pp. 1467–1472.

55. T. Yamamoto, M. Okubo, and T. Kuroki, *Trans. Inst. Fluid-Flow Mach.* **107**, 111–120 (2000).

56. M. Okubo, T. Kuroki, H. Kametaka, et al., *IEEE Trans. Ind. Applic.* **37**(5), 1447–1455 (2001).

57. N. Miyoshi, et al., Development of new concept three-way catalyst for automotive lean-burn engines, SAE Paper 950809, 1995.

58. I. Hachisuka, et al., SAE Paper 001196, 2000.

59. M. Schiavello, *Heterogeneous Photocatalysis*, Wiley, New York, 1997.

60. M. Chanon (ed.), *Homogeneous Photocatalysis*, Wiley, New York, 1997.

61. D. F. Ollis, and H. Al-Ekabi (eds.), *Proc. 1st International Conference on TiO$_2$ Photocatalyical Purification and Treatment of Water and Air*, 1993.

62. Daikin Industries, *Electric Air Cleaner*, Product catalog, Daikin Industries, LTD, Osaka, 2001 (in Japanese).

63. T. Yamamoto, M. Okubo, K. Hayakawa, et al., *IEEE Trans. Ind. Applic.* **37**(5), 1492–1498 (2001).

64. A. Chakrabarti, A. Mizuno, K. Shimizu, et al., *IEEE Trans. Ind. Applic.* **31**(3), 500–506 (1995).

65. T. Oda, T. Kato, T. Takahashi, et al., *IEEE Trans. Ind. Applic.* **34**(2), 268–272 (1998).

66. S. Masuda, S. Hosokawa, X. Tu, et al., *J. Electrostat.* **34**, 415–438 (1995).

67. J. Y. Park, I. Tomicic, G. F. Round, et al., *J. Phys. D: Appl. Phys.* **32**, 1006–1011 (1999).

68. B. M. Penetrante, and S. E. Schultheis (eds.), *Non-thermal Plasma Techniques for Pollution Control (Part B: Electron Beam and Electrical Discharge Processing)*, NATO ASI Series, Series G: Ecological Science Vol. 34, Part B, Splinger- Verlag, Heidelberg, Germany, (1993).

69. Masuda Research Inc., *Plasma Deodorization System—ADO Series*, Products catalog, Masuda Research Inc., Tokyo, 2002 (in Japanese).

70. R. Zhang, T. Yamamoto, and D. S. Bundy, *IEEE Trans. Ind. Applic.* **32**(1), 113–117 (1996).

71. C. D. Cooper and F. C. Alley, *Air Pollution Control—A Design Approach*, Waveland, 1994, pp. 151–179.

72. G. W. Penny, *Electr. Eng.* **56**, 159–163 (1937).

73. H. Lim, K. Yatsuzuka and K Asano, *J. Inst. Electrostat. Jpn.* **22**(3), 145–152 (1998) (in Japanese).

74. Y. Kawada, et al., *Proc. IEEE/IAS Annual Meeting*, 1999, pp. 1130–1135.

75. S. Jayaram, G. S. P. Castle, J. S. Chang, et al., *IEEE Trans. Ind. Applic.* **32**(4), 851–857 (1996).

76. A. Zukeran, P. C. Looy, A. Chakrabarti, et al., *IEEE Trans. Ind. Applic.* **35**(5), 1184–1191 (1999).

77. A. Mizuno, Y. Yamazaki, H. Ito, et al., *IEEE Trans. Ind. Applic.* **28**(3), 535–540.
78. S. Masuda, S. Hosokawa, X. L. Tu, et al., *IEEE Trans. Ind. Applic.* **29**(4), 774–780.
79. A. Mizuno, Y. Kisanuki, M. Noguchi, et al., *IEEE Trans. Ind. Applic.* **35**(6), 41–45 (1999).
80. H. Yoshida, Z. Marui, M. et al., *J. Inst. of Electrostat. Jpn.* **13**(5), 425–430 (1989) (in Japanese).
81. Y. Kisanuki, M. Yoshida, K. Takashima, et al., *J. Inst. of Electrostat. Jpn.* **24**(3), 153–158 (2000)(in Japanese).
82. H. Suda, T. Ueno, T. Yamauchi, et al., *Plasma Discharge Deordorizing System*, Matsushita Electric Works, Ltd. Technical Report, 2001, pp.58–63 (in Japanese).
83. M. Okubo, T. Yamamoto, T. Kuroki, et al., *IEEE Trans. Ind. Applic.* **37**(5), 1505–1511, New York, New York, (2001).
84. W. C. Hinds, *Aerosol Technology—Properties, Behavior, and Measurement of Airborne Particles*, Wiley, New York, 1982.
85. S. K. Friedlander, *Smoke, Dust, and Haze—Fundamental of Aerosol Dynamics*, Oxford University Press, New York, 2000.
86. R. G. Rice and M. E. Browning, (eds.), *Ozone: Analytical Aspect and Odor Control*, Pan American Group/International Ozone, 1976.
87. M. Kuzumoto, *J. Plasma Fusion Res.* **74**(10), 1144–1150 (1998) (in Japanese).
88. Y. Kamase, T. Mizuno, and M. Sakurai, *Ishikawajima-Harima Eng. Rev.* **40**(1), 3–6 (2000) (in Japanese).
89. Masuda Research Inc., *Ceramic Ozonizer and Small Ozonizers,* Products catalog, Masuda Research Inc., Tokyo, Japan 2002.
90. J. A. Libra and A. Saupe, *Ozonation of Water and Wastewater: A Practical Guide to Understanding Ozone and Its Application*, Wiley, New York, 2000.
91. J. J. McKetta (ed.), *Encyclopedia of Chemical Processing and Design: Wastewater Treatment with Ozone to Water and Wastewater Treatment*, Marcel Dekker, New York, New York, 1999, Vol. 66.
92. F. L. Burton and H. D. Stensel, *Wastewater Engineering: Treatment and Reuse*, McGraw-Hill, New York 2000.
93. D. H. Liu and B. G. Liptak (eds.), *Wastewater Treatment*, Lewis, New York, 2000.
94. Y. V. Basova, *J. Appl. Electrochem.* **68**(8), 639–644 (2000).
95. S. Torii, *Yuuki Denkai Gosei [Electroorganic Syntheses]*, Kodansha, LTD., 1981 (in Japanease).
96. D. R. Grymonpre, W. C. Finney, and B. R. Locke, *Chem. Eng. Sci.* **54**, 3095–3105 (2000).
97. B. Sun, M. Sato, and J. S. Clements, *Environ. Sci. Technol.* **34**, 509–513 (2000).
98. H. V. Langenhove, E. Wuyts and N. Schamp, *Water Res.* **20**(12), 1471–1476.
99. G. J. Skladany, M. A. Deshusses, J. S. Devinny, in *Odor and VOC Control Handbook* (H. J. Rafson, ed.), McGraw-Hill, New York, 1998, pp. 8-150–8-191.
100. R. F. Michelsen, in *Handbook of Air Pollution Control Engineering and Technology.* (J. C., Mycock, J. D. McKenna, and L. Theodore, eds.), CRC Boca Raton, FL, 1995, pp. 375–394.
101. Y. Tanji, T. Kanagawa, and E. Mikami, *J. Ferment. Bioeng.* **67**(4), 280–285 (1989).
102. K. Cho, M. Hirai, and M. Shoda, *J. Ferment. Bioeng.* **73**(1), 46–50 (1992).
103. A. P. Togna, and M. Singh, *Environ. Prog.* **13**(2), 94–97.
104. N. H. Jansen, and K. Rindel, *Water Sci. Technol.* **41**(6), 155–164.
105. L. Y. Wu, Y. Loo, and L. C. C. Koe, *Water Sci. Technol.* **44**(9), 295–299.
106. A. Ruokojarvi, M. T. Hartkainen, M. Olkkonen, et al., *Environ. Technol.* **21**, 1173–1180.
107. R. D. Pomeroy, *J. Water Pollut. Control. Fed.* **54**, 1541–1545.
108. T. O. Williams and R. A. Boyette, in *Proc. Control of Odors and VOC Emissions Conference,* Water Environmental Federation, Alexandria, VA, 1997, pp. 3.47–3.54.

9

Radon Pollution Control

Ali Gökmen, İnci G. Gökmen, and Yung-Tse Hung

CONTENTS

1. INTRODUCTION

Human beings are exposed to two sources of radiation in the environment: natural and man-made. Natural sources include radioactive radon, radioisotopes with a long half-life, such as potassium in the body, cosmic rays (energetic γ-rays and particles from the sun and interstellar space), and some rocks. Various sources of artificial radiation include medical X-rays, nuclear medicine for cancer treatment, and some consumer products containing radioisotopes. Natural sources of radiation account for 82% of total exposure for humans. A common radioactive element is radium, one of whose decay products, radon, poses health concerns. Radon emanates from rock, soil, and underground water as a gas. In the solar system, various radioisotopes of radon gas form from decay of radioactive uranium and thorium elements found naturally. The contributing effects from natural and man-made radiation sources on human beings is shown in Fig. 1.

Radon emanating from soil fills the atmosphere but eventually transmutes to other elements and is removed. The concentration of radon gas reaches equilibrium because of this influx and outflux of gas in the atmosphere. However, the concentration of radon may show significant variation in closed living places. Inhalation of radioactive radon gas is a threat to public health. Homes built on granite and phosphate rocks containing uranium ore may expose their owners to health risks resulting from high radon concentrations. Homes with improved thermal insulation and minimum air circulation can be even more significant candidates for radon buildup.

Several important issues related to radon are introduced in the four sections of this chapter. Section 1 discusses radon and its decay products. Section 2 presents various instrumental methods of radon measurement. Section 3 focuses on the health effects of

From: *Handbook of Environmental Engineering, Volume 2: Advanced Air and Noise Pollution Control*
Edited by: L. K. Wang, N. C. Pereira, and Y.-T. Hung © The Humana Press, Inc., Totowa, NJ

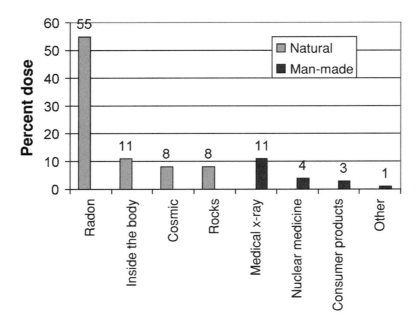

Fig. 1. Percent distribution radiation doses to humans from various sources (Cobb, C.E. (1989) Living with Radiation, *National Geographic*, **175**(4), pp 403–437).

low-level radioactivity of radon and its decay products. Finally, Section 4 discusses radon mitigation in buildings.

Radon (Rn) is a colorless, odorless, and tasteless inert gas, which decays to form a series of radioactive particles. An "isotope" of an element has the same number of protons (same atomic number) but a different number of neutrons and different mass number (total number of neutrons and protons). Some isotopes of an element may be radioactive and emit radiation from their nuclei. There are three principal kinds of radiation emission: α-, β-, γ-rays. γ-Rays are highly penetrating rays similar to X-rays. β-Rays are electrons traveling at high speeds. α-Rays are also particles; each α particle is composed of two protons and two neutrons—equivalent to a helium nucleus. Each time an α-ray is expelled from an atomic nucleus, the atom changes to a lighter, new element and the α-particle becomes a helium atom. If the new element is also radioactive, emission of radiation will continue until, eventually, stable, nonradioactive nuclides are formed.

1.1. Units of Radioactivity

The unit of radioactivity is expressed as disintegrations per second, in becquerel (Bq). The older, but commonly used unit of radioactivity is Curie (Ci). It is defined as the activity of 1 g of pure Ra-226 obtained from uranium. The activity in Ci is equal to 1 Ci$=3.7\times10^{10}$ Bq.

The decay of a radioisotope is described by an exponential law for first-order reaction:

$$A = -dN/dt = \lambda N \qquad (1)$$

Here, A is the activity (in Bq), N is the number of radioactive atoms, λ is the decay constant, related to the half-life, and t is the time. The solution of this first-order rate equation yields an exponential relation:

$$N = N_0 e^{-\lambda t} \tag{2}$$

Here, N_0 is the number of atoms present initially at time $t=0$ and N is the number of atoms present at time t. The half-life of an isotope, $t_{1/2}$, is defined as the time necessary for the initial number of radioisotopes to decay into half of the initial number; that is, when the time is equal to the half-life $t=t_{1/2}$, the number of particles N will be equal to the half of the initial number of particles, $N=N_0/2$. The relation between the decay constant λ and the half-life $t_{1/2}$ is

$$\lambda = 0.693/t_{1/2} \tag{3}$$

The specific activity of the radioisotope is defined as the activity of 1 g of pure radioisotope. The specific activity of any radioisotope can easily be calculated from Eq. (1) provided its half-life is known. As an illustration, the specific activity of 1 g of U-238 can be calculated using its half-life of 4.47×10^9 yr, which is converted into seconds in the following calculation:

$$A = \lambda N = \frac{0.693}{4.47 \times 10^9 \times 365 \times 24 \times 3600} (s^{-1}) \frac{1 \times 6.02 \times 10^{23}}{238} (atoms)$$

$$A = 1.24 \times 10^5 \; Bq/g \; U\text{-}238, \; \text{specific activity (activity of 1 g U-238)},$$

Here, 6.02×10^{23} is Avogadro's number, the number of atoms in 238 g (1 mol) of U. The activity of a trace amount of uranium in soil can be expressed in 1 kg soil if the abundance of uranium in the soil is known. The concentration of uranium in the soil is on the order of several parts per million (ppm, or gram of radioactive isotope in 1 ton of soil sample). Thus, the activity of uranium in 1 kg of soil, assuming 1 ppm uranium concentration, can be predicted as

$$A = 1.24 \times 10^5 \left(Bq/g \; U\right) \times 1 \times 10^{-6} \left(g \; U/g \; soil\right) \times 1000 \; \left(g \; soil/kg \; soil\right)$$

$$A = 124 \; Bq/kg \; soil$$

1.2. Growth of Radioactive Products in a Decay Series

Different radon isotopes are produced by the radioactive decay of radium, which results from the decay of uranium, U-238, U-235, and thorium, Th-232, which occur naturally in some rocks. Some examples of these reactions can be written as follows ($\cdots$ indicates the presence of other decay products):

Uranium series:

$$^{238}U \rightarrow \ldots {}^{226}Ra \rightarrow {}^4He + {}^{222}Rn \rightarrow \ldots {}^{206}Pb \; (stable)$$

Thorium series:

$$^{232}Th \rightarrow \ldots {}^{224}Ra \rightarrow {}^4He + {}^{220}Rn \rightarrow \ldots {}^{208}Pb \; (stable)$$

Actinium series:

$$^{235}U \rightarrow \ldots {}^{223}Ra \rightarrow {}^4He + {}^{219}Rn \rightarrow \ldots {}^{207}Pb \; (stable)$$

Table 1
Isotopes in the U-238 Decay Series

Nuclide	Half-life	Decay particle
U-238	4.5×10^9 yr	α
Th-234	24 d	β
Pa-234	1.2 min	β
U-234	2.5×10^5 yr	α
Th-230	8×10^4 yr	α
Ra-226	1.6×10^3 yr	α
Rn-222	3.8 d	α
Po-218	3 min	α
Pb-214	27 min	β
Bi-214	20 min	β
Po-214	160 μs	α
Pb-210	22 yr	β
Bi-210	5 d	β
Po-210	138 d	α
Pb-206	Stable	

The uranium series starts with U-238 and ends with Pb-206. Each radioactive element has its own decay rate. The radioisotopes, decay modes, and half-lives of each decay product of the U-238 series are shown in Table 1.

In a decay series, radioisotopes have different modes of decay and different half-lives that may change many orders of magnitude from one isotope to the next. For example, the parent radioisotope U-238 in a uranium series decays to its daughter Th-234 by α-particle emission with a half-life of 4.47×10^9 yr, but another radioisotope in the same series, Ra-226, decays to Rn-222 by α-particle emission with a half-life of only 1600 yr.

In a decay series, if the daughter has a much shorter half-life than the parent and when radioisotopes of a decay series are kept together long enough (about five half-lives of the daughter), they reach the "secular equilibrium" and the parent and daughter radioisotope activitites become equal. For example, Ra-226 isolated from all of its decay products by selective precipitation would reach equilibrium after 8000 yr (five times the half-life of Ra-226) with its parent isotope U-238, and then they both will have the same activity. Similarly, Rn-222 ($t_{1/2} = 3.8$ d) and its decay products Po-218 ($t_{1/2} = 3$ min), Pb-214 ($t_{1/2} = 27$ min), Bi-214 ($t_{1/2} = 20$ min), Po-214 ($t_{1/2} = 160$ μs) will reach the same activity after about 210 min. The change in the activity of radon and its shorter-lived decay products are illustrated in Fig. 2 in a time scale of 300 min, where initially only radon gas is permitted in a chamber passing through a filter. The activity of Rn-222 with $t_{1/2} = 3.8$ d remains nearly constant in this time interval. However, the activities of shorter-lived decay products are initially zero, but the activity of Po-218 ($t_{1/2} = 3$ min) will reach the same activity as Rn-222 in a time period less than 30 min and Po-214 ($t_{1/2} = 19.7$ min) will attain the activity of Rn-222 in a time period of about 200 min.

The thorium decay series starts with Th-232 and ends with Pb-208, and the actinium series starts with U-235 and ends with Pb-207. Different isotopes of radon, Rn-222,

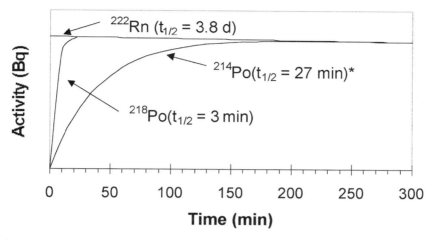

Fig. 2. The change in activity of Rn-222 ($t_{1/2}$=3.8 d) and its decay products Po-218 ($t_{1/2}$=3.1 min) and Po-214 ($t_{1/2}$=164 μs). Note: In a time scale of 300 min. Initially, radon is isolated from all its decay products (1).

Rn-220, and Rn-219, are produced in a decay series of uranium, thorium, and actinium, respectively. The half-lives of these Rn isotopes, Rn-219, Rn-220, and Rn-222, are 3.92 s, 54 s, and 3.82 d, respectively. As radon undergoes further radioactive decay, it produces a series of short-lived radioisotopes, known as radon "daughters" or "progeny." Because radon is a gas and a chemically inert element, it evolves from soil and underground water and decays to other radioactive elements in the series and finally transmutes to a stable lead isotope. The shorter half-lives of Rn-219 (3.92 s) and Rn-220 (54 s), compared to that of Rn-222 (3.82 d), make diffusion of the former into the above-ground air less probable (2). Consequently, Rn-222 is of most concern, and much of this gas can escape directly into outdoor air and contribute to the annual effective dose of ionizing radiation to humans (3). Steck et al. (4) found unusually high annual average outdoor Rn concentrations in parts of central North America and concluded that local soils may contribute to elevated outdoor Rn-222 concentrations, so such exposure should be included in epidemiological studies. Radon is potentially mobile and can diffuse through rock and soil to escape into the aboveground atmosphere. Radon formed in rocks and soils is released into the surrounding air. The amount of escaping radon varies enormously (5–7), depending on the geology (e.g., U content and its chemical form, degree of faulting), soil characteristics (e.g., permeability, moisture content), and climatic variables (e.g., temperature, humidity). Typical rates of radon release from soils throughout the world range from about 0.0002 to 0.07 Bq/(m³s). Radon production rates from any soil are extremely dependent on the geological characteristics of the soil and its underlying geological strata (8). Porous soils overlaying uranium-rich alum shales, granite, and pegmatite rocks are a particularly high risk for radon, whereas gas-impermeable soils consisting of fine sand, silt, and moist clay present a low risk (9).

Outdoors, radon emanating from the ground is quickly dispersed, and concentrations never reach levels that can be a threat to health. Whereas radon gas can disperse quickly in open air, it can enter and accumulate in dwellings as a component of soil gas drawn from the soil by mass flow driven by the pressure difference between a house and soil beneath (10). According to Baird (11), most radon that seeps into homes comes from the top meter of the soil below and around the foundations. Inside confined areas, low rates of air exchange can result in a buildup of radon and its daughters to concentrations tens of thousands of times higher than those observed outside (12). Radon concentrations within a building depend very much on both the concentration of radon in the soil surrounding the structure and the presence of entry points that allow the gas to infiltrate from outside (13). Some of the common entry points of radon into buildings include foundation joints, cracks in floors and walls, drains and piping, electrical penetrations, and cellars with earth floors (14).

2. INSTRUMENTAL METHODS OF RADON MEASUREMENT

In the mid-1980s, widespread recognition of the health threat from radon exposure created the need for a standard of competency for radon service providers. In February 1986, the US Environmental Protection Agency (EPA or Agency) established the Radon Measurement Proficiency (RMP) Program (15) to assist consumers in identifying organizations capable of providing reliable radon measurement analysis services. The Radon Contractor Proficiency Program was established in 1989 to evaluate the proficiency of radon mitigators in residences and provide information on proficient mitigators to the public. In 1991, the EPA expanded the RMP Program, adding a component to evaluate the proficiency of individuals who provide radon measurement services in the home. In 1995, these programs were consolidated to form the Radon Proficiency Program (RPP). Presently, the RPP assesses the proficiency of these individuals and organizations and grants them a listing according to their measurement or mitigation service capabilities. RPP proficiency is determined for services involved with residential settings only and does not determine proficiency for services involving schools and other large buildings, radon in water, or radon in soil. The detectors used in measurements of radon gas and its progeny are summarized in Table 2 (15).

2.1. Radon Gas Measurement Methods

2.1.1. AC: Activated Charcoal Adsorption

For this method, an airtight container with activated charcoal is opened in the area to be sampled and radon in the air adsorbs onto the charcoal granule (16,17). At the end of the sampling period, the container is sealed and may be sent to a laboratory for analysis. The gamma decay from the radon adsorbed to the charcoal is counted on 1 scintillation detector and a calculation based on calibration information is used to calculate the radon concentration at the sample site. Charcoal adsorption detectors, depending on design, are deployed from 2 to 7 d. Because charcoal allows continual adsorption and desorption of radon, the method does not give a true integrated measurement over the exposure time. Use of a diffusion barrier over the charcoal reduces the effects of drafts and high humidity.

2.1.2. CR: Continuous Radon Monitoring

This method category includes those devices that record real-time continuous measurements of radon gas. Air is either pumped or diffuses into a counting chamber. The

Table 2
Radon Gas Measurement Methods in the Checklist of the EPA

	Radon gas measurement method
1. AC	Activated charcoal adsorption
2. CR	Continuous radon monitoring
3. AT	Alpha track detection (filtered)
4. UT	Unfiltered track detection
5. LS	Charcoal liquid scintillation
6. EL	Electret–ion chamber: long term
7. ES	Electret–ion chamber: short term
8. GC	Grab radon/activated charcoal
9. GB	Grab radon/pump-collapsible bag
10. GS	Grab radon/scintillation cell
11. SC	Evacuated scintillation cell (3-d integrating)
12. PB	Pump-collapsible bag
	Radon decay product measurement
13. CW	Continuous working level monitoring
14. GW	Grab working level
15. RP	Radon progeny integrating sampling unit

Source: ref. 15.

counting chamber is typically a scintillation cell or ionization chamber (18). Scintillation counts are processed by electronics, and radon concentrations for predetermined intervals are stored in the instrument's memory or transmitted directly to a printer.

2.1.3. AT: Alpha Track Detection (Filtered)

For this method, the detector is a small piece of special plastic or film inside a small container. Air being tested diffuses through a filter covering a hole in the container. When α particles from radon and its decay products strike the detector, they cause damage tracks (19–22). At the end of the test, the container is sealed and returned to a laboratory for reading. The plastic or film detector is treated to enhance the damage tracks and then the tracks over a predetermined area are counted using a microscope or optical reader. The number of tracks per area counted is used to calculate the radon concentration of the site tested. Exposure of alpha track detectors is usually 3–12 mo, but because they are true integrating devices, alpha track detectors may be exposed for shorter lengths of time when they are measuring higher radon concentrations.

2.1.4. UT: Unfiltered Track Detection

The unfiltered alpha track detector operates on the same principle as the alpha track detector, except that there is no filter present to remove radon decay products and other α- particle emitters. Without a filter, the concentration of radon decay products decaying within the "striking range" of the detector depends on the equilibrium ratio of radon decay products to radon present in the area being tested, not simply the concentration of radon. Unfiltered detectors that use cellulose nitrate film exhibit an energy dependency that causes radon decay products that plate out on the detector not to be recorded. This phenomenon lessens but does not totally compensate for the dependency of the calibration factor on equilibrium ratio. For this reason, the EPA currently recommends that these devices not be used when the equilibrium fraction is less than 0.35 or greater than 0.60 without adjusting the calibration factor. The EPA is currently evaluating this device further to determine more precisely the effects of equilibrium fraction and other factors on performance. These evaluations will lead to a determination as to whether to finalize the current protocol or remove the method from the list of program method categories.

2.1.5. LS: Charcoal Liquid Scintillation

This method employs a small vial containing activated charcoal for sampling the radon. After an exposure period of 2–7 d (depending on design), the vial is sealed and returned to a laboratory for analysis. Although the adsorption of radon onto the charcoal is the same as for the AC method, analysis is accomplished by treating the charcoal with a scintillation fluid, then analyzing the fluid using a scintillation counter.

2.1.6. EL: Electret–Ion Chamber: Long Term

For this method, an electrostatically charged disk detector (electret) is situated within a small container (ion chamber). During the measurement period, radon diffuses through a filter-covered opening in the chamber, where the ionization resulting from the decay of radon and its progeny reduces the voltage on the electret. A calibration factor relates the measured drop in voltage to the radon concentration. Variations in electret design determine whether detectors are appropriate for making long-term or short-term measurements. EL detectors may be deployed for 1–12 mo. Because the electret–ion chambers are true integrating detectors, the EL type can be exposed at shorter intervals if radon levels are sufficiently high.

2.1.7. ES: Electret–Ion Chamber: Short Term

This method is similar to Electret-Ion Chamber for long term measurement (EL) described in Section 2.1.6, but ES detectors may be deployed for 2–7 d. Because electret–ion chambers are true integrating detectors, the ES type can be exposed at longer intervals if radon levels are sufficiently low.

2.1.8. GC: Grab Radon/Activated Charcoal

This method requires a skilled technician to sample radon by using a pump or a fan to draw air through a cartridge filled with activated charcoal. Depending on the cartridge design and airflow, sampling takes from 15 min to 1 h. After sampling, the cartridge is placed in a sealed container and taken to a laboratory where analysis is approximately the same as for the AC or LS methods.

2.1.9. GB: Grab Radon/Pump-Collapsible Bag

This method uses a sample bag composed of material impervious to radon. At the sample site, a skilled technician fills the bag with air using a portable pump and then transports it to the laboratory for analysis. Usually, the analysis method is to transfer air from the bag to a scintillation cell and perform analysis in the manner described for the grab radon/scintillation cell (GS) method in Section 2.1.10.

2.1.10. GS: Grab Radon/Scintillation Cell

For this method, a skilled operator draws air through a filter to remove radon decay products into a scintillation cell either by opening a valve on a scintillation cell that has previously been evacuated using a vacuum pump or by drawing air through the cell until air inside the cell is in equilibrium with the air being sampled; it is then sealed. To analyze the air sample, the window end of the cell is placed on a photomultiplier tube to count the scintillations (light pulses) produced when α particles from radon decay strike the zinc sulfide coating on the inside of the cell. A calculation is made to convert the counts to radon concentrations.

2.1.11. SC: Three-Day Integrating Evacuated Scintillation Cell

For this method, a scintillation cell is fitted with a restrictor valve and a negative pressure gage. Prior to deployment, the scintillation cell is evacuated. At the sample site, a skilled technician notes the negative pressure reading and opens the valve. The flow through the valve is slow enough that it takes more than the 3-d sample period to fill the cell. At the end of the sample period, the technician closes the valve, notes the negative pressure gage reading, and returns with the cell to the laboratory. Analysis procedures are approximately the same as for the GS method described above. A variation of this method involves use of the above valve on a rigid container requiring that the sampled air be transferred to a scintillation cell for analysis.

2.1.12. PB: Pump-Collapsible Bag

For this method, a sample bag impervious to radon is filled over a 24-h period. This is usually accomplished by a pump programmed to pump small amounts of air at predetermined intervals during the sampling period. After sampling, analysis procedures are similar to those for the GB method.

2.2. Radon Decay Product Measurement Methods

2.2.1. CW: Continuous Working Level Monitoring

This method encompasses those devices that record real-time continuous measurement of radon decay products. Radon decay products are sampled by continuously pumping air through a filter. A detector such as a diffused-junction or surface-barrier detector counts the α particles produced by radon decay products (23) as they decay on this filter. The monitor typically contains a microprocessor that stores the number of counts for predetermined time intervals for later recall. Measurement time for the program measurement test is approximately 24 h.

2.2.2. GW: Grab Working Level

For this method, a known volume of air is pulled through a filter, collecting the radon decay products onto the filter. Sampling time usually is 5 min. The decay products are counted using an alpha detector. Counting must be done with precise timing after the filter sample is taken.

2.2.3. RP: Radon Progeny (Decay Product) Integrating Sampling Unit

For this method, a low-flow air pump pulls air continuously through a filter. Depending on the detector used, the radiation emitted by the decay products trapped on the filter is registered on two thermoluminescent dosimeters (TLDs), an alpha track detector, or an electret. The devices presently available require access to a household electrical supply, but do not require a skilled operator. Deployment simply requires turning the device on at the start of the sampling period and off at the end. The sampling period should be at least 72 h. After sampling, the detector assembly is shipped to a laboratory where analysis of the alpha track and electret types is performed. The TLD detectors are analyzed by an instrument that heats the TLD detector and measures the light emitted. A calculation converts the light measurement to radon concentrations.

In the open environment, radon concentration varies considerably, but an average is taken as 7.4 Bq/m^3 (24). The average radon concentration is about 20 Bq/m^3 in UK homes and 50 Bq/m^3 in US homes. The ranges of all instrumental techniques are sensitive enough to measure the indoor and outdoor radon concentration. Usually, passive methods are used for monitoring average radon concentrations in dwellings, but active methods are preferred for studying the mechanisms of radon dynamics.

3. HEALTH EFFECTS OF RADON

The above-ground atmosphere is a significant source of radon exposure, although the gas is derived through the radioactive decay of U-238 and Th-232 in rock and soil minerals (25). Baird (11) states that Rn-222 by itself does not pose much danger to people because it is inert and because most of it is exhaled after inhalation. By comparison, the daughter isotopes (the progeny), Po-214 and Po-218 in particular, are electrically charged, adhere to dust particles, and are inhaled either directly or through their attachment to airborne particles (26) to cause radiation damage to the bronchial cells. Once inhaled, they tend to remain in the lungs, where they may eventually cause cancer (27,28).

It was not until the early 1970s that this potential hazard from the inhalation of radon gas and the daughter progeny in the domestic environment was first identified. In the past, contamination of air by radon and subsequent exposure to radon daughters were believed to be a problem only for uranium and phosphate miners. However, it has recently been recognized that homes and buildings far away from uranium or phosphate mines can also exhibit high concentrations of radon. Subsequently, radon and radon progeny are now recognized as important indoor pollutants (8).

Radon exposure has been linked to lung carcinogenesis in both human and animal studies. It has also been associated with the development of acute myeloid and acute lymphoblastic leukemia and other cancers (29). However, the estimation of health risks from residential radon is extremely complex and encompasses many uncertainties. Studies on smoking and nonsmoking uranium miners indicate that radon at high concentrations is a substantial risk factor for lung cancer. Based on data regarding dose–response relationships among miners, it is estimated that between 5% and 15% of lung cancer deaths might be associated with exposure to residential radon (30). The relevance of data from mines to the lower-exposure home environment is often questioned (31). Nevertheless, a meta-analysis of eight epidemiological studies undertaken (32) found that the dose–response curve associated with domestic radon exposures was remarkably similar to that observed among miners.

Ecological (geographical) study designs have been adopted by a number of recent epidemiological investigations into the health risks associated with nonindustrial radon exposures. Lucie (33) reported positive county-level correlations between radon exposure and mortality from acute myeloid leukemia in the United Kingdom, and Henshaw et al. (29) found that mean radon levels in 15 counties were significantly associated with the incidence of childhood cancers and, specifically, all leukemias. However, these reports have been met with considerable criticism because ecological designs can suffer from serious limitations (34). In particular, the effects of migration are often difficult to account for, information on potential confounding variables can be unavailable, and estimates of exposure for populations of large areas may differ greatly from actual individual doses. More refined ecological analyses, such as that undertaken by Etherington et al. (35), have reported no association between indoor radon exposure and the occurrence of cancer.

An alternative to ecological analyses is the case-control study design, in which radon exposures among individuals with cancer are compared to those of control subjects free from the disease. Most case-control studies have reported a small but significant association between radon exposure and lung cancer mortality. For example, in a recent examination of more than 4000 individuals in Sweden, Lagarde et al. (36) estimated that there is an excess relative risk of contracting lung cancer of between 0.15 and 0.20 per 100 Bq/m^3 increase in radon exposure.

The radon (progeny) concentration in indoor air is responsible for the largest contribution to the natural radiation exposure to people. This fact and the knowledge of the enhanced lung cancer rate in cohorts exposed to high radon concentrations have raised the question about the lung cancer risk for the population caused by radon in the domestic environment. The National Research Council published in its 1988 report (4th Committee on Biological Effects of Ionizing Radiation, BEIR IV) on a lung cancer risk model mainly based on epidemiological studies of lung cancer rates among underground miners (37). The cancer risk of radon among miners is greater than other population groups. In a study regarding Czech Republic miners, 1323 cytogenetic assays and 225 subjects were examined. Chromatid breaks were the most frequently observed type of aberration and the frequency of aberrant cells was correlated with radon exposure. A 1% increase in the frequency of aberrant cells was paralleled by a 62% increase in the risk of cancer. An increase in the frequency of chromatid breaks by 1 per 100 cells was followed by a 99% increase in the risk of cancer (38).

A new report (BEIR VI) entitled "Health Effects of Exposure to Radon" is a reexamination and a reassessment of all relevant data. Compared with the 1988 report, much more information was available (39). Again, the BEIR VI committee based their lung cancer risk model primarily on miner studies (empirical approach) because of principal difficulties in the dosimetric approach (lung model, atomic bomb survivors) and the biologically motivated approach (tissue growth, cell kinetic). Epidemiological studies on lung cancer and radon exposure in homes give limited information because of the overwhelming cancer risk from smoking (responsible for 90–95% of all lung cancer cases). In the development of an empirical risk model to describe rates of radon-induced lung cancer, several assumptions are needed. It is not only the shape of the exposure–response function but also the factors that influence risk that must be modeled to extrapolate the risk from the radon exposure of miners to the radon situation in homes. Usually, the radon concentrations in mines are one magnitude higher than in common houses; however, the highest concentrations in homes often reach values as high as those in mines.

The committee adopts a linear, no-threshold, relative-risk model with different weighting factors for exposures 5–14, 15–24, and more than 24 yr ago and two effect-modification factors (attained age and exposure rate indexed either in duration of exposure or in exposure rate). From the individual ERR (excess relative risk), a lifetime relative risk (LRR) and a population attributable risk (AR) are deduced. The latter indicates how much of the lung cancer burden could, in theory, be prevented if all exposures to radon were reduced to the background level of radon in the outdoor air.

During the development of the risk model, combined effects of smoking and radon were also extensively discussed. Finally, the committee preferred to use a submulti-plicative relationship to describe the synergistic effects between these carcinogens. The committee could not identify strong evidence indicative of differing susceptibility to lung carcinogens by sex. No clear indication of the effect of age on exposure could be identified and only for infants (age 1 yr) was a slightly higher risk (+8%) adopted. Despite sophisticated analyses of existing data, the committee states that, for the extreme low-exposure region mainly, the mechanistic basis of cancer induction supports the linear, no-threshold model. However, there is also the possibility of a nonlinear relation, or even a threshold below which no additional lung cancer risk exists.

Compared with the BEIR IV report, the improved risk model gives slightly higher risks for lung cancer from radon exposure. Thus, in the United States, the estimated attributable risk (AR) for lung-cancer death from domestic exposure to radon raised from approx 8% to 10% in the exposure age–duration model and even to 15% in the exposure age–concentration model. This means that a total of 15,000–20,000 lung cancer deaths per year were attributable to indoor residential radon progeny exposure in the United States. Although the models used are the most plausible to date, it must be emphasized that these numbers are derived by extrapolation from generally substantially higher exposures. Presently, there is no way to validate these estimates (39,40).

The extent of the radon problem will vary globally, and in equatorial areas where domestic conditions are usually different, the indoor level of Rn-222 and its daughters is likely to be considerably lower than in the northern regions. In the latter areas, the radon problem can be evaluated by making reference to the United Kingdom. Here, surveys have shown that 100,000 houses built on certain types of ground mostly in Cornwall and Devon and in some parts of Derbyshire, Northamptonshire, Somerset, Grampian, and the Highlands of Scotland are more likely to have high indoor radon levels. Southwest England (i.e., Cornwall and Devon) is particularly affected; 53% of UK homes were estimated to contain a concentration of Rn-222 above the action level of 200 Bq/m^3. The granites found in this region have relatively high uranium content and are suitably jointed and fractured to generate a high radon emanation rate. Varley and Flowers (41) showed that soil gas concentrations over the granites were twice that found in soils above other rocks and, as expected, homes located in granite regions had the highest indoor radon levels. It was estimated that residential radon is responsible for approx 1 in 20 cases of lung cancer deaths in the United Kingdom (approx 2000 per year). The first direct evidence for the link between residential radon and lung cancer has been published relatively recently (although not without some controversy [*see* Miles et al. (42)] and agrees with these figures (43). Working in southwest England, these authors found that the relative risk of lung cancer increased by 8% per 100 Bq/m^3 increase in the residential radon concentration.

In the United States, results of 19 studies of indoor radon concentrations are summarized (44), covering 552 single-family homes. They determined that the mean indoor concentration as 56 Bq/m³. As part of the more recent US National Residential Radon Survey, Marcinowski et al. (45) estimated an annual average radon concentration of 46.3 Bq/m³ in US homes. They also calculated that approximately 6% of homes had radon levels greater than the US Environmental Protection Agency (EPA) action level for mitigation of 148 Bq/m³. Exposure to high concentrations of radon progeny produces lung cancer in both underground miners and experimentally exposed laboratory animals. The goal of the study was to determine whether or not residential radon exposure exhibits a statistically significant association with lung cancer in a state with high residential radon concentrations. A population-based, case-controlled epidemiological study was conducted examining the relationship between residential radon gas exposure and lung cancer in Iowa females (46) who occupied their current home for at least 20 yr. The study included 413 incident lung cancer cases and 614 age–frequency-matched controls. Participant information was obtained by a mailed-out questionnaire with face-to-face follow-up. Radon dosimetry assessment consisted of five components: (1) on-site residential assessment survey; (2) on-site radon measurements; (3) regional outdoor radon measurements; (4) assessment of subjects' exposure when in another building; and (5) linkage of historic subject mobility with residential, outdoor, and other building radon concentrations. Histologic review was performed for 96% of the cases. Approximately 60% of the basement radon concentrations and 30% of the first-floor radon concentrations of study participants' homes exceeded the US EPA Action Level of 150 Bq/m³ (4 pCi/L). Large areas of western Iowa had outdoor radon concentrations comparable to the national average indoor value of 55 Bq/m³ (1.5 pCi/L). A positive association between cumulative radon gas exposure and lung cancer was demonstrated using both categorical and continuous analysis. The risk estimates obtained in this study indicate that the cumulative radon exposure presents an important environmental health hazard.

Outside of the United States Albering et al. (47) found a much higher average concentration of 116 Bq/m³ in 116 homes in the township of Visé in a radon-prone area in Belgium. In Italy, Bochicchio et al. (48) reported an average concentration of 75 Bq/m³ in a sample of 4866 dwellings and observed concentrations exceeding 600 Bq/m³ in 0.2% of homes. Yu et al. (49) recently undertook one of the relatively few studies of radon concentrations in the office environment. In 94 Hong Kong office buildings, they recorded radon concentrations similar to those that have been observed in domestic situations, with a mean of 51 Bq/m³.

As activities such as the smoking of cigarets can lead to considerably elevated levels of airborne particles, smokers are at particular risk from the inhalation of radon progeny (50). Indeed, the US EPA has estimated that the cancer risk from radon for smokers is as much as 20 times the risk for individuals who have never smoked (*see* Table 3) (51).

4. RADON MITIGATION IN DOMESTIC PROPERTIES

The most enhanced research on the radon problem has been carried out in the United States. The US EPA and the US Geological Survey have evaluated the radon potential in the United States and have developed a map to assist national, state, and local organizations to target their resources and to assist building-code officials in deciding whether radon-resistant features are applicable in new construction. The map was

Table 3
Radon Risks for Dying as a Result of Lung Cancer for Smokers and NonSmokers Exposed to Different Radon Levels During Lifetime and Remedies

Radon level (Bq/m³)	Ratio of smokers/nonsmokers out of 1000 exposed to this level over a lifetime who could die from lung cancer	What to do: Stop smoking and . . .
750	135/8	Fix your home.
375	71/4	Fix your home.
300	57/3	Fix your home.
150	29/2	Fix your home.
75	15/1	Consider fixing between 75 and 150 Bq/m³.
50	9/less than 1 ⎫	Reducing radon levels below
15	3/less than 1 ⎭	75 Bq/m³ is difficult.

Source: ref. 57.

developed using five factors to determine radon potential: indoor radon measurements, geology, aerial radioactivity, soil permeability, and foundation type. Radon potential assessment is based on geologic provinces. The Radon Index Matrix is the quantitative assessment of radon potential. The Confidence Index Matrix shows the quantity and quality of the data used to assess radon potential. Geologic Provinces were adapted to county boundaries for the Map of Radon Zones. The map can be accessed from the EPA website (www.epa.gov/iaq/radon/zonemap). EPA's Map of Radon Zones assigns each of the 3141 counties in the United States to one of three zones based on radon potential:

- Zone 1 counties have a predicted average indoor radon screening level greater than 148 Bq/m³ (4 pCi/L).
- Zone 2 counties have a predicted average indoor radon screening level between 74 and 148 Bq/m³ (2–4 pCi/L).
- Zone 3 counties have a predicted average indoor radon screening level less than 74 Bq/m³ (2 pCi/L).

Since the mid-1980s, the United States has made significant progress in reducing the risk from exposure to radon. This progress is the result of a long-term effort among the EPA, the public, nonprofit organizations, state and local governments, the business community, and other federal agencies working together. More Americans are knowledgeable about radon than at any time since the mid-1980s, when radon became a national health concern. Approximately two-thirds (66%) of Americans are generally aware of radon, and of those, three-quarters (75%, on average) understand that radon is a health hazard. Since the mid-1980s, about 18 million homes have been tested for radon and about 500,000 of them have been mitigated. Approximately 1.8 million new homes have been built with radon-resistant features since 1990. The EPA will continue to focus its efforts and those of its partners on achieving actual risk reduction through the mitigation of existing homes and the building of new homes to be radon resistant. The EPA's estimates of risk reduction are predicated upon mitigation systems being properly installed, operated, and maintained. As a result of these actions to reduce radon levels in homes through 1999, the EPA estimates that approx 350 lung cancer deaths will be prevented each year. This annual rate is expected to rise as radon levels are lowered in more new and existing homes.

The 1988 Indoor Radon Abatement Act (IRAA) required the EPA to develop a voluntary program to evaluate and provide information on contractors that offer radon control services to homeowners. The Radon Contractor Proficiency (RCP) Program was established to fulfill this portion of the IRAA. In December 1991, the EPA published "Interim Radon Mitigation Standards" as initial guidelines for evaluating the performance of radon mitigation contractors under the RCP Program. The effectiveness of the basic radon mitigation techniques set forth in the "Interim Standards" has been validated in field applications throughout the United States. This experience now serves as the basis for the more detailed and final Radon Mitigation Standards (RMS) presented in that document. A detailed document on RMS can be found on the EPA's website (www.epa.gov/iaq/radon/pubs/graphics/mitstds). The RMS provides radon mitigation contractors with uniform standards to ensure quality and effectiveness in the design, installation, and evaluation of radon mitigation systems in detached and attached residential buildings three stories or fewer in height. The RMS is intended to serve as a model set of requirements that can be adopted or modified by state and local jurisdictions to fulfill objectives of their specific radon contractor certification or licensure programs.

The American Society of Home Inspectors (ASHI) recommends that homeowners and home buyers test their current or prospective home for the presence of radon gas in indoor air. The EPA strongly recommends that steps be taken to reduce indoor radon levels when test results are 148 Bq/m^3 (4 pCi/L) or more of radon in the air. A radon mitigation system inspection checklist can be obtained from the website (www.epa.gov/radon/risk_assessment_factsheet.html).

Radon moves up through the ground to the air above and into homes through cracks and other holes in the foundation. Homes may trap radon inside, where it can build up. Any home may have a radon problem. This means new and old homes, well-sealed and drafty homes, and homes with or without basements (51). Radon from soil gas is the main cause of radon problems. Sometimes, radon enters the home through well water. In a small number of homes, the building materials can give off radon, too. However, building materials rarely cause radon problems by themselves.

Radon problem in homes can be attributed to the following sources:

1. Soil around the house
2. Cracks in solid floors
3. Construction joints
4. Cracks in walls
5. Gaps in suspended floors
6. Gaps around service pipes
7. Cavities inside walls
8. The water supply

Nearly 1 out of every 15 homes in the United States is estimated to have elevated radon levels. Although radon problems may be more common in some areas, any home may have a problem. The only way to know about the radon level in a home is to test. There are two general ways to test for radon in homes:

Short-term testing: The quickest way to test is with short-term tests. Short-term tests remain in a home for 2–90 d, depending on the device. "Charcoal canisters," "alpha track," "electret–ion chamber," "continuous monitors," and "charcoal liquid scintillation" detectors (15) are most commonly used for short-term testing. Because radon levels tend to vary from day to day and season to season, a short-term test is less likely than a long-term test to give a year-round average radon level. If one needs results quickly, however, a short-term test followed by a second short-term test may be used to decide whether to fix a home.

Long-term testing: Long-term tests remain in a home for more than 90 d. "Alpha track" and "electret" detectors are commonly used for this type of testing. A long-term test will give a reading that is more likely to indicate a home's year-round average radon level than a short-term test

The EPA recommends the following testing steps:

1. Take a short-term test. If the result is 148 Bq/m^3 (4pCi/L) or higher (0.02 working levels [WL] or higher) take a follow-up test (Step 2) to be sure.
2. Follow up with either a long-term test or a second short-term test:

   • For a better understanding of year-round average radon level, take a long-term test.
   • If the results are needed quickly, take a second short-term test.

   The higher the initial short-term test result, the more certain one can be that a short-term rather than a long-term follow-up test should be taken. If the first short-term test result is several times the action level (e.g., about 370 Bq/m^3 [10 pCi/L] or higher), one should take a second short-term test immediately.

3. If followed up with a long-term test, fix the home if the long-term test result is 148 Bq/m^3 (4 pCi/L) or more (0.02 WL or higher).

   If followed up with a second short-term test, the higher the short-term results, the more certain one can be that the home should be fixed. Consider fixing the home if the average of first and second test is 148 Bq/m^3 (4 pCi/L) or higher (0.02 WL or higher).

High radon concentrations in dwellings can be reduced to acceptable levels by using various techniques. A homeowner may wonder whether his house has a high potential of radon concentration as a result of the mineral composition of the location, building material used in construction, or the condition of house. The radon concentration can be measured to determine the seriousness of the condition. If high radon levels are found, some remedial measures can be implemented. Radon problems may be handled most efficiently during construction through some low-cost modifications to common practices. As air space is improved in new houses for better thermal insulation, radon buildup in the house may become a serious problem.

This section discusses a number of radon concentration control methods. Control measures may be classified as the following: removal of a source containing high uranium concentration around the house, modification of source material that gives off radon isolated from the indoor air, ventilation of indoor air, and air cleaning by some physical or chemical methods. Several methods are available to contractors for lowering radon levels in homes. Methods preventing radon from entering homes are preferred to methods that reduce radon levels after it has entered homes. For example, soil suction can be used to prevent radon from entering homes by drawing the radon from below the house and venting it through a pipe or pipes to the air above the house for dilution (52).

4.1. Source Removal

The soil around the outside of a house is the most common source of radon. Removal of sources is the first logical option, but it is frequently expensive, especially for existing houses. The following discussion describes source removal for both new and existing houses.

The removal of high-radium-content soil around a house requires replacement with new soil and should have low diffusion for soil gas flow. Fractional reduction of radon flux for a 10-ft cover thickness of a nonuraniferous fill material is 80% (53). The main disadvantage of this method lies in the initial capital for excavation of the site and extra earth-moving costs for replacement with the chosen fill material. If a building was constructed inadvertently from high-uranium-content material, the slabs could be replaced by a new one containing nonuraniferous material. If the fill material is not a strong source of radon, sealants may be used to close cracks and openings in the slab, or acceptable radon levels may be reached by increasing ventilation moderately. In these cases, the costs of removal must be balanced against increased ventilation costs or the costs of a sealing program.

4.2. Contaminated Well Water

Radon concentration in water is typically of the order of 100 Bq/L (several thousand picocuries per liter). Thus, use of such water could be an important radon source. Surveys of US drinking water sources indicate that 74% had radon concentrations below 100 Bq/L and only 5% had values above 400 Bq/L. The problem may be the result of deep-drilled wells and a concentration of 400 Bq/L will increase the indoor radon concentration by about 0.04 Bq/L.

Contaminated well water can be a major source of radon in both planned and existing structures. The safety limit of Ra-226 and Ra-228 (total) in drinking water is 0.18 Bq/L (5 pCi/L). The EPA published the safety limits of drinking water on its website (www.epa.gov/safewater/mcl.html). Municipal and surface water can be substituted for the contaminated well water. If the replacement of well water with a cleaner source is not possible, various water-treatment methods should be considered. Decay products of radon can be removed if water is kept in a storage tank long enough so that all short-lived products die out (54). A holding time of 19 or 31 d is required for 96.9–99.6% removal. The tank may be compartmentalized to prevent back mixing. Spray aeration of contaminated well water increases the rate of desorption of radon by increasing the surface area for mass transport across the water–air interface. The disadvantage of this method is use of exhaust ventilation for the process, as well as higher capital and operating costs.

Granulated activated carbon may be used to process radon-contaminated water. Although the carbon filter requires regeneration, the decay of radon may enhance the lifetime of the filter. Because of the low capital cost of a carbon filter, it is recommended for single-home applications. The spray aeration method is more economical for multiple dwellings (55).

4.3. Building Materials

In the past, it was thought that building materials were the principal sources of indoor radon (56). However, most recent studies have shown that with the exception of some unusual materials such as Swedish alum shale concrete, the effect of building materials on indoor radon is small. However, significant amounts of radon are emitted from materials with an elevated radium level have been found in phosphogypsum wallboards and concrete containing alum shale. Although not quite as high, elevated radium concentrations have also been found in concrete containing fly ash and some bricks (57).

4.4. Types of House and Radon Reduction

Different types of house will affect which radon-reduction system is selected. Houses are normally categorized according to the foundation design. These may include basement, slab-on-grade, which has concrete poured at ground level, or crawl space, which has a shallow unfinished space under the first floor. Different house foundation types are illustrated in Fig. 3. Some houses may have more than one foundation design feature and may require a combination of radon-reduction techniques in order to reduce radon levels below 4 pCi/L (0.15 Bq/L).

Basement and slab-on-grade houses: Four types of soil suction methods can be used to reduce radon concentration for houses that have a basement or a slab-on-grade foundation. These include subslab suction, drain tile suction, sump hole suction, or block wall suction.

Active subslab suction (also called subslab depressurization) is usually the most reliable radon reduction method. Suction pipes are inserted through the floor slab into the crushed rock or soil underneath. A fan that is connected to the pipes draws the radon gas from below the house and releases it to the outdoor air. Passive subslab suction is similar to the active subslab suction except it depends on air currents instead of a fan to draw radon gas up from below the house. Passive subslab suction is usually not as effective in reducing high radon levels compared to the active subslab suction method. For houses that have drain tiles to direct water away and form a complete loop around the foundation, suction on these drain tiles can be used to reduce radon levels. If a house with a basement has a sump pump to remove unwanted water, the sump can be capped so that it can continue to drain water and serve as the location for a radon suction pipe. Block wall suction can be used in houses with basements and hollow block foundation walls. This method can remove radon from the hollow spaces within the basement's concrete block wall and is often used together with subslab suction.

Crawl-space houses: In houses with crawl spaces, radon levels can sometimes be lowered by ventilating the crawl space passively or actively. Crawl-space ventilation reduces indoor radon levels by reducing the home's suction on the soil and by diluting the radon beneath the house. Passive, or natural, ventilation in a crawl space is achieved by opening vents or by installing additional vents. Active ventilation uses a fan to blow air through the crawl space. In colder climates, the water pipes in the crawl space need to be insulated against the cold.

A second method of crawl-space ventilation involves covering the earth floor with a heavy plastic sheet. A vent pipe and fan can be used to draw the radon from under the sheet and vent it to the outdoor air. This type of soil suction is called submembrane depressurization.

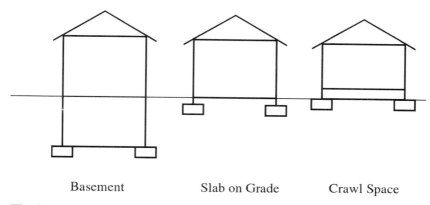

Basement   Slab on Grade   Crawl Space

Fig. 3. House Foundation Types: basement, slab-on-grade, and crawl space.

An example for crawl-space ventilation is the village of Varnhem, Sweden, where the soil is permeable gravel containing alum shale with an elevated activity of uranium (58). The crawl space was ventilated separately from the ventilation of the dwelling. The concentration of radon was measured between 122,000 and 340,000 Bq/m³ in the capillary breaking layer. However, the radon activity dropped down to 70–240 Bq/m³ after ventilation. The modification in design added approx 4% to the building cost. Moreover, there was no conflict between energy conservation and radon protective or safe design. Radon-safe construction can be made as energy efficient as conventional design. Energy conservation by increased tightness of the building and very low air-exchange rates must be discouraged because of its effect on indoor air quality and humidity even without regard to radon. In order to avoid strong negative pressure indoors, heat exchangers with balanced ventilation are preferred to exhaust ventilation with a heat exchanger.

Other types of radon-reduction methods: Other radon reduction methods applicable to houses consist of sealing, house pressurization, natural ventilation, and heat recovery ventilation. Most of these methods are either temporary measures or only partial solutions to be used in combination with other methods.

Sealing cracks and other openings in the foundation is a common element of most approaches to radon reduction. It limits the flow of radon into the home and it also reduces the loss of conditioned air, thereby making other radon-reduction techniques more effective and cost-efficient. However, it might be difficult to identify and permanently seal the places where radon is entering to the house. Normal settling of a house may open new entry routes and reopen old ones.

House pressurization uses a fan to blow air into the basement or living area from either upstairs or outdoors. Enough pressure is created at the lowest level indoors to prevent radon from entering into the house. The effectiveness of this method is limited by house construction, climate, other appliances in the house, and the occupant's lifestyle. In order to maintain enough pressure to keep radon out, doors and windows at the lowest level must be closed except for normal entry and exit.

Natural ventilation occurs in all houses to some degree. By opening doors, windows, and vents on the lower floors, one can increase the ventilation in the house. This increase in ventilation will mix radon with outdoor air and can reduce radon concentration. It can also lower indoor radon levels by reducing the vacuum effect. Natural ventilation in any type of

Table 4
Installation and Operating Cost of Radon Reduction in Homes

Technique	Typical radon reduction	Typical range of insulation costs (contractor)	Typical operating cost range (annual)*	Comments
Subslab suction (subslab depressurization)	80–99%	$800–2500	$75–175	Works best if air can move easily in material under slab.
Passive subslab suction	30–70%	$550–2250	There may be some energy penalties	May be more effective in cold climates; not as effective as active subslab suction.
Draintile suction	90–99%	$800–1700	$75–175	Works best if draintiles form complete loop around house.
Blockwall suction	50–99%	$1500–3000	$150–300	Only in houses with hollow blockwalls; requires sealing of major openings.
Sump hole suction	90–99%	$800–2500	$100–225	Works best if air moves easily to sump under slab, or if draintiles form complete loop.
Submembrane depressurization in a crawl space	80–99%	$1000–2500	$70–175	Less heat loss than natural ventilation in cold winter climates.
Natural ventilation in a crawl space	0–50%	None ($200–500 if additional vents installed)	There may be some energy penalties.	Costs variable
Sealing of radon entry routes	0–50%	$100–2000	None	Normally used with other techniques; proper materials and installation required.
House (basement) pressurization	50–99%	$500–1500	$150–500	Works best with tight basement isolated from outdoors and upper floors.
Natural ventilation	Variable	None ($200–500 if additional vents installed)	$100–700	Significant heated/cooled air loss; operating costs depend on utility rates and amount of ventilation.

(continued)

Table 4   *(Continued)*

Technique	Typical radon reduction	Typical range of insulation costs (contractor)	Typical operating cost range (annual)*	Comments
Heat recovery ventilation	25–50% if used for full house; 25–75% if used for basement	$1200–2500	$75–500 for continuous operation	Limited use; best in tight house; for full house, use with levels no higher than 8 pCi/L; no higher than 16 pCi/L for use in basement;less conditioned air loss than natural ventilation.
Water systems Aeration	95–99%	$3000–4500	$40–90	More efficient than GAC; requires annual cleaning to maintain effectiveness and to prevent contamination; carefully vent system.
Activated carbon (GAC)	85–99%	$1000–2000	None	Less efficient for higher levels than aeration; use for moderate levels (around 5000 pCi/L or less); radon byproducts can build on carbon may need radiation shield around tank and care in disposal.

Note: The fan electricity and house heating/cooling loss cost range is based on certain assumptions regarding climate, house size, and the cost of electricity and fuel. Costs may vary. Numbers based on 1991 data. http://www.epa.gov/iaq/radon/pubs/consguid.html.

house except the ventilation of a crawl space should usually be considered as a temporary radon-reduction approach because of the loss of conditioned air and related discomfort, greatly increased costs of conditioning additional outside air, and security concerns.

A heat recovery ventilator (HRV), also called an air-to-air heat exchanger, can be used to increase ventilation. It operates by using the heated or cooled air being exhausted to warm or cool the incoming air. HRVs can be used to ventilate all or parts of the house, although they are more effective in reducing radon concentrations if confined to the basement. If properly balanced and maintained, HRVs ensure a constant degree of ventilation throughout the year. They also can improve air quality in houses that have other indoor pollutants. However, there might be a significant increase in the heating and cooling costs with an HRV.

Comparison of radon-reduction installation and operating costs are compared in Table 4 (59). This table was prepared by the EPA using 1991 data. Although the costs may vary, it gives an idea about typical radon reduction for the investment made.

REFERENCES

1. G. Friedlander, J. W. Kennedy, E. S. Macias, et al., *Nuclear and Radiochemistry*, Wiley, New York, 1981.
2. E. Steinnes, in *Geomedicine* (J. Lag, ed.), CRC, Boca Raton, FL, 1990, pp. 163–169.
3. R. H. Clarke and T. R. E. Southwood, *Nature* **338**, 197–198 (1989).
4. D. J. Steck, R. W. Field, and C. F. Lynch, *Environ. Health Perspect.* **107**, 123–127 (1999).
5. C. Bowie and S. H. U. Bowie, *Lancet* **337**, 409–413 (1991).
6. N. R. Varley and A. G. Flowers, *Environ. Geochem. Health* **15**, 145–151 (1993).
7. R. L. Jones, *Environ. Geochem. Health* **7**, 21–24 (1995).
8. B. LeHvesque, D. Gauvin, R. McGregor, et al., *Health Phys.* **72**(6), 907–914 (1997).
9. IARC (International Agency for Research on Cancer), *Man-made Mineral Fibres and Radon*, IARC Monographs on the Evaluation of the Carcinogenic Risk of Chemicals to Humans, Vol. 43, International Agency for Research on Cancer, Lyon, 1988.
10. G. Sharman, *Environ. Geochem. Health* **14**, 113–120 (1992).
11. C. Baird, *Environmental Chemistry*, Freeman, New York, 1998.
12. H. U. Wanner, *IARC Sci. Publ.* **109**, 19–30 (1993).
13. W. Jedrychowski, E. Flak, J. Wesolowski, et al., *Central Eur. J. Public Health* **3**(3), 150–160 (1995).
14. K. K. Nielson, V. C. Rogers, R. B. Holt, et al., *Health Phys.* **73**(4), 668–678 (1997).
15. *EPA Radon Proficiency Program (RPP) Handbook*, US Environmental Protection Agency, Washington, DC, 1996; available from www.epa.gov/iaq/radon/handbook.
16. M. Kawano and S. Nakatani, in *The Natural Radiation Environment* (J. A. S. Adams and W. M. Lowder, eds.), 1964, pp. 291–312, University of Chicago Press.
17. K. Megumi and T. Mamuro, *J. Geophys. Res.* **77**, 3051–3056 (1972).
18. D. K. Talbot, J. D. Appleton, T. K. Ball, et al. A comparison of field and laboratory analytical methods for radon site investigation, *J. Geochem. Explor.* **65**(1), pp. 79–90. (1998).
19. S. A. Durrani and R. Ilic, (eds.), *Solid State Nuclear Track Detection*, Pergamon, Oxford, 1997, p. 304.
20. S. A. Durrani, *Radiati. Measure.* **34**, 5–13 (2001).
21. A. Canabo, F. O. Lopez, A. A. Arnaud, et al., *Radiati. Measure.* **34**, 483–486 (2001).
22. V. S. Y. Koo, C. W. Y. Yip, J. P. Y. Ho, et al., *Appl. Radiati. Isotopes* **56**, 953–956 (2002).
23. K. Jamil, F. Rehman, S. Ali, et al., *Nucl. Instrum. Methods Phys. Res. A* **388**, 267–272 (1997).
24. J. I. Fabrikant, *Health Phys.* **59**(1), 89 (1990).
25. P. W. Abrahams, *Sci. Total Environ.* **291**, 1–32 (2001).
26. B. S. Cohen, *Health Phys.* **74**(5), 554–560 (1998).
27. P. Polpong and S. Bovornkitti, *J. Med. Assoc. Thailand* **81**(1), 47–57 (1998).
28. A. P. Jones, *Atmos. Environ.* **33**, 4535–4564 (1999).
29. D. L. Henshaw, J. P. Eatough, and R. B. Richardson, *Lancet* **335**, 1008–1012 (1990).
30. K. Steindorf, J. Lubin, H. E. Wichmann, et al., *Int. J. Epidemiol.* **24**(3), 485–492 (1995).
31. J. H. Lubin, L. TomaHsek, C. Edling, et al., Estimating lung cancer mortality from residential radon using data for low exposures of miners, *Radiati. Res.* **147**(2), 126–134 (1997).
32. J. H. Lubin and J. D. Boice, *J. Natl. Cancer Inst.* **89**(1), 49–57 (1997).
33. N. P. Lucie, *Lancet* **2**(8654), 99–100 (1989).
34. S. P. Wolff, *Nature* **352**(6333), 288 (1991).
35. D. J. Etherington, D. F. Pheby, and F. I. Bray, *Eur. J. Cancer* **32**(7), 1189–1197 (1996).
36. F. Lagarde, G. Pershagen, G. Akerblom, et al., *Health Phys.* **72**(2), 269–276 (1997).
37. The National Research Council, 4th Committee on Biological Effects of Ionizing Radiation, *BEIR IV Report*, National, Washington, DC, 1988.
38. Z. Smerhovsky, K. Landa, P. Rössner, et al., *Mutat. Res.* **514**, 165–176 (2002).

39. BEIR VI Report (2000) National Research Council, 6th Committee on Biological Effects of Ionizing Radiation, *BEIR VI Report* National Academic Press, Washington, DC, 2000.
40. H. Friedmann, *Eur. J. Radiol.* **35**, 221–222 (2000).
41. N. R. Varley and A. G. Flowers, *Radiat. Prot. Dosim.* **77**, 171–176 (1998).
42. D. Miles, J. O. O'Brien, and M. Owen, *Br. J. Cancer* **79**, 1621–1622 (1999).
43. S. Darby, E. Whitley, P. Silcocks, et al., *Br. J. Cancer* **78**, 394–408 (1988).
44. A. V. Nero, M. B. Schwehr, W. W. Nazaro, et al., *Science* **234**(4779), 992–997 (1986).
45. F. Marcinowski, R. M. Lucas, and W. M. Yeager, *Health Phys.* **66**(6), 699–706 (1994).
46. R. W. Field, D. J. Steck, B. J. Smith, et al., *Sci. Total Environ.* **272**, 67–72 (2001).
47. H. J. Albering, J. A. Hoogewer, and J. C. Kleinjans, *Health Phys.* **70**(1), 64–69 (1996).
48. F. Bochicchio, G. Campos-Venuti, C. Nuccetelli, et al., *Health Phys.* **71**(5), 741–748 (1996).
49. K. N. Yu, E. C. M. Yung, M. J. Stokes, et al., *Health Phys.* **75**(2), 159–164 (1998).
50. S. E. Hampson, J. A. Andrews, M. E. Lee, et al., *Risk Anal.* **18**(3), 343–350 (1998).
51. US Environmental Protection Agency, *Consumer's Guide to Radon Reduction*; US EPA Office of Air and Radiation, Washington, DC, 1992; available from www.epa.gov/iaq/radon/pubs/consguid.
52. US EPA, *How to Reduce Radon Levels in Your Home*, Office of Radiation and Indoor Air Report No. 402-K92-003, US EPA Office of Air and Radiation, Washington, DC, 1985.
53. J. E. Fitzgerald, et al., *A Preliminary Evaluation of the Control of Indoor Radon Daughter Levels in New Structures*, EPA Office of Radiation Programs, Report No. EPA-520/4-76-018.
54. A. P. Becker, *Evaluation of Waterborne Radon Impact on Indoor Air Quality and Assessment of Control Options*, PB84246404 (1984).
55. G. W. Reid, S. Hataway, and P. Lassovszwy, *Health Phys.* **48**(5) (1985).
56. United Nations Scientific Committee on the Effects of Atomic Radiation, *Sources and Effects of Ionizing Radiation*, Annex B: Natural sources of radiation UNIPUB, New York, 1977.
57. R. A. Wadden and P. A. Scheff, *Indoor Air Pollution* Wiley, New York, 1983.
58. S. Ericson and H. Schmied, in *Modified Design in New Construction Prevents Infiltration of Soil Gas That Carries Radon* (in Radon and its Decay Products: Occurrence, Properties and Health Effects, ACS Symp. Series No 331. P. H. Hopke, ed.), American Chemical Society, Washington, DC, 1987, p. 526.

10

Cooling of Thermal Discharges

Yung-Tse Hung, James Eldridge, Jerry R. Taricska, and Kathleen Hung Li

CONTENTS

1. INTRODUCTION

The discharge of water at elevated temperatures is often described as *thermal pollution*. It is produced by industries such as electric power plants, pulp and paper mills, chemical facilities, and other process industries that use and subsequently discharge water. Even if the discharged water is merely elevated in temperature, allowing it to return to streams, rivers, lakes, and other waters can dramatically alter the native environment.

Often, an elevated water temperature will be detrimental to native species of plants and animals. In such instances, alternate, invasive, and possibly undesirable species will be supported by the elevated water temperature. As water warms, the solubility of oxygen decreases. In addition to promoting competitive species of plants and animals, warmed waters may lead to aesthetic and odor problems if anaerobic conditions are created. For these and many other reasons, the discharge of elevated process water is widely regulated by governmental agencies in developed nations.

The fundamental principle of the evaporation of water to achieve cooling is widely utilized in nature. In the human body, sweat glands produce perspiration (water), which cools the body by evaporation. The fact that this cooling is more effective during a 95°F day in Arizona than an afternoon of the same temperature in Florida is widely understood. This concept is illustrated by the following example.

Problem

Given a pail full of water, change the water in the pail into vapor.

From: *Handbook of Environmental Engineering, Volume 2: Advanced Air and Noise Pollution Control*
Edited by: L. K. Wang, N. C. Pereira, and Y.-T. Hung © The Humana Press, Inc., Totowa, NJ

Solution

Several possible methods are possible to achieve this goal:

1. Let the pail stand until all of the water in the pail has evaporated.
2. Obtain a second, empty pail and pour the water into the empty pail. Repeat this process again and again between the pails. In this way, one will intuitively increase the water–air interaction, so that evaporation of the water will be enhanced.
3. If possible, blow air over the surface of the water in the pail. A simple house fan will suffice. Again, intuitively, evaporation of the water is enhanced and the moving air will carry away the water vapor as fresh air is pushed across the water.
4. As in the previous mention of Arizona versus Florida, low-humidity air will support the desired evaporation of water much more efficiently than high-humidity air.

The purpose of this chapter is to discuss cooling ponds and cooling towers. Essentially, a cooling pond is choice 1 in the previous example. A cooling tower combines choices 2–4.

Cooling ponds are the simplest choice to lower the temperature of a warm-water discharge stream and may be the least costly alternative. Such a solution to a water-cooling problem relies entirely upon contact of the pond water surface with the surrounding environment. As such, the warm water will slowly approach the desired outlet temperature. Ponds are designed to be either a *recirculating* or *a once-through* type.

A cooling pond may be a good choice if ample room for such a pond is available and land costs are low at a given location. The slow process most often requires a very large pond surface, perhaps several acres, for proper heat transfer to occur. Also, if they are located in a rural area and if the quality of the water being cooled is fairly good, such man-made ponds can offer recreational uses in addition to serving to lower the water temperature. An important note regarding cooling ponds is that this method of lowering the water temperature is most likely to limit evaporative losses of water versus a cooling tower.

However, if the needed land area for a pond is not available and/or the slow cooling of the water is unacceptable, a cooling tower will be the design choice. Cooling towers are classified as *natural* or *mechanical draft*, as well as *wet* or *dry* types.

2. COOLING PONDS

2.1. Mechanism of Heat Dissipation (Cooling)

A cooling pond permits a sufficient buildup of the process water so that the needed heat exchange can occur between the water and the surrounding environment. The warmed water entering the pond will lose and gain heat as it passes through the pond by the combined mechanisms of conduction, convection, radiation, and evaporation. Numerous factors affect the rates of these various mechanisms as well as the overall rate of heat transfer. If the heat inputs are equal to heat outputs, the pond will be at an equilibrium condition and the surface temperature of the pond will be at some constant value.

The heat exchange between the pond and the surrounding environment will be governed by the temperature difference between the atmosphere and the pond surface, the surface area of the pond, meteorological conditions such as rain or sun, wind speed, humidity, geographical location, and so forth. Obviously, these conditions will vary depending on both time of day as well as time of year.

The various heat inputs and outputs have been identified and presented in typical ranges (1). These values are as follows:

Inputs

1. q_s = short-wave solar radiation
   Range of q_s from 1085 to 7588 kcal/d-m² or 400 to 2800 Btu/d-ft²
2. q_a = long-wave solar radiation
   Range of q_a from 6504 to 8672 kcal/d-m² or 2400 to 3200 Btu/d-ft²

Outputs

3. q_{sr} = reflected portion of q_s
   q_{sr} ranges from 109 to 542 kcal/d-m² or 40 to 200 Btu/d-ft²

4. q_{ar} = reflected portion of q_a
   q_{ar} ranges from 190 to 325 kcal/d-m² or 70 to 120 Btu/d-ft²

5. q_{br} = long-wave black radiation from water surface
   q_{br} ranges from 6504 to 9756 kcal/d-m² or 2400 to 3600 Btu/d-ft²

6. q_e = evaporative heat loss
   q_e ranges from 5420 to 21680 kcal/d-m² or 2000 to 8000 Btu/d-ft²

Input or Output: Dependent on Circumstances

7. q_c = conductive–convective heat loss
   q_c ranges from −867 to +1084 kcal/d-m² or −320 to +400 Btu/d-ft²

The net heat flux across the air–water interface (pond surface) can now be expressed as the difference between inputs and output:

$$q_t = (q_s + q_a) - (q_{sr} + q_{ar} + q_{br} + q_e) + q_c \qquad (1)$$

Note that as indicated above value for conductive–convective heat loss can be either positive of negative.

As stated previously, at equilibrium q_t is zero. The surface temperature of the pond at this condition is called the *equilibrium temperature*. Also, note that the values for q_{br}, q_e and q_c are dependent on the pond surface temperature. The variables q_s, q_a, q_{sr} and q_{ar} are independent of pond surface temperature. The discharge of a heated effluent into the pond will add another heat quantity to the input side of Eq. (1), q_d.

As the result of changing weather conditions, the surface temperature of most bodies of standing water (be these natural ponds, and lakes or man-made cooling ponds) approaches the equilibrium temperature asymptotically. The equilibrium temperature of such bodies of water is normally calculated on the basis of average daily temperatures.

2.2. Design of Cooling Ponds

Often the design of a cooling pond is dictated by a common rule of thumb. One such design rule widely accepted for cooling pond design is 1–2 ac/MW of pond capacity or 200–400 kcal of heat will be dissipated per hour per square meter of pond surface area. Such rules of thumb should always be double checked with sound engineering principles.

Local weather patterns and land formations are obvious limitations that must be considered, as such factors will affect the rate of heat transfer possible from the pond's surface. Thus, the size of the pond needed for the required cooling should also be considered (2).

Basic design equations are presented here for both recirculating cooling ponds as well as for once-through-type ponds. The effects of longitudinal mixing and short circuiting of both types of cooling pond are discussed. Seasonal and geographical considerations are also considered, as well as other miscellaneous considerations regarding cooling pond design.

2.2.1. Evaluation of Heat Dissipation of Completely Mixed and Recirculating Cooling Ponds

The primary parameter when designing a recirculating cooling pond is the amount of water surface area required to dissipate the excess heat of the water that enters the pond. An example would be a cooling pond at a thermal electrical power generation site. In the design, a difference will exist above the equilibrium surface temperature of the pond. The surface area of the pond needed to dissipate the surplus heat of the water in the pond is directly related to the mean surface heat transfer capacity of the pond (1).

2.2.1.1. CAPACITY OF A COMPLETELY RECIRCULATING POND TO DISSIPATE ADDED HEAT LOADS

The ability of a pond to dissipate a heat load from an influent has been demonstrated (1) to be described as

$$q = KA\,(T_s - E) \tag{2}$$

where q is the heat dissipated (kcal/d), K is the heat transfer coefficient at the pond surface (kcal/d - m^2-°C), A is the surface area of the pond (m^2), T_s is the surface temperature of the pond (°C), and E is the equilibrium temperature of the pond at zero added heat load (°C). Equation (2) may also be expressed as

$$T_s = q/KA + E \tag{3}$$

Now writing the equation in terms of heat flux, q_{rj} kcal/d-m^2, Eq. (3) becomes

$$T_s = q_{rj}/K + E \tag{4}$$

Equation (4) expresses the relationship between the pond surface and the rise of the pond temperature above the equilibrium temperature. Equation (2) or some modification of this equation is used in cooling pond design to predict pond surface temperature.

2.2.1.2. SURFACE HEAT TRANSFER COEFFICIENT

An expression has been derived for the surface heat transfer coefficient, K, by means of a balance of the rates of heat exchange at the pond surface. This expression can be manipulated and several approximations are made to arrive at the following definition of K (3):

$$K = 92.5 + (C_1 + \beta)\,f(U) \tag{5}$$

where K is the surface heat transfer coefficient (Kcal/d-m^2-°C), 92.5 is the back radiation, C_1 is the Bowen's conduction–evaporation coefficient (0.468 mm Hg °C), β is the slope of the saturated vapor pressure curve between the dew point temperature and the

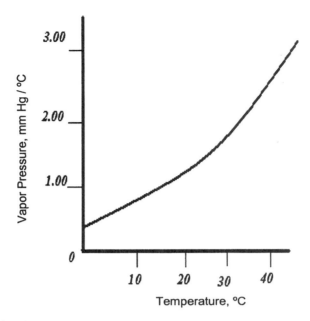

Fig. 1. Plot of the slope of the vapor pressure curve for water versus. the average dew point and the pond water surface temperature.

pond surface temperature (mm Hg °C), and $f(U)$ is the evaporative wind speed function (Kcal/d-m^2-mm Hg).

Equation (5) defines the surface heat transfer coefficient in terms of three components. These three components approximately represent the sum of heat dissipation from back radiation, conduction, and evaporation.

2.2.1.3. RELATIONSHIP FOR THE SLOPE OF THE SATURATED VAPOR PRESSURE CURVE, β

Various values of β are determined by one of two methods; see vapor pressure tables or a vapor pressure chart such as that in Fig. 1:

$$\beta = (e_s - e_a)/(T_s - T_d) \qquad (6)$$

where β is the slope of the vapor pressure curve (mm Hg/°C), e_s is the saturated vapor pressure of water at the pond surface (mm Hg), e_a is the air vapor pressure (mm Hg), T_s is the surface temperature of the pond water (°C), T_d is the dew point temperature (°C), and T_{avg} is the average temperature $[(T_d + T_s)/2]$ (°C).

2.2.1.4. RELATIONSHIP FOR THE WIND SPEED FUNCTION, $f(U)$

Brady et al. (3) used curve fitting, plotting techniques, and multiple regression analysis of data collected at three cooling pond sites to develop an empirical relationship to predict the wind speed function. In metric units, this expression states

$$f(U) = 189.8 + 0.784\ U^2 \qquad (7)$$

where $f(U)$ is the wind speed function (kcal/d-m^2) and U is the wind speed (km/h). It is important to note that Eq. (7) predicts that the wind speed function, $f(U)$, is not sensitive to relatively low wind velocities. This has the effect of reducing the accuracy of wind speed data at low wind velocities. This also reduces the sensitivity of the heat transfer coefficient in calm wind conditions. A possible explanation of the apparent insensitivity of K to calm wind conditions is related to the cooling effect of vertical convection currents in the atmosphere (3).

2.2.1.5. RELATIONSHIP FOR EQUILIBRIUM TEMPERATURE, E

An expression of pond equilibrium temperature has been derived by Brady et al. by examining the balance of heat-exchange rates at the pond surface. This expression can be manipulated, and by several approximations, E is expressed as (3)

$$E = T_d + (H_s/K) \qquad (8)$$

where E is the equilibrium temperature (°C), H_s is gross solar radiation (kcal/d-m^2), K is the heat transfer coefficient (kcal/d-m^2 °C), and T_d is the dew point temperature (°C). A test of Eq. (8) was made by evaluation of values of E directly with data for dew point temperature and gross solar radiation at three recirculating cooling pond sites. Plots of this evaluation revealed that Eq. (8) was normally accurate within a few degrees Celsius if the heat transfer coefficient was greater than 135 kcal/d-m^2-°C (3).

2.2.2. Prediction of Surface Temperature, T_s, of a Completely Mixed Recirculating Cooling Pond

The surface heat transfer coefficient K and the equilibrium temperature E can be determined using known weather conditions and Eqs. (5)–(8). An iterative approach is used to account for the fact that the average surface temperature (T_s) is not known initially. The first estimate of T_s is corrected in successive approximations. Required weather conditions are the dew point temperature, gross solar radiation, and wind speed. With these values known, T_s can be calculated for a specific heating load (3).

2.2.2.1. CALCULATION OF COOLING POND EXIT TEMPERATURE

Problem

Given the following information, calculate the cooling pond exit temperature:

Application:	Fossil-fuel-fired power plant
Hot water source:	Plant condensers
Type of cooling pond:	Completely recirculating
Surface area of pond, A:	4×10^6 m^2
Heat load:	8×10^9 kcal/d
Hot water flow rate:	9255 L/s
Location:	Southern United States
Prevailing weather condition:	July
Mean dew point, T_d:	23°C
Mean solar radiation:	5750 kcal/m^2-d
Mean wind speed:	13.5 km/h

Solution

Estimate the surface water temperature T_s to be 36°C, and with the given mean dew point temperature, T_d, of 23°C (given in conditions above), calculate T_{avg}:

$$T_{avg} = (T_d + T_s)/2 = (36 + 23)/2 = 29.5°C$$

Using Fig. 1 with the average temperature to determine the slope of the vapor pressure curve, $\beta = 1.77$ mm Hg °C. At the given wind speed $U = 13.5$ km/h and using Eq. (7), determine

$$f(U) = 189.8 + 0.784 \; U^2 \tag{7}$$
$$f(U) = 189.8 + 0.784 \; (13.5)^2 = 323.6 \; \text{kcal/m}^2 \; \text{d}$$

The surface heat coefficient K is now determined from Eq. (5):

$$K = 92.5 + (C_1 + \beta) f(U) \tag{5}$$
$$K = 92.5 + (0.468 + 1.77) \; 323.6 = 816.5 \; \text{kcal/m}^2 - \text{d} - °C$$

Using the given solar radiation $H_s = 5750$ kcal/m²-d, the equilibrium temperature E can now be determined from Eq. (8):

$$E = T_d + \left(H_s / K \right) \tag{8}$$
$$E = 23 + \left(5750 / 816.5 \right) = 23 + 7.0 = 30°C$$

The surface temperature is now determined using Eq. (3):

$$T_s = q / KA + E \tag{3}$$
$$T_s = \left\{ 8 \times 10^9 / \left[(816.5)(4 \times 10^6) \right] \right\} + 30.0$$
$$T_s = 2.5 + 30.0 = 32.5°C \text{ versus the assumed value of } 36°C$$

Repeating this procedure using 33°C as T_s, from Fig. 1, $\beta = 1.65$, and previously calculated value for $f(U)$ of 323.6 kcal/m²-d, the recalculated K and E are

$$K = 777.9 \text{ kcal/m}^2\text{-d-}°C \text{ and } E = 30.4°C$$

and

$$T_s = q / KA + E \tag{3}$$
$$T_s = \left\{ 8 \times 10^9 / \left[(777.9)(4 \times 10^6) \right] \right\} + 30.4$$
$$T_s = 2.6 + 30.4 = 33.0°C$$

Because this calculated value for T_s equals the assumed value, no additional iterations are needed.

2.2.2.2. DESIGN CHART

Figure 2 will assist in this type of cooling pond design calculation (3). This figure gives the values for the heat transfer coefficient K (in kcal/m²-d-°C) at the intersection of the wind speed U (in km/h) and average temperature T_{avg} (in -°C). The controlling local weather conditions may be found in standard references (4).

2.2.3. Longitudinal Mixing Effects on Recirculated Cooling Pond Performance

In addition to dissipating heat from water flowing from an industrial process, the lowest possible intake temperature of the water re-entering the process is a prime objective of proper operation of the pond if that water is to be reused in the process. Dissipating heat and lowering the temperature of the water recirculated back to the process are not necessarily complementary objectives.

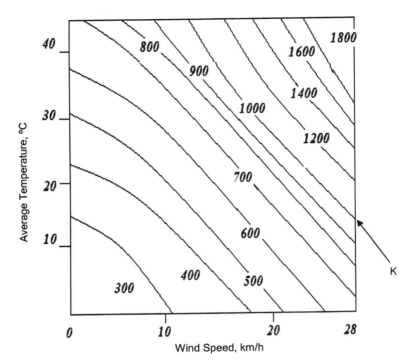

Fig. 2. Values of heat transfer coefficients in terms of wind speed and average temperature.

Pond exit temperature (process intake temperature) at a given set of weather conditions is directly related to the amount of longitudinal mixing that takes place within the cooling pond. Also affecting this temperature is the uniformity of flow through the pond. Short-circuit flows of water within the pond will result in direct transport of warm water entering the pond to the pond discharge point. Such short-circuit flow will return water to the process at an elevated temperature.

If insufficient longitudinal mixing occurs, a number of techniques are available to alleviate the condition. Transverse baffles placed in the pond will increase the length of the flow path of water in the pond. Vertical skimmer walls may be placed in the pond to minimize wind-driven short circuits. This effect of wind-driven water currents is also minimized by a high length-to-width ratio of pond size. Relatively shallow pond design will minimize vertical water current formation in the pond.

2.2.4. Evaluation of Heat Dissipation in Once-Through Cooling Ponds

Various mechanisms of heat transfer of warm water have been described previously. Various models exist that allow for prediction of water discharge temperature from such cooling ponds. The needed pond surface area may be predicted using such models for a given temperature change of the water in the pond.

Edinger and Geyer (1) developed a method that predicts the design parameters of a cooling pond with a good balance of accuracy versus facility. This method is also summarized in a government publication (5). The model assumes complete (ideal) mixing in the depth and width of the water flow. No longitudinal mixing of the water, in the flow direction, is assumed. Further assumptions are constant cross-sectional flow and constant flow velocity. The absence of flow channel curvature is also assumed in this model. In this way, the water entering the pond is modeled as completely mixed at the point of entry into the pond. The warm water then transfers heat to the surroundings exponentially as it proceeds as a "slug" of warm water passing through the pond. These assumptions and approximations allow for a mathematical expression that predicts the surface temperature as a function of longitudinal distance of the pond. In differential form, the net rate of heat exchange is

$$dq_t/dt = -K(T_s - E) \tag{9}$$

where dq_t/dt is the net rate of surface heat exchange (kcal/m²-d), K is the heat transfer coefficient (kcal/m²-°C), T_s is the pond surface temperature (°C), and E is the equilibrium temperature (°C).

This heat exchange causes the water temperature to fall in relation to distance of water travel in the cooling pond. This may be expressed mathematically:

$$-K(T_s - E) = \rho C_p du(\delta T/\delta X) \tag{10}$$

where ρ is the density of water (1.0 g/cm³), C_p is the specific heat of water (1.0 cal/g°C), d is the average depth of the flow path (m), u is the average flow velocity (m/d), $\delta T/\delta X$ is the longitudinal temperature gradient (°C/m), and X is the distance from pond entry, (m). Integration of Eq. (10) with the boundary condition of $T = T_x$ at $X = X$ and that $T = T_0$ at $X = 0$ results in the expression

$$e^{-\alpha} = (T_x - E)/(T_0 - E) = -KX/e^{\rho C_p du} \tag{11}$$

where $\alpha = KX/\rho C_p du = KA/\rho C_p Q$ and Q = volumetric flow rate, m³/d. With the previously discussed weather conditions, the cooling pond area required for a given heat load may be calculated. or alternatively, the exit temperature of water leaving a cooling pond of a known surface area may be determined.

2.2.4.1. CALCULATION OF POND EXIT TEMPERATURE

In the previous example, the following values were found to be appropriate:

$$K = 778.5 \text{ kcal/m}^2\text{-d-°C}$$

$$C_p = 1.0 \text{ kcal/kg-°C}$$

and

$$\rho = 1.0 \text{ g/cm}^3 = 1000 \text{ kg/m}^3$$

The volumetric flow, Q, in terms of m³/d is determined as follows:

$$Q = 9,255 \text{ L/s} = (9225)(60)(60)(24)/1,000 = 8.0 \times 10^3 \text{ m}^3/\text{d}$$

Assume that the inlet temperature of water entering the power plant condensers is 28°C. If the water is warmed 10 °C as it passes the condensers, the temperature of the water entering the cooling pond can be assumed to be 38°C (T_0). Therefore, the following data apply. As shown in Eq. (11), the value for α can be determined as follows:

$$\alpha = KA/\rho C_p Q$$

$$\alpha = (778.5)(4 \times 10^6)/[(1000)(1.0)(8 \times 10^5)] = 3.89$$

Substituting values for E, T_0 and α into Eq. (11) and solving for the temperature water, T_x, exiting the cooling pond,

$$e^{-\alpha} = (T_x - E)/(T_0 - E) \tag{11}$$
$$e^{-3.89} = (T_x - 30.4)/(38 - 30.4)$$
$$T_x = 30.56°C$$

The effect of pond surface area on the exit temperature of the water may be evaluated by repeating this calculation for various values for the area of the pond (A). Alternatively, exit water temperature may be used to determine the required cooling pond surface area.

2.2.5. Relationship of a Completely Mixed to a Totally Unmixed Pond Once-Through Cooling Ponds

The purpose of once-through cooling ponds, both completely mixed and completely unmixed, is twofold: to dissipate heat from water rejected from some industrial process and to achieve the lowest possible effluent temperature of water exiting the pond. Both results are directly related to both longitudinal mixing and short-circuiting currents present within the pond.

2.2.5.1. COMPARISON BETWEEN NET PLANT TEMPERATURE RISE OF A COMPLETELY MIXED POND AND A COMPLETELY UNMIXED POND

The ratio of the net plant temperature rise in the water exiting the pond from a completely mixed versus a completely unmixed pond has been determined (1) to be

$$(T_{sm} - E)/(T_{su} - E) = e^\alpha/(1 + \alpha) \tag{12}$$

where T_{sm} is the completely mixed cooling pond's actual surface temperature (°C) and T_{su} is the completely unmixed cooling pond's actual surface temperature (°C).

At fixed flow rate, Q, surface area, A, and similar weather conditions, the completely unmixed pond will yield faster cooling of the water compared to the completely mixed pond. This is a consequence of the warm water in the unmixed pond being unmixed initially or otherwise diluted with cooler water upon entry into the pond. The driving force of cooling, $T_s - E$, is maintained at the highest possible value.

2.2.5.2. AREA OF COMPLETELY UNMIXED COOLING POND NEEDED TO PROVIDE IDENTICAL SURFACE TEMPERATURE AS A COMPLETELY MIXED COOLING POND

The area of a completely unmixed cooling pond needed to provide the identical surface temperature as a completely mixed cooling pond can be determined using Eq. (11). In this instance, $T_{sm} = T_{su}$ so $T_{sm} - E = T_{su} - E$ as well; thus,

$$(T_{sm} - E)/T_{su} = e^\alpha/(1 + \alpha_m) = 1 \tag{13}$$

where α_m is the α evaluated for the completely mixed cooling pond and, α_u is the α evaluated for the completely unmixed cooling pond. The actual difference between α_m and α_u is the area of each respective cooling pond. The flow rate, Q, is identical for each pond. Additionally, the specific heat of the water, C_p, the water density, ρ, and the surface heat transfer coefficient, K, are the same for both or either ponds. Therefore, the ratio of the two required areas for the two types of cooling pond can be determined.

2.2.6. Effect of Seasonal Weather Variations on Cooling Pond Requirements and Performance

The equilibrium temperature, E, for most locations inside the continental United States reaches a low in early January and then reaches a high 6 mo later in early July. The pattern of variation of E is actually quite smooth for most locations. The difference between seasonal high and low is more distinct at more northern (higher latitude) locations (4).

The surface heat-exchange coefficient, K, varies in a relatively inversely proportional relationship to E. The highest heat-exchange coefficients are realized in summer and, lowest in winter. The variation of K is not as regular as the variation of E as the result of seasonal wind variations. As a result, the maximum value of K occurs in some locations (coastal, mountains, etc.) at times of the year other than midsummer.

2.2.7. Effect of Geographic Location on Cooling Pond Performance

As mentioned above, geographic locations will affect both E and K. At higher latitudes, solar radiation is reduced and significantly affects the equilibrium temperature of a cooling pond. Topography (hills, flat plains, etc.) will influence predominant wind speeds as well as wet bulb temperature and, hence, influence K. As a result, within the continental United States, the best conditions for cooling pond performance exist in the southern Great Plains between the Mississippi River and the Rocky Mountains (Kansas, Oklahoma, north Texas, eastern Colorado, etc.). Favorable conditions for cooling pond performance also exist along the Atlantic seaboard between Cape Cod and the Gulf of Mexico. Poor conditions for cooling pond performance will be found in the Great Basin region between the Rocky mountains and Sierra Nevada Mountains (Nevada and Utah). It should be stressed that local "microclimate" and local topographic conditions must be considered when determining cooling pond performance (4).

2.2.8. Other Factors Related to Cooling Pond Design

Topics pertinent to the design of cooling ponds are presented in the following subsections:

2.2.8.1. DIKES AND BAFFLES

A cooling pond has traditionally been constructed by building dikes and/or other embankments to form a man-made pond. Therefore, cooling ponds are normally constructed where the local topography is amenable to such construction. If a pond must be excavated, construction costs will escalate (4).

2.2.8.2. EVAPORATION OF WATER

In some locations, particularly southern and western, evaporative loss of water from a cooling pond can be a serious problem. Such water must be replenished, and if construction of a cooling pond is proposed for arid or semiarid regions, sources of substitute water must be available.

2.2.8.3. INFILTRATION

This term refers to entry and movement of water through soils. If present, infiltration may affect the design and location of a cooling pond for two reasons: (1) If recharging of the local water table takes place, that water table may rise in elevation and affect surrounding land uses; (2) groundwater contamination may occur.

2.2.8.4. FOGGING

A disadvantage of cooling ponds is the potential for fog formation. If a fog forms and obscures local visibility, this can be a problem, particularly if a highway or an airport runway is nearby.

2.2.8.5. RECREATION

As previously mentioned, if the water being cooled in the pond is of sufficiently good quality, boating, fishing, and perhaps even swimming recreational opportunities may be an auxiliary benefit of pond construction.

2.2.8.6. ECONOMICS

If sufficient land is available at a reasonable cost and the local topography is amenable to construction needs, a cooling pond may be a good solution to a cooling problem, as the pond will cost less to construct than a tower.

3. COOLING TOWERS

To treat large volumes of warm water from an industrial process such as a coal or nuclear power plant, a cooling pond is not practical because of the enormous size of the pond needed. Indeed, the term *lake* would be more appropriate to describe such a body of water. As a result, a *cooling tower* is the practical solution to the cooling needs of most large industrial plants. Such towers are often massive in size and visible from a long distance. Note the size of the cooling system in Fig. 3. The design of cooling towers is now a specialized niche of environmental engineering. Therefore, it is important for engineers to understand the basic design of such towers, their advantages and disadvantages, their operational considerations, and their environmental impact.

Fig. 3. The cooling system at the CHPP-5 (TEC-5) heating and power plant in Kharkiv, Ukraine. (Courtesy of Dr. Gennadiy K. Voronovsleiy, General Director.)

3.1. Mechanism of Heat Dissipation in Cooling Towers

The normal operation of a cooling tower involves both heat and mass transfer. Excellent explanations of the material and heat balances involved have been given in detail elsewhere (6). For the purposes of this discussion, the relative temperature and humidity gradients present within a tower are given in Fig. 4.

In this example, the top of the tower in Fig. 4a contains water at a higher temperature versus the dry bulb temperature of the air. The water is therefore being cooled by both evaporation and transfer of sensible heat from the water to the air. As evaporation of water is also taking place at the air–water interface, the humidity will change as well.

Possible relative conditions at the bottom of the cooling tower are sketched in Fig. 4b. Now, the water temperature is above the wet bulb temperature but below the dry bulb temperature. As the water is being cooled, the temperature at the water–air interface is lower than the bulk temperature of the water. A temperature gradient will exist in the direction of the water–air interface. This is a consequence of the fact that the air temperature is higher than the temperature at the water-air interface. Sensible heat is being transferred from both water and air to the interface. The sum of these two heat transfers will equal the latent heat flow from the interface into the air as the consequence of evaporation.

3.2. Types of Towers

Cooling towers are normally classified as natural or induced draft types. Additionally, towers are either wet or dry types (7).

Induced (also called mechanical) draft towers employ a fan to move air through the tower. Natural draft towers rely upon the difference in density between the warm air exiting a tower and the cool air inside the tower to act as the driving force for air movement. Most towers at large industrial locations have direct contact between air and water and are hence considered to be "wet" towers. Some towers, however, are designed to keep the air and water separate and, hence, are considered to be "dry" towers.

Dry cooling towers are used in specialized applications where temperatures are very high. Convection is the primary mechanism of heat transfer. Because dry towers are markedly less efficient than wet towers, they are not discussed at length here.

Common types of cooling tower are presented in Figs. 5–13 and are described in the following subsections.

3.3. Natural Draft Atmospheric Cooling Towers

The flow scheme for this design of cooling tower is presented in Figs. 5. Air enters the tower through open louvers on the sides of the tower. As the air passes through the tower, it comes into contact with water that is sprayed downward in the tower. Wind currents primarily drive the horizontal airflow through the tower. Such towers are sometimes open, containing only the falling water spray.

An alternate design of this type of tower uses packing (or fill) with an open design. Splashboards, triangular PVC rods, or an open polyethylene design are among the several possibilities for the fill design. The basic purpose of these various fills is the same: to increase the water–air contact surface with minimum added resistance to both airflow and water flow.

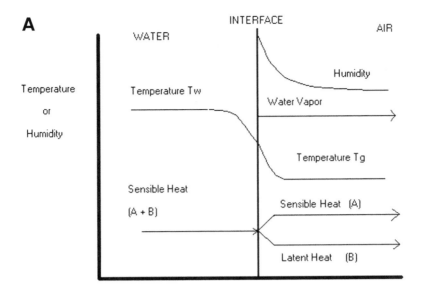

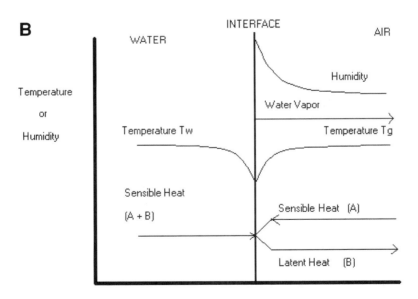

Fig. 4. **(a)** Relative temperature and humidity conditions at the top of a typical cooling tower; **(b)** relative temperature and humidity conditions at the bottom of a typical cooling tower.

With the exception of large hyperbolic cooling towers that are a special case (such as those in Fig. 3), this type of cooling tower is normally used for smaller installations such as hospitals, office buildings, or other relatively small facilities. This type of cooling tower is typical of refrigeration units.

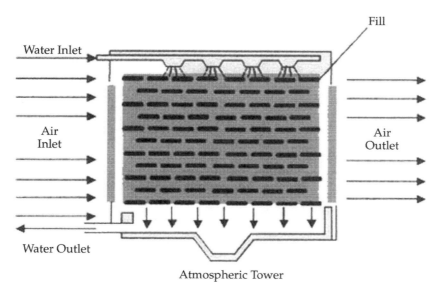

Fig. 5. Natural draft atmospheric cooling tower.

Advantages and disadvantages of this type of cooling tower have been previously described (6) as follows:

Advantages

No moving or other mechanical parts
Relatively little maintenance
No recirculation or air required

Disadvantages

Comparatively large sizes needed for a given heat load
High initial capital cost
High pumping costs
Must have unobstructed location to take advantage of available winds
Cooling dependent on wind direction and velocity
Often had excessive drift loss

The last disadvantage is of concern. Drift droplets will likely contain chemical impurities and, as such, will be considered to be airborne emissions by governmental regulatory agencies (8,9).

Ozone treatment of cooling tower water has recently been suggested by the US Department of Energy as a method of treating cooling tower water that saves costs and reduces the need for chemical additives (10).

3.4. Natural Draft, Wet Hyperbolic Cooling Towers

As seen in Fig. 3, this type of tower is named for the distinctive hyperbolic shape of the stack through which air passes after contacting warm water sprayed over packing inside the lower part of the tower. Often massive in size, hyperbolic cooling towers are often 150–180 m (490–515 ft) high and 100–135 m (330–450 ft) in diameter. Such towers

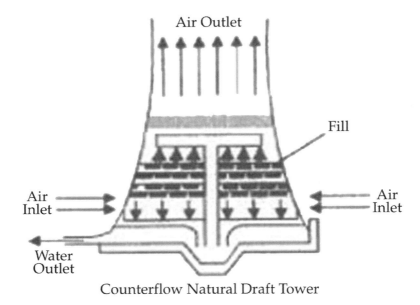

Counterflow Natural Draft Tower

Fig. 6. Counter flow natural draft cooling tower.

are often made from reinforced concrete. Airflow through the tower is driven by the density difference between the heated (so lighter) air leaving the tower and the colder (heavier) air outside the tower. Wind velocity blowing across the tower opening may additionally aid this airflow; however, if the tower is correctly designed, the operation of the cooling tower will not be dependent on wind flows. This type of cooling tower design is illustrated in Figs. 6 and 7.

The bottom 3–6 m (10–20 ft) contains fill (packing) that supports extended surface-to-surface contacts of air and water in both the counterflow and crossflow designs. In a counterflow design, warm water is distributed by spray nozzles over the fill material and flows downward as the air passes upward. If crossflow design is utilized, warm water flows downward over the fill as air passes in a 90° plane to the water flow. The center of the tower is open and the airflow here turns upward to exit the tower.

Counterflow design supports more efficient heat transfer, as the coolest water will always be in contact with the coolest air and the minimum amount of fill is required. However, the geometry of the crossflow design promotes improved air–water contact, and at the same capacity as the counterflow tower, the crossflow tower will operate at a lower head loss.

The purpose of the fill (packing) material, as mentioned above, is to support and enhance surface-to-surface contact of air and water to facilitate heat transfer. Transfer of heat may be modeled based on a thin film of air surrounding a water droplet, as in Fig. 8. This movement of heat can be modeled using the Merkel equation (11,12):

$$KaV/L = \int_{T_1}^{T_2} dT/(h_w - h_a) \qquad (14)$$

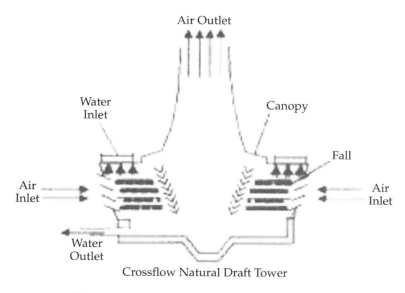

Fig. 7. Cross flow natural draft cooling tower.

where *KaV/L* is the tower characteristic, *K* is the mass transfer coefficient (lb water/ft^2 or kg water/ m^2), *a* is the contact area/tower volume (ft or m), *V* is the active cooling volume/plan area (ft or m), T_1 is the hot water temperature (°F or °C), T_2 is the cold water temperature (°F or °C), *T* is the bulk water temperature (°F or °C), h_w is the enthalpy of the air–water mixture at bulk water temperature (J /kg dry air or Btu/lb dry

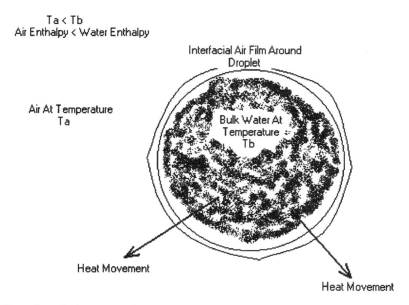

Fig. 8. Film of cool air surround a warm droplet of water, with movement of heat from the water to the air.

air), and h_a is the enthalpy of the air–water vapor mixture at wet bulb temperature (J/kg dry air or Btu/lb dry air).

Further derivation of this equation is possible to determine the tower characteristic and can be found in standard reference texts. Here, it is used to demonstrate that the function of fill or packing is to maximize the surface of water available to the air to promote the most efficient heat transfer possible.

Natural draft hyperbolic cooling towers are attractive because of their relative lack of mechanical and electrical components. They can accommodate large quantities of water and are relatively efficient cooling units. However, the height of such units, which is needed for proper draft, is objectionable from a public relations viewpoint. The plume of condensed water vapor leaving a large hyperbolic tower is often viewed as pollution by the general public.

3.5. Example 1

Select the proper wet bulb temperature for a proposed cooling tower in southern California. When designing a cooling tower, the highest possible wet bulb temperature must be determined. This will ensure that the cooling tower will operate at all times. The Los Angeles airport has the following maximum wet bulb temperatures (4):

- 67°F is exceeded during 2% of summer hours
- 69°F is exceeded during 0.5% of summer hours
- 71°F is exceeded during 0.1% of summer hours

Depending on how critical the application is, the designer may wish to design to a slightly higher temperature than 71°F. Note that at 0.1% of the summer hours, if 71°F is used for design purposes, the tower may be expected to fail over 20 h, or approx 1 d:

$$(90 \text{ d/summer})(24 \text{ h/d})(0.01) = 21.6 \text{ h}$$

Of course, these hours of poor performance will most likely be spread over the entire summer period, so the amount of safety factors needed is at the discretion of the design engineer.

3.6. Hybrid Draft Cooling Towers

Two hybrid types of draft cooling tower are seen in Figs. 9 and 10. Although still considered to be draft-type cooling towers, both designs employ a fan to assist airflow through the towers. The large size of the cooling tower limits the actual size of the fan that can be used, so the design of the tower narrows to a point where the fan is installed and then the tower widens after the fan as the air exits the tower.

The advantage of these hybrid designs is that no packing or fill products are used. The lack of the packing to force additional water–air contact requires the fan to make up for this lost efficiency in heat transfer. The fan also adds to maintenance and operational costs of the tower. These designs do reduce initial capital outlay, as the towers will be considerably smaller than true natural draft cooling towers.

3.7. Induced (Mechanical) or Forced Draft Wet Cooling Towers

As opposed to the hybrid induced draft towers just discussed, cooling towers that employ a fan to drive air through a packed cooling tower are known as mechanical or induced draft type. Often, a bank of such cooling towers are built together to satisfy the

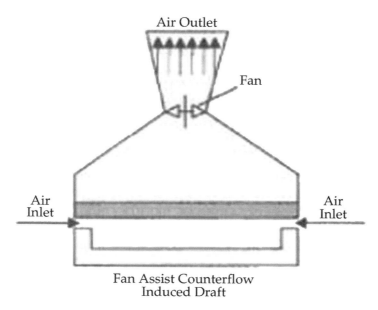

Fig. 9. Fan-assist (draft or mechanical type) counterflow induced draft cooling tower.

total cooling requirement, as the practical fan size limits the size of each individual cooling tower. Both counterflow and crossflow air paths may be used.

In Figs. 11 and 13, it is seen that the fan is located behind the packed section of the tower. From the exit section of the tower, the fan is able to pull the air through the packing. The designs in Figs. 12 and 14 place the fan at the entrance to the tower to push air through the packed section and then out of the tower. The cooling towers in Figs. 11 and

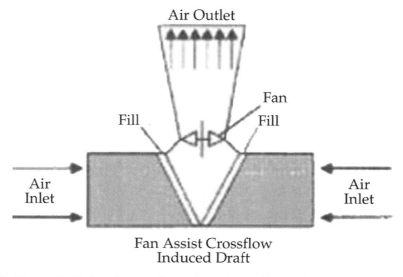

Fig. 10. Fan-assist (induced or mechanical type) crossflow induced draft cooling tower.

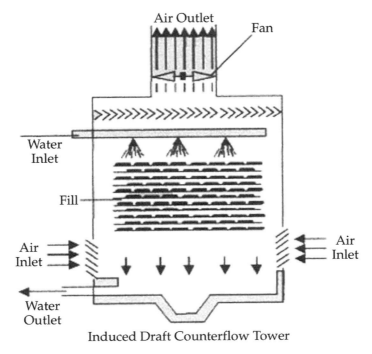

Fig. 11. Induced draft (mechanical) counterflow cooling tower.

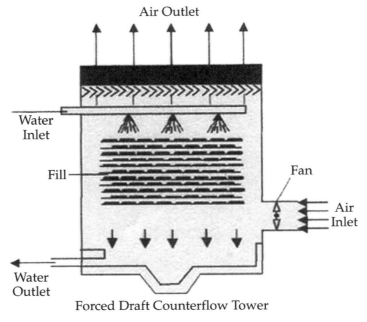

Fig. 12. Forced (mechanical) draft counterflow cooling tower.

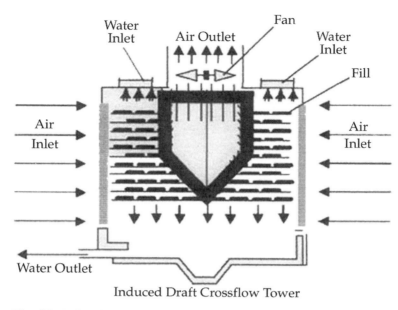

Fig. 13. Induced draft (mechanical) crossflow cooling tower.

13 are called *induced draft* and the designs in Figs. 12 and 14 are called *forced draft* cooling towers.

The Induced draft design normally utilizes larger fans than forced draft. This allows for higher air exit velocities and lessens problems with recirculation as a result. Air distribution is often superior using induced draft versus forced draft design. As the fan is in the exit in an induced draft cooling tower, this design permits the warm, moist air to pass through the fan, increasing corrosion problems.

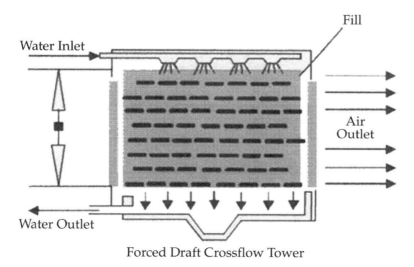

Fig. 14. Forced draft (mechanical) crossflow cooling tower.

Both induced and forced draft cooling towers can be noisy. Both types will also obviously consume large amounts of electric power during normal operations and have more maintenance considerations than natural draft cooling towers. These designs are practical for rooftops. Also, if sufficient land is available, a bank of either type of design will not be the dominant feature of the landscape as with large natural draft hyperbolic cooling towers.

3.8. Cooling Tower Performance Problems

This discussion is intended to be brief, as many good sources of operational advice are now available, particularly on the Internet.

Some of the causes of poor performance in a cooling tower (13–20) and suggested corrective actions are given in the following subsections.

3.8.1. Scale Formation

In a cooling tower, water evaporates and potentially enables mineral deposits to form a "scale" on surfaces. If such a scale does form, it acts as a barrier to heat transfer and the performance of the cooling tower will degrade. Possible sources of such scale are hard water and/or inadequate blow-down (21).

If hard water is present, water softeners combined with adjustment of the blow-down rate will likely prevent further formation of scale. The advice of a reputable industrial water-treatment service company may prove valuable in such a situation.

Removing existing scale deposits can be difficult. Hard-water deposits generally will dissolve in a mild acid backwash. However, if scale is thick and/or has had the opportunity to dry, it may have to be manually chipped off surfaces.

3.8.2. Clogged Spray Nozzles

Again, hard water can cause deposits to form in the spray nozzles that introduce the water into the cooling tower. Algae or other solids may also cause deposits. Spray nozzles should be inspected on a regular basis to verify that the proper flow pattern of water into the tower is present. If partial clogging occurs, the flow of water over the fill will be uneven and water–air contact surface will be lost. Consequently, heat transfer and tower performance will suffer. A proper water–treatment program recommended by an experienced water–treatment company should be followed.

3.8.3. Poor Airflow

Poor airflow through the tower will degrade tower performance. Various causes can partially interrupt airflow: Debris in the tower exit, entrance, or the packing will cause unwanted and undesirable air currents to form that will short circuit parts of the tower. Loose fan and motor mountings, poor fan alignment, damaged fan blades, gear box problems, and other problems can all contribute to poor airflow as well. Thus, regular maintenance and visual inspection of the cooling tower are important.

3.8.4. Poor Pump Performance

If pumps (recirculation, blow-down, etc.) are not performing properly, then water flow through the tower will be degraded as will tower performance. Again, preventive maintenance is preferable to an emergency shutdown. Perhaps most instructive are case studies of problem cooling tower installations (22–24).

3.8.5. Choosing the Correct Cooling Tower

As previously mentioned, the Internet offers a window to a wide source of technical information. As information is readily found on the World Wide Web, this chapter will not direct readers to specific sites for design advice. Using several search engines with proper key phrases will return several excellent results to consult regarding a cooling tower project. The basic data to gather at the start of such a project (25) are as follows:

1. The heat load on the tower
2. Range (temperature difference of water in versus water out)
3. Approach (temperature difference, water leaving versus air entering wet bulb temperature),
4. Approximate tower size

As can be seen, if items 1–3 are known, one will be ready to consult with the engineering staffs of potential suppliers of a cooling tower to determine item 4.

If the wet bulb temperature is not known, a quick estimate can be made by dipping a thermometer into the water and then reading the lowest temperature the thermometer registers as the water evaporates. This temperature is the wet bulb temperature. Several readings in the environmental conditions for which the cooling tower is proposed should be taken. Pertinent articles regarding cooling tower design (26) frequently appear in various trade magazines as well.

3.8.6. Legionnaires' Disease

Legionnaires' disease is a form of pneumonia. The bacterium that causes this disease will flourish in cooling tower water if conditions are favorable (18). As no amount of drift elimination will prevent submicron particles of water from leaving a cooling tower, it is important to stress that a proper water control program be followed for normal cooling tower operation. As previously noted, a qualified water-treatment service company should be engaged to ensure that such bacterial growth is prevented in the operations of a cooling tower. To prevent airborne Legionnaires' and other diseases, the cooling water can be treated by specially formulated chemicals (18), filtration, copper ionization, ultraviolet (15), ozone (10,14–17), or dissolved air–ozone flotation (19). Controlling hydrogen sulfide emissions from cooling tower water has been discussed by Nagl (20).

NOMENCLATURE

a	Contact area/tower volume (1/ft or 1/m)
A	Surface area of pond (ft or m^2)
C_1	Bowen's conduction–evaporation coefficient (0.468 mm Hg °C)
C_p	Specific heat of water (1.0 cal/g °C)
d	Average depth of flow path (ft or m)
dq_t/dt	Net rate of surface heat exchange (kcal/m^2-d)
E	Equilibrium temperature (°C)
E_0	Equilibrium temperature of the pond at zero added heat load (°C)
e_a	Air vapor pressure (mm Hg)
e_s	Saturated vapor pressure of water at pond surface, (mm Hg)
$f(U)$	Evaporative wind speed function (kcal/d-m^2-mm Hg)
h_a	Enthalpy of air–water vapor mixture at wet bulb temperature (J/kg dry air or Btu/lb dry air)

h_w	Enthalpy of air–water mixture at bulk water temperature (J/kg dry air or Btu/lb dry air)
H_s	Gross solar radiation, (kcal/d-m^2)
K	Surface heat coefficient or heat transfer coefficient at the pond surface (kcal/d-m^2-°C)
K	Mass transfer coefficient (lb water/h ft^2)
KaV/L	Tower characteristic
q	Heat dissipated (kcal/d)
q_a	Long-wave solar radiation (kcal/d-m^2 or 400 Btu/d-ft^2)
q_c	Conductive–convective heat loss (kcal/d-m^2 or 400 Btu/d-ft^2)
q_d	Discharge of a heated effluent (kcal/d-m^2 or 400 Btu/d-ft^2)
q_e	Evaporative heat loss (kcal/d-m^2 or 400 Btu/d-ft^2)
q_{rj}	Heat flux (kcal/d-m^2)
q_s	Short-wave solar radiation (kcal/d-m^2 or 400 Btu/d-ft^2)
q_t	Net heat flux across the air–water interface (kcal/d-m^2 or 400 Btu/d-ft^2)
q_{ar}	Reflected portion of q_a (kcal/d-m^2 or 400 Btu/d-ft^2)
q_{br}	Long-wave black radiation from water surface (kcal/d-m^2 or 400 Btu/d-ft^2)
q_{sr}	Reflected portion of q_s (kcal/d-m^2 or 400 Btu/d-ft^2)
Q	Volumetric flow rate (m^3/d)
T	Bulk water temperature (°F or °C)
T_{avg}	Average temperature [$(T_d+T_s)/2$], (°C)
T_d	Dew point temperature (°C)
T_s	Surface temperature of the pond (°C)
T_x	Boundary condition of T at $X=X$
T_0	Boundary condition of T at $X=0$
T_{sm}	Completely mixed cooling pond's actual surface temperature (°C)
T_{su}	Completely unmixed cooling pond's actual surface temperature (°C)
T_1	Hot water temperature (°F or °C)
T_2	Cold water temperature (°F or °C)
u	Average flow velocity (m/d)
U	Wind speed (km/h)
V	Active cooling volume/plan area (m or ft)
X	Distance from pond entry (m)
α	Mixing coefficient
α_m	α evaluated for the completely mixed cooling pond
α_u	α evaluated for the completely unmixed cooling pond
β	Slope of the saturated vapor pressure curve between the dew point temperature and the pond surface temperature (mm Hg °C)
ρ	density of water (1.0 g/cm^3)
$\delta T/\delta X$	Longitudinal temperature gradient, (°C/m)

GLOSSARY

Approach. The difference in temperatures of water temperature exiting the tower and the dry bulb temperature of the air entering the tower (ΔT, water out vs dry bulb air).

Basin.   The bottom of the tower, water is collected here after passing the tower.

Blow-down.   Some of the water collected in the basin is exited from the tower rather than recirculated within the tower. This prevents build up of solids in the water. (*Note:* The same term is used in wet scrubbing, as the objective is also the same.)

Cooling range.   The temperature difference between hot water entering the tower and cooled water exiting the tower.

Drift.   Also known as *windage;* small droplets of water, which are carried out of the tower by the airflow, normally expressed as a percent of total water flow rate.

Drift eliminators.   Baffles, or other specialized collection products, placed in the air exit path to capture water droplets. (*Note:* Per the previous mention of scrubbing, again, mist elimination in scrubbing is a similar application.)

Dry bulb temperature.   Air temperature, measured by standard means.

Heat load. Amount of heat exchanged within the tower between water and air, expressed per unit of time. It can also be expressed as the product of the quantity of water circulated within the tower per unit time and the cooling range.

Make-up water.   Water that is added to the tower to replenish water losses due to evaporation, drift, blow-down, and other miscellaneous water losses.

Packing.   Also called *fill;* this is a product placed within the tower to promote uniform water flow distribution and to enhance water–air contacts. (*Note:* Again, note the similarity to wet scrubbing applications.)

Performance.   Normally expressed as the amount of cooling for a given quantity of water at a given wet bulb temperature.

Relative humidity.   The ratio of the partial pressure of water vapor in the air to the vapor pressure of water at the temperature of the air, which contains the water vapor.

Wet bulb temperature.   The temperature measured by an ordinary thermometer as a thin film of water on the thermometer is evaporated into a surrounding air stream. (*Note:* Recall the example earlier in this discussion of cooling in Arizona versus Florida. If the air surrounding the thermometer is already saturated with water, no evaporation and hence no cooling of the water can take place. In such an instance, the wet and dry bulb temperatures are equal.)

ACKNOWLEDGMENT

This chapter is dedicated to Dr. William Shuster, a former chairman and professor of Chemical and Environmental Engineering at the Rensselaer Polytechnic Institute in Troy, New York. Dr. Shuster has mentored over 50 doctoral students from around the world and is well known in the field. The authors of this chapter salute Dr. Shuster for his lifelong contributions to chemical engineering education.

REFERENCES

1. J. E. Edinger, and J. C. Geyer, *Publication No. 65-902*, Edison Electric Power Institute, 1965.
2. L. D. Berman, *Evaporative Cooling of Circulating Water*, Pergamon, New York, 1961.
3. D. K. Brady, W. L. Graves, Jr., and J. C. Geyer, *Publication No. 69-901*, Edison Electric Power Institute, 1969.
4. US Department of Commerce, *Climactic Atlas of the United States*, US Government Printing Office, Washington, DC, 1968.

5. US Department of the Interior, *FWPCA, Industrial Waste Guide on Thermal Pollution*, US Government Printing Office, Washington, DC, 1968.
6. K. K. McKelvey, and M. E. Brooke, *The Industrial Cooling Tower*, Elsevier, Amsterdam, 1959.
7. Editor, *Power* (1973).
8. US EPA, in *Compilation of Air Pollutant Emission Factors, AP-42, Vol. 1, Sec. 13.4*, 5th ed., Federal Register, 2002.
9. US EPA, In *National Pollutant Discharge Elimination System—Proposed Rules*, Federal Register, 2002.
10. US Department of Energy, *Federal Technology Alerts*: Ozone treatment for cooling towers, December, 1995.
11. R. C. Rosaler, *The Standard Handbook of Plant Engineering*, 2nd ed., McGraw-Hill, New York, 1995.
12. D. W. Green, (ed.), *Perry's Chemical Engineers' Handbook*, 6th ed., McGraw-Hill, New York, 1984.
13. Washington State University Cooperative Extension Energy Program, Technical Brief WSUEEP98013 (revised) (1998).
14. WQA, *Ozone: A Reference Manual*, Water Quality Association, Washington, DC (2002).
15. J. Roseman, *Water Qual. Products* **8**(4), 14–16 (2003).
16. L. K. Wang, *Pretreatment and Ozonation of Cooling Tower Water, Part I*, US Department of Commerce, National Technical Information Service, Springfield, VA, 1984.
17. L. K. Wang, *Pretreatment and Ozonation of Cooling Tower Water, Part II*, US Department of Commerce, National Technical Information Service, Springfield, VA, 1984.
18. L. K. Wang, *Prevention of Airborne Legionnaires' Disease by Formulation of a New Cooling Water for Use in Central Air Conditioning Systems*, US Department of Commerce, National Technical Information Service, 1984.
19. L. K. Wang, *Proceedings of the Seventh Mid-Atlantic Industrial Waste Conference*, 1985, pp. 207–216.
20. G. Nagl, *Environ. Technol.* **9**(7), 19–24 (1999).
21. W. W. Shuster, in *Handbook of Environmental Engineering, Volume 4, Water Resources and Natural Control Processes* (L. K. Wang, and N. C. Pereira, eds.), Humana, Totowa, NJ, 1986, pp. 107–138.
22. W. Barber, *Nucleonics Week* (June 4, 1998).
23. G. Thurman, *Power Mag.* (2001).
24. R. Swanekamp, *Power Mag.* (2001).
25. J. Terranova, *HPAC Eng.* (2001).
26. J. Katzel, *Plant Eng.* (2000).

11
Performance and Costs of Air Pollution Control Technologies

Lawrence K. Wang, Jiann-Long Chen, and Yung-Tse Hung

CONTENTS

1. INTRODUCTION

1.1. Air Emission Sources and Control

In general, air toxics are hazardous air pollutants (HAPs) that cause cancer or other human health effects. One hundred ninety compounds are specifically identified in the Clean Air Act (CAA) amendments of 1990 as air toxics that the US Environmental Protection Agency (US EPA) must investigate and regulate. Air emission control is one of important tasks of the US EPA (1–23).

The top 14 HAPs identified by the US EPA are toluene, formaldehyde, methylene chloride, methyl chloroform, ethylene, *m*-xylene, benzene, *o*-xylene, perchloroethylene, *p*-xylene, chlorobenzene, acetic acid, trichlorotrifluoroethane, and trichloroethylene (22,23).

Most HAPs can be classified as volatile organic compounds (VOCs), semivolatile organic compounds (SVOCs), particulate matters (PMs), and pathogenic micro-organisms (19–26). HAP control will be very active in the 21st century on several fronts: new regulations, the Maximum Achievable Control Technology (MACT) hammer,

From: *Handbook of Environmental Engineering, Volume 2: Advanced Air and Noise Pollution Control*
Edited by: L. K. Wang, N. C. Pereira, and Y.-T. Hung © The Humana Press, Inc., Totowa, NJ

and residual risk. Each presents issues for industrial plant compliance in the forth-coming years(27).

Air emission controls can be divided into (1) controls for point sources of emissions and (2) controls for area sources of emissions. Point sources include stacks, ducts, and vents from industrial plants and from remediation technologies such as air stripping, soil vapor extraction, thermal desorption, and thermal destruction. Add-on emission controls usually can be added readily to point sources. Area sources include lagoons, landfills, spill sites, and remediation technologies such as excavation. Air emission con-trols for area source controls are generally more difficult to apply and less effective than controls for point sources. Some emission sources such as solidification/stabilization, bioremediation, and storage piles may be either point or area sources of emissions. Area sources can be converted to point sources using enclosures or collection hoods (3,8,10,12,28,29).

1.2. Air Pollution Control Devices Selection

There are many air pollution control devices (APCDs) available (1,3–5,8–11,30–35). Each APCD has relative advantages and disadvantages and no single control option will always be the best choice for air pollution control or site remediation technology. Selection criteria for APCDs include (a) demonstrated past use of the control technology for the specific application of interest, (b) ability meet or exceed the required average capture and/or control efficiency, (c) compatibility with the physical and chemical prop-erties of the waste gas stream, (d) reliability of control equipment and process, (e) capital cost of control equipment, (f) operating costs of system (including byproducts disposal or regeneration costs), and (g) permitting requirements.

The information in this chapter is intended to be used to screen potential APCD options and used in conjunction with detailed engineering evaluations, vendor data, and feasibility studies to select air pollution control technologies.

The cost-effectiveness of APCDs is very process- and site-specific. In general, a con-trol system is designed or modified for each specific application; so, in theory, any desired removal or control efficiency can be achieved. In practice, a trade-off exists between removal or control efficiency and cost (26,28,29,36–48).

2. TECHNICAL CONSIDERATIONS

2.1. Point Source VOC Controls

Various APCDs for controlling VOC from point sources of air emission are evaluated and summarized in Table 1. Carbon adsorption, thermal oxidation, catalytic oxidation, condensation, biofiltration, ultraviolet, and emerging control technologies are intro-duced in volume 1 of this handbook series (48). Only the internal combustion engines (ICEs) and membrane process are briefly introduced in below.

The principle of operation of a control device that incorporates an ICE is to use a con-ventional automobile or truck ICE as a thermal incinerator. The major components include the standard automobile or truck engine, supplemental fuel supply (propane or natural), carburetor and off-gas lines from a point emission control device (adsorbent bed, catalytic converter, etc.). ICEs may be used for VOC control from any point source where the airstream must meet certain air quality criteria (2–12).

Table 1
Typical Required Emission Stream and Contaminant Characteristics for Point Sources VOC Controls

Control	Concentration (ppm)	Flow rate (scfm)	Applicable organics	Inorganic/corrosive contaminants	PM damage	Temperature (°C)	Humidity	Variable stream	Pressure drop (in. H_2O)
Carbon adsorption	<10,000	Regenerable: <500,000; Disposal: <1,500	Most compounds with 50<MW<200	Some inorganics and metals can be controlled	PM must be removed	<120 (most effective around 38)	<50%	Variable flow or concentration present no problems	23–125
Thermal oxidation (incineration)	<25% LEL	Packaged: <50,000 custom to 100,000 and above	Most organics acceptable Formation of acid gases or PICs possible	Extra controls may be required for halogenated organics and sulfur	PM not a problem	Combustion at 760–870	High humidity streams may reduce Btu content	Low flow rates require supplementary fuel	4 (a 70% efficient heat exchanger may have a drop of 15)
Catalytic oxidation (incineration)	<25 % LEL >10 Btu/scf may overheat catalyst	Packaged: <50,000	Halogenated organics usually a problem	Corrosives and metals can degrade catalyst	PM can clog catalyst	260–593	Increasing humidity increases required supplemental heat	Variable flow required dilution/ supplemental fuel; Might overheat catalyst	6 (a 70% efficient heat exchanger may have a drop of 15)
Condensers	>5,000	< 2,000	BP > 38°C preferred	Corrosive gases a problem	PM not a problem	As close to dew point of HAPs as possible	Humid streams can ice over refrigeration unit	Changes in VOC content, concentration, or temperature can reduce condensation	Minimal
Internal combustion engines	<25 % LEL	<1,000/unit (200 typical)	All combustible organics	Corrosive gases unacceptable; will not treat metals or other combustibles	PM may cause fouling	Must be less than combustion temperature	High humidity stream may reduce Btu content	Variable flowrate requires supplementary fuel; excessive loading dumps untreated hydrocarbons	Pressure increases across device (power is generated)
Soil beds/ biofilters	1,000 (pilot system)	600–90,000	Most organics acceptable	SO_2, NO_x, H_2S can be controlled	PM can plug soil pores	38: optimal (highly sensitive)	90% RH required to prevent drying out bed	Variable flow or concentration may impact microbe culture	2–3 (pilot system)
Emerging/ miscellaneous controls (membrane, UV)	Membrane: >300, Low UV: 100 in pilot studies		Some organics acceptable	Corrosive gases may degrade membrane	PM may be problem for membrane	Unknown effects	Unknown	Unknown	Varies

Source: ref. 28.

Membrane filtration is an emerging APCD for removing VOCs and SVOCs from waste gas streams. The membrane module acts to concentrate the toxic organic compounds by being more permeable to organic constituents than air. The imposed pressure difference across a selective membrane drives the separation of the VOCs and SVOCs from the waste gas streams.

2.2. Point Source PM Controls

Various APCDs for controlling PMs from point sources of air emissions are evaluated and summarized in Table 2. Fabric filtration (baghouse), wet scrubbers, dry scrubbers, electrostatic precipitators (ESPs), quench chambers, Venturi scrubbers, and operational controls are all introduced in detail in other chapters of this handbook series. Only the high efficiency particulate air (HEPA) filter is briefly described in this subsection.

The HEPA filters are commonly used in medical and environmental facilities requiring 99.9% or greater PM removal. HEPA filters can be used as a PM polishing step in ventilation and air conditioning systems for buildings undergoing asbestos or lead paint removal, for enclosures, or with solidification and stabilization mixing bins at a remediation site. The major components of a PM control system employing HEPA filters include the following: (1) HEPA filters, (2) filter housing, (3) duck work, and (4) fan.

2.3. Area Source VOC and PM Controls

Various APCDs for controlling VOCs and PMs from area sources of air emissions are evaluated and summarized in Table 3. Table 4 indicates the ranges of removal efficiency (RE) for point source PM controls. The applicable APCDs, which are briefly described in the following section, include covers, foams, wind screen, water sprays, operational controls, enclosures, and collection hoods (28,29).

Covers control emissions of contaminated particulate matter and VOCs/SVOCs by physically isolating the contaminated media from the atmosphere. Some cover materials (e.g., sawdust and straw) are tilled into the contaminated soil or contaminated media as an anchoring mechanism.

Modified fire-fighting foams are commonly used to control PMs/VOCs/SVOCs emissions during the remediation of hazardous waste sites or hazardous spill sites. Suppressing of PM/VOC/SVOC is accomplished by blanketing the emitting source (liquid, slurry, soil, or contaminated equipment) with foam, thus forming a physical barrier to those HAP emissions. Some foams are "sacrificial," meaning that the chemicals compromising the foam will react with specific VOCs/SVOCs and further suppress their emissions.

Wind screens can be used to reduce PM emissions from storage piles, excavation sites, and other area sources. The principle is to provide an area of reduced wind velocity that allows settling of the large particles and reduces the particle flux from the exposed surfaces on the leeward side of the screen. In addition, wind screens reduce moisture loss of a contaminated medium (such as soil), resulting in decreased VOC/SVOC and PM emissions.

The control mechanism of water sprays is the agglomeration of small particles with large particles or with water droplets. Also, water will cool the surface of a contaminated medium, such as soil. Typically, water is applied with mobile water wagons or fixed perforated pipes.

Table 2
Typical Required Emission Stream and Contaminant Characteristics for Point Source PM Controls

Control	Loading (grains/dscfm)	Flow rate (scfm)	PCDD/PCDF, PICs control	Inorganic/corrosive contaminants	PM size	Inlet temperature (°C)	Variable stream	Pressure drop (in. H$_2$O)
Fabric filter (baghouse)		Any rate	Effectiveness disputed	Various fabrics susceptible to various corrosive compounds; no "sticky" PMs	>0.3 μm (varies with fabric)	Above stream dew point, but <300 (special fabrics up to 1200)	Not a problem	<20 (6–8 optimum)
Wet scrubbers; packed tower, ionizing wet	>200 for 90% or better RE	1,000–200,000	Effectiveness disputed	Acid gases, IWS, metals	>10μm	Insensitive for PMs	Not a problem	12–32
Dry scrubbers DSI, SDA	0.02 grains/dscfm	Better for >1,000	SDA may be effective when used with baghouse	Acid gases, heavy metals		DSI: optimum efficiency at 15 from saturation SDA: to 1000	Not a problem	Small
Electrostatic precipitator	>2 g/scf	<200,000	No	Not effective for resistive particles	0.01–70 μm	30–60- above stream dew point	Sensitive to loading	0.4
Quench chambers	Any	Any rate	No	Can remove 50% acids,depending on liquor used		Incineration temperature	Sensitive to loading	
Venturi scrubber	Any	<80,000	Effectiveness disputed	HCl, SO$_2$, NO$_x$, HCN can be controlled	85–99% RE for PM > 0.5 μm	30–60 above stream dew point	Not a problem	<50
Operational controls	Varies	Any rate	Yes	N/A	N/A			
HEPA filter	Low	Any rate			>0.03 μm	Dew point of stream	Not a problem	

Source: ref. 28.

389

Table 3
Typical Required Emission Source and Contaminant Characteristics for Area Source VOC and PM Controls

Control	Medium	Wind	Site Surface	Equipment required	Duration of control effectiveness	Sunlight/temperature	Precipitation
Covers: soil, organics, synthetics, paving, road carpet/slag	Soil, liquid, or sludge	Little effect	Level surface is best	Graders, front end loaders, other heavy equipment	Months	UV, weather may degrade synthetic and organic covers	May wash away soil cover
Foams	Soil, liquid, or sludge	Application difficult in winds > 10 mph	Some foams unsuited for steep slopes	Pump, foaming nozzle, water, application unit	From a few hours up to several months	Increased temperature will decrease useful life of foam	Can wash away some foams
Wind screen	Soil, liquid, or sludge	High winds may cause damage	Any	Supports	Easily damaged, otherwise no limit	No effect	No effect
Water sprays	Soil	Wind increases evaporation rate	Not used near surface waters	Nozzles, water wagon, pump	Effective until water evaporates	Heat decreases control time	Improves efficiency
Water sprays with additives	Soil	Wind reduces life of spay	Not used near surface waters	Nozzles, water wagon, pump	Effective until water evaporates	Heat decreases control time, freezing weather may preclude application	Improves efficiency
Operational controls	Soil, liquid, or sludge	No effect	No effect	Varies	Varies	No effect	May improve efficiency
Enclosures	Soil, liquid, or sludge	No effect	No effect	Structure	Months to years	Not effect	No effect
Collection hoods	Soil, liquid, or sludge	High winds may necessitate use of wind screen	Level surface best	Hood, ducting, blower	Indefinite	Effectiveness may decrease somewhat as temperature rises	No effect

Source: ref. 28.

Table 4
Ranges of % RE for Point Source PM Controls

	Fly ash	PCDD/PCDF	Acid gases	<10 μm	>10 μm	Metals
Baghouses	—	Entrained fraction removed		99+[a]	99+[a]	90–95[b]
Wet scrubbers	—	—	95–99+	Low	—	40–50[b]
Venturi scrubbers	—	—	99	80–95	80–95	Variable
Dry scrubbers	—	90–99+	95–99+	99+	99+	95–99[b]
ESP	99+	98 with SDA	—	99[c]	99[c]	85–99[b]
Quench Chambers	—	—	50	—	—	—
HEPA filters	—	Entrained fraction removed	—	99.9+[d]	99+[d]	—

[a]Except for "sticky" particles.
[b]Lower removal efficiency for mercury.
[c]For resistive particles.
[d]With high pressure drop.
Source: refs. 26 and 28.

Operational controls are those procedures and practices inherent to most air emission control projects that can be instituted to reduce VOC/SVOC/PM emissions. These may include: (1) cleaning practices, (2) seasonal scheduling, (3) vehicle speed control, (4) storage pile orientation, (5) excavation practices, (6) dumping practices, and (7) soil/materials handling practices.

Enclosures provide a physical barrier between the emitting area and the atmosphere and, in essence, convert an area source to a point source HAP emission. Prior to releasing the air trapped within the enclosure, conventional point source controls are employed to control VOC/SVOC/PM emissions.

Collection hoods are commonly used to capture VOCs/SVOCs/PMs emitted from small-area sources and route those emissions to appropriate APCDs. In practice, hoods are designed using the capture velocity principle, which involves the creation of an airflow sufficient to remove the contaminated air after the emitting source.

2.4. Pressure Drops Across Various APCDs

The total system pressure drop is the summation of duct pressure drop, stack pressure drop, APCD #1 pressure drop, APCD #2 pressure drop, APCD #3 pressure drop, and so forth. The assumed pressure drops across various APCDs are shown in Table 5. Equation (1) can be used to calculate the total system pressure drop, ΔP_t (in. H$_2$O).

$$\Delta P_t = \Delta P_{duct} + \Delta P_{stack} + \Delta P_{device\#1} + \Delta P_{device\#2} + \Delta P_{device\#3} + \Delta P_{device\#n} \quad (1)$$

3. ENERGY AND COST CONSIDERATIONS FOR MINOR POINT SOURCE CONTROLS

3.1. Sizing and Selection of Cyclones, Gas Precoolers, and Gas Preheaters

Gas conditioning equipment includes those components that are used to temper or pretreat the gas stream to provide the most efficient and economical operation of the downstream control devices. Preconditioning equipment, installed upstream of the control devices, consists of mechanical dust collectors, wet or dry gas coolers, and gas

Table 5
Assumed Pressure Drops Across Various Components

System component	Pressure drop (in. H_2O)
Stack	0.6
Ductwork	0.6
Thermal incinerator	4.1
Heat exchanger	2
Catalytic incinerator	6
Absorber	Variable[a]
Carbon adsorber	6
Condenser	3
Fabric filter	6
Electrostatic precipitator	0.5
Venturi scrubber	ΔP_v

[a]Use Eq. (1) to determine the pressure drop.

preheaters. Where the control device is a fabric-filter system or electrostatic precipitator, mechanical dust collectors are required upstream if the gas stream contains significant amounts of large particles (3).

Gas stream pretreatment equipment can be installed upstream of the control device (i.e., cyclones, precoolers, and preheaters) and enable the emission stream to fall within the parameters specified by the downstream process equipment manufacturers. The best solution for compliance with Clean Air Act regulations may be one or a combination of technologies. Accordingly, cyclones and other airstream pretreatment equipment become very important to a complete air emission control systems (1,3–18).

Mechanical dust collectors, such as cyclones, are used to remove the bulk of the heavier dust particles from the gas stream. These devices operate by separating the dust particles from the gas stream through the use of centrifugal force. The efficiency of a cyclone is determined by the entering gas velocity and the diameter at the cyclone inlet. The cyclone inlet area can be calculated from Eq. (2). In this equation, d is the critical particle size in (μm). The critical particle size is the size of the smallest particle the cyclone can remove with 100% efficiency. Therefore, simply select a critical particle size and then calculate the appropriate cyclone dimensions which will remove 100% of all particles that size and larger.

$$A_{cyc} = 3.34 \left[Q_{e,a} \left(r_p - D_G \right) / \mu \right]^{1.33} (d)^{2.67} \qquad (2)$$

where $Q_{e,a}$ is the actual emission stream flow rate (acfm), r_p is the density of the particle (lb/ft^3), D_G is the density of the emission stream (lb/ft^3), μ is the emission stream viscosity (lb-ft/s), and d is the critical particle size (μm)

Gas stream coolers can be wet or dry. Dry-type coolers operate by radiating heat to the atmosphere. Wet-type coolers (spray chambers) cool and humidify the gas by the addition of water sprays in the gas stream; the evaporating water reduces the temperature of the gas stream. A third method of cooling is through the addition of dilution air. The applications and operational conditions of gas stream coolers are introduced in Chapter 10.

Gas preheaters are used to increase the emission stream temperature. Condensation causes corrosion of metal surfaces and is of particular concern in fabric-filter applications where moisture can cause plugging or "blinding" of the fabric pore. Therefore, gas preheaters can be used to elevate the temperature of an emission stream above its dew point. The temperature of the emission stream should be 50–100°F above its dew point if the emission stream is to be treated (i.e., PM collected) by a downstream ESP or a fabric filter (26).

3.2. Sizing and Selection of Fans, Ductworks, Stacks, Dampers, and Hoods

Other auxiliary process equipment include but are not limited to fans, ductworks, stacks, dampers, and hood, which are self-explanatory. Figure 1 shows the components of a typical hood exhaust system, which includes the hood, duct, cyclone cleaners, and a fan (28,29). Detailed procedures and examples for designing fans, ducts and so forth can be found in Chapter 6 of this book.

Hoods are commonly used to capture PMs/VOCs emitted from small-area sources (e.g., waste stabilization/solidification mixing silos, bioremediation reactors) and route those emissions to appropriate APCDs. Three hood designs that are commonly used are depicted in Fig. 2. The selection of hood type will be dependent on the emitting source characteristics (e.g., source area and accessibility, emitting air velocity, surrounding air currents) and the required capture efficiency. Major components of a hood exhaust system are depicted in Fig. 1.

Hoods can be used to capture PM/VOC emissions from exsitu waste stabilization/ solidification mixing silos and bioremediation reactors. The use of a hood will be contingent upon access to the emitting source and upon the area of the emitting source. As the distance between a source and hood increases, so does the required total volumetric flow rate of air into the hood to maintain a given capture efficiency. Because the cost of most air pollution control equipment is proportional to the volumetric flow rate, a point is reached where it is not economically feasible to use a hood. The emitting

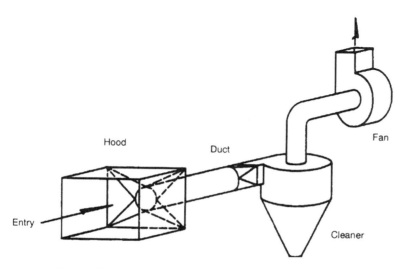

Fig. 1. Components of a hood exhaust system. (From ref. 7.)

source area will impact the hood size required, and for canopy and capturing hoods, it will impact the airflow rate required to maintain a given capture efficiency. The advantages/disadvantages of using a hood to capture PM/VOC emissions are outlined in Table 6. Parameters that influence the capture efficiency of hood exhaust systems are given in Table 7.

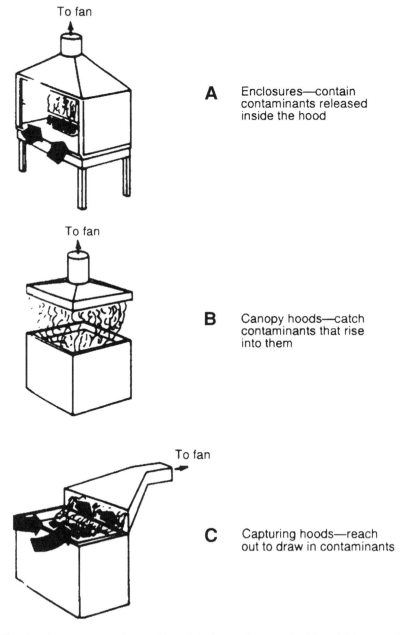

Fig. 2. Three commonly used hood designs. (From refs. 28 and 33)

Table 6
Advantages/Disadvantages of Hoods to Capture PM/VOC Emissions

Advantages	Disadvantages
90–100% PM/VOC capture efficiencies are possible.	Emission source must be accessible to hood.
Much data available regarding selection and design of hoods.	Contaminant diluted by airflow into hood. This can affect air pollution control efficiencies (e.g., carbon adsorption, incineration).
Conversion of an area source into a point source. Most air pollution control equipment is designed for point sources.	Power cost may be high as a result of required capture velocity and head loss through ductwork. Use of hoods is practical for small-area sources only. Hoods subject to corrosion (e.g., acid gases, lime).

Source: US EPA.

Hood PM/VOC capturing efficiencies can be as high as 90–100%. However, PM/VOC control efficiencies are functions of both the hood capture efficiency and the air pollution control equipment removal efficiency. Hood exhaust systems designs are based on the hood aspect ratio (width/length of hood), the required capture velocity (v), and the distance of the furthest point of the emitting source from the hood centerline (X). Ranges of capture velocities required as a function of surrounding air turbulence and the emitting source are listed in Table 8. The velocities obtained from Table 8 can then be used in hood design. Various hood designs are shown in Table 9. For a more thorough presentation on hood design, see ref. 4. The following are design equations of various hoods:

1. Slot hood

$$Q = 3.7LVX \tag{3}$$

Table 7
Parameters That Affect Hood Capture Efficiencies

Parameter	Comment
Distance between hood and farthest point of emitting source	As this distance increases, for a given volumetric flow rate into the hood, the capture efficiency decreases.
Volumetric flow rate into the hood	As the volumetric flow rate into the hood increases, the capture efficiency increases.
Surrounding air turbulence	As the surrounding air turbulence increases, the required volumetric flow rate into the hood increases to maintain a given capture efficiency.
Hood design	Hood designs are tailored to specific types of emitting sources. For example, canopy hoods are designed to collect emissions from heated open-top tanks.

Source: US EPA

Table 8
Range of Capture Velocities

Condition of dispersion of contaminant	Examples	capture velocity (fpm)
Released with practically no velocity into quiet air	Evaporation from tanks; degreasing, etc.	50–100
Released at low velocity into moderately still air	Spray booths; intermittent container filling; low-speed conveyor transfers; welding; plating; pickling.	100–200
Active generation into zone of rapid air motion	Spray painting in shallow booths; barrel filling; conveyor loading; crushers.	200–500
Released at high initial velocity into zone of very rapid air motion	Grinding; abrasive blasting.	500–2000

Note: In each category, a range of capture velocity is shown. The proper choice of value depends on several factors:

Lower end of range	Upper end of range
1. Room air current minimal or favorable to capture	1. Disturbing room air current
2. Contaminants of low toxicity or of nuisance value only	2. Contaminants of high toxicity
3. Intermittent, low production	3. High production, heavy use
4. Large hood, large air mass in motion	4. Small hood, local control only

2. Flanged hood

$$Q = 2.8\,LVX \tag{4}$$

3. Plain opening

$$Q = V\left(10X^2 + A\right) \tag{5}$$

4. Flanged opening

$$Q = 0.75V\left(10X^2 + A\right) \tag{6}$$

5. Booth

$$Q = VA = VWH \tag{7}$$

6. Canopy

$$Q = 1.4\,PVD \tag{8}$$

where X is the centerline distance to point x in the emissions plume (ft), L is the length (ft), W is the width (ft), H is the height (ft), D is the distance between hood and source (ft), A is the area (ft^2), Q is the flow rate, (ft^3/min), P is the perimeter of hood (ft), and V is the velocity at point x (ft/min)

3.3. Cyclone Purchase Costs

Cyclones are used upstream of particulate control devices (e.g., fabric filters, ESPs) to remove larger particles entrained in a gas stream. Equation (9) yields the cost of a carbon steel cyclone with a support stand, fan and motor, and a hopper or drum to collect the dust:

Table 9
Hood Design [U.S. EPA]

	Description	Aspect ratio (W/L)
	Slot	0.2 or less
	Flanged slot	0.2 or less
	Plain opening	0.2 or greater and round
	Flanged opening	0.2 or greater and round
	Booth	To suit work
	Canopy	To suit work

$$P_{cyc} = 6,520 A_{cyc}^{0.9031} \qquad (9)$$

where P_{cyc} is the cost of cyclone, (August 1988 $) and A_{cyc} is the cyclone inlet area, (ft^2 [0.200 ft$^2 \le A_{cyc} \le$ 2.64 ft^2]). The cost of a rotary air lock for the hopper or drum is given by

$$P_{ral} = 2,730 A_{cyc}^{0.965} \qquad (10)$$

where P_{ral} is the cost of the rotary air lock (August 1988 $) and A_{cyc} is the cyclone inlet area (ft^2 [0.350 ft$^2 \le A_{cyc} \le$ 2.64 ft^2])

3.4. Fan Purchase Cost

In general, fan costs are most closely correlated with fan diameter. The readers are referred to Chapter 6 for detailed fan design. Equations (11)–(13) can be used to obtain fan prices. Costs for carbon steel fan motor ranging in horsepower from 1 to 150 hp are provided in Eqs. (14) and (15). Equation (12) or (13) is used in conjunction with Eq. (14) or (15), respectively.

The cost of a fan is largely a function of the fan wheel diameter, d_{fan}. The wheel diameter is related to the ductwork diameter through use of manufacturer's multirating tables. The readers should be able to obtain the fan wheel diameter for a given ductwork diameter by consulting the appropriate multirating table or by calling the fan manufacturer.

For a centrifugal fan consisting of backward curved blades including a belt-driven motor and starter and a static pressure range between 0.5 and 8 in. of water, the cost as a function of fan diameter in July 1988 dollars is provided by

$$P_{fan} = 42.3 d_{fan}^{1.20} \tag{11}$$

where P_{fan} is the cost of the fan system (July 1988 \$) and d_{fan} is the fan diameter (in. [12.25 in. $\le (d_{fan} \le 36.5$ in.]).

The cost of a fiber-reinforced plastic (FRP) fan, not including the cost of a motor or starter, is provided by

$$P_{fan} = 53.7 d_{fan}^{1.35} \tag{12}$$

where P_{fan} is the cost of the fan without a motor or starter, (April 1988 \$) and d_{fan} is the fan diameter (in. [10.5 in. $\le d_{fan} \le 73$ in.]). The cost of a motor and starter as obtained in Eq. (14) or (15) should be added to the fan cost obtained in Eq. (12).

A correlation for a radial-tip fan with welded, carbon steel construction and an operating temperature limit of 1000°F without a motor or starter is provided by

$$P_{fan} = a_f \times (d_{fan})^{b_f} \tag{13}$$

where P_{fan} is the cost of the fan without motor or starter (July 1988 \$), and a_f, b_f are coefficients, and d_{fan} is the fan diameter (in.). The values for the parameters a and b are provided in Table 10.

The cost of fan motors and starters is given in Eq. (14) or (15) as a function of the horsepower requirement. The cost obtained from either of these equations should be added to the fan cost obtained in Eq. (12) or (13). For low-horsepower requirements,

$$P_{motor} = 235 \, hp^{0.256} \tag{14}$$

where P_{motor} is the cost of the fan motor, belt, and starter (February 1988 \$) and hp is the motor horsepower ($1 \le hp \le 7.5$).

Table 10
Equation (13) Parameters

Parameter	Group 1	Group 2
Static pressure (in.)	2–22	20–32
Flow rate (acfm)	700–27,000	2,000–27,000
Fan wheel diameter (in.)	19.125–50.5	19.25–36.5
a_f	6.41	22.1
b_f	1.81	1.55

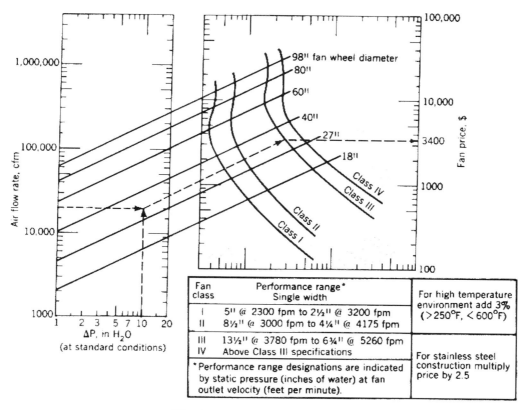

Fig. 3. Fan price. (From ref. 30.)

For high-horsepower requirements,

$$P_{\text{motor}} = 94.7 \, \text{hp}^{0.821} \tag{15}$$

where P_{motor} is the cost of the fan motor, belt, and starter (February 1988 $) and hp is the motor horsepower ($7.5 \le \text{hp} \le 250$). More, but different, fan purchase cost data are provided in refs. 26 and 30.

The fan purchase cost (*see* Fig. 3) is a function of the flow rate moved by the fan and the pressure drop (ΔP) across the control system. The fan is assumed to be located downstream of the final control device in the control system. Therefore, the fan capacity must be based on the final control device's exit gas flow rate at actual conditions ($Q_{fg,a}$). The control system pressure drop (ΔP) is the total of the pressure drops across the various control system equipment, including the stack and ductwork. Table 5 presents conservative pressure drops across specific control system components that can be used if specific data are unavailable.

Using the actual flow rate and total parameters, we can obtain the fan purchase cost from Fig. 3. Fans are categorized into classes I to IV according to control system pressure drop. Guidelines are presented in Fig. 3 for determining which class of fan to use. There is some overlap between the classes. The lower-class fan is generally selected because of cost savings. To estimate the cost of a motor for the fan, multiply the fan cost by 15%. (*Note*: The fan and motor costs are included in the cost curves for thermal incinerators and packaged carbon adsorbers.)

3.5. Ductwork Purchase Cost

The cost of ductwork for a HAP control system is typically a function of material (e.g., PVC, FRP), diameter, and length. To obtain the duct diameter requirement as a function of the emission stream flow rate at actual condition ($Q_{e,a}$) use Eq. (16), which assumes a duct velocity (U_{duct}) of 2000 ft/min:

$$d_{duct} = 12\left[(4/\pi)\left(Q_{e,a}/U_{duct}\right)\right]^{0.5} = 0.3028 Q_{e,a}^{0.5} \qquad (16)$$

The cost of PVC ductwork (in $/ft) for diameters between 6 in. and 24 in. is obtained using

$$P_{PVCD} = a_d\left(d_{duct}\right)^{b_d} \qquad (17)$$

where P_{PVCD} is the cost of PVC ductwork, ($/ft [August 1988 $]), d_{duct} is the duct diameter (in. [factor of 12 in./ft above]), $a_d = 0.877$ (6 in. $\leq d_{duct} \leq 12$ in.) or 0.0745 (14 in. $\leq d_{duct} \leq 24$ in.), and $b_d = 1.05$ (6 in. $\leq d_{duct} \leq 12$ in.) or 1.98 (14 in. $\leq d_{duct} \leq 24$ in.).

For FRP duct having a diameter between 2 and 5 ft, the following equation can be used to obtain the ductwork cost:

$$P_{FRPD} = 24 D_{duct} \qquad (18)$$

where P_{FRPD} is the cost of FRP ductwork ($/ft [August 1988 $]) and D_{duct} is the duct diameter (ft). Note that the duct diameter is in units of feet for this equation.

It is more difficult to obtain ductwork costs for carbon steel and stainless-steel construction because ductwork using these materials are almost always custom fabricated. For more information on these costs, consult other sources (1,3).

The ductwork purchase cost is typically proportional to the ductwork weight, which is a function of (1) the material of construction, (2) length, (3) diameter, and (4) thickness. Carbon steel ducts are normally used for noncorrosive flue gases at temperatures below 1150°F. Stainless-steel ducts are generally used with gas temperatures between 1150°F and 1500°F or if the gas stream contains corrosive materials. Figures 4 and 5 present purchase costs for carbon steel and stainless-steel ducts, respectively. It is assumed that the major portion of ductwork is utilized to transport the emission stream from the process to the control system; therefore, the flow rate, $Q_{e,a}$, of the emission stream at actual conditions is used to size the ductwork.

3.6. Stack Purchase Cost

It is difficult to obtain stack cost correlations because stacks are usually custom fabricated. Smaller stacks are typically sections of straight ductwork with supports. However, the cost of small (e.g., 50–100 ft) FRP stacks can be roughly estimated as 150% of the cost of FRP ductwork for the same diameter and length. Similarly, the cost of small carbon steel and stainless-steel stacks is also approx 150% of the cost of corresponding ductwork (1,3).

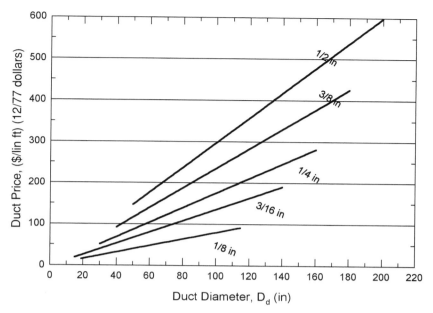

Fig. 4. Carbon steel straight-duct fabrication price at various thicknesses. (From ref. 30.)

For larger stacks (200–600 ft), the cost is typically quite high, ranging from $1,000,000 to $5,000,000 for some applications. Equation (19) and Table 11 can be used to obtain costs of large stacks:

$$P_{\text{stack}} = aH_{\text{stack}}^{b} \tag{19}$$

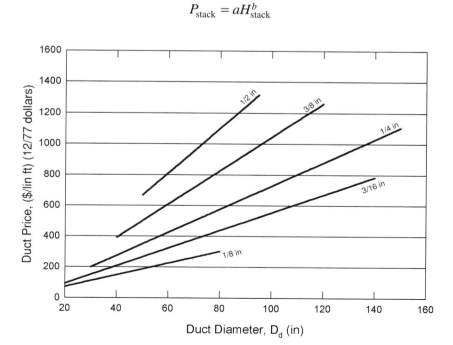

Fig. 5. Stainless-steel straight-duct fabrication price at various thicknesses. (From ref. 30.)

Table 11
Parameters for Costs of Large Stacks

Lining	Diameter (ft)	*a*	*b*
Carbon steel	15	0.0120	0.811
316L Stainless steel	20	0.0108	0.851
Steel in top	30	0.0114	0.882
Section	40	0.0137	0.885
Acid resistant	15	0.00601	0.952
Firebrick	20	0.00562	0.984
	30	0.00551	1.027
	40	0.00633	1.036

Source: ref. 1.

where P_{stack} is the total capital cost of large stack(10^6 \$), H_{stack} is the stack height (ft), and *a* and *b* are coefficients. (*see* Table 11).

The stack purchase cost is a function of: (1) the material of construction, (2) stack height, (3) stack diameter, and (4) stack thickness. In addition, minimum stack exit velocities should be at least 1.5 times the expected wind velocity; for instance, in the case of 30-mph winds, the minimum exit velocity should be at least 4000 ft/min. For purposes of this handbook, the stack cost is estimated with respect to the final control device's exit gas flow rate at actual conditions. Figures 6 and 7 present purchased costs for unlined, carbon steel stacks.

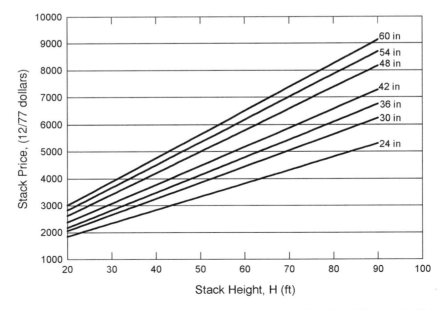

Fig. 6. Carbon steel stack fabrication price for a 0.25-in. plate. (From ref. 30.)

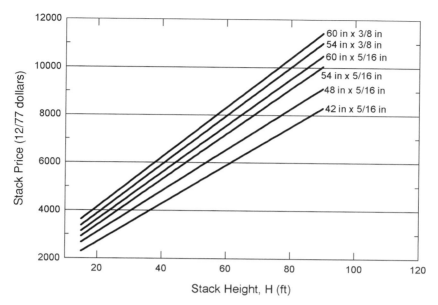

Fig. 7. Carbon steel stack fabrication price for 5/16-in. and 3/8-in. plates. (From ref. 30.)

Without specific information, assume the following items to simplify the costing procedures:

1. The stack is constructed with 0.25-in.-thick carbon steel plate.
2. The stack height equals 50 ft.
3. The stack diameter is calculated using a stack exit velocity of 4000 ft/min. Therefore,

$$D_{stack} = 12 \left(\frac{4}{\pi} \times \frac{Q_{fg,a}}{U_{stack}} \right)^{0.5} = 0.2141 \left(Q_{fg,a} \right)^{0.5} \tag{20}$$

where D_{stack} is the stack diameter (in.), $Q_{fg,a}$ is the flue gas flow rate at actual conditions, (acfm), and U_{stack} is the velocity of the gas stream in the stack (ft/min).

3.7. Damper Purchase Cost

Dampers are commonly used to divert airflow in many industrial systems. Two types of damper are discussed: backflow and two-way diverter valve dampers. The cost of backflow dampers for duct diameters between 10 and 36 in. is given by

$$P_{damp} = 7.46 d_{duct}^{0.944} \tag{21}$$

where P_{damp} is the cost of the damper (February 1988 $) and d_{duct} is the ductwork diameter (in.).

The cost of a two-way diverter valve for ductwork diameters between 13 and 40 in. are given by:

$$P_{divert} = 4.84 d_{duct}^{1.50} \tag{22}$$

where P_{divert} is the cost of the two-way diverter valve (February 1988 $) and d_{duct} is the ductwork diameter (in.).

4. ENERGY AND COST CONSIDERATIONS FOR MAJOR POINT SOURCE CONTROLS

4.1. Introduction

The auxiliary APCDs discussed in Section 3 include cyclones, gas precoolers, gas preheaters, fans, ductworks, stack, dampers, and hoods, which are required pretreatment and collection means for almost all point source air pollution control projects. This section provides generalized evaluation procedures for a given major add-on HAP control system, which can be one or a combination of the following: (1) thermal incinerator, (2) heat exchanger, (3) catalytic incinerator, (4) carbon adsorber, (5) absorber (scrubber), (6) condenser, (7) fabric filter, (8) electrostatic precipitator, or (9) Venturi scrubber.

The auxiliary APCDs will always be needed for point source air emission controls. Usually, only one or two major APCDs will be the add-on HAP control units.

4.2. Sizing and Selection of Major Add-on Air Pollution Control Devices

Selection of one or more major add-on APCDs for a specific air emission control project will be decided based on both the technical feasibility and the economical feasibility of using the intended APCDs. The readers are referred to other related chapters for a specific major APCD and Tables 1 and 2 for their technical evaluation and comparison.

4.3. Purchased Equipment Costs of Major Add-on Air Pollution Control Devices

This subsection provides generalized procedures for estimating capital and annualized costs for a given add-on HAP control system. (*Note*: The calculation of the cost of HAP waste disposal is outside the scope of this handbook; however, this cost must be included in any rigorous control cost estimation.) The procedures are presented in a step-by-step format and illustrated at each step with cost calculations.

The major equipment purchased cost (i.e., the cost of the major components that comprise the control system) is related to a specific equipment design parameter and can be expressed either analytically or graphically. Gathering current costs from vendors was beyond the scope of this project and necessitates use of dated cost data compiled by others. In general, the cost estimates may be escalated using the *Chemical Engineering* Fabricated Equipment cost indices (36), partially reported in Table 12.

If more recent cost data are available, they should be substituted for the cost curve data presented. These cost curves should not be extrapolated beyond their range. The cost data presented in these figures were obtained from cost information published in US EPA reports. Using the specific value for the design variable, obtain purchased costs form the specific cost curve for each major control system component. Presented in the following subsections are brief descriptions of the equipment costs included in each HAP control cost curve.

4.3.1. Thermal Incinerator Purchase Costs

The cost curve for thermal incinerators (*see* Fig. 8) includes the fan plus instrumentation and control costs, in addition to the major equipment purchase cost. The heat-exchanger cost and other auxiliary equipment cost (*see* Section 3) should also be included. More thermal incinerator costs for comparison with bio-oxidation costs are reported by Boswell (17).

Table 12
Chemical Engineering Equipment Index

Date	Index	Date	Index	Date	Index
Feb. 1990	389.0	May 1988	369.5	Aug. 1986	334.6
Jan. 1990	388.8	Apr. 1988	369.4	July 1986	334.6
Dec. 1989	390.9	Mar. 1988	364.0	June 1986	333.4
Nov. 1989	391.8	Feb. 1988	363.7	May 1986	334.2
Oct. 1989	392.6	Jan. 1988	362.8	Apr. 1986	334.4
Sept. 1989	392.1	Dec. 1987	357.2	Mar. 1986	336.9
Aug. 1989	392.4	Nov. 1987	353.8	Feb. 1986	338.1
July 1989	392.8	Oct. 1987	352.2	Jan. 1986	345.3
June 1989	392.4	Sept. 1987	343.8	Dec. 1985	348.1
May 1989	391.9	Aug. 1987	344.7	Nov. 1985	347.5
Apr. 1989	391.0	July 1987	343.9	Oct. 1985	347.5
Mar. 1989	390.7	June 1987	340.4	Sept. 1985	347.2
Feb. 1989	387.7	May 1987	340.0	Aug. 1985	346.7
Jan. 1989	386.0	Apr. 1987	338.3	July 1985	347.2
Dec. 1988	383.2	Mar. 1987	337.9	June 1985	347.0
Nov. 1988	380.7	Feb. 1987	336.9	May 1985	347.6
Oct. 1988	379.6	Jan. 1987	336.0	Apr. 1985	347.6
Sept. 1988	379.5	Dec. 1986	335.7	Mar. 1985	346.9
Aug. 1988	376.3	Nov. 1986	335.6	Feb. 1985	346.8
July 1988	374.2	Oct. 1986	335.8	Jan. 1985	346.5
June 1988	371.6	Sept. 1986	336.6	Dec. 1984	346.0

[a](2, 30)

Note: CE Equipment Index = 437.4 in April 2000; CE Equipment Index = 273.7 in December 1979; CE Equipment Index = 226.2 in December 1977.

Source: refs. 2 and 36.

4.3.2. Heat-Exchanger Purchase Costs

If the HAP control system includes a heat exchanger, the cost (*see* Fig. 9) is part of the major equipment purchase cost and, thus, must be added. The remaining auxiliary equipment (ductwork and stack) purchase costs and costs of freight and taxes must be added to obtain the total purchased cost.

4.3.3. Catalytic Incinerator Purchase

The cost curve for catalytic incinerators (*see* Fig. 10) provides the cost of an incinerator less catalyst. Catalyst costs [$2750 per cubic foot in June 1985 (26)] and the cost of a heat exchanger, if applicable (*see* Fig. 9), must be added to obtain the major equipment purchase cost. All auxiliary equipment (ductwork, fan, and stack) purchase costs, the cost of instrumentation and controls, and freight and taxes must be added to obtain the total purchase cost. Boswell reported catalytic incinerator's costs in comparison with bio-oxidation costs (17).

4.3.4. Carbon Adsorber Purchase Costs

Two cost curves are presented for carbon adsorbers: Fig. 11 for packed carbon adsorbers and Fig. 12 for custom carbon adsorbers. The cost curve for packaged carbon

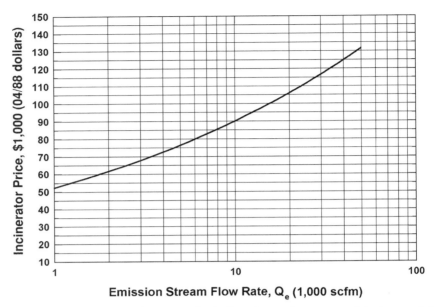

Fig. 8. Price for thermal incinerators, including fan and motor, and instrumentation and controls costs. (From ref. 34.)

adsorbers includes the fan plus instrumentation and control costs, in addition to the major equipment purchase cost. The cost of the remaining auxiliary equipment (ductwork, and stack) as well as costs of freight and taxes must be added to obtain the total purchase cost. The cost curve for custom carbon adsorbers does not include the cost of

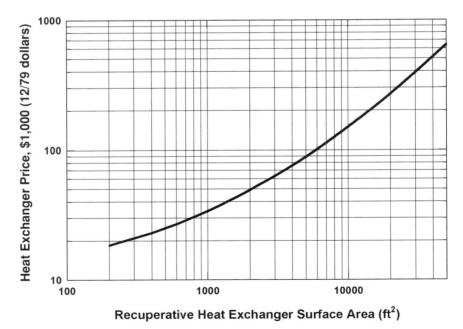

Fig. 9. Price for thermal oxidation recuperative heat exchangers. (From ref. 30.)

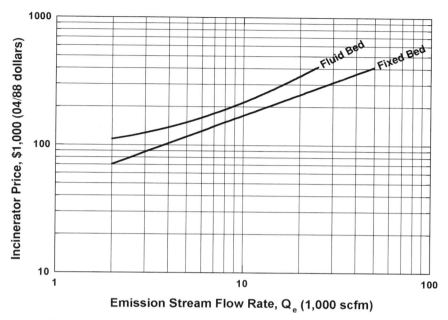

Fig. 10. Price for catalytic incinerators, less catalyst. (From ref. 34.)

carbon (part of the major equipment purchased cost), however, it does include the cost of instrumentation and controls. The cost of carbon is $1.80 to $2.00 per pound in 1991 (26). All auxiliary equipment (ductwork, fan, and stack) purchase costs and freight and taxes must be added to obtain the total purchase cost.

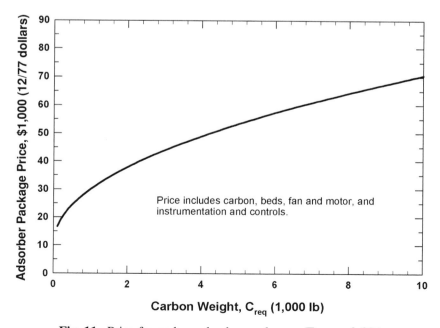

Fig. 11. Price for carbon adsorber packages. (From ref. 32.)

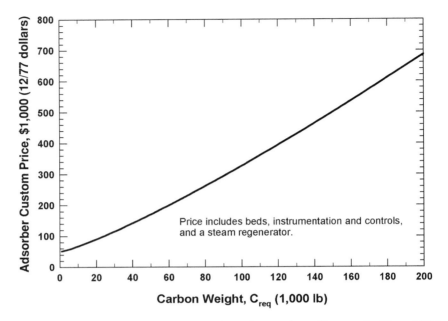

Fig. 12. Price for custom carbon adsorbers, less carbon. (From refs. 28 and 32.)

4.3.5. Absorber Purchase Costs

The cost curve for absorbers (*see* Fig. 13) does not include the cost of packing, platforms, and ladders. The cost of platform and ladders (*see* Fig. 14) and packing (*see* Table 13) must be added to obtain the major equipment purchase cost. All auxiliary equipment (ductwork, fan, and stack) purchase costs, the cost of instrumentation and controls, and freight and taxes must be added to obtain the total purchase cost.

4.3.6. Condenser Purchase Costs

The cost curve for condensers (*see* Fig. 15) yields the total capital cost for cold-water condenser systems. For systems needing a refrigerant (ethylene glycol), the applicable cost from Fig. 16 must be added to the cost obtained Fig. 15. Because a total capital cost is determined, no additional cost estimates are necessary; therefore, proceed to Section 6 to calculate annualized operating costs. The cost of a refrigerant (ethylene glycol) is estimated to be $0.31 per pound for June 1985 (30).

4.3.7. Fabric Filter Purchase Costs

The cost curve for a negative pressure fabric filter (*see* Fig. 17) does not include the cost of bags (*see* Table 14), which depend on the type of fabric used. This cost must be added to obtain the major equipment purchase cost. All auxiliary equipment (ductwork, fan, and stack) purchase costs, the cost of instrumentation and controls, and freight and taxes must be added to obtain the total purchase cost.

4.3.8. Electrostatic Precipitator Purchase Costs

The cost curve presented in Fig. 18 provides the major equipment purchase cost for an insulated electrostatic precipitator. All auxiliary equipment (ductwork, fan, and

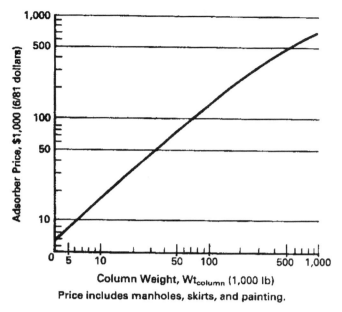

Fig. 13. Prices for adsorber column. (From ref. 1, 9, 36.)

stack) purchase cost, the cost of instrumentation and controls, and freight and taxes must be added to obtain total purchase cost.

4.3.9. Venturi Scrubber Purchase Costs

The cost curve for Venturi scrubbers (*see* Fig. 19) includes the cost of instrumentation and controls, in addition to the major equipment purchase cost. This cost curve

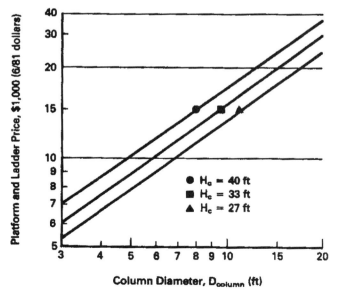

Fig. 14. Prices for adsorber platform and ladders. (From ref. 1, 9, 36.)

Table 13
Price of Packing for Absorber System

Packing Type and Material	Cost/ft² (6/81 dollars)			
Packing diameter (in.)	1	1.5	2	3
Pall rings				
Carbon steel	24.3	16.5	15.1	—
Stainless steel	92.1	70.3	0.8	—
Polypropylene	21.9	14.8	13.8	—
Berl saddles				
Stoneware	28.1	21.7	—	—
Porcelain	34.5	25.6	—	—
Intalox saddles				
Polypropylene	21.9	—	13.6	7.0
Porcelain	19.4	14.8	13.3	12.2
Stoneware	18.2	13.3	12.2	11.0
Packing rings				
Carbon steel	30.3	19.8	17.0	13.9
Porcelain	13.2	10.6	9.7	8.1
Stainless steel	109.0	82.6	22.9	—

Source: ref. 26.

is based on a Venturi scrubber constructed from 1/8-in. carbon steel. Figure 20 is used to determine if 1/8-in. steel is appropriate for a given application (use the high curve). If thicker steel is required, Fig. 21 presents a price adjustment factor for various steel thickness; this factor is used to escalate the cost obtained from Fig. 19. In addition,

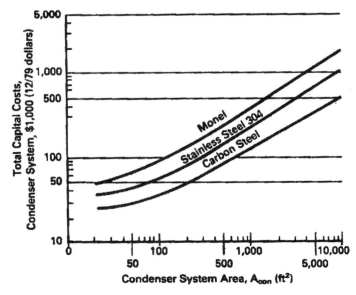

Fig. 15. Total capital costs for cold water condenser systems. (From ref. 37.)

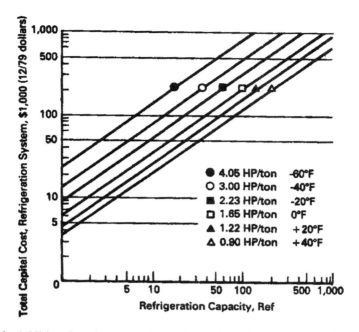

Fig. 16. Additional capital costs for refrigerant condenser systems. (From ref. 37.)

if stainless steel is required, multiply the scrubber cost estimate by 2.3 for 304L stainless steel or by 3.2 for 316L stainless steel. Costs of all auxiliary equipment (ductwork, fan, and stack) and freight and taxes must be added to obtain the total purchased cost.

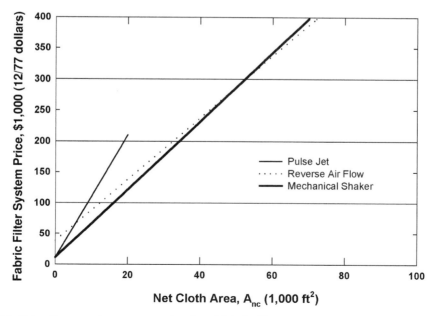

Fig. 17. Price for negative-pressure, insulated fabric-filter system, less bags. (From ref. 30.)

Table 14
Bag Prices

Class	Type	12/77 Dollars/gross ft^2						
		Dacron	Orlon	Nylon	Nomex	Glass	Polypropylene	Cotton
Standard	Mechanical shaker <20,000 ft^2	0.40	0.65	0.75	1.15	0.50	0.65	0.45
Standard	Mechanical shaker >20,000 ft^2	0.35	0.50	0.70	1.05	0.45	0.55	0.40
Standard	Pulse jet[a]	0.60	0.95	—	1.30	—	0.70	—
Custom	Mechanical shaker	0.25	0.35	0.45	0.65	0.30	0.35	0.40
Custom	Reverse air	0.25	0.35	0.45	0.65	0.30	0.35	0.40

[a]For heavy felt, multiply source by 1.5.
Source: Data from refs. 26 and 30.

5. ENERGY AND COST CONSIDERATIONS FOR AREA SOURCE CONTROLS

5.1. Introduction

Energy and cost information about various control technologies used to control emissions from area sources is presented in this section. The control technologies generally are applicable to the control of all classes of air contaminants, including VOCs, SVOCs, PM, and metals associated with PM. The specific control technologies addressed in this section are covers and physical barriers, foams, wind barriers, water sprays, water sprays with additives, operational controls, enclosures, collection hoods, and miscellaneous controls.

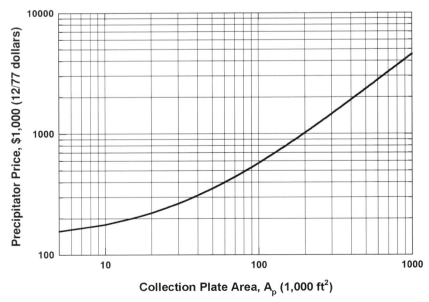

Fig. 18. Price for insulated electrostatic precipitators. (From ref. 30.)

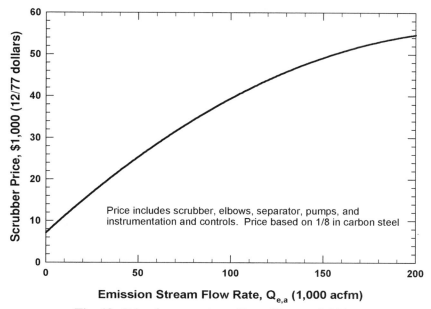

Fig. 19. Price for venturi scrubbers (From ref. 30.).

Emissions from area sources are more difficult to measure, model, and control than emission from point sources. The sources may be several acres in size and the concentration of emissions in the source/atmospheric boundary layer is generally very low. Therefore, the types of control suitable for point sources are not applicable to area sources. Two general control approaches exist for area sources: (1) collect the emissions in a hood or enclosure and route the airstream to a point source control device and (2) prevent the emissions from occurring. The first approach is merely a conversion of the area source to a point source and is the most suitable for batch or in situ remediation processes such as solidification/stabilization and bioremediation. The second approach is primarily suited for materials handling operations such as excavation.

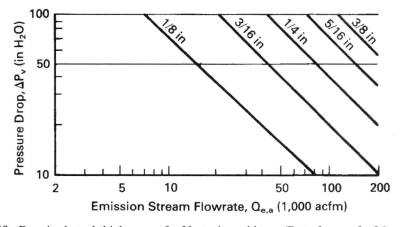

Fig. 20. Required steel thicknesses for Venturi scrubbers. (Data from refs. 26 and 30.)

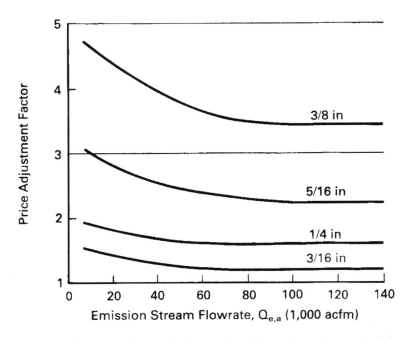

Fig. 21. Price adjustment factors for Venturi scrubbers. (Data from refs. 26 and 30.)

5.2. Cover Cost

The amount (depth, thickness, etc.) of cover material required to achieve a given control efficiency is not well defined in the literature. However, there are general sizing guidelines reported in the literature. Cost estimates of implementing cover-based VOC/PM control measures are presented in Table 15. Caution should be exercised when using these cost estimates because costs are highly dependent on the site characteristics, labor costs, weather conditions, and the availability of specific cover materials at each site.

Table 15
Cost (1992 $) of Implementing Cover-Based Area Control Measures

Cover material	Equipment[a]	Labor/materials
Backfill dirt	2.0 m^3	15/m^3
Clay	1.0 m^3	15/m^3
Road base, road carpet, and gravel[b]	3–6 m	4–10/m
Asphalt, road base[c]	6–12/m	200–300/m
Wood fibers with plastic[d]	0.5/m^2	0.5/m^2
Polymer sheeting	1.0/m^2	1.0/m^2

[a]Assumes material not already on site.
[b]7.5 m wide and 0.15 m gravel.
[c]7.5 m wide and 0.10 m asphalt.
[d]Wood fiber depth not stated.
Source: Data from refs. 6 and 14.

Table 16
Foam Costs

Foam	Brand name	Cost ($/m^3)
Temporary	FX9162	1.10
	AC6545	0.54–0.86
Long term	FX 9162/9161	3.80
	AC904	1.30–1.94
	AC912	1.94–2.70
	AC918	2.70–3.77
	AC930	3.77–5.28

Source: Data from refs. 12–15.

5.3. Foam Cost

Costs for various foam types are given in Table 16. These costs are a function of the area to be treated at the application depths recommended by the manufacturer. Costs for foam application units range from $8000 to $12,000 per month for manifold application units (including bulk storage tanks) and $3250 to $7750 for hand-line application units. Small 3M application units can be rented for about $660 per week; about $500 of ancillary equipment is also required (1992 dollars).

5.4. Wind Screen Cost

Capital costs for wind screens vary with the type of control desired (VOC or PM) and the operation requiring control (e.g., inactive sites, excavation, etc.). Costs as a function of pollutant to be controlled and operation requiring control are outlined in Table 17.

5.5. Water Spray Cost

For mobile water spray systems, capital costs are estimated to be $23,000/water wagon per year, with operating and maintenance (O&M) costs (fuel, water, labor, and truck maintenance) estimated to be $44,000/water wagon per year. Furthermore, the number of water wagons required can be estimated by assuming that a single truck applying 1 L/m^2 can treat roughly 1 mile^2/h, (approx 11,000 m^2). Capital and O&M costs for fixed water systems will vary with the type of emission source to be controlled (e.g., "truck-out," excavation, loading operations) and the amount of plumbing required.

Table 17
Wind Screen System Costs

Type of pollutant	Control	Cost
VOC	Inactive surface impoundment	$0.7–1.4/m^2 impoundment area
Particulate matter	Inactive sites[a]	$0.6–13/m^2 of inactive area
	Storage piles[b]	$721/m^2 of pile area
	Excavation[c]	$11/m^2 of excavation area

[a]Minimum price assumes chain link fence is available to secure wind screen (valid for small areas). Cost per linear meter for wind screen is about $40, not including support structure.

[b]Assumes conical-shaped storage pile roughly 10 m in diameter.

[c]Assumes 60-m-diameter excavation site and 1.8-m-high wind screen around two-thirds of site.

Source: ref. 8.

5.6. Water Additives Costs

Water additives costs include the costs associated with water spray systems and also include the cost of additives and storage tanks for the additives. Storage tank costs will vary depending on the size of the operation, the water/additive application rate, and the time between deliveries of additive. The dilution ratio, application rate, and frequency must be determined to predict the cost per square foot. Some additive costs by product name and classification are as follows:

1. Hygroscopic salt = $0.02–0.10/ft^2
2. Bitumens/adhesives = $0.15–0.32/ft^2
3. Surfactant = $0.002/ft^2

5.7. Enclosure Costs

Enclosures range in size from 30 ft in diameter to 130 ft wide × 62 ft tall × unlimited length. For self-supported structures wider than 60 ft, footings may be required. Prior to erecting an enclosure, the site may require grading so that the slope is less than 3%. The costs of air-supported and self-supported enclosures are as follows:

1. Air-supported enclosure = $5.5/m^2-month rent + unknown O&M
2. Self-supported enclosure = $19/m^2-month rent + $48/m^2 O&M

The costs presented do not include the costs of gas collection/treatment systems.

5.8. Hood Costs

Hood exhaust systems designs are based on the hood aspect ratio (width/length of hood), the required capture velocity (v), and the distance of the furthest point of the emitting source from the hood centerline (x).

The costs of hood exhaust systems are highly dependent on the volumetric flow rate, the length of ducting required, the hood/ducting materials of construction required (e.g., carbon steel, stainless steel), hood size, and fan size required to move the air. An example of a hood exhaust system cost breakdown is presented in Table 18.

5.9. Operational Control Costs

For a target control efficiency, the operational practices/procedures required can generally be determined. Operational practices/procedures that are amenable to this approach are as follows:

1. Road cleaning practices
2. Seasonal scheduling
3. Vehicle speed control
4. Excavation practices
5. Dumping practices

Quantification of PM emission controls achievable for soil loading practices and storage pile geometry/orientation is not possible. However, guidelines are available for each of these operational control measures.

For the majority of the operational control measures presented in this section, the cost is negligible, with the exception of road cleaning equipment and possibly seasonal scheduling. The cost of seasonal scheduling will vary with season primarily because of

Table 18
Hood Exhaust System Cost Estimate

Equipment	Applicable dimensions	Cost
Canopy hood	3/16-in.-thick carbon steel, 10 ft in diameter	$2400
Ductwork	100 ft of 1-ft-diameter, 16-gage carbon steel straight duct	$1300
	Four 1-ft diameter, 16-gage carbon steel 90° elbows.	$1750
Radial-tip fan	Moves 11,000 acfm at 10 in. H_2O with a 45.5-in. wheel diameter.	$7700
	Total cost:	$13,150

Source: ref. 8.

labor costs and equipment availability. Cost for street cleaning practices are estimated to be $140 per day per street cleaner and $66 per day per crew. The use of larger excavation equipment to minimize emissions will increase costs to some extent.

6. CAPITAL COSTS IN CURRENT DOLLARS

In this handbook, the total capital cost includes only manufacturing area costs; therefore, it excludes offsite costs. The total capital cost of a control system is the sum of direct costs, indirect costs, and contingency costs.

Direct costs include the total purchase equipment cost (i.e., the major equipment purchase cost plus the auxiliary equipment purchase cost), instrumentation and controls, freight and taxes, and installation costs (i.e., foundation and supports, erection and handling, electrical, piping, insulation, and painting). (*Note*: The summation of the total purchased equipment cost, the cost of instrumentation and controls, and freight and taxes is defined as the total purchase cost.)

Indirect costs consist of in-house engineering design and supervision costs, architect and engineering contractor expenses, and preliminary testing costs. An example of contingency costs is the penalties incurred for failure to meet completion dates or performance specifications.

The capital cost estimation procedure presented in this handbook is for a factored or "study" estimate. Usual reliability for a study type estimate is ±30%. To determine the total capital cost by a factored cost estimate, a reliable estimate of the total purchase cost is calculated and predetermined factors are applied to determine all other capital cost elements.

The procedure to estimate the total capital cost is as follows: (1) Obtain the total purchased equipment cost by estimating the purchased cost of major and auxiliary equipment; (2) estimate the cost of instrumentation and controls plus freight and taxes as a percentage of the total purchased equipment cost; (3) estimate the total purchase cost by adding items 1and 2; and (4) estimate total capital cost by applying a predetermined cost factor to the total purchase cost.

Table 19 shows how the capital cost in current dollars can be estimated for an intended APCD system. Table 20 presents the capital cost elements and factors (26). The following are the footnotes to Table 19.

Table 19
Estimate of Capital Costs in Current Dollars

Cost elements	Figure of table cost	Escalation factor (current FE/base FE) (*see* Table 12)	Current cost
1. Major equipment purchase cost			
Thermal Incinerator[a]	$ ———— ×	(———— / ————)	= $ ————
Heat Exchanger[b]	$ ———— ×	(———— / ————)	= $ ————
Catalytic incinerator[c]	$ ———— ×	(———— / ————)	= $ ————
Catalyst,[c] V_{cat} = ——ft^3 ×	————$/ft^3$	(———— / ————)	= $ ————
Carbon adsorber[d]	$ ———— ×	(———— / ————)	= $ ————
Carbon,[d] C_{req} = ——lb ×	————$/lb	(———— / ————)	= $ ————
Absorber[e]	$ ———— ×	(———— / ————)	= $ ————
Platforms and ladders[e]	$ ———— ×	(———— / ————)	= $ ————
Packing,[e] V_{pack}=——ft^3 ×	————$/ft^3$	(———— / ————)	= $ ————
Condenser[f]	$ ———— ×	(———— / ————)	= $ ————
Refrigerant[f]	$ ———— ×	(———— / ————)	= $ ————
Fabric filter[g]	$ ———— ×	(———— / ————)	= $ ————
Bags,[g] A_{tc} = ——ft^2 ×	————$/ft^2$	(———— / ————)	= $ ————
ESP[h]	$ ———— ×	(———— / ————)	= $ ————
Venturi scrubber[i]	$ ———— ×	(———— / ————)	= $ ————
Design factors[i]	———— ×		= $ ————
	(Thickness Factor)	(Composition Factor)	
		Subtotal	= $ ————
2. Auxiliary equipment purchase cost			
Ductwork[j]	$ ——×—— ×	(——/——)	= $ ————
	(Length)		
Fan[k]	$ ———— ×	(——/——)	= $ ————
Motor[l]	$ ———— ×	0.15	= $ ————
	(Fan Current Cost)		
Stack[m]	$ ———— ×	(——/——)	= $ ————
		Subtotal	= $ ————
3. Pretotal purchase equipment cost	Item 1 Subtotal + Item 2 Subtotal		$ ————
Adjustments[n]	(Item 3)×(−0.091)		$ ————
4. TOTAL purchase equipment cost	Item 3 + Adjustments		$ ————
5. Instrumentation and controls	10% of Item 4		$ ————
6. Freight and taxes	8% of Item 4		$ ————
7. TOTAL Purchase Cost	Item 4 + Item 5 + Item 6		$ ————
8. TOTAL CAPITAL COSTS	F^o×(Item 7), where F=		$ ————

Source: ref. 26.

(a) Thermal incinerator: Figure 8 includes fan plus instrumentation and control costs for thermal incinerators, in addition to the major equipment purchase cost. Additional auxiliary equipment (ductwork and stack) purchase costs and costs of freight and taxes must be added to obtain the total purchase cost.

(b) Heat exchangers: If the HAP control system requires a heat exchanger, obtain the cost from Fig. 9, escalate this cost using the appropriate factor, and add to the major equipment purchase cost.

(c) Catalytic incinerator: Figure 10 provides the cost of a catalytic incinerator, less catalyst costs. The "table" catalyst cost is estimated by multiplying the volume of catalyst required (V_{cat}) by the catalyst cost factor ($\$/ft^3$) found in the literature. Catalyst costs, all auxiliary

Table 20
Capital Cost Elements and Factors[a]

				Control technique			
				Thermal and			
		Venturi	Fabric	catalytic			
Cost Elements	ESP	scrubbers	filter	incinerators	Adsorber	Absorber	Condensers
DIRECT COSTS							
Purchased							
equipment cost[b]	1.00	1.00	1.00	1.00	1.00	1.00	1.00
Other direct costs							
Foundation and	0.04	0.06	0.04	0.08	0.08	0.12	0.08
supports							
Erection and handling	0.50	0.40	0.50	0.14	0.14	0.40	0.14
Electrical	0.08	0.01	0.08	0.04	0.04	0.01	0.08
Piping	0.01	0.05	0.01	0.02	0.02	0.30	0.02
Insulation	0.02	0.03	0.07	0.01	0.01	0.01	0.10
Painting	0.02	0.01	0.02	0.01	0.01	0.01	0.01
Total Direct Cost	1.67	1.56	1.72	1.30	1.30	1.85	1.43
INDIRECT COSTS							
Engineering							
and supervision	0.20	0.10	0.10	0.10	0.10	0.10	0.10
Construction and							
filed expenses	0.20	0.10	0.20	0.05	0.05	0.10	0.05
Construction fee	0.10	0.10	0.10	0.10	0.10	0.10	0.10
Start up	0.01	0.01	0.01	0.02	0.02	0.02	0.02
Performance test	0.01	0.01	0.01	0.01	0.01	0.01	0.01
Model study	0.02	—	—	—	—	—	—
Total Indirect Cost	0.54	0.32	0.42	0.28	0.28	0.32	0.28
CONTINGENCY[c]	0.07	0.06	0.07	0.05	0.05	0.07	0.05
TOTAL[d]	2.27	1.94	2.21	1.63	1.63	2.24	1.76

[a]As fractions of total purchased equipment cost. They must be applied to the total purchased equipment cost.

[b]Total of purchase costs of major equipment and auxiliary equipment and others, which include instrumentation and controls at 10%, taxes and freight at 8% of the equipment purchase cost.

[c]Contingency costs are estimated to equal 3% of the total direct and indirect costs.

[d]For retrofit applications, multiply the total by 1.25.

Source: ref. 26.

equipment (ductwork, fan, and stack) purchase costs, the cost of instrumentation and controls, and freight and taxes must be added to obtain the total purchase cost.

(d) Carbon adsorber: Figure 11 (packaged carbon adsorber systems) includes the cost of carbon, beds, fan and motor, instrumentation and controls, and a steam regenerator. Additional auxiliary equipment (ductwork, and stack) purchase costs and costs of freight and taxes must be added to obtain the total purchase cost. Figure 12 (custom carbon adsorber systems) includes beds, instrumentation and controls, and a steam regenerator, less carbon. The "table" carbon cost for custom carbon adsorbers is estimated by multiplying the weight of carbon required (C_{req}) by the carbon cost factor ($/lb) found in the literature. Costs of carbon, all auxiliary equipment (duct, fan, stack) purchase costs, and freight and taxes must be added to obtain the total purchase cost.

(e) Absorber: Figure 13 does not include the cost of packing, platforms, and ladders. The cost of platforms and ladders (see Fig. 14) and packing must be added to obtain the major purchased equipment cost. The "table" packing cost is estimated by multiplying the volume of packing required (V_{pack}) by the appropriate packing cost factor found in Table 13. All auxiliary equipment (ductwork, fan, and stack) purchase costs and costs of freight and taxes must be added to obtain the total purchase cost.

(f) Condenser systems: Figure 15 yields total capital costs for cold-water condenser systems. For systems needing a refrigerant, the applicable cost from Fig. 16 must be added to obtain the total capital costs. In either case, the escalated cost estimate is then placed on the TOTAL CAPITAL COSTS line.

(g) Fabric filter systems: Figure 17 gives the cost of a negative-pressure, insulated baghouse. The curve does not include bag costs. The "table" bag cost is estimated by multiplying the gross cloth area required by the appropriate bag cost factor found in Table 14. Bag costs, all auxiliary equipment (ductwork, fan, and stack) purchase costs, the cost of instrumentation and controls, and freight and taxes must be added to obtain the total purchase cost.

(h) Electrostatic precipitators: Figure 18 provides the cost for an insulated ESP. All auxiliary equipment (duct, fan, and stack) purchase costs, the cost of instrumentation and controls, and freight and taxes must be added to obtain the total purchase cost.

(i) Venturi scrubber: Figure 19 includes the cost of instrumentation and controls in addition to the major equipment purchase cost. This cost curve is based on a Venturi scrubber constructed form (1/8)-in. carbon steel. Figure 20 is used to determine if (1/8)-in. steel is appropriate for a given application (use the higher curve). If thicker steel is required, Fig. 21 yields an adjustment factor for various steel thicknesses; this factor is used to escalate the cost obtained from Fig. 19. In addition, if stainless steel is required, multiply the scrubber cost estimate by 2.3 for 304L stainless steel or by 3.2 for 316L stainless steel. Costs of all auxiliary equipment (ductwork, fan, and stack) and freight and taxes must be added to obtain the total purchase cost.

(j) Ductwork: Figure 4 gives the cost of straight ductwork made of carbon steel for various thicknesses, based on the required duct diameter. Figure 5 gives the cost of straight ductwork made of stainless steel for various thicknesses, based on the required diameter. Preliminary calculations are necessary to estimate ductwork costs.

(k) Fan: Figure 3 gives the cost of a fan based on the gas flow rate at actual conditions and the HAP control system pressure drop (in in. H_2O). The applicable fan class is also based on the HAP control system pressure drop. Calculation of the total system pressure drop is required.

(l) The cost of a motor is estimated as 15% of the fan cost.

(m) Stack: Figure 6 gives the cost of a carbon steel stack at various stack heights and diameters. Figure 7 gives the price of a stainless-steel stack at various stack heights and diameters. Preliminary calculations are necessary to estimate stack costs. For both figures, use the curve that best represents the calculated diameter.

(n) For thermal incinerators, carbon adsorbers, and Venturi scrubbers, the purchase cost curve includes the cost for instrumentation and controls. This cost (i.e., the "Adjustment") must be subtracted out to estimate the total purchased equipment cost. This is done by adding the Item 1 subtotal and the Item 2 subtotal and multiplying the result by −0.091. This value is added to the total purchased equipment cost. For all other major equipment, the "Adjustment" equals zero.

(o) Obtain factor F from "TOTAL" line in Table 20.

7. ANNUALIZED OPERATING COSTS

7.1. Introduction

The annualized cost of an air pollution control system can be divided into direct operating costs, indirect operating costs, and credits. In this handbook, the inflation effect on costs is not considered, annualized costs are assumed to be constant in real dollars, and the total annualized cost is estimated on a before-tax basis.

7.2. Direct Operating Costs

The direct operating costs consist of utilities, operating labor charges, maintenance charges, and replacement parts and labor charges.

1. Utilities (i.e., fuel, electricity, water, steam, and materials required for the control system) are annual costs that vary depending on the control system size and operating time. They are calculated using gas stream characteristics and control equipment capacity data.
2. Operating labor costs consist of operator labor and supervision, whereas maintenance costs consist of maintenance labor and materials. The direct operating costs are established by estimating annual quantities of utilities consumed and operator and maintenance labor used and by applying unit costs to these quantities. The annual quantities of utilities and labor requirements are assumed to be proportional to the annual operating hours for the control system. Operating labor supervision and maintenance materials are taken as %ages of the operator and maintenance labor costs.
3. Costs of replacement parts are estimated as applicable, and the cost of replacement labor is assumed to equal the cost of replacement parts.

Table 20 presents June 1985 unit costs for utilities, operator labor, and maintenance labor as well as cost factors for other direct operating cost elements.

The procedures used to estimate direct operating costs (including utilities, direct labor, maintenance, and replacement costs) and indirect operating costs (including overhead, property tax, insurance, administration, and capital recovery cost) were taken from US EPA. These unit costs and cost factors are applied to estimated quantities of utilities consumed, labor expended, and parts used to obtain total direct operating costs.

If a given control system contains two or more control devices, the direct operating costs must be calculated for each device and summed. The capital recovery cost for a multiple control devices system should be calculated using a weighted-average capital cost factor.

Unless specified, use 8600 h per year, 8 h per shift, and 24 h per day, as necessary, to estimate the annual costs for utilities consumed, operator labor, and maintenance labor.

7.2.1 Utility Requirement: Electricity Requirement

The utility requirements for a control system are obtained from each component's design calculations. Use the costing information in Tables 21–23 to estimate the total utility costs. A procedure to estimate fan electricity costs is provided below, as these costs are applicable to all control techniques. Table 24 is a summary table for estimation of annualized costs in current dollars. As Tables 21–24 are important for estimation of annualized costs, they are further explained in detail.

Table 21 is further explained by the following footnotes to the table:

(a) The readers are referred to Tables 22 and 23 to estimate utility costs and replacement costs for each HAP control technique.
(b) Maintenance materials include operating supplies (e.g., lubrication, paper).
(c) CRF = capital recovery factor. For an average interest rate of 10%, the CRF for specific control devices are as follows:
     ESP and fabric filter: CRF = 0.117 (based on 20-y life span)
     Venturi scrubber, thermal and catalytic incinerators, adsorber, absorber, and condenser: CRF = 0.163 (based on 10-y life span).

Table 22 is explained by the following footnotes:

Table 21
Unit Costs to Calculate Annualized Cost

Cost Elements	Unit costs/factor
Direct Operating Costs	
1. Utilities[a]	
a. Natural gas	$0.00425/ft^3
b. Fuel oil	$1.025/gal
c. Water	$0.0003/gal
d. Steam	$0.00504/lb
e. Electricity	$0.059/kWh
f. Solvent	As applicable
2. Operating labor	
a. Operator labor	$11.53/hour
b. Supervision	15% of operator labor
3. Maintenance	
a. Labor	$11.53/hour
b. Materials[b]	100% of maintenance labor
4. Replacement	
a. Parts	As applicable
b. Labor	100% of replacement parts
Indirect Operating Costs	
1. Overhead	80% of 2a + 2b + 3a + 4a
2. Property tax	1% of Total Capital Cost
3. Insurance	1% of Total Capital Cost
4. Administration	2% of Total Capital Cost
5. Capital recovery	(CRF[c]) × Total Capital Cost
Credits	As applicable

Note: See text for additional information.
Source: ref. 26.

Table 22
Utility/Replacement Operating Costs for HAP Control Techniques

HAP control device	Utilities/replacement parts
Thermal incinerator	Natural gas or fuel oil; Electricity (fan)
Catalytic incinerator	Catalyst cost (V_{cat}); Natural gas or fuel oil; Electricity (fan)
Carbon adsorber systems	Carbon (C_{req}); Steam Cooling water; Electricity (fan)
Absorber systems	Absorbent (water or solvent); Electricity (fan)
Condenser system	Refrigerant; (Ref); Electricity (fan)
Fabric-filter system	Bags (A_{tc}); Electricity (fan + control device)
Electrostatic precipitator	Electricity (fan + control device)
Venturi scrubbers	Water; Electricity (fan)

Note: See Text for additional information.

Table 23
Additional Utility Requirements

Fuel requirement for incinerators, (ft³)
  (*Note*:The design sections for thermal and catalytic incinerators are developed under the assumption that natural gas is used as the supplementary fuel. Fuel oil could be used, however, the use of natural gas is normal industry practice. If fuel oil is used, the following equation can be used by replacing Q_f with the fuel oil flow rate in units of gallons per minute. The product of the equation then equals gallons of fuel oil.)
  Fuel requirement = 60 Q_f×HRS, where Q_f is the supplementary fuel required (scfm) and HRS is the annual operating hours (h). (*Note*: Use 8600 h unless otherwise specified.)

Steam requirement for carbon adsorber (lb)
  (*Note*: Assume 4 lb of steam required for each of recovered product.)
  Steam requirement = 4(Q_{rec})×HRS, where Q_{rec} is the quantity of HAP recovered (lb/h).

Cooling water requirement for carbon adsorber (gal)
  (*Note*: Assume 12 gal of cooling water required per 100 lbs steam.)
  Water requirement = 0.48(Q_{rec})×HRS

Absorbent requirement for absorbers (gal)
  (*Note*: Assume no recycle of absorbing fluid [water or solvent].)
  Absorbent requirement = 60(L_{gal})×HRS, where L_{gal} is the absorbing fluid flow rate (gal/min).

Water requirement for venturi scrubbers (gal)
  (*Note*: Assume 0.01 gal water is required per actual cubic feet of emission stream.)
  Water requirement = 0.6 $Q_{e,a}$×HRS, where $Q_{e,a}$ is the emission stream flow rate into scrubber (acfm).

Baghouse electricity requirement (kWh)
  (*Note*: Assume 0.0002 kW are required per square feet of gross cloth area.)
  Baghouse electricity requirement = 0.0002 A_{tc}×HRS, where A_{tc} is the gross cloth area required (ft²).

ESP electricity requirement (kWh)
  (*Note*: Assume 0.0015 kW are required per square feet of collection area.)
  ESP electricity requirement = 0.0015 A_p×HRS, where A_p is the collection plate area (ft²).

Note: See text for additional information.
Source: ref. 26.

Table 24
Estimate of Annualized Costs in Current Dollars

Cost elements	Units costs/factors	Annual expenditure	Current dollars
Direct Operating Costs			
1. Natural gas	$0.00425/ft^3	×_____ft^3	= $ _____
2. Fuel oil	$1.025/gal	×_____gal	= $ _____
3. Water	$0.003/gal	×_____gal	= $ _____
4. Steam	$0.00504/lb	×_____lb	= $ _____
5. Electricity	$0.059/kWh	×_____kWh	= $ _____
6. Solvent	$ _____per gal	×_____gal	= $ _____
7. Replacement Parts	As applicable		$ _____
8. Replacement Labor	100% of Line 7		$ _____
9. Operator Labor	$11.53/h	×_____h	= $ _____
10. Supervision labor	15% of Line 9		$ _____
11. Maintenance Labor	$11.53/h	×_____h	= $ _____
12. Maintenance Materials	100% of Line 11		$ _____
Subtotal	Add Items 1–12		$ _____
Indirect Operating Costs			
14. Overhead	80 % of Sum of Lines 8–11		$ _____
15. Property tax	1% of Total Capital Cost[a]		$ _____
16. Insurance	1% of Total Capital Cost[a]		$ _____
17. Administration	2% of Total Capital Cost[a]		$ _____
18. Capital recovery	(CRF)×Total Capital Cost[a], where CRF =		$ _____
19. Subtotal	Add Items 14–18		$ _____
20. Credits	As applicable		$ _____
Net Annualized Costs	Items 13+Item 19–Item 20		$ _____

Note: See text for additional information.
[a]Total capital cost from Line 8 of Table 19.

(a)   The readers are referred to Table 21 for utility unit costs, Sections 4.3.2–4.3.4, Table 14 for replacement part unit costs, and Table 12 for FE cost indices.
(b)   See Table 23 for additional utility requirement.
(c)   Annualized replacement catalyst costs are calculated as

$$\text{Annualized cost} = \frac{V_{cat}\left(ft^3\right)\times \$/ft^3}{3 \text{ yr}}(\text{current FE / Base FE}) \tag{23}$$

(d)   Annualized replacement carbon costs are calculated

$$\text{Annualized cost} = \frac{C_{req}\left(lb\right)\times \$/lb}{5 \text{ yr}}(\text{current FE / Base FE}) \tag{24}$$

(e)   Refrigerant replacement is the result of the system leaks; however, the loss rate of refrigerant is very low and varies for every unit. Therefore, assume that the cost of refrigerant replacement is negligible.
(f)   Annualized replacement bag costs are calculated as

$$\text{Annualized cost} = \frac{A_{tc}\left(ft^2\right)\times \$/ft^2}{2 \text{ yr}}(\text{current FE / Base FE}) \tag{25}$$

Table 24 is further explained by the following footnotes:

(a)   The readers are referred to Section 7.2.2 for the costs of natural gas, fuel oil, water, steam, and solvent.
(b)   The readers are referred to Section 7.2.1 for the electricity costs.
(c)   As applicable.
(d)   Total capital cost can be obtained from Line 8 of Table 19.

The annualized electricity requirement of fan, baghouse, and ESP can be calculated by the following equations.

(a)   Fan electricity requirement (FER)

$$\text{FER (kWh)} = 0.0002\left(Q_{fg,a}\right) \times \Delta P \times \text{HRS} \tag{26}$$

where $Q_{fg,a}$ is the actual flue gas flow rate (acfm), ΔP is the total HAP control system pressure drop (in. H_2O) (*see* Table 5), HRS is the annual operating hours (hr) (*Note*: use 8600 unless otherwise specified).

(b)   Baghouse electricity requirement, (BER),
      (*Note*: assume 0.0002 kW are required per square feet of gross cloth area)

$$\text{BER (kWh)} = 0.0002\left(A_{tc}\right) \times \text{HRS} \tag{27}$$

where A_{tc} is the gross cloth area required (ft^2).

(c)   ESP electricity requirement, (EER) (*Note*: assume 0.0015 kW are required per square feet of collection area)

$$\text{EER (kWh)} = 0.0015\left(A_p\right) \times \text{HRS} \tag{28}$$

where A_p is the collection plate area (ft^2).

(d)   Annual electricity requirement (AER)

$$\text{AER (kWh)} = \text{FER} + \text{BER} + \text{EER} \tag{29}$$

7.2.2. Utility Requirement: Fuel, Steam, Absorbent, and Water Requirements

The design sections for thermal and catalytic incinerators are developed under the assumption that natural gas is used as the supplementary fuel. Fuel oil could be used; however, the use of natural gas is normal industry practice. If fuel oil is used, eq. (30) can be used by replacing Q_f with the fuel oil flow rate in units of gallons per minute. The resultant product of the equation (gallons of fuel oil required) is then used on Line 2 of Table 24.

(a)   Fuel requirement for incinerators (Line 1 or Line 2, Table 24)

$$\text{Fuel Requirement}\left(\text{ft}^3\right) = 60\left(Q_f\right) \times \text{HRS} \tag{30}$$

where Q_f is the supplementary fuel required, (scfm), and HRS is the annual operating hours, (h) (*Note*: use 8600 unless otherwise specified).

(b)   Steam requirement for carbon adsorber (Line 4, Table 24) (*Note*: assume 4 lb of steam required for each lb of recovered product)

$$\text{Steam Requirement(lb)} = 4\left(Q_{rec}\right) \times \text{HRS} \tag{31}$$

where Q_{rec} is the quantity of HAP recovered (lb/h).

(c) Cooling water requirement for carbon adsorber (Line 3, Table 24) (*Note*: assume 12 gal of cooling water required per 100 lb steam)

$$\text{Water Requirement (gallon)} = 0.48 \left(Q_{\text{rec}} \right) \times \text{HRS} \tag{32}$$

(d) Absorbent requirement for absorbers (Line 3, Table 24) (*Note*: assume no recycle of absorbing fluid [water or solvent])

$$\text{Absorbent Requirement (gallon)} = 60 \left(L_{\text{gal}} \right) \times \text{HRS} \tag{33}$$

where L_{gal} is the absorbing fluid flow rate (gal/min).

(e) Water requirement for Venturi scrubbers (Line 3, Table 24) (*Note*: assume 0.01 gal of water required per acf of emission stream)

$$\text{Water Requirement (gallon)} = 0.6 \left(Q_{e,a} \right) \times \text{HRS} \tag{34}$$

where $Q_{e,a}$ is the emission stream flow rate into scrubber (acfm).

7.2.3. Replacement Parts Annualized Costs

(a) Annualized catalyst replacement costs (Line 7, Table 24). Over the lifetime of a catalytic incinerator, the catalyst is depleted and must be replaced (assume catalyst lifetime is 3 yr).

$$\text{Annual Catalist Cost, } \$ = \left(\text{Catalyst Current Cost} \right) / 3 \tag{35}$$

(b) Annualized carbon replacement costs (Line 7, Table 24). Over the lifetime of a carbon adsorber, the carbon is depleted and must be replaced (assume carbon lifetime is 5 yr):

$$\text{Annual Carbon Cost, } \$ = \left(\text{Carbon Current Cost} \right) / 5 \tag{36}$$

(c) Annualized refrigerant replacement costs. Refrigerant in a condenser needs to be replaced periodically because of system leaks; however, the loss rate is typically very low. Therefore, assume that the cost of refrigerant is negligible.

(d) Annualized bag replacement costs (Line 7, Table 24). Over the lifetime of a fabric filter system, the bags become worn and must be replaced (assume bag lifetime is 2 yr).

$$\text{Annual Bag Cost, } \$ = \left(\text{Bag Current Cost} \right) / 2 \tag{37}$$

The bag current cost can be obtained from Table 14.

7.3. Indirect Operating Costs

7.3.1. Introduction

The indirect operating costs include overhead costs, property tax, insurance, administration costs, and the capital recovery costs. Overhead costs are estimated as a %age of operating labor costs. Property tax, insurance, and administrative costs are estimated as a %age of the total capital cost. The capital recovery cost is estimated as the product of the capital recovery factor and the total capital cost. The factor for capital recovery costs (the total of annual depreciation and interest on capital) is determined from the expected life of the control device and the interest rate at which the capital is borrowed. The expected life of a given control device depends on the type of control application, maintenance service, and operating duty. For costing purposes, pre-established expected life values are used.

Some control technologies recover the HAPs from a given emission stream as a salable product. Therefore, any cost credits associated with the recovered material must be deducted from the total annualized cost to obtain the net annualized cost for the system.

The amount, purity, and commercial value of the recovered material determine the magnitude of credits.

7.3.2. Capital Recovery Factor

Estimate the overhead costs as 80% of the direct labor cost (the summation of operating labor and supervision labor costs) and the maintenance labor cost. The property tax estimate is calculated as 1% of the total capital cost, insurance is 1% of the total capital cost. Estimate the capital recovery cost portion of the fixed capital charges by multiplying the total capital cost by a capital recovery factor. The capital recovery factor (CFR) is calculated as follows:

$$CFR = \left[i(1+i)^n\right]\Big/\left[(1+i)^n - 1\right] \tag{38}$$

where i is the interest rate on borrowed capital (decimal) and, n is the control device life (yr).

For the purpose of this handbook, an interest rate of 10% is used. Table 25 contains data on expected control device life (n). Calculated capital recovery factors at 10% interest rate are 0.163 and 0.117 for 10- and 20-yr control device lifetimes, respectively. If more than one control device is used by the control system, use a weighted-average capital recovery factor. A weighted-average capital recovery factor (CRF_w) is determined as follows:

$$CRF_W = CRF_1\left[PC_1/(PC_1 + PC_2)\right] + CRF_2\left[PC_2/(PC_1 + PC_2)\right] \tag{39}$$

where CRF_1 is the capital recovery factor for control device 1, CRF_2 is the capital recovery factor for control device 2, PC_1 is the purchased equipment cost for control device 1, and, PC_2 is the purchased equipment cost for control device 2.

7.3.3. Calculation of Capital Recovery Factor, Annualized Operator Labor, and Annualized Maintenance Labor

(a)   Calculation of capital recovery factor (CRF) (Line 18, Table 24)

$$CFR = \left[i(1+i)^n\right]\Big/\left[(1+i)^n - 1\right] \tag{38}$$

Table 25
Estimated Labor Hours per Shift and Average Equipment Lift

Control device	Labor Requirements (h/shift)		Average equipment life (yr)
	Operator labor	Maintenance labor	
Electrostatic precipitator	0.5–2	0.5–1	20
Fabric filter	2–4	1–2	20
Venturi scrubber	2–8	1–2	10
Incinerator	0.5	0.5	10
Adsorber	0.5	0.5	10
Absorber	0.5	0.5	10
Condenser	0.5	0.5	10

Source: ref. 26.

where i is the interest rate on borrowed capital (decimal) (use 10% unless otherwise specified) and, n is the control device life, (yr).

(b)   Calculation of annualized operator labor (OL) (Line 9, Table 24)

$$OL \text{ (hr)} = (HRS)(\text{operator hours per shift}) / (\text{operating hours per shift}) \qquad (40)$$

(*Note*: Obtain operator hours per shift value from Table 25.)

(c)   Calculation of annualized maintenance labor (ML) (Line 11, Table 24)

$$ML \text{ (hr)} = (HRS)(\text{maintenance hours per shift}) / (\text{operating hours per shift}) \qquad (41)$$

(*Note*: Obtain maintenance hours per shift value from Table 25.)

8. COST ADJUSTMENTS AND CONSIDERATIONS

8.1. Calculation of Current and Future Costs

For purposes of this handbook, auxiliary equipment cost is defined to include the cost of fans, ductwork, stacks, dampers, and cyclones (if necessary), which commonly accompany control equipment. These costs must be estimated before the purchased equipment cost (PEC) can be calculated. Costs for auxiliary equipment were obtained from refs 1, 3, 9, 12, 26, 28, 30, and 34. Readers are referred to other chapters of this handbook and other references for primary, secondary, and tertiary APCD costs (12–17, 25,26,28–35,37,38).

If equipment costs must be escalated to the current year, the *Chemical Engineering* (CE) Equipment Index can be used. Monthly indices for 5 yr are provided in Table 12 (2,36). The following equation can be used for converting the past cost to the future cost, or vice versa.

$$\text{Cost}_b = \text{Cost}_a \times \left(\text{Index}_b / \text{Index}_a\right) \qquad (42)$$

where Cost_a is the cost in the month-year of a ($), Cost_b is the cost in the month-year of b ($), Index_a is the CE Fabricated Equipment Cost Index in the month-year of a and Index_b is the CE Fabricated Equipment Cost Index in the month-year of b.

It should be noted that although the CE Fabricated Equipment Cost Indices (2,36) are recommended here for Index_a and Index_b, the ENR Cost Indices (37,39) can also be adopted for updating the costs. Cost data for construction and O&M have originated from a variety of reference sources and reflect different time periods and geographic locations. Values presented in this handbook have been converted to a specific month–year (constant dollar) base except where noted.

8.2. Cost Locality Factors

In addition to adjusting to a constant dollar base, cost indexes, such as those previously described, are used to perform economic analyses, adjust to current dollars, and make cost comparisons. However, such indexes, when applied to the several components of construction or operation and maintenance costs, will only adjust the data on a national average basis.

In order to arrive at a more accurate cost figure than one that results from the use of the national average indexes alone, the locality factor can be applied to an estimated cost or cost index. The use of locality factors, which have been calculated from generally available statistics, permits the localizing of national average cost data for construction labor, construction materials, total construction cost, O&M labor costs, and power costs. The factor for labor and materials are given in Table 26 and those for power costs are given in Table 27.

8.3. Energy Conversion and Representative Heat Values

Whenever various forms of energy are interconverted, there will be some loss resulting from inefficiencies. For example, whenever electrical energy is converted to mechanical energy, some of the energy is lost as heat energy in the motor. Similarly, if an engine operating on a Carnot cycle has a source temperature of 1100°F (1560°R) and a receiver temperature of 500°F (960°R), the efficiency is only $(1.0 - 960/1560)$ or 38.5%. Because no heat engine can be more efficient than a Carnot engine, it is clear that this is the maximum efficiency for these source and receiver temperatures.

The efficiency of pumps and blowers is usually in the range of 70–80% so that mechanical energy can be converted to hydraulic energy with no more than about 30% loss. Similarly, mechanical and electrical energy can be converted from one form to the other with a loss of less than 10%. On the other hand, the conversion of heat energy to mechanical energy necessitates the wasting of roughly two-thirds of the heat energy. For

Table 26
Cost Locality Factors

	Construction[a]			O&M Labor[b]
	Labor	Materials	Total	
Atlanta	0.66	1.03	0.79	0.77
Baltimore	0.90	0.95	0.92	0.79
Birmingham	0.66	1.02	0.79	0.79
Boston	1.12	0.90	1.04	0.97
Chicago	1.25	1.10	1.20	1.02
Cincinnati	1.10	1.05	1.08	0.98
Cleveland	1.19	1.01	1.13	1.05
Dallas	0.63	0.83	0.70	0.92
Denver	0.76	1.07	0.87	1.00
Detroit	1.17	0.98	1.10	1.32
Kansan City	0.97	1.25	1.07	0.88
Los Angeles	1.17	1.16	1.17	1.32
Minneapolis	0.93	1.03	0.97	1.21
New Orleans	0.88	1.05	0.94	0.66
New York	1.43	0.91	1.24	1.14
Philadelphia	1.23	1.00	1.15	1.05
Pittsburgh	1.05	0.96	1.02	0.97
St. Louis	1.29	0.99	1.18	0.83
San Francisco	1.23	0.96	1.13	1.13
Seattle	1.16	0.91	1.07	1.21
National Index Values	1.00	1.00	1.00	1.00

[a]Calculated from EPA Sewage Treatment Plant and Sewer Construction Cost Index Third Quarter 1979.
[b]US Department of Commerce Bureau of Census, City Employment in 1976, GE76 No. 2 July 1977. Based on average earnings by city of noneducation employees (40).

Table 27
Power Cost Locality Factor[a]

Region	Power cost factor
New England	1.31
Mid-Atlantic	1.18
Northeast Central	1.10
Northwest Central	0.98
South Atlantic	0.94
Southeast Central	0.98
Southwest Central	0.87
Mountain	0.79
Pacific	0.86
US average	1.00

[a]Basis: BLS, September 1979, Producers Price Index.

example, if electrical energy is converted to heat energy, 1 kWh will generate about 3413 Btu of heat. However, if heat energy is used to generate electrical energy in a modern coal fired power plant, about 10500 Btu of heat energy is needed to generate 1 kWh; this is a conversion efficiency of only 32.5%. Typical energy conversion percentage efficiencies (%) are as follows:

1. Heat to mechanical ≤ 38.5%
2. Heat to electrical ≤ 32.5%
3. Mechanical to electrical > 90%
4. Mechanical to hydraulic 70–80%
5. Electrical to mechanical > 90%
6. Electrical to heat approx 100%
7. Electrical to hydraulic 65–80%

Representative heat values of common fuels are as follows:

1. Anthracite coal = 14,200 Btu/lb coal
2. Digester gas = 600 Btu/ft^3
3. Fuel oil = 140,000 Btu/gal
4. Lignite coal = 7400 Btu/lb Coal
5. Liquefied natural gas (LNG) = 86,000 Btu/gal
6. Municipal refuse (25% moisture) = 4200 Btu/lb
7. Natural gas = 1000 Btu/ft^3
8. Propane gas = 2500 Btu/ft^3
9. Waste Paper (10% moisture) = 7600 Btu/lb
10. Wastewater Sludge = 10,000 Btu/lb dry VS

8.4. Construction Costs, O&M Costs, Replacement Costs, and Salvage Values

The construction costs incurred by the project represent single-payment costs that occur at certain times throughout the planning period. The single-payment present-worth factor (sppwf) is used to determine the present-worth cost and is determined by:

$$\text{sppwf} = \frac{1}{(1+i)^n} \tag{43}$$

where i is the interest rate and n is the number of interest periods.

The O&M cost includes both constant and variable costs. The constant O&M cost is based on the flow rate at the beginning of the planning period. The variable O&M cost represents the difference between the O&M cost at the flow rate in the final year of the planning period and the constant O&M cost identified by the flow rate at the beginning of the planning period. The uniform-series present-worth factor (uspwf) is used to convert the constant annual O&M cost to a present-worth cost:

$$\text{uspwf} = \frac{(1+i)^n - 1}{i(1+i)^n} \tag{44}$$

The facility replacement cost identifies the cost required to extend the useful life of equipment to the end of the planning period. This is computed when a capital item has a service life of less than the remaining years in the planning period and is computed by

$$\text{Replacement Cost} = \frac{\text{Planning Period} - \text{Remaining Service Life}}{\text{Service Life}} \times \text{Capital Value} \tag{45}$$

Capital value is the capital that would be required today to completely replace the facility. This is a single-payment cost, with present worth computed using the factor sppwf.

Finally, the salvage value represents the value remaining for all capital at the end of the planning period:

$$\text{Salvage Value} = \frac{\text{Service Life} - \text{Years to Planning End}}{\text{Service Life}} \times \text{Capital Value} \tag{46}$$

Capital value is the initial investment (or cost to replace today). This is a negative cost, with the present-worth value computed using the factor sppwf.

9. PRACTICE EXAMPLES

Example 1

Assume an emission stream actual flow rate of 1000 acfm, a particle density of 30 lb/ft^3, and emission stream density of 0.07 lb/ft3, an emission stream viscosity of 1.4×10^{-5} lb/ft-s, and a critical particle size of 20 μm. Determine the cyclone inlet area for the purpose of sizing and cost estimation.

Solution

Using Eq. (2), the cyclone inlet area is

$$A_{\text{cyc}} = 3.34 \left[1000(30 - 0.07)/1.41 \times 10^{-5} \right]^{1.33} \times \left[20 \times 10^{-6} \right]^{2.67}$$
$$A_{\text{cyc}} = 2.41 \text{ ft}^2$$

Example 2

Assume an emission stream actual flow rate 1000 acfm, a n particle density of 30 lb/ft^3, an emission stream density of 0.07 lb/ft^3, an emission stream viscosity of 1.41×10^{-5} lb/ft-s, and a critical particle size of 20 μm. Determine the following:

1. The August 1988 cost of the cyclone body
2. The August 1988 cost of the rotary air lock
3. The August 1988 cost of the total cyclone system

4. The February 1990 cost of the total cyclone system
5. The April 2000 cost of the total cyclone system

Assume that the April 2000 CE Fabricated Equipment Index is 437.4 (*see* Table 12).

Solution

In Example 1, the cyclone inlet area has been calculated to be 2.41 ft^2 using Eq. (2).

1. The August 1988 cost of a cyclone is then obtained from Eq. (9) as follows:

$$P_{cyc} = 6,520(2.41)^{0.9031}$$
$$P_{cyc} = \$14,400$$

2. The August 1988 cost of a rotary air lock for this system is given by Eq. (10):

$$P_{ral} = 2,730(2.41)^{0.0965}$$
$$P_{ral} = \$2,970$$

3. The August 1988 cost of a cyclone is the sum of these two costs, or $17,400.
4. The February 1990 cost of a cyclone system is given by Eq. (42). The CE Fabricated Equipment Indexes for August 1988 and February 1990 are 376.3 and 389.0, respectively.

$$Cost_b = Cost_a(389.00 / 376.30)$$
$$Cost_b = \$2,970(389.00 / 376.30)$$
$$Cost_b = \$3,070.24$$

5. The April 2000 cost when the index = 437.40 is

$$Cost_b = \$2,970(437.40 / 376.30)$$
$$= \$3,452.24$$

Example 3

Determine the fan costs in July 1988 and in the future when the *Chemical Engineering* Fabricated equipment cost index is projected to be 650. Assume the required static pressure equals 8 in. of water with a fan diameter of 30 in.

Solution

Equation (11) can be used to obtain the fan cost as follows:

1. The July 1988 fan cost:

$$P_{fan} = 42.3(30)^{1.2}$$
$$P_{fan} = \$2,510$$

2. The future fan cost when CE Fabricated equipment cost index is 650. The July 1988 index is 363.7. Equation (42) can be used for the calculation.

$$Cost_b = \$2,510(650 / 363.7)$$
$$= \$4,485.84$$

Example 4

Determine the required FRP duct diameter assuming a duct velocity (U_{duct}) of 2000 ft/min and an actual air emission rate ($Q_{e,a}$) of 15,300 acfm.

Solution

d_{duct} is obtained using Eq. (16):

$$d_{\text{duct}} = 12\left(\frac{4}{\pi}\frac{Q_{e,a}}{U_{\text{duct}}}\right)^{0.5}$$
$$= 12\left(\frac{4}{\pi}\frac{15,300}{2000}\right)^{0.5}$$
$$= 37.4 \text{ in. or } 3.12 \text{ ft}$$

Example 5

Determine the cost of a 50-ft FRP duct ($d_{\text{duct}} = 3.12$ ft) when the CE Fabricated equipment cost index reaches 437.4 in April 2000 and when the same cost index reached 700.

Solution

1. The August 1988 cost of FRP ductwork can be calculated using Eq. (18):

$$P_{\text{FRPD}} = 24 \times 3.12 = \$74.88 / \text{ft}$$

Thus, for a 50-ft length, the August 1998 cost of ductwork is $50 \times 74.88 = \$3744$
2. The April 2000 cost when the CE equipment cost index is 437.4 is

$$\text{Cost}_b = 3,744 \times (437.4 / 376.3)$$
$$= \$4,351.9$$

3. The future cost when the cost index reaches 700 is

$$\text{Cost}_b = 3,744 \times (437.4 / 376.3)$$
$$= \$6,964.66$$

Example 6

Assume that a 50-ft duct length and 3.12-ft diameter of FRP ductwork will be required. Determine the stack size and its future cost when the CE Fabricated equipment cost index reaches 700.

Solution

Assume a 50-ft FRP stack is required. The cost of this stack is approx 150% the cost of an equal length of ductwork. From the case given in Example 5, the cost of 50 ft of FRP ductwork is $3744. The FRP stack cost in August 1988 is $1.5 \times \$3744 = \$5616$. The future FRP stack cost when the cost index reaches 700 will be $5616 (700/376.3) = \$10,447$.

Example 7

Assume that a two-way diverter valve is required for a duct of diameter 37 in. Determine the cost of the valve for the following conditions.

1. In February 1988
2. In the future when the CE Fabricated equipment cost index reaches 700

Solution

1. The cost of two-way diverter valve in February 1988 (cost index = 363.7) can be calculated using Eq (22).

$$P_{damp} = 4.84(37)^{1.50} = \$1,090$$

2. The future cost of the valve when the cost index reaches 700 will be

$$Cost_b = \$1,090 \times (700 / 363.7)$$
$$= \$2,097.88$$

Example 8

Design a flanged slot hood assuming that the air emission stream flow rate is 18,600 ft^3/min for spray painting operation in shallow booths. The centerline distance to point (X) in the emission plume is 10 ft.

Solution

1. From Table 8, the ranges of capture velocities for spray painting operation in shallow booths is 200–500 fpm. The average capture velocity of 350 fpm is chosen for design.
2. From Table 9, the W/L ratio for flanged slot is 0.2 or less. The W/L ratio of 0.2 is chosen to fit the room.
3. From Eq. (4) for the flanged slot, the following calculations are presented:

$$Q = 2.8 \, LVX$$

$$18,600 \text{ ft}^3 / \min = 2.8L \times 350 \text{ ft} / \min \times 10 \text{ ft}$$
$$L = 1.9 \text{ ft. Select 2 ft for the length of hood.}$$
$$W = 0.2 \, L = 0.4 \text{ ft. Select 0.4 ft for the width of hood.}$$

Example 9

The flue gas flow rate at actual conditions when exiting a heat exchanger is approx 40,000 acfm. Assume that there is no other available specific data for the stack. Calculate the stack diameter. Use the cost curve in Fig. 6 to estimate the stack cost. The CE Fabricated equipment index can be found in Table 12.

Solution

The actual gas flow rate exiting the heat exchanger (flue gas flow rate) is 40,000 acfm. Therefore, using Eq. (20), the stack diameter is

$$D_{stack} = 0.2140(40,000)^{1/2} = 43 \text{ in.}$$

With the stack diameter known, use the appropriate curve in Fig. 6 (use the closest curve: 42 in.) to estimate the stack cost as follows:

$4,500 \times (347/226.2) = \$6,903$ (Note: 12/77 dollars escalated to reflect 6/85 dollars)

$4,500 \times (437.4/226.2) = \$8,701$ (Note: 12/77 dollars escalated to reflect 4/00 dollars)

Example 10

A fan is carrying an airflow through an incinerator, a heat exchanger, ductwork, and stack. The flow rate exiting the heat exchanger is approx 40,000 acfm. The fan price curve (*see* Fig. 3) and the pressure drop information (*see* Table 5) are available. Consult Table 12 and CE Fabricated equipment cost indexes. Determine the fan cost and the motor cost.

Solution

For this example case, the fan and motor costs (*see* Fig. 3) are included in the thermal incinerator cost curve; however, these costs can be calculated separately. The total pressure drop across the control system is 7.3 in. H_2O (obtained from summing the values from Table 5 for the incinerator, heat exchanger, ductwork, and stack). The flow rate exiting the heat exchanger ($Q_{fg,a}$) is 40,000 acfm. The pressure drop from the guidelines on Fig. 3 indicates that a class II fan (the lower class fan) is appropriate. The estimated fan and motor costs are as follows:

1. Fan cost

$$\$5,000 \times (347.0 / 226.2) = \$7,670$$

(*Note*: 12/77 dollars escalated to reflect 6/85 dollars.)
or

$$\$5,000 \times (437.4 / 226.2) = \$9,668$$

(*Note*: 12/77 dollars escalated to reflect 4/00 dollars.)

2. Motor cost=Fan cost×0.15
   =$7,670×0.15=$1,150.5 in June 1985
   =$9,668×0.15=$1,450.2 in April 2000

Example 11

The emission stream flow rate at actual conditions is approx 16,500 acfm. The ductwork is assumed to be 100 ft in length and made of 3/16-in.-thick plate. The emission stream contains no chlorine or sulfur compounds (noncorrosive) and has a gas temperature of 960°F. The carbon steel straight-duct fabrication price (*see* Fig. 4) is available. Determine the duct diameter and the April 2000 duct price. Use the CE Fabricated equipment index (*see* Table 12).

Solution

In this example case, as no specific data on the ductwork are available, use the above assumptions to estimate the cost of the ductwork. The duct diameter is estimated using Eq. (16):

$$D_{duct} = 0.3028 \times (16,500)^{1/2} = 39 \text{ in.}$$

As the emission stream contains no chlorine or sulfur and the gas temperature is 960°F, carbon steel ductwork is used. The cost of the ductwork is estimated using Fig. 4:

$$(\$52 / \text{ft}) \times 100 \text{ ft} \times (437.4 / 226.2) = \$10,055$$

(*Note*: 12/77 dollars escalated to reflect 4/00 dollars.)

Example 12

The example thermal incinerator system case consists of an incinerator with a combustion chamber volume (V_c) of approx 860 ft^3 and a primary heat exchanger with a surface area (A) of 4200 ft^2. Figures 8 and 9 are available. Consult Table 12. Determine the prices of incinerator and its heat exchanger.

Solution:

From the cost data presented in Figs. 8 and 9, June 1985 and April 2000 cost estimates are obtained as follows:

1. Incinerator plus instrumentation and control costs

$$\$98,000 \times (347.0 / 226.2) = \$150,336$$

(*Note*: 12/77 dollars escalated to reflect 6/85 dollars.)
or

$$\$98,000 \times (437.4 / 226.2) = \$189,501$$

(*Note*: 12/77 dollars escalated to reflect 4/00 dollars.)

2. Heat-exchanger cost

$$\$85,000 \ (347.0/273.7) = \$107,764$$

(*Note*: 12/79 dollars escalated to reflect 6/85 dollars.)
or

$$\$85,000 (437.4 / 273.7) = \$135,839$$

(*Note*: 12/79 dollars escalated to reflect 4/00 dollars.)

Example 13

Recommend a few references from which an environmental engineer may purchase fans, carbon adsorbers, fabric filters, ducts, stacks, and cyclones.

Solution

Many manufacturers and suppliers of the fans, carbon adsorbers, fabric filters, ducts, stacks, and cyclones can be found from the literature (41–43).

1. *Pollution Engineering*, 2000–2001 Buyer's Guide, Vol. 32, No. 12, November 2000.
2. *Environmental Protection*, 2003 Buyer's Guide, Vol. 14, No. 2, March 2003.
3. *Water Engineering and Management*, 2003 Annual Buyer's Guide, Vol. 149, No. 12, December 2002.

Example 14

Recommend a few reference sources from which an environmental engineer may purchase heat exchangers, air preheaters, motors, coolers, packings, condensers, solvents, surfactants, gas membrane filters, catalytic incinerators, and catalytic products.

Solution

The following are four excellent reference sources (44–48):

1. *Chemical Engineering*, Buyer's Guide 2001, Vol. 107, No. 9, August 2000.
2. *Environmental Technology*, 2000 Resource Guide, Vol. 9, No. 6, July 2000.
3. *Environmental Protection*, 2002 Executive Forecast, Vol. 13, No. 1, January 2002.
4. *Air pollution Control Engineering*, Humana Press, Totowa, NJ, 2004.

NOMENCLATURE

A	Area (ft^2)
A_{con}	Condenser system area (ft^2)
A_{cyc}	Cyclone inlet area (ft^2)
A_p	Collection plate area (ft^2)
A_{tc}	Total cloth area of bag filter (ft^2)
a, b	Coefficients in Table 11
a_d	0.877 ($6 < d_{duct} < 12$ in.) or 0.0745 ($14 < d_{duct} < 24$ in.)

a_f, b_f	Coefficients in Table 10
B	Dust outlet diameter
BER	Baghouse electricity requirement (kWh)
b_d	1.05 ($6 \leq d_{duct} \leq 12$ in.) or 1.98 ($14 \leq d_{duct} \leq 24$ in.)
C_{rep}	Replacement carbon (lb)
Cost_a	Cost in the month-year of a ($)
Cost_b	Cost in the month-year of b ($)
CRF	Capital recovery factor
CRF_w	Weighted average capital recovery factor
D	Distance between hood and source (ft)
d_{duct}	Duct diameter (in.)
d_{fan}	Fan diameter (in.)
D_{column}	Absorber column diameter (ft)
D_G	Density of emission stream (lb/ft^3)
D_o	Gas outlet diameter
D_{stack}	Stack diameter
EER	ESP electricity requirement (kWh)
ESP	Electrostatic precipitator
FE	CE (*Chemical Engineering*) Fabricated equipment cost index
FER	Fan electricity requirement (kWh)
H	Height (ft)
hp	Motor horsepower ($1 < \text{hp} < 7.5$)
HRS	Annual operating hours
H_{stack}	Stack height (ft)
ΔH	Pressure drop expressed as number of inlet velocity heads
i	Interest rate
Index_a	CE Fabricated Equipment Cost Index in the month-year of a
Index_b	The CE Fabricated Equipment Cost Index in the month-year of b
L	Length (ft)
L_{gal}	Absorbing fluid flow (gal/min)
ML	Annual maintenance labor (hr)
OL	Annual operator labor (hr)
n	Number of interest periods
P	Perimeter of hood (ft)
PC	Purchased equipment cost ($)
P_{cyc}	Cost of cyclone ($)
P_{damp}	Cost of damper ($)
P_{divert}	Cost of two-way diverter valve ($)
P_{fan}	Cost of fan system ($)
P_{FRPD}	Cost of FRP ductwork ($)
P_{motor}	Cost of fan motor, belt, and starter ($)
P_{PVCD}	Cost of PVC ductwork ($/ft)
P_{ral}	Cost of rotary air lock ($)
P_{stack}	Total capital cost of large stack ($)
ΔP	Pressure drop expressed as static pressure head

Q	Flow rate (ft^3/min)
Q_f	Supplementary fuel required (scfm)
$Q_{e,a}$	Actual emission stream flow rate (acfm)
$Q_{fg,a}$	Flue gas flow rate at actual conditions (acfm)
Q_{rec}	Recovered HAP (lb/hr)
sppwf	Single payment present worth factor
uspwf	Uniform series present worth factor
U_{stack}	Velocity of gas stream in stack (ft/min)
V	Velocity at point x (ft/min)
V_{cat}	Catalyst volume (ft^3)
W	Width (ft)
Wt$_{column}$	Absorber column weight (1000 lbs)
X	Centerline distance to point x in emission plume (ft)
μ	Emission stream viscosity (lb-ft/s)

REFERENCES

1. W. M. Vatavuk, *Chem. Eng.* **97**, 126–130 (1990)
2. Editor, *Chemical Engineering Equipment Indices*, McGraw-Hill, New York, 2001.
3. B. G. Liptak, (ed.), *Environmental Engineers' Handbook, Volume II: Air Pollution,* Chilton, Radnor, PA, 1974
4 ACGIH. *Industrial Ventilation*, 16th ed. American Conference of Government Industrial Hygienists, Lansing, MI, 1980.
5. E. Aul, Personal communication to Barry Walker of Radian Corporation, 1992.
6. K. Smit, Sr., *Means Site Work Cost Data*, R. J. Grant, NY, 1992.
7. NIOSH. *The Industrial Environment—Its Evaluation and Control*, Report No. HSM-99-71-45. Washington, DC, 1973.
8. US EPA, *Engineering Bulletin—Control of Air Emissions form Material Handling*, Report No. EPA/540/2-91/023. US Environmental Protection Agency, Cincinnati, OH, 1991.
9. W. Vatavuk, *Estimating Costs of Air Pollution Control*, Lewis, Chelsea, MI, 1990.
10. G. A. Vogel, *J. Air Pollution Control Assoc.* **35**, 558–566, (1985).
11. Sprung Instant Structures Inc., Personal communication from Grant Cleverley to Barry Walker of Radian Corporation. Sprung, Allentown, PA, 1992.
12. US EPA, *Method for Estimating Fugitive Particulate Emissions from Hazardous Waste Sites*, Report No. EPA/600/2-87/066 (NTIS PB87-232203), US Environmental Protection Agency, Cincinnati, OH, 1987.
13. US EPA, *Evaluation of the Effectiveness of Chemical Dust Suppressants on Unpaved Roads*, Report No. EPA/600/2-871/112. US Environmental Protection Agency, Research Triangle Park, NC, 1987.
14. US EPA, *Control of Open Fugitive Dust Sources*,Report No. EPA-450/3-88-008, US Environmental Protection Agency, Research Triangle Park, NC, 1988.
15. US EPA, *Dust and Vapor Suppression Technologies for Use During the Excavation of Contaminated Soils, Sludges, or Sediments. Land Disposal, Remediation Action, Incineration, and Treatment of Hazardous Waste*, Report No. EPA/600/9-88/021, US Environmental Protection Agency, Cincinnati, OH, 1988.
16. L. K. Wang, *Water and Air Quality Control with a Coil Type Filter*, PB80-195-662, US Department of Commerce, National Technical Information Service, Springfield, VA,1980.
17. J. Boswell, *Chem. Eng. Prog.*, **98**, 48–53, (2002).

18. L. K. Wang, *An Investigation of Asbestos Content in Air for Eagleton School*, PB86-194-172/AS, US Department of Commerce, National Technical Information Service, Springfield, VA, 1984.
19. S. Turner, *Environ. Protect.* Vol. **12**, 68–73,2001.
20. B. Martha, D. Day, and B. Wight, *Environ. Protect.* Vol. **12**, 51–62, (2001).
21. M. H. S. Wang and L. K. Wang, in *Handbook of Environmental Engineering. Vol. 1. Air and Noise Pollution Control*, Humana Press, Totowa, NJ, 1979, pp. 271–353.
22. L. K. Wang, *Standards and Guides of Air Pollution Control, Volume 1*, PB88-181094/AS, US Department of Commerce, National Technical Information Service, Springfield, VA, 1987.
23. L. K. Wang, *Standards and Guides of Air Pollution Control, Volume 2,* PB88-181102/AS, US Department of Commerce, National Technical Information Service, Springfield, VA,1987.
24. L. K. Wang and L. Kurylko, *Prevention of Airborne Legionairs' Disease by Formulation of a New Cooling Water for Use in Central Air Conditioning Systems*, PB85-215317/AS, US Department of Commerce, National Technical Information Service, Springfield, VA, 1984.
25. R. L. Pennington, *Environ. Technol.* **5**, 18–25, (1996).
26. US EPA, *Control Technologies for Hazardous Air Pollutants*, Report No. EPA/625/6-91/014, US Environmental Protection Agency, Washington, DC, 1991.
27. B. S. Forcade, *Environ. Protec.* **14**(1), 22–25, (2003).
28. US EPA, *Control of Air Emissions from Superfund Sites*, Report No. EPA/625/R-92/012, US Environmental Protection Agency, Washington, DC, (1992).
29. US EPA, *Organic Air Emission from Waste Management Facilities*, Report No. EPA/625/R-92/003, US Environmental Protection Agency, Washington, DC, 1992.
30. US EPA, *Capital and Operating Costs of Selected Air Pollution Control Systems*, Report No. EPA/450/5-80-002, US Environmental Protection Agency, Washington, DC, 1978.
31. W. M. Vatavuk, and R. B. Neveril, *Chem. Eng*, **89**, 129–132 (1982).
32. W. M. Vatavuk, and R. B. Neveril, *Chem. Eng*, **90**, 131–132 (1983).
33. C. D.Cooper, and F. C.Alley, *Air Pollution Control: A Design Approach*, Waveland , Prospect Heights, IL, 1990.
34. US EPA, *OAQPS Control Cost Manual*, Report No. EPA 453-96-001, US Environmental Protection Agency, Research Triangle Park, NC, 1996.
35. US EPA, *Organic Chemical Manufacturing. Vol. 5: Adsorption, Condensation, and Absorption Devices*, Report No. EPA 450/3-80-077, US Environmental Protection Agency, Washington, DC, 1980.
36. Editor, *Chem. Eng.* **107**, 410 (2000).
37. J. C. Wang, D. B. Aulenbach, and L. K. Wang, *Clean Production* (K. B. Misra, ed.), Springer-Verlag, Berlin, 1996, pp. 685–720.
38. Editor, *Public Works* **133**, (2002).
39. Editor, *Engineering News Record. ENR Cost Indices*, McGraw-Hill Publishing, New York, 2001.
40. US Department of Commerce, *City Employment in 1976*, Report No GE7-No. 2, US Department of Commerce, Washington, DC, 1977.
41. Editor, *Pollut. Eng*, **32**(12), (2000).
42. Editor, *Environ. Protect.* **14**(2), (2003).
43. Editor, *Water Eng. Manage.* **149**(12), (2003).
44. Editor, *Chem. Eng*, **107**(9), (2000).
45. Editor, *Environ. Technol.* **9**(6), (2000).
46. Editor, *Environ. Protect.* **13**(1), (2002).
47. US EPA, *Construction Cost Indexes*, US. Environmental Protection Agency, Washington, DC, 1979.
48. L. K. Wang, N. C. Pereira, and Y. T. Hung (eds). *Air Pollution Control Engineering*, Humana Press, Totowa, NJ, 2004.

APPENDIX:

CONVERSION FACTORS

Multiply	By	To obtain
Acres	43,560	ft^2
Atmospheres	29.92	in. mercury
Atmospheres	33.90	ft water
Atmospheres	14.70	psi
Btu	1.055	kJ
Btu	777.5	ft-lb
Btu	3.927×10^{-4}	hp-h
Btu	2.928×10^{-4}	kW-h
Btu/lb	2.326	kJ/kg
ft^3	28.32	L
ft^3	0.03704	yd^3
ft^3	7.481	gal
ft^3/s	0.6463	mgd
ft^3/s	448.8	gpm
yd^3	0.765	m^3
°F	$0.555 \times (°F-32)$	°C
ft	0.3048	m
gal	3.785	L
gal, water	8.345	lb, water
gpd/ft^2	0.04074	m^3/m^2-d
gpm	0.06308	L/s
gpm/ft^2	0.06790	L/m^2-s
hp	0.7457	kW
hp	42.44	Btu/min
hp	33.00	ft-lb/min
hp-h	2.685	MJ
in	25.4	mm
lb (mass)	0.4536	kg
mil gal	3,785	m^3
mgd	3,785	m^3/d
ppm (by weight)	1.000	mg/L
psi	6.895	kN/m^2
sq ft	0.0929	m^2
tons (short)	907.2	kg

Note: Energy conversion in practice should take into account the efficiencies of using heat energy to produce an electrical power of 1 kW-hr, the Btu required is $1/(2.928 \times 10^{-4})(0.325) = 10,508$, but not $1/(2.928 \times 10^{-4}) = 3415$, which does not include the actual heat to electrical energy conversion efficiency.

12
Noise Pollution

James P. Chambers

Contents

1. INTRODUCTION

Noise is playing an ever-increasing role in our lives and seems a regrettable but ultimately avoidable corollary of current technology. The trend toward the use of more automated equipment, sports and pleasure craft, high-wattage stereo, larger construction machinery, and the increasing numbers of ground vehicles and aircraft has created a gradual acceptance of noise as a natural byproduct of progress. Indeed, prior to 1972 the only major federal activity in noise control legislation was a 1968 amendment to the Federal Aviation Act, whereby the FAA was directed to regulate civil aircraft noise during landings and takeoffs, including sonic booms.

Nevertheless, various noise-monitoring studies and sociological surveys in recent years have indicated the need for noise abatement. Noise pollution is thus another environmental pollutant to be formally recognized as a genuine threat to human health and the quality of life. The fundamental insight we have gained is that noise may be considered a contaminant of the atmosphere just as definitely as a particulate or a gaseous contaminant. There is evidence that, at a minimum, noise can impair efficiency, adversely affect health, and increase accident rates. At sufficiently high levels, noise can damage hearing immediately, and even at lower levels, there may be a progressive impairment of hearing.

This chapter is descriptive. It deals with the sources, characteristics, and effects of noise, describes methods for the measurement and analysis of noise, and lists some of the guidelines that are used to control the problem.

From: *Handbook of Environmental Engineering, Volume 2: Advanced Air and Noise Pollution Control*
Edited by: L. K. Wang, N. C. Pereira, and Y.-T. Hung © The Humana Press, Inc., Totowa, NJ

2. CHARACTERISTICS OF NOISE

For all practical purposes, noise may be defined as unwanted sound; therefore, noise characteristics are essentially sound characteristics. Sound waves propagate through an elastic medium at a speed intrinsic to that material. In a gaseous medium such as air, sound waves produce significant changes in the density of the air, which, in turn, produce pressure changes. The parameter lending itself to quantification is *sound pressure*, the incremental variation in pressure above and below atmospheric pressure. In engineering terms, the acoustic pressure can be viewed as the gage pressure.

The standard US atmosphere has as an ambient pressure 101,300 N/m² (pascals, Pa). The human ear can detect sound pressures ranging from as low as 2×10^{-5} N/m², the threshold of hearing, to over 200 N/m², the threshold of pain. This wide range has prompted the use of a logarithmic scale to express sound pressures. The decibel (dB) is a dimensionless unit used to express the *sound pressure level* (SPL or L_p); the term "level" is used to emphasize the fact that a logarithm of a ratio is being expressed. More specifically, the sound pressure level is defined as

$$\text{SPL} = 10 \, \log\left(p^2 / p_{\text{ref}}^2\right)$$
$$= 20 \, \log\left(p / p_{\text{ref}}\right) \text{ decibels (dB)} \qquad (1)$$

where p is the measured root-mean-square sound pressure (N/m²) and p_{ref} is the reference sound pressure, 2×10^{-5} N/m². It is useful to note that the reference pressure is the threshold of hearing such that 0 dB corresponds to the limit of hearing. Noise measurements are, quite simply, sound measurements, and the term "noise level" is often the word used synonymously with sound pressure level.

Comfort requires that the sound level, from all sources, should be of the order of 65 dB or less (i.e., sound with a root-mean-square pressure of 3.56×10^{-2} N/m²). Some typical noise levels are as follows (1):

100–110	Jet fly-by at 300 m (1000 ft)
90–100	Power mower
80–90	Heavy truck 64 km/h (40 mph) at 15 m (50 ft), food blender (at receiver), motorcycle at 15 m (50 ft)
60–70	Vacuum cleaner (at receiver), air conditioner at 6 m (20 ft)
40–50	Quiet residential–daytime
20–30	Wilderness

Noise levels in general have increased over the years and some authorities hold that average noise levels in cities have increased at about 1 dB per year for the last 30 yr.

The sound pressure level represents the magnitude of a noise source and is one of the characteristics that can assess whether a given noise is considered to be annoying. There are other characteristics, both intrinsic to the noise and its context, that dictate whether people will consider it to be annoying (2):

1. Frequency content or bandwidth
2. Duration
3. Presence of pure tones or transients
4. Intermittency
5. Time of day
6. Location (or activity)

The above factors introduce much subjectivity into noise pollution characterization, and various rating schemes have been devised by psychoacousticians and researchers that are meant to correlate with the annoyance-related characteristics of a noise signal. More will be said about this in Section 6.

3. STANDARDS

The Noise Control Act of 1972 became Public Law PL 92574 in October of that year. Under the Act, the Environmental Pollution Agency (EPA) had to develop criteria identifying the effects of noise on public health and welfare in all possible noise environments and to specify the noise reduction necessary for protection with an adequate margin of safety. The EPA's basic "Identification of Levels" document (3) was published in March 1974 and it concluded that virtually all of the population is protected against lifetime hearing loss when annual exposure to noise, averaged on a 24-h daily level, is less than or equal to 70 A-weighted decibels (dBA) (See Section 6 for discussion on A-weighted decibels.) This noise-level goal forms the initial base of the long-range federal program designed to prevent the occurrence of noise levels associated with the adverse effect on public health and welfare. Even so, noise levels in excess of 55 dBA can cause annoyance. The federal government's regulatory development and related activity is aimed at the annoyance-type noises that pervade the community. These noises in the approximate order of importance, especially to urban communities, are (1) surface transportation noise, (2) aircraft noise, (3) construction equipment and industrial noise, and (4) residential noise.

Although states and municipalities retain primary responsibility for noise control, they often rely on EPA recommended limits of noise levels and exposures. Presently, industry is governed by noise regulations adopted by OSHA (Occupational Safety and Health Administration), which sets noise exposure limits at an employee's location for environments of steady noise, mixed noise, and impact noise. For steady noise (i.e., noise at a constant dBA level over a period of time), a maximum exposure of 90 dBA (about the sound level emitted from a loud engine) for an 8-h day is prescribed, with a halving of exposure time for each additional 5-dBA increment.

Table 1 presents permitted exposure times for various noise levels. For mixed or varying-level noise, the exposure may not exceed a daily noise dose (D_t) of unity, as expressed in Eq. (2):

Table 1
Permissible Steady Level Noise Exposure

Sound level (dBA)	Time permitted (h-min)	Sound level (dBA)	Time permitted (h-min)
85	16–0	102	1–31
89	9–11	104	1–9
90	8–0	106	0–52
92	6–4	108	0–40
94	4–36	110	0–30
96	3–29	112	0–23
98	2–50	114	0–17
100	2–0	115	0–15

Source: ref. 4.

$$D_t = D_1/T_1 + D_2/T_2 + D_3/T_3 + \ldots + D_n/T_n < 1 \qquad (2)$$

where D_n is the actual duration of exposure at noise level n and T_n is the noise exposure limit for noise level n from Table 1.

Impact noises are generated by machines such as drop hammers and punch presses and exposure to such noises must not exceed a 140-dB peak sound pressure level. The peak sound pressure level also determines the maximum number of impacts per day that an employee may be exposed to, as indicated in Table 2.

Table 2
Permissible Impact Noise Exposure

Peak SPL (dBA)	Impacts/day
140	100
130	1,000
120	10,000
110	100,000

Source: ref. 4.

If an employee is exposed to both steady noise and impulsive noise throughout the day, the combined effect can be handled quite simply. For predictave purposes one needs to treat the ratio of the number of impacts N_n at a given peak sound pressure level to the maximum number of impacts allowed at that level and add this fraction to the steady-level calculation. The combined fractions from all sources should not exceed unity. Furthermore, a hearing conservation program must be implemented that will include, at least, an annual audiometric test for employees exposed to noise levels greater than 85 dBA for 8 h or whose noise dosage D_t meets or exceeds 0.5. Such a plan protects the workers by monitoring potential deterioration of their hearing and protects employers from unwarranted claims of damaged hearing prior to employment. For measurement purposes, the use of a sound level meter with A weighting filter and slow time response (Section 6) functionally incorporates both steady and impulsive noise.

Example

A group of factory workers are subjected to the following sound levels daily.

Location	Noise	Time
Tool crib	85 dBA	8–11 AM
Press room	92 dBA	11–12 noon
HVAC room	85 dBA	1–4 PM
Turbine room	94 dBA	4–5 PM

In addition, they are exposed to 12 impulsive events from various sources that have a sound level of 130 dB (peak). The workers are off-site for lunch from 12 noon until 1 PM. Determine if it is permissible (i.e., safe, from an acoustic/ hearing standpoint) for the workers to work in this environment.

From Eq. (2), the noise dosage from all events is $D = 3/16 + 1/6.06 + 3/16 + 1/4.6 + 12/1000$ = 0.772. Therefore, the workers can work in this environment, but they should have their hearing checked periodically because their exposure is greater than 0.5.

4. SOURCES

In trying to identify the various sources of noise, one immediately thinks of the din that characterizes modern cities. In fact, a major emphasis has been placed on community sound studies in urban areas (5–10). This owes to the demonstrable fact that urban areas are generally noisier than rural areas, and because larger numbers of people live in urban areas, where they are presumably affected by the noise, the benefits may be expected to be proportionally larger. Urban noise levels are a complex mixture of noise from transportation, factories, industries, machines, and people. Basically, noise sources can be grouped into three types: transportation, industrial, and residential.

Transportation sources of noise are comprised principally of automotive and aircraft noises; motorcycles, scooters, and snowmobiles should also be considered. A main contributor to transportation noise is automotive traffic. At speeds in excess of 60 miles/h (mph), tire noises are most discernible, whereas at lower speeds, engine noises tend to dominate. The road gradient can also have an effect on vehicular noise emission; for example, a 5% road gradient adds about 3 dBA to truck noise, whereas the effect on cars is usually insignificant. Noise levels increase as the number of vehicles and average speed increases. Aircraft noises have been the source of nuisance complaints from the public for a long time. Here again, various factors, such as the amount of aircraft activity, flight paths, takeoff, and approach and landing procedures, determine the amount of noise contributed to the total level. For example, the reduction in community noise from a plane at an altitude of 3000 ft as opposed to 1500 ft (prior to entering its glide slope) can be as much as 9 dBA.

Some industrial operations and equipment are significant noise sources. Principal examples are machinery or machine tools, pneumatic equipment, high-speed rotating or stamping operations, and duct, fan, and blower systems. Typical noise levels for operating personnel may be quite high. Noise levels of 105–115 dBA are encountered in grinding polycarbonates and other tough plastics; industrial wood saws emit noise levels of 100–105 dBA depending on the type of wood being cut; noise levels of 100–110 dBA are common with lathe operations; and structure-borne noises from gear housings can vary between 92 and 105 dBA. In some cases, the personnel exposure time is small, perhaps 10 min for a quick equipment check. In other cases, a full 8-h day may be spent in the vicinity of the noise. Community exposure to such noises would, of course, depend on the proximity to the noise sources, and ambient noise levels in residential areas could be affected by more than 10 dBA.

Residential sources, both indoor and outdoor, may not seem so significant at first. However, when one considers air conditioners, lawn mowers, power saws, dishwashers, kitchen and laundry appliances, television, stereos, pets, and children, the overall severity of these sources cannot be ignored. Furthermore, the simple increase in the numbers of tools, cars, gadgets, and appliances used by modern industrial societies can create a substantial noise burden.

5. EFFECTS

Sound is of great value to mankind. It warns of danger and appropriately arouses and activates all of us. It allows us the advantages of music and speech. It can calm or excite us; it can elicit our joy or sorrow. However, irrelevant or excessive sound becomes noise and is undesirable. People react to noise through its effect on the nervous system, and at this point, a certain amount of subjectivity and value judgment enters our considerations; for example, not all people react to noise in the same way. A lawn mower and motorcycle may emit an equivalent sound level, but a certain portion of the population may find one to be inoffensive and the other to be annoying. At the high and low ends of the noise-level scale, the effects on humans are obvious; for example, at 30 dBA, noise is not an annoyance, whereas at 120 dBA, it is definitely annoying to the point of producing physical discomfort in all hearers. It is at the in-between values of noise level that humans show varying susceptibility to it.

Effects of noise include physiological and annoyance types. In the former category, there is evidence indicating that exposure to noise of sufficient intensity and duration can permanently damage the inner ear, with resulting permanent hearing loss. Loss of sleep from noise can increase tension and irritability; even during sleep, noise can lessen or diminish the relaxation that the body derives from sleep. In the annoyance category, noise can interfere with speech communication and the perception of other auditory signals; the performance of complicated tasks can be affected by noise. Noise can adversely affect mood, disturb relaxation, and reduce the opportunity for privacy (2). In all of the above ways, noise can detract from the enjoyment of out environment and can affect the quality of human life.

6. MEASUREMENT

An effective noise abatement program is difficult to establish without an adequate survey and assessment of the noise problem. However, attempting to quantify ambient noise levels can be a tedious and frustrating undertaking. Unlike air and water pollution measurements, noise measurements must include subjective as well as objective factors; that is, a straightforward physical measurement of noise magnitude must be augmented with subjective loudness and annoyance-related factors. This complication has given rise to a multitude of units, rating scales, and measurement schemes (10). Nevertheless, there are some basic elements that must be considered with regard to the magnitude of noise and its frequency and temporal distribution. These elements will be considered in the following paragraphs along with some of the more prominent noise measurement parameters.

Noise levels are commonly measured by a hand-held instrument called a sound-level meter that gives either a single-number evaluation of the time-varying pressure in decibels or a spectral breakdown of the signal. The most vital part of a sound meter is the microphone, and an important measure of microphone performance for noise surveys is its directional response to sound. When noise comes from many different directions (owing to multiple sources and reflections from walls, ground, etc.), the measuring microphone must respond identically to the various noises regardless of the angle of incidence.

The sound pressure level is a purely objective quantification of noise based on the measured physical property, sound pressure. The effect of noise on humans, however, depends not only on its magnitude but also on its frequency content because the ear is not equally sensitive to noise (and its loudness) at all frequencies in the audible range of 20–20,000 Hz. Attempts to characterize the frequency response of the human ear by subjective methods have given rise to psychoacoustic data, which, in turn, have been used to develop frequency correction factors. Thus, a frequency-weighting system was derived according to which some frequencies were emphasized more than others. This system yields a single-number rating of the noise, representing noise levels in a manner similar to the subjective impression of the human ear. This particular weighting system is designated scale "A" and readings using this system are expressed as A-level decibels or dBA. Sound-level meters are available that allow the sound to pass through an electronic A-weighting network, thus yielding a single number that approximates the response of a human ear to the sound. The A-scale places less emphasis on low-frequency sound (below 500 Hz), and provides more weight to annoying middle- and high-frequency sounds (500–4000 Hz). In practice, regulations are set limiting the maximum permissible level of A-weighted sound that may be emitted from a source. An alternate weighting scale, C-weighting, was developed to incorporate the human response to loud and typically lower-frequency sound sources such as explosions. Because the use of dBC is typically in niche applications, the use of the more utilized dBA will be considered here.

An A-weighted sound-level measurement is the least complex noise evaluation system. It is adequate, perhaps, to quantify human response to a noise, but it does not give any information on how various frequency components contribute to a particular noise dBA level. This type of frequency information is most useful when designing a noise control system. Because absorption materials and other noise control products exhibit different noise attenuation characteristics at different frequencies, choosing the proper materials and devices must be based on a frequency analysis of the noise source. For such an analysis, an instrument called the octave band analyzer is most commonly used. As its name implies, this instrument separates the noise frequency spectrum into contiguous frequency bands one octave in width and it measures the sound pressure level in each of the bands. Some modern sound-level meters incorporate octave band measurements.

An octave is the interval between two sounds having a basic frequency ratio of 2; that is, the upper cutoff frequency is twice the lower cutoff frequency and the center frequencies are progressively doubled for each octave. In noise studies, the center frequencies are 31.5, 63, 125, 250, 500, 1000, 2000, 4000, and 8000 Hz. The center frequency of each octave band is the geometric mean, or the square root, of the lower and upper cutoff frequencies; that is, $f_0 = \sqrt{f_1 f_2}$, where f_0 is the center frequency in Hertz, f_1 is the lower cutoff frequency in Hertz, and f_2 is the upper cutoff frequency in Hertz. Table 3 illustrates the center, lower, and upper frequencies for each of the octave bands in the range of human hearing.

The sound pressure level versus frequency information provided by the octave band analyzer usually enables one to identify the dominant noise bands and thereby select the proper control materials. There are, however, certain noise control cases (e. g., sound reduction of machinery noise) in which narrower-band analyzers become necessary. In such instances, so-called narrow-band (or spectrum) analyzers are used. Half-octave analyzers have an upper cutoff frequency of $\sqrt{2}$ times the lower cutoff frequency; third-

Table 3
Octave Band Lower–Center–Upper Frequencies

Lower cutoff	Center frequency	Upper cutoff
22	31.5	44
44	63	88
88	125	176
176	250	352
352	500	706
706	1,000	1,414
1,414	2,000	2,828
2,828	4,000	5,656
5,656	8,000	11,312

Source: ref. 4.

octave analyzers have an upper cutoff frequency equal to the cube root of 2, or 1.26, times the lower cutoff frequency; tenth-octave analyzers have an upper cutoff frequency equal to the tenth root of 2, or 1.07, times the lower cutoff frequency.

The definition of center frequency still applies to narrow-band analysis. Consequently a table of center frequencies and frequency ranges (as in Table 3) may be constructed for any of the above fractional octave analyzers. For example, the lowest band of the one-third octave analyzer covers the range from 22 to 28 (22 × 1.26) Hz and the center frequency is 25 Hz. The next band covers the range from 28 Hz to 35 (28 × 1.26) Hz and the center frequency is 31.5 Hz, and so on into higher-frequency bands.

In addition to magnitude and frequency, noise can also have a temporal or time-varying character. This additional dimension of time establishes the need for supplementary equipment to record temporal variations in sound pressure levels. A temporal parameter of great value in determining noise control in indoor spaces is the reverberation time. The reverberation time (RT) of a space is defined as the time required for the sound pressure level to decay 60 dB. The usual equipment required to measure the RT for noise control purposes consists of an impulsive sound source, a sound-level meter, and a recording device. The RT is calculated from the sound decay curve based on the measurement of slope.

Temporal distribution is particularly useful for determining and expressing noise exposure in urban areas, where the noise levels fluctuate considerably in the course of a 24-h day. One way of evaluating temporal characteristics of noise is by expressing noise levels (L) represented by L_x, where x is the maximum percent of the time that a specified dBA level may be exceeded. Thus, L_1 may be read as a noise level that is exceeded only 1% of the time—a very high noise level indeed; on the other hand, L_{95} may be regarded as background noise that is exceeded 95% of the time. L_{50} corresponds to a temporal median noise level.

L_x for a particular community may be determined either by a sufficiently advanced sound-level meter or direct acquisition and manipulation with a digital computer. The results are shown as a curve whose ordinate is the percent of time a sound level h is exceeded and the abscissa is the sound level (in dBA).

Another way of expressing the temporal behavior of community noise is by the equivalent sound level (L_{eq}) as shown in Eq. (3). This is a single-number noise descriptor whose mathematical definition for a time interval t_1 to t_2 is

$$L_{eq} = 10 \log\left[\frac{1}{t_2 - t_1} \int_{t_1}^{t_2} \frac{p^2(t)}{p_{ref}^2(t)}\, dt\right] \qquad (3)$$

where $p(t)$ is the time-varying A-weighted sound pressure (in N/m^2) and p_{ref} is the reference root-mean-square sound pressure of 2×10^{-5} N/m^2.

L_{dn} is yet another descriptor of community noise and is called the day–night average sound level, as shown in Eq. (4). Here, data are analyzed as in the case of L_{eq} except that a 10-dBA penalty is applied to nighttime levels, with nighttime being defined as the period between 10 PM and 7 AM. The minimum sampling period for the evaluation of L_{dn} is 24 h, and the formula used is

$$L_{dn} = 10 \log\left\{\frac{1}{24}\left[15\left(10^{L_d/10}\right) + 9\left(10^{(L_n+10)/10}\right)\right]\right\} \qquad (4)$$

where L_d is L_{eq} during daytime hours (7 AM to 10 PM) and L_n is L_{eq} for nighttime hours (10 PM to 7 AM). The L_{dn} obtained from the above equation may be corrected for seasonal, background noise levels and the presence of pure tones or impulses (10).

There are several additional noise descriptors in addition to the above that tend to characterize the "noisiness" or annoyance of sound and are based on a fair amount of subjectivity. These metrics include the perceived noise level (PNL), the effective perceived noise level (EPNL), and speech articulation index (AI) (10).

Most of the measurements and parameters described above have been designed to characterize ambient noise levels and community exposure to noise. Such measurements aid in formulating legislation and standards and in devising community-related noise control programs. In contrast to community noises, there are industrial noises within factories, workshops, and so forth that must be monitored in order to determine compliance with OSHA noise regulations. Such acoustical measurements are meant to evaluate employee exposure to work-related noises and require different measuring techniques.

For steady-level noise surveys, measurements are performed with an A-weighted sound-level meter utilizing the slow response setting, and a comparison is then made with Table 1 to check for compliance or violation. When noise is not at a constant level or when an employee's tasks take him or her from one area to another of differing sound levels, the daily noise dose can be computed from Eq. (2). Thus, a sound-level meter and stopwatch could be used to provide the D_n and T_n of Eq. (2). However, such procedures may grow time-consuming and distracting to the worker. An easier way is to use a noise dosimeter. The dosimeter is a light, compact instrument that can be carried in a pocket and allows continual, unobtrusive monitoring of noise-exposure levels. The instrument constantly measures and records noise-exposure, and at the end of the work shift it will indicate the percentage of allowable exposure received by the individual. Thus, it continually and automatically computes the dosage/time formula of Eq. (2). For extremely loud impulsive sources the use of a sound-level meter with a special accessory called the peak-hold circuit can be useful.

In order that measurement accuracy is ensured, acoustical instruments such as sound-level meters and dosimeters must be calibrated regularly. Calibration is required by OSHA before and after each day of use. If measurements are continuous over a period of hours, periodic checks on calibration are recommended. These calibration checks are necessary to obtain valid data. Calibrators called pistonphones are available that allow a rapid field calibration of acoustical instruments. Also, when purchasing instruments, it is worthwhile to ensure that the instruments are amenable to field calibration. Having to return an instrument to the factory for calibration can be time-consuming and expensive.

Hearing conservation programs to monitor sound responses of employees are also part of the noise measurement program. Hearing tests are performed on employees with the aid of an audiometer. Basically, the employees listen through headphones to test tones generated at various frequencies and the employees respond to what they hear. Such tests, carried out annually, detect changes in the employees' hearing ability.

In order that noise measurements are valid for legal purposes, they and the devices that make these measurements must meet certain standards that were developed by the American National Standards Institute (ANSI). Indeed, if action against an alleged violation is contemplated, meter and recorder construction, calibration, and use must conform strictly to ANSI standards; if not, the quality and validity of the tests and data will come into question.

In the above paragraphs, we have tried to present some of the more salient features of noise measurement and instrumentation. The technical literature abounds with descriptions of various noise studies and measurement techniques, and the interested reader would do well to consult these references (5–16).

7. CONTROL

There are essentially three approaches to noise reduction and control. The first of these is to control noise at the source. If the source is sufficiently quiet, the rest of the problem is essentially solved. Source control can be achieved by careful consideration of noise control during the design of new products. Thus, adequate mufflers to control intake and exhaust noises and absorptive enclosures and design modifications to engines can result in quieter industrial operations and automobiles. Improved rib design can result in reduced tire noise. Similar considerations apply to jet aircraft. Noisy sources can be housed in enclosures whose performance depends on the type of enclosure material used. Source noise control can also be undertaken as a retrofit measure, but this may be more expensive and could result in performance compromises. Examples of retrofit noise reduction efforts are seen in the aircraft industry. Also, in many machines, there can be found metal-to-metal contacts; these may be replaced with softer material or a cushioning element can be introduced between the metal parts. A new course of action that has arisen recently is active noise/vibration control that attempts to reduce radiated sound levels by means of either injecting sound near the source to force destructive interference or modifying the radiation efficiency of the source.

One variation of the source control theme is operation oriented, in that effective noise control may be achieved by introducing alternative methods of performing an operation. One can see that noisy operations at night incur the 10-dB penalty in the LDN metric of Eq. (4), whereas the same operation performed during the day would not. This approach is being followed by major airlines in their normally scheduled flights by following certain noise-abatement takeoff and landing procedures at major airports. Similar practices may be introduced wherever applicable in industry.

When the desired amount of noise reduction cannot always be achieved by good acoustic design at the noise source, the next best solution is the modification or alteration of the noise path between the source and the receiver. Rerouting or relocating noise sources is an example of path modification and is best applied in the planning stage of highways and airports. In many plants, noisy and quieter equipment is dispersed throughout the plant and it may be found feasible to concentrate the noisier equipment in a special limited area where effective noise control procedures may then be introduced.

Another method of path modification is to interpose barriers between the source and receiver. Such a "shielding" is useful in attenuating highway noise levels imposed on nearby areas. Absorbent-lined telephone booths are good examples of barriers being used to reflect or absorb noise into enclosed spaces. More sophisticated versions of this approach are used to combat noise intrusions into buildings.

Usually, the source control step coupled with path modification should result in an adequate noise reduction whereby the individual is subjected to no more than an acceptable noise level. However, this is not always possible, as in the case of factories and workshops, where noise levels may be high in spite of adequate controls. In such cases, the third approach to noise control is that of personal protection or control at the receiver. Either the individual's exposure to noise levels must be limited to dosage levels (as in the OSHA specifications) by limiting time and dosage level, or by further protection being afforded through the wearing of devices such as ear plugs or head phones. In the case of residential or community noise control, replacement windows or doors (with better fitting seals) may be needed on older homes to reduce interior noise in houses.

One method of noise control that does not quite fall into any of the above three categories is the concept of land-use planning. The noise levels for proposed airport, highway, and building sites can often be reasonably predicted. It is therefore the responsibility of designers, planners, and builders to assess the compatibility of proposed land use with the acoustic environment (17). Establishing land-use patterns that separate the most objectionable noise sources from noise-sensitive areas (by means of acoustical zoning, noise contours, and community noise-source inventories) is an appealing solution to a substantial part of the urban and community noise problem. The widespread implementation of land-use planning depends on the availability of highly simplified tests and screening procedures (such as EPNL and L_{dn}) that enable persons with no background in acoustics to determine the potential severity of a noise problem and thereby to assess the acoustic suitability of proposed sites. The obvious advantages of land-use planning is that it prevents noise from becoming a problem in the first place. The equally obvious limitation of land-use planning is that it does nothing to improve the existing noise situation.

It should be borne in mind that in order to achieve effective and economic noise control, a complete study and analysis of the noise problems is essential. An attempt to characterize a noise problem by using a single-decibel-level reading may not always result in an effective solution. Such single-level readings, although appealing and easily

understood, are subject to error in overlooking the major importance of decibel levels
in each frequency band, particularly when there are pure tone components present (see
Section 6). Annoyance is strongly dependent on dB level in each frequency band, and
noise control techniques are equally dependent on the decible level in each frequency
band. Thus, identifying noise problems with regard to both magnitude and frequency is
an important step toward intelligent noise control (10). Noise control is unique in that the
solution of a noise problem ordinarily will not create other environmental problems, as
is often the case with air, water, and solid waste disposal methodologies. An improperly
controlled noise problem will always manifest itself as a noise problem only.

Much of noise-abatement technology is within the existing state of the art and needs
only the proper incentives to be applied. OSHA and other governmental regulations
have already and will continue to provide that incentive, and practical techniques will be
demonstrated and documented to serve as guides for future noise-abatement programs.

Chapter 13 will deal more specifically with the subject of noise control and abatement and will provide design examples for both indoor and outdoor noise abatement.

REFERENCES

1. L. E. Kinsler, A. R. Frey, A. B. Coppens, et al., (2000) *Fundamentals of Acoustics*, 4th ed.,
   Wiley, New York, 2000.
2. M. J. Crocker (ed.), *Handbook of Acoustics*, Wiley–Interscience, New York, 1998.
3. US Environmental Protection Agency, *Information on Levels of Environmental Noise
   Requisite to Protect Public Health and Welfare with Adequate Margin of Safety*, Report No.
   EPA/550/9-74-004, US Environmental Protection Agency, Washington, DC, 1974.
4. L. K. Wang and N. C. Pereira (eds.), *Handbook of Environmental Engineering Volume 1, Air
   and Noise Pollution Control*, Humana, Clifton, NJ, 1979.
5. US Department of Transportation, *A Community Noise Survey of Medford, Massachusetts*,
   DOT Report No. DOT-TSG-OSR-72-l, US Department. of Transportation, Washington, DC,
   1972.
6. M. Simpsonand and D. Bishop, *Community Noise Measurements in Los Angeles. Boston
   and Detroit*, Report No. 2078, Bolt, Beranek and Newman, Cambridge, MA, 1971.
7. US Environmental Protection Agency, *Community Noise*, EPA Report No. NTID 300.3, US
   Environmental Protection Agency, Washington, DC, 1971.
8. Massachusetts Port Authority, *The Results of Four Years of Noise Abatement at Boston's
   Logan International Airport*, Massport's Noise Abatement Office, Boston, 1981.
9. J. M. Fields, *An Updated Catalog of 318 Social Surveys of Residents' Reactions to
   Environmental Noise* (1943–1989) Report No. NASA TM-187553, National Aeronautics
   and Space Administration, Washington, DC, 1991.
10. C. E. Wilson, *Noise Control*, Harper & Row, New York, 1989.
11. F. G. Haag, *J. Sound Vibr.* **10**(12), 15–17 (1977).
12. M. M. Prince, S. J. Gilbert, R. J. Smith, et al., *J. Acoust. Soc. Am.* **113**(2), 871–880 (2003).
13. D.L. Sheadel *Proceedings of the 94th Annual Meeting of the Air and Waste Management
    Association*, 2001, paper 01–686.
14. E. A. G. Shaw, *Noise Control Eng. J.* **44**(3), 109–119 (1996).
15. C. J. Jones, and J. B. Ollerhead, *Proceedings of Euro-Noise '92*, pp. 119–127.
16. S. Fidell, K. Pearsons, L. Silvati, et al., *J. Acoust. Soc. Am.* **111**(4), 1743–1750 (2002).
17. D. L. Sheadel, *Proceedings of the 92nd Annual Meeting of the Air and Waste Management
    Association*, 1999, paper 99–574.

13
Noise Control

James P. Chambers and Paul Jensen

Contents

1. INTRODUCTION

Most people think acoustics applies only to rooms with special functions, such as concert halls or churches. Actually, any space has acoustical qualities, and if these qualities are inappropriate, the utility of the space may be compromised. Normally, noise problems are associated with sounds that people can hear. However, ultrasonic and infrasonic sounds can also produce psychological effects and, under certain conditions, definite physiological effects.

Several examples are appropriate and will serve to illustrate the range of acoustical problems often encountered. Most of us are aware that some spaces must be quiet to be useful. If intrusive noise were present in a bedroom, school study hall, or library reading room, for example, it would be difficult to use that facility as it was intended. Sometimes, even "quiet" can be inappropriate. A sports arena or nightclub would be dull places to visit if no noise were present. There is a middle ground too. Open plan office spaces need to be moderately quiet to allow for pleasant working conditions, but not so quiet that private conversation would be impossible.

From: *Handbook of Environmental Engineering, Volume 2: Advanced Air and Noise Pollution Control*
Edited by: L. K. Wang, N. C. Pereira, and Y.-T. Hung © The Humana Press, Inc., Totowa, NJ

The acoustic properties of these spaces, and indeed of any space, are determined by the geometry of the space and of the materials within it. The factors governing the behavior of sound within spaces are well understood and can be used to ensure that the spaces function well acoustically. This function falls into the professional disciplines of architectural acoustics or noise control. Because noise control rests on the factors affecting the behavior of sound, it is pertinent to discuss some of the physics involved.

2. THE PHYSICS OF SOUND

2.1. Sound

Sound is a disturbance that propagates through an elastic medium (air, water, etc.) at a speed characteristic of that medium. Noise and its control can encompass a wide range of mediums such as underwater noise from ocean traffic or unwanted vibrations in mechanical structures. For the purposes of this text, noise will refer to airborne disturbances unless otherwise noted. When a sound source in air vibrates, it causes the air to oscillate, which, in turn, produces extremely small changes in the pressure of the surrounding air. The pressure waves spread out like ripples on a pond when a stone is dropped into it, except that sound waves fill the whole volume of air, whereas ripples are confined to the surface of the pond.

2.2. Speed of Sound

In a free field, sound propagates with the velocity c defined by

$$c = 20.05\sqrt{T_K} \ (m/s) \tag{1}$$

or

$$c = 49.03\sqrt{T_R} \ (f/s)$$

where T_K and T_R are the temperature in Kelvin and Rankine, respectively (1).

A simpler formula for the velocity of sound in air sufficiently accurate at normal temperatures, 0–30°C, is

$$c = 331 + 0.6T_C \ (m/s) \tag{2}$$

where T_C is the temperature in centigrade.

Example 1

Determine the speed of sound at 20°C (68°F) in both metric and English units.

Solution:

The Kelvin temperature is $T_K = 273.2 + 20 = 293.2$ K and the Rankine temperature is $T_R = 459.7 + 68 = 527.7°$R

The speed of sound c is then

$$c = 20.05\sqrt{293.2} = 343 \ m/s$$

or

$$c = 49.03\sqrt{527.7} = 1125 \ ft/s$$

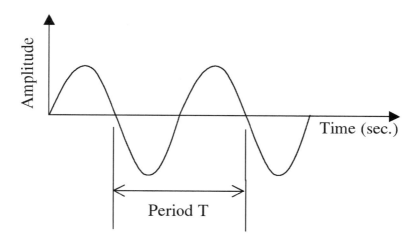

Fig. 1. Harmonic oscillation of pressure.

2.3. *Sound Pressure*

Sound waves produce changes in the density of the medium (air) as they travel through it. These changes in air density cause pressure fluctuations around the ambient static pressure. If air particles oscillate in a harmonic mode (sinusoidally varying with time), sound pressure will also change harmonically and cause a pure tone.

For a pure tone, the sound pressure p can be described as

$$p = a \, \sin(\omega t) = a \, \sin(2\pi f t) \text{ Pascals} \tag{3}$$

where a is the amplitude in Pascals, ω is the angular frequency in radians per second, t is the time in seconds, and f is the frequency in hertz.
The angular frequency is defined as

$$\omega = 2\pi f \text{ rad/s} \tag{4}$$

Figure 1 shows a pure tone oscillation, although pure tones do not often exist in nature. Even musical instruments do not produce pure tones. Instead, the sounds they emit consist of a fundamental tone and a number of harmonics. The harmonics occur at integer multiples of the fundamental frequency and they confer on an instrument its special character. A buzz saw, too, is rich in harmonics.

In most situations, the disturbances created in the air that we call sound cannot easily be expressed mathematically in time and space. This results from the fact that over an extended portion of a vibrating surface creating the airborne disturbance, some portions are compressing the surrounding air while other portions are causing rarefactions. This results in a phase difference in the pressure in either time or space. Complicating the situation further, physical boundaries cause reflections and allow the disturbances to interact with each other. Fortunately, the sound pressure in a sound field will in most cases vary in a random fashion, and statistical techniques can therefore be used to deal with the phenomenon.

The field of acoustics and noise control has nearly uniformly adopted the metric system throughout and, as such, the unit used for measuring sound pressure is the Pascal (Pa). In earlier years, the units bar, microbar, and dyne per square centimeter were used to measure sound pressure.

The following conversion factors apply:

$1 \text{ bar} = 10^5 \text{ Pa}$
$1 \text{ μbar} = 10^{-6} \text{ Pa} = 1 \text{ dyn/cm}^2$
$1 \text{ dyn/cm}^2 = 10^{-1} \text{ Pa}$
$1 \text{ Pa} = 1 \text{ N/m}^2$

The sound pressure in a sound wave can be measured with a microphone. The electric signal generated by the microphone is typically amplified and recorded onto an oscilloscope or other recording medium. A detailed picture of the sound wave can be produced by an oscilloscope; however, sufficient information can usually be obtained by a continuous display of the instantaneous value of the sound pressure as a function of time only. Consequently, it is possible to use simpler equipment such as a sound-level meter, available at electronics stores, rather than an oscilloscope to analyze noise.

2.4. Frequency

The frequency of a sound indicates the number of cycles performed in 1 s:

$$f = 1/T \text{ Hz} \qquad (5)$$

where T is the period of one full cycle. The unit for frequency is the hertz (Hz):

$1 \text{ Hz} = 1 \text{ cycle/sec} = 1 \text{ cps}$
$1000 \text{ Hz} = 1 \text{ kilohertz} = 1 \text{ kHz}$

The audible frequency range to humans, 16–20,000 Hz, has been divided into a series of octave bands and one-third (1/3) octave bands. Just as with an octave on a piano keyboard, an octave in sound analysis represents the frequency interval between a given frequency and twice that frequency. The interval is identified by the center frequency, representing the geometric mean of the bounds of that interval. The internationally agreed upon 1000-Hz center frequency determines the center frequencies of the remaining bands. The center frequencies and approximate cutoff frequencies are listed in Table 1.

2.5. Wavelength

The wavelength λ is equal to the distance the oscillations have propagated in the time period T:

$$\lambda = cT = c/f \qquad (6)$$

This shows that the wavelength is inversely proportional to the frequency. In the audio frequency range (16–20,000 Hz), the low frequencies have wavelengths of several meters (or feet), whereas the wavelengths for the high frequencies are only a few centimeters (or fractions of an inch).

Example 2

Determine the wavelength of a 125-Hz and an 8000-Hz tone at 20°C (68°F) in both metric and English units.

Table 1
Center and Approximate Frequency Limits for Octave and One-Third Octave Bands Covering the Audio Frequency Range

Octave			One-Third Octave		
Lower limit (Hz)	Center frequency (Hz)	Upper limit (Hz)	Lower limit (Hz)	Center frequency (Hz)	Upper limit (Hz)
11	16	22	14.1	16	17.8
			17.8	20	22.4
			22.4	25	28.2
22	31.5	44	28.2	31.5	35.5
			35.5	40	44.7
			44.7	50	56.2
44	63	88	56.2	63	70.8
			70.8	80	89.1
			89.1	100	112
88	125	177	112	125	141
			141	160	178
			178	200	224
177	250	355	224	250	282
			282	315	355
			355	400	447
355	500	710	447	500	562
			562	630	708
			708	800	891
710	1,000	1,420	891	1,000	1,122
			1,122	1,250	1,413
			1,413	1,600	1,778
1,420	2,000	2,840	1,778	2,000	2,239
			2,239	2,500	2,818
			2,818	3,150	3,548
2,840	4,000	5,680	3,548	4,000	4,467
			4,467	5,000	5,623
			5,623	6,300	7,079
5,680	8,000	11,360	7,079	8,000	8,913
			8,913	10,000	11,220
			11,220	12,500	14,130
11,360	16,000	22,720	14,130	16,000	17,780
			17,780	20,000	22,390

Source: Data from refs. 1 and 2.

Solution

$$125 \text{ Hz:} \quad \lambda = \frac{343}{125} = 2.74 \text{ m}$$
$$\lambda = \frac{1125}{125} = 9 \text{ ft}$$

$$8000 \text{ Hz}: \quad \lambda = \frac{343}{8000} = 0.0043 \text{ m} = 4.3 \text{ cm}$$

$$\lambda = \frac{1125}{8000} = 0.14 \text{ ft} = 1.7 \text{ in.}$$

2.6. rms Sound Pressure

Most common sounds consist of a rapid, irregular series of positive-pressure disturbances (compressions) and negative-pressure disturbances (rarefactions) measured from
the static pressure. Typically, there is no net increase in the ambient pressure as a result
of the presence of acoustic disturbances except in certain exotic problems such as sound
produced by explosions. Thus, the mean value of any series of sound pressure disturbances is not meaningful, as there are as many compressions as rarefactions and the net
acoustic pressure is zero.

The root-mean-square (rms) sound pressure yields a nonzero value to describe the pressure disturbance, and this is more meaningful. Physically, the rms value is indicative of
the energy density of the disturbance. Mathematically, the rms value is obtained by squar-
ing the sound pressures at any instant of time and then integrating over the sample time
and averaging the results. The rms value is then the square root of this time average:

$$p_{rms} = \sqrt{\lim_{T \to \infty} \frac{1}{2T} \int_{-T}^{T} p(t)^2 dt} \tag{7}$$

2.7. Sound Level Meter

The rms value of the sound pressure can be measured by a sound-level meter that can
typically display overall sound levels or octave band levels in either a linear format or
with various weighting functions (dBA, dBC, etc.). Specific details on sound-level
meters are covered by various ANSI and IEC standards (3,4).

2.8. Sound Pressure Level

The sound pressures that are normally measured with a sound-level meter cover an
extremely large range. The sound pressure of the faintest sound a human ear can hear
is equivalent to about 2×10^{-5} Pa and sound pressures that can be measured close to
jet engines are equivalent to about 2×10^2 Pa. The ratio of these two sound pressures
is 10^7. In order to handle in a simple fashion such a large measurement range, a logarithmic measurement scale is used. There are other benefits to using such a scale
because the human response to sensations (experiences, sounds, fragrances, pains, etc.)
corresponds to a logarithmic intensity scale rather than to a linear scale.

The sound pressure level then is a logarithmic ratio L_p defined as

$$L_p = 10 \log \frac{p_{rms}^2}{p_{ref}^2} = 20 \log \frac{p_{rms}}{p_{ref}} \tag{8}$$

where p_{rms} is the sound pressure of interest (in Pa) and p_{ref} is a reference sound pressure (in Pa) usually chosen as the limit of hearing of 20 μPa. The unit for the sound
pressure level, SPL or L_p, is the decibel (dB) (5).

Table 2
Relation Between Sound Pressure and Sound Pressure Level (SPL or L_p)

Sound pressure (p)	Sound pressure level (L_p)
p_{ref}	0 dB
$1.12p_{ref}$	1 dB
$1.26p_{ref}$	2 dB
$2p_{ref}$	6 dB
$3.16p_{ref}$	10 dB
$10p_{ref}$	20 dB
$100p_{ref}$	40 dB
$10,000p_{ref}$	80 dB
$1,000,000p_{ref}$	120 dB

Source: ref. 1.

The relationship between sound pressure and sound pressure level (with 20 μPa as the reference sound pressure) is shown in Table 2.

Example 3

Determine the sound pressure level for sound pressures of $p = 1$ Pa and $p = 1$ atm (1.013×10^5 Pa) (reference to 20 μPa).

Solution:

$$L_p = 20\log\frac{1}{20\times10^{-5}} = 20\log\frac{10^5}{2} = 94 \text{ dB}$$
$$L_p = 20\log\frac{1.103\times10^5}{20\times10^{-5}} = 20\log\frac{1.103\times10^{10}}{2} = 194 \text{ dB}$$

2.9 Loudness

Sound waves cause the membrane in the ear to vibrate, and these vibrations are transmitted through interconnected small bones to the inner ear, the cochlea. Thousands of hair cells in the cochlea retransmit the sound information to the brain through nerves. A young unspoiled ear can hear pure tones if they have sound pressure levels as a function of frequency as illustrated in the lowest curve in Fig. 2.

For example, as can be seen in Fig. 2, the threshold of hearing at 1000 Hz is about 4 dB. It can also be seen in Fig. 2 that the ear is most sensitive to sound in the frequency range 300–6000 Hz. The sensitivity drops at lower and higher frequencies. The family of curves in Fig. 2 (Fletcher and Munson curves) (6,7) indicate the sound pressure level a tone must have in order to be as loud as the 1000-Hz tone with a sound pressure level as indicated on the curve. A 20-Hz tone must have a sound pressure level about 70 dB higher than a tone at 1000 Hz in order for a person just to hear the tone. The curves also indicated the loudness—the subjective interpretation of the magnitude of sound—for pure tones. The units describing loudness are called phons. By definition, the phon is equal to the sound pressure level (in dB) reference to 20 μPa of an equally loud 1000-Hz tone. Pain will occur when the loudness exceeds 120 phons.

An approximate measurement of the loudness of a pure tone can be made by using a sound-level meter with a weighting network. The weighting network approximates an

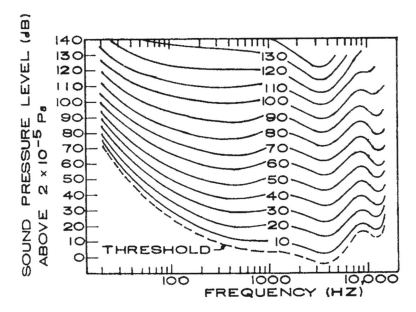

Fig. 2. Normal equal loudness contours for pure tones. (From ref. 6.)

average loudness curve. Most sound-level meters are equipped with weighting networks, A, B, and C (4). The sensitivity of these networks is shown in Fig. 3.

Initially the intent was that for loudnesses in the 30–60-phon range, the A-weighting should be used, whereas in the 60–90-phon range, the B-weighting applies, and for loudnesses above 90 phons, the C-weighting is appropriate.

Because the Fletcher and Munson curves are based on pure tones, measurement of a complicated noise spectrum by a sound-level meter often gives results that vary from the loudness measured in accordance with the definition of the phon. This occurs because the brain sums the loudness of the individual components of the spectrum dif-

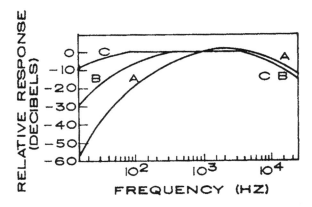

Fig. 3. Frequency response of A-, B-, and C-weighting networks. (From ref. 4.)

Table 3
Sample Sound Pressure Levels (dBA)

Sound pressure level (dBA)	Event
100–110	Jet fly at 300 m (1000 ft)
90–100	Power mower
80–90	Heavy Truck 64 km/h (40 mph) at 15 m (50 ft),food blender (at receiver), motorcycle at 15 m (50 ft)
60–70	Vacuum cleaner (at receiver), air conditioner at 6m (20 ft)
40–50	Quiet residential—daytime
20–30	Wilderness

Source: ref. 7.

ferently from the straight summation performed by the sound-level meter. The sound-level meter gives a value that approximates the rms value of the sound signal, but the function of the brain is much more complicated. For that reason, it is not strictly correct to say that measurements made with a sound-level meter indicate accurately the loudness in phons. All that the reading indicates is the sound level with the A-, B-, or C-weighting. Table 3 shows some typical sound levels of common sound sources in dBA, where dBA indicates that the A-weighting network has been used.

2.10. Sound Power Level

The sound power radiated by a sound source covers an extremely wide range. It is more convenient to express the sound power level L_w using a logarithmic scale based on an internationally selected sound power as a reference. Thus,

$$L_w = 10\log(W/W_0)\text{dB} \qquad (9)$$

where $W_0 = 10^{-12}$ watts (W) (5). For many practical problems, the sound power, W, can be expressed as

$$W = \frac{P_{rms}^2}{\rho c}\, dA \qquad (10)$$

where ρ is the ambient density, c is the speed of sound, and dA is the area around the source. One can note that as one moves away from a pointlike source, p_{rms} decays as $1/r^2$, but the sphere (or hemisphere) dA that encloses the source increases as $4\pi r^2$ so that W remains constant. It is for this reason that much noise legislation is worded for power rather than pressure so that an accurate portrait of the noise from a source can be properly specified and verified.

2.11. Sound Energy Density

The sound energy density D is the energy that arises from the sound field present in a small volume of the air in a room. The relation between the space-average mean-square sound pressure and the space-average sound energy density is

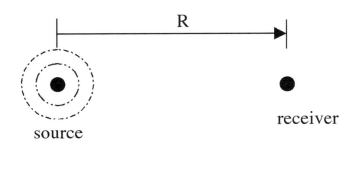

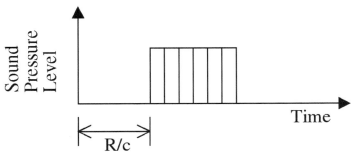

Fig. 4. Sound field out-of-doors. (From ref. 1.)

$$D = \frac{P_{\text{rms}}^2}{\rho c^2} \tag{11}$$

where ρ is the density of the air (kg/m^3).

3. INDOOR SOUND

3.1. Introduction

The acoustical environment in a room depends greatly on the size, shape, and other properties of the confining walls, floor, and ceiling. Before examining how the properties of the room affect the sound field, it is pertinent to examine a simpler situation. What happens when a sound source such as a pistol is fired out-of-doors in the absence of any nearby reflecting surfaces as is seen in Fig. 4. The sound waves will travel through the atmosphere from the pistol to the listener (which acousticians typically call the receiver), who will only hear one crack. The farther the receiver is away, the later the crack will arrive and the weaker it will be. The direct sound wave is the only sound heard.

The sound wave arrives at time t:

$$t = R/c \tag{12}$$

where R is the distance between the source and the receiver and c is the speed of sound.

If there is a large wall or building at a distance near the firing position as evidenced in Fig. 5, the receiver will hear a different sound. The first sound wave will be from the

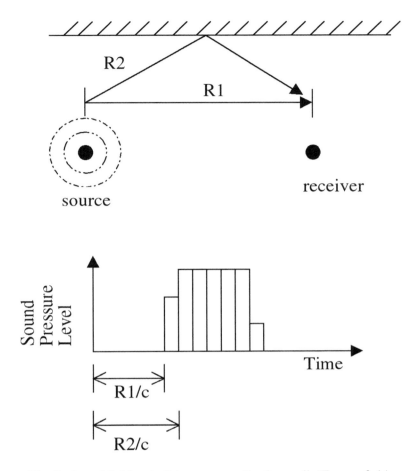

Fig. 5. Sound field out-of-doors near reflecting wall. (From ref. 1.)

direct arrival and the second sound wave from the reflected arrival off of the wall. The two sound waves, will arrive at

$$t_1 = R_1/c \tag{13}$$

and

$$t_2 = R_2/c \tag{14}$$

The human ear is able to recognize individual sound impulses if they are separated by a time period of about 50 ms. If the time difference between the direct sound and the reflected sound is greater, an echo will be heard; if the time difference is less, only one modified crack will be heard, as evidenced in Fig. 5.

Example 4

Determine the time a receiver will hear a pistol shot when the distance between the pistol and the receiver is 200 m and there is a building behind the pistol, 40 m away. Make the calculation for the two temperature conditions −10°C and 25°C.

Solution:

$-10°C$:

$$c = 20.05\sqrt{T_K} = 20.05\sqrt{263.2} = 325 \text{ m/s}$$

$$t_1 = \frac{200}{325} = 0.615 \text{ s}$$

$$t_2 = \frac{200}{325} = 0.872 \text{ s}$$

$25°C$

$$c = 20.05\sqrt{T_K} = 20.05\sqrt{298.2} = 346 \text{ m/s}$$

$$t_1 = \frac{200}{346} = 0.578 \text{ s}$$

$$t_2 = \frac{280}{346} = 0.810 \text{ s}$$

The reflected sound wave is weaker than the direct sound ways for two reasons:

1. It will have traveled farther than the direct sound wave, and the sound intensity decreases with distance.
2. Some sound energy will be lost in the process of reflection.

It is often advantageous to think of the reflected sound as coming from the mirror image of the sound source. This can be done if the reflecting surface is large compared to the wavelength of the sound and if any irregularity on the reflecting surface is small compared to the wavelength.

If the pistol is fired in an enclosed space and the receiver is inside the room, he or she will then receive a number of impulses one after the other. In a room, however, the individual impulses tend to blend together because they occur in such rapid succession. The individual impulses are difficult to separate for a second reason to. We can imagine that the reflections are caused by fictitious mirror sound sources of first, second, and higher order, with the higher orders representing successive reflections. The higher the order of the mirror image, the weaker the sound source because source energy will be lost every time the sound wave reflects.

3.2. Sound Buildup and Sound Decay

Until now, only impulsive sound has been discussed. What happens when sounds of longer duration are present? Figure 4 can again be used to examine the outdoor situation previously described. If a sound source is placed in a position where it emits sound for 6 s, a receiver in position R will receive sound for a period of 6 s. However, the greater the distance between source and receiver, the weaker the sound and the later the receiver starts hearing the sound. If there is a large wall near either the sound source or the receiver, a different picture will be seen. As shown in Fig. 5, the receiver will initially only hear the direct sound. Then a weaker reflected sound will arrive, and the sound level will increase. After 6 s, the direct sound will disappear and only the reflected sound will be heard until that also disappears.

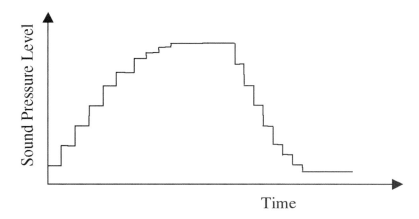

Fig. 6. Buildup and decay of sound field in a room. (From ref. 1.)

The changes in the sound field in an enclosed space, when a continuous sound source is turned on and off, are shown in Fig. 6. The situation is similar to the outdoor example, only many more reflecting surfaces are present.

In the first half of Fig. 6, every step up results from the arrival of a new reflection. Expressed another way, every step up corresponds to the contribution from one of the mirror images. The farther away the mirror images, the later the contribution arrives.

The individual steps on the curve gradually become smaller and smaller because the mirror images lie farther and farther away and the intensity of the sound wave gets reduced at each reflection. A stationary or steady-state situation represented by the horizontal portion of the curve in Fig. 6 is rapidly reached.

When the sound source is switched off, the sound level drops off, following a similar step function because the contribution from the individual mirror images gradually disappears. The time it takes to reach the stationary situation and the time it takes from the moment the sound source is switched off until the sound has disappeared depend solely on the acoustical properties of the room. Thus, the rise time or the decay time can be utilized to characterize a room acoustically. It should be noted that Fig. 6 is presented for qualitative purposes and that actual build up and decay curves may be smoother or quite spikey indeed based on the geometry and contents of the room such as furniture or people.

The equations used to determine the acoustical properties are usually developed according to statistical techniques; thus, several equations exist, each differing slightly as a result of the different statistical assumptions made. However, all equations are based on the energy balance that will always exist.

Before developing the equations determining the sound buildup and sound decay in a room, it will be necessary to calculate how much sound energy will be incident on a wall element dS in 1 s. To do so, consider any volume element dV at a distance R from the wall element dS at such a direction that the perpendicular to dS makes the angle θ with R, as seen in Fig. 7.

Because the surface area of the sphere (radius R) centered at dS is $4\pi R^2$ and because the projection of dS on the surface of the sphere is $dS\cos\theta$, then the fraction

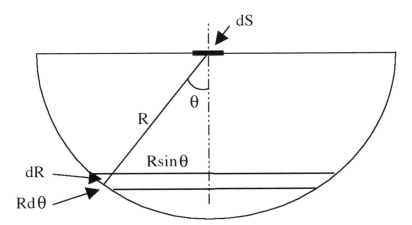

Fig. 7. Figure for use in the calculation of sound energy incident on a wall element per second. (From ref. 1.)

$(dS \cos \theta)/4\pi R^2$ of the total sound energy passing through the infinitesimal volume toward the wall $(D \, dV)$ will be incident on dS. (It is assumed that the sound energy in the volume element dV will propagate equally well in all directions.)

All volume elements dV located within a hemisphere of radius c and centered at dS will transmit sound energy to the wall element dS during a 1-s time period. Thus, τ, the sound energy incident on the wall element dS, can be obtained by integrating over the hemisphere.

For the volume element dV, we will use the rotational element at a distance R from dS at the angle θ between R and $dS's$ perpendicular. From Fig. 7, it can be seen that $dV = 2r\pi \sin\theta \, d\theta \, R \, dR$

$$
\begin{aligned}
\tau &= \int_0^c \int_0^{\pi/2} \frac{dS \cos\theta}{4\pi R^2} D 2\pi R \sin\theta \; d\theta \, dR \\
&= \int_0^c \int_0^{\pi/2} \frac{1}{2} D dS \, \cos\theta \; \sin\theta \; d\theta dR \\
&= \frac{1}{4} Dc dS
\end{aligned}
\tag{15}
$$

The absorption coefficient α_n for a wall element S_n is defined as the proportion of incident sound energy that is absorbed. This means that the expression for absorbed energy is $(1/4)Dc\alpha_n S_n$. The energy balance is

$$\text{Added energy} - \text{Absorbed energy} = \text{Change in energy}$$

Mathematically, this becomes

$$
W - \frac{1}{4} Dc \sum \alpha_n S_n = V \frac{dD}{dt}
\tag{16}
$$

where W is the emitted sound power in (watts), D is the energy density (watts-s/m^3), c is the speed of sound (m/s), V is the room volume (m^3) and α is the absorption coefficient defined as the percentage of impinging energy absorbed on any given wall element of area S.

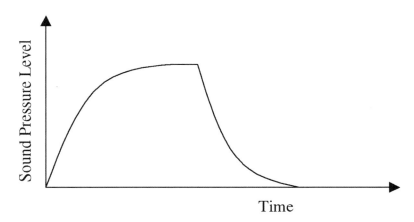

Fig. 8. Sound buildup and decay in room (exponential curve). (From ref. 1.)

Statistical techniques must be applied to the middle term of the above equation that expresses the amount of energy absorbed per unit time. If we assume that at any time the sound field is diffuse (that the energy intensity is everywhere equal), then during sound buildup and decay, the solution to Eqn. (16) is as follows:

Sound buildup:

$$D = D_0(1 - \exp(-cAt/4V)) \text{ (W-s/m}^3) \tag{17}$$

Sound decay:

$$D = D_0 \exp(-cAt/4V) \text{ (W-s/m}^3) \tag{18}$$

where D_0 represents the energy density once equilibrium is reached and A is the total absorption in the room equal to $\Sigma \alpha_n S_n$. During the stationary (equilibrium) period,

$$D_0 = 4W/cA \text{ (W-s/m}^3) \tag{19}$$

In the next section, the way in which these equations are put to use will be amplified, but before doing so, it is pertinent to make some additional comments. First, examine Fig. 8, which represents both the sound buildup and decay according to the exponential expressions just developed.

These curves approximate the physical process involved; however, they do not represent the way we perceive the process. The ears seldom hear the sound buildup but do hear the sound decay. Because the human perception is more strongly related to the sound decay, the main emphasis has been placed on that phenomenon, (i.e., sound decay). The second comment pertains to the conditions under which the diffuse sound field assumption is valid.

3.3. Diffuse Sound Field

Only large rooms, where the number of normal modes of vibration is large, can be considered to have a diffuse sound field. In the diffuse sound field, the average energy density will be the same at all points. This means that there would be no net flow of power in any direction. Because we will always have a net flow of power away from a source to the places where the energy is absorbed, we will actually never have a diffuse

Table 4
Values for Air Absorption, 4*m*.

Relative humidity(%)	Temperature (°C)	4*m* (dB/1000 m)							
		63 Hz	125 Hz	250 Hz	500 Hz	1000 Hz	2000 Hz	4000Hz	8000 Hz
25	15	0.2	0.6	1.3	2.4	5.9	19.3	66.9	198
	20	0.2	0.6	1.5	2.6	5.4	15.5	53.7	180.5
	25	0.2	0.6	1.6	3.1	5.6	13.5	43.6	153.4
	30	0.1	0.5	1.7	3.7	6.5	13.0	37.0	128.2
50	15	0.1	0.4	1.2	2.4	4.3	10.3	33.2	118.4
	20	0.1	0.4	1.2	2.8	5.0	10.0	28.1	97.4
	25	0.1	0.3	1.2	3.2	6.2	10.8	25.6	82.2
	30	0.1	0.3	1.1	3.4	7.4	12.8	25.4	72.4
75	15	0.1	0.3	1.0	2.4	4.5	8.7	23.7	81.6
	20	0.1	0.3	0.9	2.7	5.5	9.6	22.0	69.1
	25	0.1	0.2	0.9	2.8	6.5	11.5	22.4	61.5
	30	0.1	0.2	0.8	2.7	7.4	14.2	24.0	68.4

Note: The evaluation of 4m in linear terms can be found from the above values divided by 4340.
Source: ref. 9.

sound field. However, the concept of a diffuse sound field is useful in rooms with not too much absorption and where the measurement positions are not close to either highly absorption surfaces or the sound source.

3.4. Reverberation Time

The relationship expressed in the preceding discussion is incorporated into the concept of reverberation time. The reverberation time *t* in a room is defined as the time it takes the sound pressure level to decrease to a value 60 dB below its original value. The pioneer of room acoustics. W.C. Sabin (8) developed this definition and derived the following formula based on the preceding equations.

$$t = 0.163 \frac{V}{A + 4mV} \text{ metric}$$

(20)

$$t = 0.049 \frac{V}{A + 4mV} \text{ English}$$

where *V* is the room volume (m³[ft³]), *A* is the total absorption ($\Sigma \alpha_n S_n$), and *m* is an attenuation coefficient relating to the absorption. The 4*m* term is often expressed in units of dB/1000 m to highlight its relative influence on sound levels. For use in Eq. (20), the values in Table 4 must be divide by 4340 to provide 4*m* in units of 1/m.

The total absorption *A* is defined as

$$A = \Sigma \alpha_n S_n = \alpha_1 S_1 + \alpha_2 S_2 + \cdots$$

(21)

where α_1, α_2, $\cdots$ are the absorption coefficients of the different wall elements of surface area S_1, S_2, $\cdots$, respectively. *A* has the unit m²-Sabin (ft²-Sabin) if the unit for *S* is m² (ft²).

3.5. *Optimum Reverberation Time*

From an examination of the reverberation time formula, it can be seen that the total absorption in a room can be calculated from measured reverberation time. Optimum values exist for the reverberation time for various functional spaces. Thus, if a room is to be modified to change its function, the change in the amount of absorption materials in the room can be determined. Similarly, if a space is to be designed, knowledge of the total amount of absorption within the room can be used to calculate reverberation time and hence determine whether the room will be acoustically satisfactory. Changes can be made in the materials used if necessary. Table 5 presents some examples of optimum reverberation time for different functional spaces.

Table 5
Reverberation Times(s) for Some Common Spaces

Room size (m^3)	Speech or lecture hall	Opera house	Churches
100	0.65	0.85	1.15
1000	0.8	1.25	1.5
10,000	0.95	1.6	1.8
100,000	1.1	1.9	2.2

Source: ref. 1.

3.6. *Energy Density and Reverberation Time*

It is important to recognize, however, that changing the reverberation time also affects energy density. The energy density in a room (away from the individual sound sources, where the sound field is more likely to be diffuse) is directly proportional to the reverberation time. From Eqs. (19) and (20),

$$D_0 = 4W/cA \tag{22}$$

and

$$t = 0.16 \ V/A \ (\text{assuming } m = 0) \tag{23}$$

The equation representing the relationship can be determined to be

$$D_0 = \frac{W_t}{13.6V} \tag{24}$$

where c is taken as 340 m/s.

This formula clearly shows that the energy density in the room can only be reduced (and a noise-reduction achieved) either by reducing the sound power emitted to the room or by a reduction of the reverberation time—the last being equivalent to an increase in the total absorption.

In rooms, such as industrial areas, where short reverberation times are desirable, the energy density will be low also. In such a case, there is no problem, because both characteristics are desirable. In music rooms, however, where high energy density and medium reverberation time are assets, some compromise may be necessary.

3.7. Relationship Between Direct and Reflected Sound

It was determined that the energy density in the steady-state situation was

$$D_0 = 4W/cA \text{ W-s/m}^3 \tag{19}$$

Close to the sound source, however, the energy density will be larger than D_0 because of the contribution D_d from the sound source. Assuming the sound source radiates evenly in all directions, then

$$D_d = W/c4\pi R^2 \text{ W-s/m}^3 \tag{25}$$

The ratio of D_d to D_0 becomes

$$D_d/D_0 = A/16\pi R^2 \tag{26}$$

and as long as that ratio is greater than 1, we are in the direct sound held. This means that D_d is dominant when

$$R^2 \leq A/16\pi R^2 \tag{27}$$

This distance R can be used to determine how far away from a machine an operator must be in order to achieve a noise reduction in a room solely through the use of absorption materials. The total energy density in a room is given by

$$D = D_0 + D_d = \frac{4W}{cA} + \frac{W}{c4\pi R^2} = \frac{W}{c}\left[\frac{4}{A} + \frac{1}{4\pi R^2}\right] \tag{28}$$

The sound pressure level, SPL reference to 20 µPa, can be determined by inserting Eq. (28) into Eq. (11) and utilizing Eq. (8):

$$\text{SPL} = \text{PWL} + 10\log\left[\frac{1}{4\pi R^2} + \frac{4}{A}\right] \tag{29}$$

where

$$\text{PWL} = 10\log W/W_0 \text{ [see Eq. (9)]}$$

Example 5

Determine the distance R for which the direct energy density is the main contributor in two different rooms.

> Room 1:   Volume $V = 300$ m^3
>   Reverberation time $t = 1.0$ s
>
> Room 2:   Volume $V = 20,000$ m^3
>   Reverberation time $t = 2.5$ s

Solution

Room 1:

$$A = \frac{0.16 \times V}{t} = \frac{0.16 \times 300}{1.0} = 48 \text{ m}^2\text{-Sabin}$$

$$R^2 \leq \frac{48}{16\pi} = 0.95 \text{ m}$$

$$R \leq 1 \text{ m}$$

Room 2:

$$A = \frac{0.16 \times V}{t} = \frac{0.16 \times 20000}{2.5} = 1280 \ \text{m}^2\text{-Sabin}$$

$$R^2 \leq \frac{1280}{16\pi} = 25.5 \ \text{m}$$

$$R \leq 5 \ \text{m}$$

4. SOUND OUT-OF-DOORS

4.1. Sound Propagation

One major difference between sound inside an enclosed space and sound outdoors is that the effect of multiple reflecting surfaces outdoors is usually not as significant as indoors.

At a distance R from a point source, the energy density will be equal to

$$D = \frac{W}{c \times 4\pi R^2} \tag{30}$$

The sound pressure level SPL reference to 20 µPa can be determined by inserting Eq. (30) into Eq. (11) and utilizing Eq. (8):

$$\text{SPL} = \text{PWL} - 10 \log 4\pi R^2 \ \text{dB} \tag{31}$$

Implicit in this equation is the inverse square law, which states that for each doubling of distance the sound pressure level will drop off by 6 dB. This is only true if there are no winds, no temperature gradients, and no reflecting surfaces.

4.2. Wind and Temperature Gradients

Another major difference between sound inside an enclosed space and sound outdoors is the influence of meteorology. Under normal circumstances, the wind velocity will change as a function of distance above the ground. If the wind velocity changes with the height above the ground, the sound velocity is also changed. This causes the sound waves to be turned toward the ground in the downwind direction and away from the ground in the upwind direction, as indicated in Fig. 9.

A temperature change above the ground also influences the propagation of the sound waves. If the temperature increases with height, the sound waves will be turned toward

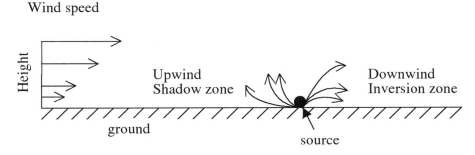

Fig. 9. The influence of wind direction and wind velocity on sound propagation. (From ref. 1.)

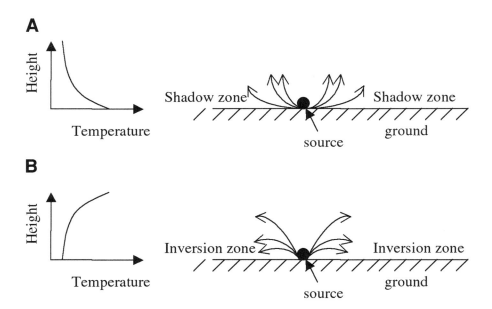

Fig. 10. The influence of temperature change on sound propagation: **(A)** increasing temperatures by height; **(B)** decreasing temperatures by height. (From ref. 1.)

the ground although the opposite is true for temperatures, which decrease with height, as shown in Fig. 10.

4.3. Barriers

The propagation of sound outdoors is affected by land forms such as mountains, hills, dikes, and so forth. These natural as well as artificial barriers can be effective in reducing received sound, provided that the barriers have no holes and that they block the direct path of sound from the source to the receiver.

Four primary parameters determine how effective a barrier is in reducing sound: (1) source-to-barrier distance, (2) barrier-to-receiver distance, (3) the height of the barrier and (4) the length of the barrier. Figure 11 shows a method that can be used to estimate the reduction obtainable from a barrier (10). To use the figure, the four parameters involved are combined into a single descriptor—the increase in sound path length. This descriptor is then related directly to sound reduction. The method is limited to barriers that can be regarded as rigid and infinitely long. Figure 11 gives the average noise reduction that can be obtained for broadband noise. For low frequencies the attenuation is lower, and for high frequencies the attenuation is higher.

As shown in Fig. 11, the source-to-receiver distance is D. With the barrier in place, the shortest source-to-receiver distance becomes $a + b$. The difference $(a + b) - D$ is the parameter δ, the sound path length difference. For a situation in which $\delta = 1.0$, for example, the sound reduction for an infinitely long barrier would be about 17 dB when metric units are used and about 12 dB when English units are used.

Recent research has focused on absorptive and partial length barriers (9). A more flexible, and complicated, treatment of barriers comes from Kurze and Anderson (11,12). In this

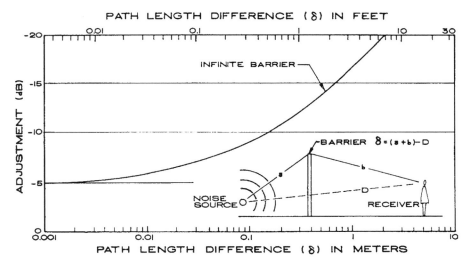

Fig. 11. Average noise reduction for an acoustic barrier of infinite length. (From refs. 1 and 10.)

model, four parameters combine into one variable, the Fresnel number N, to determine how effective a barrier is in reducing sound. They are (1) the source-to-barrier distance, (2) the barrier-to-receiver distance, (3) the height of the barrier and (4) the frequency of interest. Figure 12 is a sketch of the barrier problem to be utilized in Eq. 32 that describes the insertion loss IL or the change in sound level before and after the barrier is added:

$$N = \frac{2f(A+B-C)}{c}$$

$$\text{IL} = 0, \qquad N < -0.1916 - 0.635b'$$

$$\text{IL} = 5(1+0.6b') + 20\log\frac{\sqrt{-2\pi N}}{\tan\sqrt{-2\pi N}}, \qquad -0.1916 - 0.635b' \leq N \leq 0$$

$$\text{IL} = 5(1+0.6b') + 20\log\frac{\sqrt{2\pi N}}{\tanh\sqrt{2\pi N}}, \qquad 0 < N < 5.03 \tag{32}$$

$$\text{IL} = 20(1+0.15b'), \qquad N \geq 5.03$$

where b' is 0 for a wall and 1 for a berm (earthen barrier). This method is also limited to barriers that can be regarded as infinitely long but it has the advantage of being frequency dependent. One can check whether a barrier is sufficiently long to qualify as infinite by looking at the flanking paths as a separate diffraction problem. If the insertion loss of the flanking path is sufficiently larger than the path over the top, it can be neglected.

5. NOISE REDUCTION

5.1. Absorptive Materials

A reduction of the sound pressure level in a room can often be efficiently accomplished by reducing the sound power emitted by the sound source. By reducing the

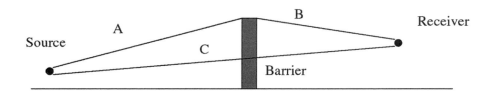

Fig. 12. Geometry for calculation of insertion loss using the method of Kurze and Anderson. (From ref. 12.)

emitted sound power, both the direct and the reflected sound will be reduced. However, it is often not appropriate to modify the sound source because modification may have adverse effects, such as reduced efficiency, interference with access to the machine, and so forth (13–15). In such cases, the energy density can be changed by modifying the sound field. One commonly used modification is obtained by installing absorption materials at strategic places such as the walls and ceiling of the room. The beneficial result of this treatment is reduction of the intensity of the reflected sound (15,16).

In Section 3, we developed the formulas governing energy density, sound power, reverberation time, room volume, and total absorption. In this section, we will consider how much noise reduction can be achieved by the use of absorptive materials.

Let us look at an example where an existing room has a total absorption of A_0. If S square meters of absorptive material is installed with an absorption coefficient α, then the total absorption after the installation will be

$$A = A_0 + \alpha S \text{ m}^2\text{-Sabin} \tag{33}$$

However, only the surface exposed to the incident sound determines the absorption, so if the new material covers any of the older material

$$A = A_0 + S (\alpha - \alpha_0) \text{ m}^2\text{-Sabin} \tag{34}$$

where α_0 is the absorption coefficient for the wall that will be covered with the absorptive treatment from Eq. (19). The energy density of the reflected sound will be reduced by

$$\Delta = 10 \log A/A_0 \text{ dB} \tag{35}$$

Since noise is a combination of direct and reflected signals, the noise reduction that can be achieved is highly dependent on the initial amount of absorption in the space. In most practical installations, it will not be possible to achieve more than 5–10 dB noise reduction. A 10-dB reduction is equivalent to approximately half of the sound loudness, rated subjectively.

By placing absorption materials on room surfaces, the direct sound will not be reduced; therefore, machine operators who are located a few feet from the noisy machine will receive little benefit. However, the contribution to their noise exposure from adjacent machines will be reduced.

Example 6

Determine the noise reduction of the reflected sound that can be obtained in a $20 \times 25 \times 35 \text{ m}^3$ space where the reverberation time is 4.0 s and the existing ceiling material having an absorption coefficient of 0.25 is exchanged with a material having an absorption coefficient of 0.90.

Solution

$$V = 20 \times 25 \times 35 = 17,500 \text{ m}^3$$
$$t = \frac{0.16 \times 17,500}{A_0}, t = 4.0 \text{ s}$$
$$A_0 = \frac{0.16 \times 17,500}{4} = 700 \text{ m}^2\text{-Sabin}$$
$$A = A_0 + S(\alpha - \alpha_0)$$
$$A = 700 + 25 \times 35 \times (0.90 - 0.25)$$
$$A = 700 + 570 = 1270 \text{ m}^2\text{-Sabin}$$

The noise reduction Δ is

$$\Delta = 10 \log \frac{1270}{700} = 10 \log 1.82$$
$$\Delta = 2.6 \text{ dB}$$

When using absorption materials in an attempt to reduce the energy density of the reflected sound, it is beneficial to place the materials as close to the sound source as possible. Good results often occur when a noisy machine is placed near a wall covered with an effective absorption material. Another alternative would be to cover the machine with a barrier furnished with absorptive material. A canopy with absorptive material can be effective if it is suspended close to the machine. It is important to recognize that the absorption coefficient varies with frequency, so that before a selection is made, it will be necessary to know the frequency content of the disturbing noise. In most practical situations, noise-reduction techniques should be aimed at the middle- and high-frequency regions (500–8000 Hz), which is the range where human perception is the greatest. Sounds in this frequency range are typically the most tiring, cause the greatest task interference, and are the most damaging to hearing.

The acoustics of a room are not affected by whether the sound energy is absorbed by a real or an imaginary surface. In this respect, an open window is a very effective sound absorber because it acts as a sink for all of the arriving sound energy. For real surfaces, the sound energy can either be converted to heat or transmitted as sound waves in the building structure. Sound-absorbing materials can, on the basis of the physical process involved, be separated into two main groups: (1) the porous absorbers and (2) the resonant absorbers.

5.1.1. Porous Absorbers

Porous absorbers are complicated structures often manufactured from glass or mineral fibers pressed into boards held together by suitable adhesives. Owing to the complexity of the construction details of the porous materials (such as length of channels, side branches, irregularity in shape of fibers, etc.), it is difficult to predict accurately the absorption values from calculations although great strides have been made in recent years (17–19). In general, though, the acoustical properties of these materials depend on the air particles within the materials being set into motion (oscillating) when a sound

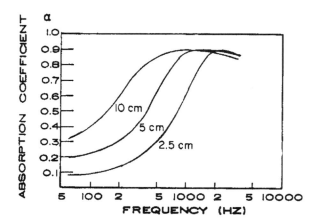

Fig. 13. Absorption coefficients of porous materials with 2.5, 5, and 10 cm thicknesses. (From ref. 1.)

wave strikes the material. Furthermore, this motion, resisted by viscous forces near the surfaces of the fibers, results in part of the sound energy being transformed into heat.

The porosity of the materials is of great importance. If the porosity is low, then the sound waves will have difficulty in penetrating the material, and most of the sound will be reflected from the surface. If the porosity is large, the sound wave will easily penetrate the material and the reflection from the surface will be very small. This, however, does not necessarily mean that the absorption will be great. Substantial absorption will be obtained only if the sound wave that penetrates into the material is reduced to a great extent before it leaves the material.

One valid generalization about porous absorbers is that the first maximum of absorption will occur when the thickness of the porous material equals one-quarter of a wavelength of the incident sound, for example, if $d = \lambda/4$. Therefore, if it is important to obtain high absorption values at low frequencies, the thickness of the material must be large. However, low-frequency absorption can be obtained with thin materials having high flow resistance values as long as they are installed with an airspace between them and the wall.

Figure 13 shows how the absorption coefficient theoretically should change when the thickness is changed. The first maximum on the curves should occur at a frequency having a wavelength equivalent to four times the thickness of the material. However, remember that the complexity of these porous materials requires that one should only use data that have been obtained experimentally.

If the absorption material is covered with thin plastic film to prevent contamination or spilling, then the absorption coefficient will change. The greatest change will take place at high frequencies, at which the film will be impermeable to sound, thus decreasing the absorption coefficient. The least change will occur when the film is no thicker than 50 μm (1–2 mil).

5.1.2. Resonance Absorbers

Resonance absorbers can be thought to consist of a simple mechanical system containing a mass m and a stiffness s (20). Each resonance absorber will thus possess a frequency at which sound energy impinging on it will be absorbed. The formula for the resonance frequency of such a system (1) is

$$f_0 = \frac{1}{2\pi}\sqrt{\frac{s}{m}} \text{ Hz} \tag{36}$$

5.1.2.1. MEMBRANE ABSORBERS

A membrane absorber consists of an impermeable plate placed at a distance from a wall. The air in the cavity between the plate and the wall acts as a spring with a characteristic stiffness. The resonance frequency for such a system (1) is

$$f_0 = 600/\sqrt{md} \text{ Hz} \tag{37}$$

where m is the weight of the front plate (in kg/m^2) and d is the depth of the cavity (in m). In English units, the resonance frequency will be

$$f_0 = 1340/\sqrt{md} \text{ Hz} \tag{38}$$

where m is the weight of the front plate (in lb/ft^2), and d is the depth of the cavity (in ft) (1).

The above formula does not consider the stiffness of the front plate itself and it naturally must be included if it is significant. The stiffness of a panel or plate depends on the support system utilized. This is especially true if the distance between the wall and the plate is large, because the air has little stiffness and the plate stiffness becomes more important. If the distance between panel or plate supports is small, the panel or plate has great stiffness.

The resonance frequency of a panel system is normally in the range 100–400 Hz for absorbers with panel thicknesses of 5–15 mm (¼–⅝ in.) and cavity depths of 10–25 mm (⅜–1 in.). Figure 14 shows a typical membrane absorber and associated absorption coefficient curve.

When a sound wave strikes a membrane absorber, the front plate starts to oscillate. The oscillations are the greatest if the incoming sound frequency is close to the resonance frequency of the membrane absorber. The absorption of a membrane absorber can be increased by filling the cavity behind the plate with a porous absorber. However, the maximum absorption that can he obtained that membrane absorbers seldom exceeds a coefficient of 0.5.

5.1.2.2. HELMHOLTZ RESONATORS

A Helmholtz resonator is an enclosed space with stiff walls, having only a single opening to a room, normally called the throat. Such a resonator can be represented by a simple mechanical system consisting of an oscillating mass m (the "plug" of air in the throat) resisted by a spring with a stiffness s (the air in the enclosed space).

A Helmholtz resonator, like any mechanical system, has a resonance frequency determined by

$$f_0 = \frac{c}{2\pi}\sqrt{\frac{S}{l_a V}} \text{ Hz} \tag{39}$$

where c is the speed of sound (in m/s), S is the area of the throat opening (in m^2), V is the volume of the enclosed space (in m^3), and l_a is the adjusted length of the throat (in m) (1).

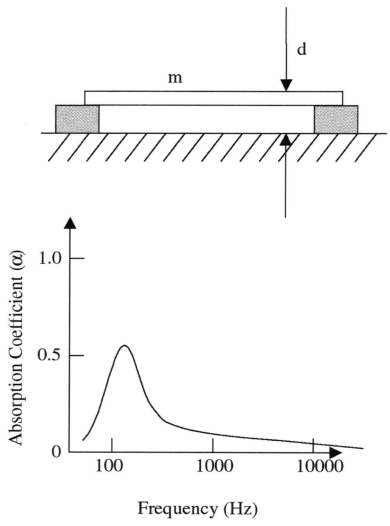

Fig. 14. Qualitative description of absorption coefficients for a membrane absorber as a function of frequency. (From ref. 1.)

For most practical purposes, the adjusted length of the throat is determined (7) by

$$
\begin{aligned}
l_a &= l + 1.7a \quad \text{flanged} \\
l_a &= l + 1.4a \quad \text{unflanged}
\end{aligned}
\tag{40}
$$

where a is the diameter of the throat and l is the length of the throat. An opening consisting of a circular hole in the thin wall of a resonator will have an adjusted length of approx $1.6a$ (7).

The formula for the resonance frequency is only valid if the dimensions of the resonator are small compared to the resonance frequency wavelength and if the dimensions of the throat are small compared to the enclosed space. A Helmholtz resonator absorbs sound effectively only in the vicinity of the resonance frequency. The absorption can be large provided that the internal damping of the resonator is minimal; filling the enclosed space with a porous material will increase damping. For optimum performance, the Helmholtz resonator should be empty and the walls of the enclosed space should be very stiff.

The highest absorption that can be obtained with a Helmholtz resonator (1) is

$$A_0 = \frac{\lambda^2}{2\pi} \tag{41}$$

The Helmholtz resonator is very selective, so that the absorption falls off rapidly as the sound frequency shifts away from the resonance frequency. Helmholtz resonators are often utilized to solve special problems, such as reducing standing waves in music rooms or concert halls.

5.1.2.3. PERFORATED PANE/ABSORBERS

If the front plate in a membrane absorber is perforated, one may visualize the perforated plate as consisting of a large number of small Helmholtz resonators with a common cavity. The common cavity can be imagined to be divided into a number of smaller spaces by fictitious walls so that each hole in the front plate has one space behind it.

Most standard perforated plates contain between 6% and 25% open area. Most holes are made circular in shape with diameters between 1 and 4 mm. The thickness of the front plates depends on the material and usually varies between 1 and 10 mm.

The resonance frequency for such a system is given by (21)

$$f_0 = \frac{c}{2\pi} \sqrt{\frac{p}{100dl_a}} \text{ Hz} \tag{42}$$

where p is the perforation percentage, d is the distance between the plate and the wall (in m), and l_a is the adjusted length of the throat (in m).

This formula shows that there is greater latitude in the possible resonance frequency for which the perforated plate absorber can be designed than for the single Helmholtz resonator. Practical experience has shown that the perforated panel absorbers have the additional benefit of possessing good absorption over a wider frequency range than a single Helmholtz resonator, especially if porous material is placed behind the plate.

If the open area of the perforations exceeds about 25%, then the Helmholtz resonator part of the acoustic properties can be neglected and the acoustic properties can be predicted solely on the basis of the porous material behind the plate.

5.1.2.4. SLOT RESONATORS

A slot resonator is a variation of a Helmholtz resonator, the only differences being that one dimension of the cavity space is much larger than the other and the opening to the space is a slot. The resonance frequency for a slot resonator is much more complicated than for a Helmholtz resonator; the procedure will not be discussed here.

5.2. Nonacoustical Parameters of Absorptive Materials

Absorptive materials should always be thought of as an integral part of a building design—not as an afterthought. In this way, their appearance can be used positively to provide the most pleasing environment.

Absorptive materials should be protected from abuse, especially if located within 2 m from the floor, and should be noncombustible and flame retardant and should not shed or collect dust. The acoustical properties should not change as a result of maintenance and cleaning. Porous materials cannot be painted because the absorptive characteristics will be completely ruined; they need to be cleaned on a regular basis or replaced if they become unattractive.

5.3. Absorption Coefficients

Design values for absorption coefficients should be obtained from qualified laboratory measurements. Reputable laboratories follow the guidelines of the International Standard Organization recommendations for measuring the reverberation time and calculating the absorption coefficients (22). Such standardizations have been found to be necessary to obtain reliable and consistent coefficients.

One comment should be made regarding reported absorption coefficients. They are all derived from measurements made under idealized conditions as near as possible to diffuse sound fields. A diffuse sound, however, is rarely obtainable in real life. The net result is that absorption coefficients in practice will be effectively lower than given. The laboratory values still have their use because they permit comparisons between different products that have been tested under similar conditions. Values for most commercially available absorption materials can be found from vendors and manufacturers (1,9,12). Materials do not have to be "acoustical" to possess absorption coefficients.

Some typical absorption coefficient values are shown in Table 6 for common construction materials. As can be seen, even these materials have acoustical properties.

6. SOUND ISOLATION

6.1. Introduction

Sound isolation refers to the process of insulating an area adjacent to or near a sound source in order to achieve an acceptable acoustical environment in the insulated area. An acceptable acoustical environment can mean that people within the insulated area are not seriously disturbed by intrusive sounds; in the opposite sense, it could mean that people are confident that they cannot be overheard by others outside the room—they have acoustical privacy.

When dealing with room acoustic problems, it is important to know the amount of sound energy removed from an area. When dealing with sound isolation, it is important to know how the sound energy leaves one area and enters another. Airborne sound can travel from one room to another in several ways:

1. Through a dividing wall (by causing the wall to vibrate and reradiate sound on the other side)
2. Through cracks, holes, and so forth
3. By means of flanking paths (other structural paths or through a ceiling plenum)

Direct transmission through the dividing wall is often responsible for most of the intrusive sound.

Table 6
Absorption Coefficients at Various Frequencies for Some Common Construction Materials

	Frequency (Hz)					
	125	250	500	1000	2000	4000
Wood flooring	0.15	0.11	0.10	0.07	0.07	0.07
Linoleum flooring	0.02	0.03	0.03	0.03	0.03	0.02
Padded carpet	0.11	0.14	0.20	0.33	0.52	0.82
Brickwall	0.01	0.01	0.02	0.02	0.03	0.03
Brick (unglazed)	0.03	0.03	0.03	0.04	0.05	0.07
Concrete	0.01	0.01	0.02	0.02	0.02	0.04
Plaster, gypsum	0.13	0.15	0.02	0.03	0.04	0.05
Windows	0.4	0.3	0.2	0.17	0.15	0.1
Doors	0.18	0.12	0.10	0.09	0.08	0.07
Drapes	0.06	0.1	0.38	0.63	0.7	0.73
Water surface	0.008	0.008	0.013	0.015	0.02	0.025
Rock wool, 2.5 cm	0.09	0.23	0.53	0.72	0.75	0.77
Rock wool, 5.0 cm	0.20	0.53	0.74	0.78	0.75	0.77
Rock wool, 10.0 cm	0.68	0.84	0.82	0.78	0.75	0.77

Source: Data from refs. 1 and 9.

6.2. Transmission Loss

When dealing with room acoustic problems, it is important to know the amount of sound energy that will be absorbed, although it is less important to know what percentage of the absorbed sound will be converted to heat or will be transmitted away from the room through openings or other paths. When it comes to sound isolation, however, it definitely is of importance to know how much sound energy will be transmitted through the building's structural elements.

Transmission loss is a term describing the sound-attenuating properties of a material or system. The transmission loss (TL) of a wall or slab construction is defined as the difference between the sound power impinging on the surface in the source room and the sound power radiated from the surface in the receiving room. In logarithmic form, this is expressed as

$$TL = 10 \log W_1/W_2 \, dB \qquad (43)$$

This can be written as

$$TL = 10 \log 1/\tau \, dB \qquad (44)$$

where τ is the transmission coefficient. The transmission loss can only be measured accurately when great care has been taken to eliminate the indirect transmission paths.

6.2.1. Measurement of Transmission Loss

In order to calculate the transmission loss of a material, it is necessary to make measurements in properly designed rooms separated by the material for which the TL quantification is desired. When making measurements of transmission loss of structures,

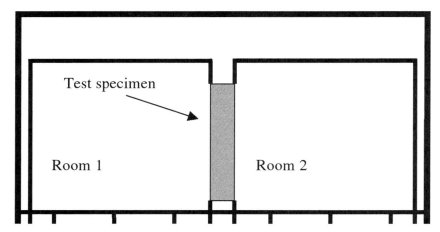

Fig. 15. Reverberation rooms for measurement of transmission loss. (From ref. 1.)

it is of utmost importance that the sound transmission between source and receiver room primarily takes place through the wall of interest (23). Transmission of sound through floor, side walls, and so forth must be negligible. This can he accomplished by constructing the two rooms on separate foundations and inserting resilient layers—such as sheets of cork, glass fiber, or mineral wool—where the two rooms abut each other. Some rooms have a separate frame for the installation of the test specimen between the two rooms, as shown, in Fig. 15.

The sound power which impinges on a surface S can in the stationary situation be determined by

$$W_1 = \frac{1}{4} cSD_1 \tag{15}$$

Recall that the energy density in the steady-state situation could be expressed by

$$D_2 = 4W_2/cA_2 \tag{19}$$

The energy density in the diffuse field also relates to the rms sound pressure:

$$p_{1,\mathrm{rms}}^2 = \rho c^2 D_1$$
$$p_{2,\mathrm{rms}}^2 = \rho c^2 D_2 \tag{11}$$

Substituting these formulas in the formula for the transmission loss, we get

$$\mathrm{TL} = 20\log\left(p_{1,\mathrm{rms}}/p_{2,\mathrm{rms}}\right) - 10\log(A_2/S) \tag{45}$$

$$\mathrm{TL} = \mathrm{SPL}_1 - \mathrm{SPL}_2 - 10\log(A_2/S) \tag{46}$$

where SPL_1 and SPL_2 are the sound pressure levels in the source and receiver room, respectively. Measurements of the SPL can be directly made, but A_2 should be calculated from reverberation time data. Although some time was spent in the preceding sections detailing methods for calculating A determining it from the reverberation time removes any uncertainty in the assumed values for α.

6.2.2. Calculation of Transmission Loss

There are occasions when estimates of specific wall construction TL are needed. A calculating procedure, rather than a measurement procedure, is needed for these occasions and is discussed in the following subsections.

6.2.2.1. TRANSMISSION LOSS OF A SINGLE-THICKNESS WALL

An accurate calculation of the transmission loss of even a simple single wall is extremely difficult to perform and is beyond the intent of this text. If some assumptions are made, then it is possible to derive a simplified formula that approximates actual transmission loss fairly well. The following assumptions are made:

1. Only plane waves are incident on the walls.
2. The wall radiates plane waves.
3. The wall is infinitely large.
4. No energy losses occur inside the wall.
5. The surfaces in the receiving room are totally absorptive.

The simplified formula is (12,24)

$$TL = 20 \log f + 20 \log m - C \qquad (47)$$

where f is the frequency (in Hz), m is the weight of the wall (in kg/m^2), and C is a constant depending on the units used (for metric and English units, the values are 43 and 29, respectively). If one relaxes assumption 5 and has a diffuse field impinging on the incident wall, the transmission loss is lowered by 5 dB to account for the possible varying angles of incidence (12). The formula indicates that the TL will increase by 6 dB for each doubling of frequency and doubling of surface weight (or thickness for a constant material). This relationship is normally called the "mass law."

Figure 16 shows a graphic presentation of Eq. (47) (dashed line) and an empirically determined curve (solid line). As can be seen from Fig. 16, the empirically determined curve is lower than the theoretical curve. It can also be seen that for walls with surface weight less than about 200 kg/m^2 (40 lb/ft^2), each doubling of the weight only results in a TL increase of approx 5 dB, whereas above 200/kg m^2, the ''mass law'' is valid again.

One of the assumptions made was that the wall was only affected by plane waves normal to the wall. This, however, is not realistic. Sound waves will strike the wall from all directions, at all angles. An incident sound wave just grazing the wall will deform the wall and cause bending modes to be generated. At some frequency, dependent on physical properties of the wall, the bending modes will be easily excited. This is the critical frequency fc. At the critical frequency, the transmission loss theoretically becomes zero. The critical frequency fc can be determined from

$$f_c = \frac{c^2}{2\pi} \sqrt{\frac{m}{B}} \ \text{Hz} \qquad (48)$$

where c is the speed of sound (in m/s), m is the surface weight (in kg/m^2), and B is the bending stiffness (in Nm) (1).
The bending stiffness B can be determined from

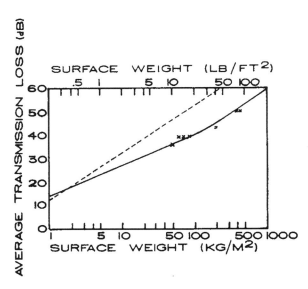

Fig. 16. Average transmission loss in the frequency range 100–3000 Hz as a function of the surface weights of walls. (From ref. 1.)

$$B = \frac{E \times d^3}{12 \times (1 - \sigma^2)} \qquad (49)$$

where d is the thickness of the wall (in m), E is Young's modulus (in Pa), and σ is Poisson's ratio.

By utilizing the relationship $m = d\rho$, where ρ is the density of the wall, the critical frequency becomes

$$f_c = \frac{c^2}{2\pi d} \sqrt{\frac{12\rho(1 - \sigma^2)}{E}} \text{ Hz} \qquad (50)$$

Poisson's ratio will be equal to about 0.3 for many structural materials. Attention to the critical frequency is important because, if the sound to be isolated is significant at that frequency, alternate construction may be needed.

Example 7

Determine the critical frequency for a 6-mm-thick window pane. The density of glass is about 2500 kg/m³. Young's modulus is 3.5×10^{10} kg/m-s²

Solution

$$f_c = \frac{340^2}{2\pi \times 0.006} \sqrt{\frac{12 \times 2500 \times (1 - 0.09)}{3.5 \times 10^{10}}}$$

$$f_c = 2720 \text{ Hz}$$

The transmission loss curves for a 6-mm-thick window is shown in Fig. 17 (as calculated according to the above discussion). Note in Fig. 17 that the transmission loss

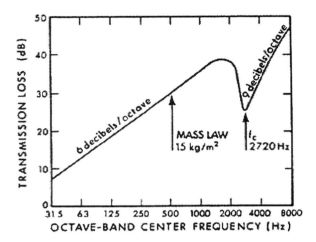

Fig. 17. Transmission loss curves for a 6-mm-thick window as a function of frequency. (From ref. 1.)

at the critical frequency is not zero, as is suggested by theory. Laboratory studies have been undertaken and they seem to indicate that the minimum TL should be around 25 dB.

6.2.2.2. TRANSMISSION LOSS OF DOUBLE WALL

The transmission loss of a double wall depends on the following:

1. The material between the individual walls
2. The thickness of the cavity between the individual walls
3. The structural ties between the individual walls
4. The materials of which the walls are made
5. The thickness of each wall

In the idealized situation where the two wall sections are completely independent of each other, the transmission loss at very low frequencies will be approximately equivalent to the TL of a single wall with the same total surface weight. However, at the resonance frequency of the wall, the transmission loss will be less than expected. The resonance frequency for a double wall is determined by

$$f_r = \frac{c}{2\pi}\sqrt{\frac{(m_1+m_2)\times\rho}{m_1\times m_2\times d}} \text{ Hz} \tag{51}$$

where m_1, and m_2 are the surface weight (in kg/m²), ρ is the density of the air (1.2 kg/m³), and d is the distance between wall elements (in m) (1).

If the two wall elements are of equal density and thickness, the formula for the resonance frequency becomes

$$f_r = \frac{c}{2\pi}\sqrt{\frac{2\rho}{m_1 d}} \text{ Hz} \tag{52}$$

The transmission loss at frequencies lower than f_r will be lower than that of a single wall with total weight equal to (m_1, +m_2). At higher frequencies however, the increase in transmission loss will be greater than the characteristic 6 dB per octave for a single wall. It is therefore important to keep the resonance frequency as low as possible.

In order to obtain maximum utilization of the individual wall elements (to get as high a transmission loss as possible), it will also be necessary to place absorptive material in the cavity between the two wall elements. This can be accomplished by placing thermal isolation fiberglass blankets in the cavity. It is necessary to cover only half the wall area in order to increase the transmission loss by 3–5 dB. The function of the blankets is to reduce standing waves that may be set up between the walls, rather than to reduce the direct sound transmission.

6.3. *Noise Reduction*

The noise reduction (NR) is the difference in decibels between the sound pressure levels in two rooms. The NR accounts for all sound paths and thus is more inclusive than the TL.

The NR can be expressed as

$$\text{NR} = \text{TL} - 10 \log S/A_2 - (C_1 + C_2) \text{ dB} \tag{53}$$

where S is the area of the common wall (in m^2), A_2 is the total absorption in the receiving room (in m^2-Sabin). (1), C_1 is a correction that depends on air leaks, and C_2 is a correction that depends on flanking transmission, which will be described in subsection 6.3.2 (C_1 and C_2 are zero for no air leaks and no flanking transmission).

6.3.1. *Air Leaks*

No aspect of a wall construction requires more attention to detail than ensuring that no air path leaks exist. The seriousness of air leaks can be illustrated by the following. A wall with an air leak can be considered as consisting of two elements with different transmission losses. The sound power transmitted by the two elements with a common incident sound power is

$$W_{\text{trans}} = \tau_a S_a + \tau_w S_w \tag{54}$$

where τ_a and τ_w are average transmission coefficients for the air leak and for the wall, respectively S_a and S_w are the areas of the air leak and the wall, respectively (in m^2). The composite transmission loss is then

$$\text{TL} = 10 \log \frac{S_a + S_w}{\tau_a S_a + \tau_w S_w} \tag{55}$$

If an infinitely "good" wall (no transmission at all, $\tau = 0$) is placed between two rooms—and there is no flanking transmission—the formula indicates that the maximum NR between the two spaces will be no greater than 10 dB if 10% of the wall area is not closed off ($\tau = 1$). If 1% of the wall area is not closed off, the noise reduction will be 20 dB. If only 0.1% of the wall area is not closed off, the noise reduction will be 30 dB. If only 0.01% of the wall area is not closed off, the noise reduction will not exceed 40 dB.

A typical wall between apartments may have an area of 30–40 m^2. In such a wall, 0.01% of the wall area may be the equivalent of an 8–10-m-long crack no greater than 0.5 mm. In order to achieve sound isolation between apartments, it will be necessary to have an average NR of about 50 dB. If that much sound isolation is provided, then the 70–80 dBA sound levels caused by such common sources as radios and television sets will be reduced to a sound level of about 20–30 dBA in an adjacent apartment. A sound level in the 20–30 dBA range is equivalent to a typical ambient sound level: thus, the intrusion will be minimal.

6.3.2. Flanking Transmission

A flanking path is any other path the sound can take in getting from one room to another than directly through the separating wall (the direct path). Although a sound path through an air leak is also a flanking path by the definition, it has been discussed separately because of the ease of handling those transmission problems. In this sub section, the remaining types of flanking transmission are considered.

The determination of the actual value of the flanking transmission is one of the least investigated and possibly one of the least understood. One explanation for this lack of understanding is that in situations where no more than 45–50 dB NR is required, the contribution from flanking transmission is minimal. In the situation where higher NR values are necessary, precautions have to be taken to reduce the flanking transmission in addition to increasing the TL of the dividing wall.

Flanking transmission can generally be reduced by breaking the structural vibration pathway. This can be accomplished by physically separating the source side from the receiver side. Sufficient separation is often obtained by ensuring that a small space exists between the side walls (or floor and/or ceiling) and the dividing wall. In such cases, the space is filled with a resilient material such as cork, glass fiber, or mineral wool. These materials are used to eliminate the undesirable attributes of the space and, fortunately, they do not short circuit the vibration pathway.

In more critical cases, the source room walls or ceilings may be treated with resiliently mounted skins or suspension ceilings. In the most severe cases, the entire source area may be "detached" from the main structure by "floating" the space on resilient material.

6.4. Noise Isolation Class (NIC)

The Noise Isolation Class is a metric that is used to specify a desired sound isolation between two spaces (9). It is determined by measurements of the sound pressure level in each of the spaces, say on opposite sides of a door. Much like the noise reduction (NR), it incorporates all transmission paths in a de facto manner since it is a direct measurement of the space. It is evaluated by using a broad band, omnidirectional source in one space to ensonify that space and then measuring the sound pressure level on both sides of the interface between the spaces to determine the loss of sound level. It is important to not measure too close to either surface so as to not preferentially measure any given sound path. If possible, a 1 m spacing between interface and measurment location should suffice. Sound pressure levels in one-third octave bands between 125 Hz and 4000 Hz are used to evaluate the NIC. The NIC of the interface is determined by comparing the measured loss values to a contour generated by specifying a loss at 500 Hz and then correcting it by the following values at the other frequencies of interest:

Freq	125	160	200	250	315	400	500	630	800	1k	1250	1.6k	2k	2.5k	3.15k	4k
Correction	−16	−13	−10	−7	−4	−1	0	1	2	3	4	4	4	4	4	4

This value at 500 Hz is adjusted upward or downword and the NIC is determined when either (1) any individual predicted one-third octave value exceeds the actual measured one-third octave loss by 8 dB or (2) the sum total of predicted values exceed the measured values by 32 dB (deficiencies only). The lower, or more conservative, value of NIC is to be used.

Example 8

Sound levels were measured on opposite sides of a doorway between two rooms and the following values were obtained. Determine the NIC of the door.

Freq (Hz) 1/3 Oct. Bd.	Source	Receiver
125	65.5	42.5
160	61.7	43.5
200	55.3	34.1
250	48.7	29.8
315	49.0	25.8
400	43.6	21.7
500	49.4	23.4
630	44.4	21.2
800	49.9	24.5
1000	50.2	23.3
1250	49.1	21.6
1600	48.1	23.8
2000	52.8	33.6
2500	55.1	33.5
3150	55.2	28.5
4000	54.7	24.6

Solution

At a value of NIC = 24, the measured loss at 2 kHz is 19.2 dB while the predicted loss is 28 dB (24 + 4 correction). Thus, the predicted level exceeds the measured level by more than 8 dB. Note that the 32 dB criterion is exceeded at NIC = 25, so the lower value is used to describe the door's effectiveness.

7. VIBRATIONS

7.1. Introduction

Structural vibrations are always present even though sensitive instruments may be needed to detect those caused by minuscule forces such as footfalls or distant traffic. When vibrations are more severe, they can be sensed. If the vibrations occur in the low part of our audibility range, we may not be able to distinguish between feeling and hearing them.

Vibrations can become a problem in a number of ways. For example. they can cause unwanted motion in equipment. Little motion can be tolerated for sensitive scientific equipment such as electron microscopes. More motion can be tolerated for industrial equipment, but each has its limits. Continuous vibrations can cause metal fatigue. Sufficiently intense vibrations can cause physiological disorders in personnel handling equipment such as grinders or chain saws. Vibrations can cause noise. Because vibrations are of such significance, it is of interest to determine how vibratory energy can be prevented from traveling through and within building structures and how equipment can be insulated from vibrations.

7.2. *Vibration Isolation*

Vibrations are prevented from reaching machinery or passing from the machinery to surrounding structures by incorporating specially designed devices into their supports. These devices, collectively termed "vibration isolators", act to reduce the intensity of the vibrations that pass through them.

Most machinery vibration problems deal with harmonic oscillations caused by some cyclical function of the machine (gear rpm, machinery imbalance, etc.) The vibratory response of the isolating system (i.e., the forces transmitted through the system) is shown in . Figure 18 shows the transmissibility plotted as a function of the ratio of forcing frequency, ω, to the resonance frequency, ω_0, of the isolator. The resonance frequency ω_0 of the isolator is

$$\omega \cong \sqrt{s/M} \tag{56}$$

where s is the stiffness of the isolator (in N/m) and M is the weight of the machine supported by the isolator (in kg) (1).
The transmissibility (ε) is defined as

$$\text{Transmissibility} = \text{transmitted force/impressed force} \tag{57}$$

It can be shown that if there is no damping in the support system, then the formula can be expressed by

$$\varepsilon = \frac{1}{1 - \left(\omega / \omega_0 \right)} \tag{58}$$

when ω/ω_0 becomes zero, the transmissibility equals 1.0 and no isolation will be obtained. If ω/ω_0 equals 1.0, the theoretical amplitude of the transmitted force goes to infinity, because the driving frequency equals the natural frequency of the system. Therefore, if it is desired to reduce the magnitude of the driving force, the transmissibility must be less than 0 dB. This occurs if the driving frequency is greater than $\omega_0 \sqrt{2}$ as determined from the mathematical derivation. In practice, the driving frequency must be two or four times greater than the resonance frequency of the vibration isolator.

There is a family of curves shown in Fig. 18 because the actual transmissibility is dependent on energy losses that occur within the isolator. Collectively, these losses are expressed in terms of the loss factor η, which is directly proportional to the damping coefficient ξ of the isolator:

$$\eta = \xi / \omega_0 M \tag{59}$$

There is a convenient relationship between the resonance frequency of the isolator and its static deflection that enables us easily to choose an appropriate device on the basis of static deflection. This is fortunate for two reasons. First, the static deflection of the isolator may affect the physical stability of the machinery isolated (especially during

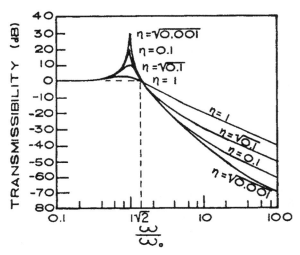

Fig. 18. Transmissibility as a function of loss factor. (From ref. 1.)

startup or shutdown) and, second, physical constraints often control the situation. The relationship occurs because the static deflection is determined by

$$x = Mg/s \tag{60}$$

where g is the gravity acceleration (in m/s²). Recall that

$$\omega_0 = 2\pi f_0 \tag{4}$$

$$\omega_0 = \sqrt{s/M} \tag{56}$$

then we get

$$\omega_0 = 2\pi f_0 = \sqrt{s/M} = \sqrt{g/x} \tag{61}$$

$$f_0 = \frac{1}{2\pi}\sqrt{g/x} \tag{62}$$

and because $g = 9.81$ m/s², f_0 becomes

$$f_0 = \frac{1}{2\sqrt{x}} \tag{63}$$

where x is the static deflection (in m). In English units,

$$f_0 = \frac{3.13}{\sqrt{x}} \tag{64}$$

Corresponding values of static deflection and resonance frequency are shown as follows:

X (m)	0.001	0.003	0.01	0.03	0.1
f_0 (Hz)	16	9	5	3	1.6

The resonance frequency for the foundation of the support system should preferably be less than one-third that of the driving frequency.

8. ACTIVE NOISE CONTROL

The bulk of the work presented here has examined passive noise control by means of either blocking sound transmission from the source or into the receiver or by absorbing it along its propagation route. An alternate means of noise control is active noise control whereby energy is added to a system in an attempt to reduce the net pressure at a receiver. There are two main avenues of this research that can be loosely described as either sound cancellation or source reduction. The first methodology attempts to measure sound that is propagating past a receiver, say in the tail pipe of a car, and to inject noise of an equal magnitude and opposite phase in order to cancel the propagating sound wave. The second methodology attempts to alter the vibration pattern of a structure in an attempt to force the structure into vibration modes that posses lower radiation efficiencies. The treatment of this relatively new line of research is beyond the scope of this text, but the reader is directed to refs. 25–27.

9. DESIGN EXAMPLES

9.1. Indoor Situation

In an industrial plant the noise from one machine was disturbing all operators in the 12-m × 17-m × 5-m (40 ft × 50 ft × 15 ft) production area. The machine was unattended for most of the day and the suggestion was made to enclose it completely.

Because the machine is unattended during most of the day, an enclosure seems a reasonable approach. The physical demands of the situation must be examined before the type of enclosure is chosen:

> Does the enclosure need inspection windows?
> Will an access door be necessary?
> What are the material requirements'?

In this case we have the following,

> No inspection windows are necessary
> A 0.5-m × 0.5-m access door is needed

Material requirements are as follows:

- Easy cleaning
- Low cost
- Ease of assembly and disassembly
- Structural viability

On the basis of the material requirements, plywood is selected as the material for the enclosure.

The acoustical requirements can now be examined. First, we will select a noise criterion. In many cases, people will select the ambient noise, with the machine turned off, as a criterion, or select a sound level that bears a relationship to hearing damage. The latter will often be chosen if the ambient criterion places too great strains on the construction of the noise control features.

Table 7
Sound Pressure Levels in Decibels at Octave Band Center Frequencies

	Frequency (Hz)					
	250	500	1000	2000	4000	8000
Ambient	66	54	50	44	40	36
Machine noise	92	100	104	107	103	99
Criterion	91	86	82	80	80	81
Noise reduction		14	22	27	23	18
Reverberation time (s)	4.1	2.7	2.0	1.8	1.7	1.2

Source: ref. 1.

In order to determine the criterion, ambient measurements were made with the machine turned off; next, sound pressure level measurements were made at a distance of 2 m (approx 6 ft) from the sound source. In addition to these measurements, we also made reverberation time measurements for the purpose of determining the acoustical characteristics of the processing area. The results of the measurements are shown in Table 7. Because the ambient is extremely low, a noise criterion relating to the prevention of hearing damage is selected and noted in Table 7.

Using Eq. (20), we can calculate the total sound absorption in the production room:

$$t = 0.163 \frac{V}{A + 4mV} \text{ metric}$$

$$t = 0.049 \frac{V}{A + 4mV} \text{ english}$$

(20)

The total absorption $A + 4mV$ (in m²-Sabin) becomes

40 60 82 90 96 136

It is now possible to calculate the power level (PWL) emitted by the machine by using Eq. (29). The vaule in the brackets in the equation becomes

0.030 0.027 0.025 0.024 0.024 0.023

and 10 log becomes (in dB)

−15 −16 −16 −16 −16 −16

and the PWL finally becomes (dB)

107 116 120 123 119 115

In this particular situation, the material selected is plywood about 1/2-in.-thick, which as indicated in Fig. 16, appears to meet the acoustical criterion (7 kg/m²; 28-dB transmission loss, average [500 Hz]). After the plywood has been painted, it is easily cleanable. The cost of plywood makes this solution economically attractive. Easy assembly is achieved through the use of machine screws, and it is structurally viable in the necessary size.

Can the enclosure be constructed by the plywood alone or do we need absorption inside? To answer this question, we need to calculate the average sound pressure level in the enclosure, with finished dimensions of $2.0 \times 2.0 \times 1.5$ m^3. A determination must also be made of the transmission loss as a function of frequency.

From Table 6, we can select the absorption coefficient (α) for plywood (use door)

0.12 0.10 0.09 0.08 0.07 0.06 (estimated)

Because the wall area is around 20 m^2, the total absorption (A) becomes

2.5 2.0 1.8 1.6 1.4 1.2 m^2-Sabin

An average distance away from the sound source inside the enclosure will be about 0.5 m, so the value in the bracket in Eq. (29) becomes

1.9 2.3 2.5 2.8 3.2 3.6
and log becomes
3 4 4 4 5 6
and the PWL is
110 120 124 127 124 121

By using Fig. 16, we find that the average transmission loss of 1/2-in. plywood is about 28 dB. (The average transmission loss does for most homogeneous materials correspond to the transmission loss at 500 Hz.)

Next, we need to determine the critical frequency f_c. Young's modulus for plywood is 0.675×10^{10} Pa. The critical frequency f_c can be determined from Eq. (50):

$$f_c = \frac{340^2}{2\pi 0.012} \sqrt{\frac{12 \times 600 \times (1 - 0.09)}{0.675 \times 10^{10}}} \text{ Hz} \qquad (50)$$

$$f_c = 1510 \text{ Hz}$$

The total transmission loss spectrum can now be determined—remember that the transmission increases by about 6 dB for doubling of frequency. Figure 19 shows the transmission loss as a function of frequency. The transmission loss (in dB) that can be determined from that curve is

22 28 33 28 37 46

Because no gasketing is envisioned between individual wall panels and around the access door, the full potential of the plywood will not be obtained. For calculation purposes, it is expected that all cracks that may be in the enclosure will amount to about 0.01 m^2.

It will, therefore, be necessary to calculate the composite transmission loss. In order to do this, we first use Eq. (44) to determine the transmission coefficients (τ) for the plywood panels:

6.2 1.6 0.5 1.6 0.2 0.25 ($\times 10^{-4}$)

and τ for the cracks is 1.0.
Then, we use Eq. (55) to determine the composite transmission loss:

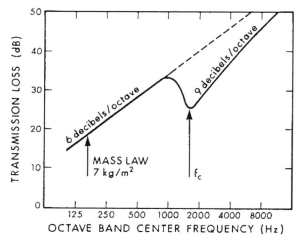

Fig. 19. Transmission loss of ½-in. plywood. (From ref. 1.)

$$\text{TL} = 10 \log \frac{S_a + S_w}{\tau_a S_a + \tau_w S_w} \tag{55}$$

and the results (in dB) become

22 27 30 27 31 33

The NR of the enclosure can now be determined using Eq. (53):

$$\text{NR} = \text{TL} - 10 \log S/A_2 - (C_1 + C_2) \text{ dB} \tag{53}$$

C_1 can be considered zero because cracks were made part of the TL, and because there is no flanking, $C_2 = 0$. The NR (in dB) becomes

25 32 36 34 38 41

This NR is then subtracted from the SPL calculated to be inside the enclosure.

By subtracting the NR from the calculated inside SPL, we get the outside SPL_{out}:

SPL_{out} 85 88 88 93 86 80 dB
Criterion   91 86 82 80 80 81 dB

This result shows clearly that this "bare" enclosure cannot meet the criterion in the 500–4000 Hz frequency range and that absorption inside the enclosure will be necessary. We then decided to pad all enclosure walls (approx 20 m²) with 2.5-cm-thick rockwool. From Table 6, we get the absorption coefficients (α)

0.23 0.53 0.72 0.75 0.77 0.80 (estimated)

and by utilizing Eq. (21), we get A_2 (in m²-Sabin):

4.6 10.6 14.4 15.0 15.4 16.0

To calculate the reduction in the sound pressure level inside the enclosure, we utilize the absorption characteristics of the "bare" plywood, A_2, and Eq. (35):

$$\Delta = 10 \log A/A_0 \text{ dB} \tag{35}$$

The NR (Δ) (in dB) becomes

3 7 9 10 10 11

Because the outside SPL will drop with the equivalent amount as the inside, the outside SPL becomes

| SPL_{out} | 82 | 81 | 79 | 83 | 76 | 69 dB |
| Criterion | 91 | 86 | 82 | 80 | 80 | 81 dB |

The expected SPL, therefore, will only exceed our criterion in the 2000-Hz octave band by about 3 dB.

In addition to lowering the inside SPL, the rockwool has a second benefit. It will reduce the area of the open crack to perhaps one-tenth of what was expected in the "bare" enclosure. This means that the composite TL of the enclosure will be greater than previously determined.

Using a crack area of 0.001 m^2, we get a composite TL (in dB) of the enclosure walls of

22 28 32 28 36 41

which differs from the previous calculated composite TL (in dB) as shown

— +1 +2 +1 +5 +8

This means that the calculated outside SPL_{out} can be reduced by these amounts:

| SPL_{out} | 82 | 80 | 77 | 82 | 71 | 61 dB |
| Criterion | 91 | 86 | 82 | 80 | 80 | 81 dB |

Once again, we see that our criterion can only be expected to be exceeded in the 2000-Hz octave band (about 2 dB). The decision was made to build the enclosure, and after the installation, the measurements made outside the enclosure (SPL_{meas}, in dB) were

85 80 76 75 70 63

and the difference (in dB) from the calculated result becomes

+3 — −1 −7 −1 +2

The only major difference was in the 2000-Hz octave band, the octave band containing the critical frequency. The measurements showed that the dip in the transmission loss at that frequency may not have been as pronounced as expected and that the TL curve might look somewhat like the curve shown in Fig. 20. Beyond that, the correlation between calculations and measurements is extremely good.

9.2. Outdoor Situation

Noise, an increasing harassment in our daily life, is largely the result of modern technology. However,what technology created—noise pollution—technology can correct.

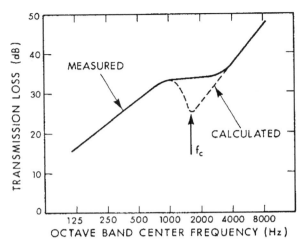

Fig. 20. Transmission loss of ½ in. plywood (corrected). (From ref. 1.)

Noise can be controlled, and much of the technology required for solving most noise problems is available today.

As more people have responded to noise intrusions, various government bodies have developed some type of noise regulation governing environmental noise. The implementation of noise standards is by no means uniform across the United States. A good review of various ways to handle noise ordinances varying from subjective nuisance to objective defined levels can be found in ref. 28. A typical implementation is the use of the day-night level (L_{DN}):

$$L_{DN} = 10\log\left[\frac{1}{24}\left(15\times10^{L_{eqD}/10} + 9\times10^{\left(L_{eqN}+10\right)/10}\right)\right] \qquad (65)$$

where the L_{eqD} is the equivalent sound level during daylight hours of 7 AM to 10 PM and the L_{eqN} is the nighttime level from 10 PM to 7 AM. This equivalent level can be the average of several spot measurements or, more typically, a running average from a recording sound-level meter. The +10 added to the L_{eqN} represents a penalty for nighttime noise because there is a greater expectation of quiet at night. Although values vary across municipalities, typical design values suggest L_{DN} less than 55 dBA for outdoor situations and L_{DN} less than 45 dBA for indoor situations to avoid complaints.

The following example reflects a community's recognition of a noise intrusion from a neighboring industrial plant. Our problem was to determine the likely causes of the noise complaint and develop solutions for its control. Figure 21 shows a layout of the plant, as well as the location of the surrounding residential area. Measurements of noise were performed both close to equipment and out in the community. To determine the severity of the noise problem, it is necessary to have a noise criterion. At the time of the problem, the state in which this situation existed had no quantified regulation, just a common nuisance law. For this example, the Illinois Noise Pollution Control Regulation was adopted to control community noise levels. Under the Illinois regulation, allowable noise emission levels vary, depending on whether the source may be categorized as an industrial,

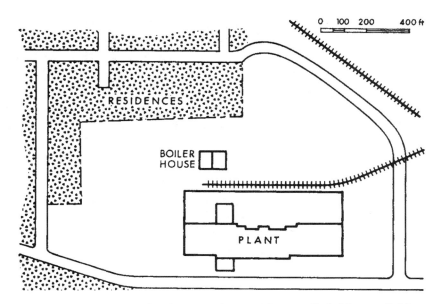

Fig. 21. Plant layout for the exemplary problem studied. (From ref. 1.)

commercial, or residential. Emission criteria are dependent also on the category of the area that receives these noise levels.

Briefly, the Illinois coding system identified industrial land areas as Class "C" land, commercial areas as Class "B" land, and residential areas as Class "A" land. According to the Illinois regulation, the plant in this example was categorized as Class "C" land and the community as Class "A" land.

Figure 22 indicates the criterion that applies when a Class "C" land (i.e., the plant) radiates noise to an adjacent Class "A" land. The regulation distinguishes between daytime (7 AM to 10 PM) and nighttime (10 PM to 7 AM) criteria, and because the plant is operating on a round-the-clock basis, it is the more severe nighttime criterion that was selected. Figure 22 is used for evaluating the noise radiated by the plant, and a violation exists if the noise measured in the community exceeds the criterion curves.

Although the criterion curve is necessary to evaluate whether or not there is a violation of the noise code, it is also useful in identifying (1) those noise sources that are major contributors to the violation and (2) the amount of noise reduction needed to comply with the regulation. The range of sound pressure level measured in the community, as well as the criterion curve, is shown in Fig. 23 and it is quite obvious that a violation exists. More than 90 data samples were measured and analyzed in the course of this project; a representative sample is included in Table 8.

Figure 24 shows the layout of the plant with some of the noise sources indicated. With the measurement made 5 ft from the cooling tower and using Eq. (31), we can determine the sound pressure level expected approx 250 ft away.

$$SPL_5 = PWL - 10\log 4\pi\, 5^2$$

$$SPL_{250} = PWL - 10\log 4\pi\, 250^2$$

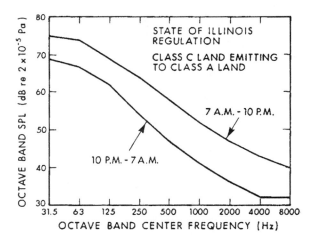

Fig. 22. Noise criterion. (From ref. 1.)

Subtraction results in

$$\mathrm{SPL}_5 - \mathrm{SPL}_{250} = 10\log 4\pi\, 250 - 10\log 4\pi\, 5^2 = 20\log 250/5$$
$$\mathrm{SPL}_5 - \mathrm{SPL}_{250} = 20\log 50 = 34\,\mathrm{dB}$$
$$\mathrm{SPL}_{250} = \mathrm{SPL}_5 - 34\,\mathrm{dB}$$

Figure 25 shows the sound pressure level measured 5 ft from the cooling tower, the calculated SPL at 450 ft solely contributed by the cooling tower, and the measured SPL at that position. The nighttime criterion is also shown. This figure shows without any doubt that the cooling tower is a major contributor to the sound pressure level measured at that location.

The protrusion of the criterion curve indicates the required noise reduction. These facts raise a question: Will the plant be able to meet the noise criterion if the cooling

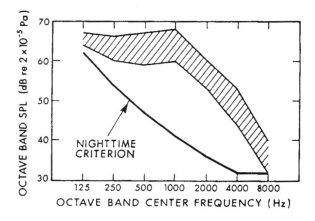

Fig. 23. Range of noise level in the community studied. (From ref. 1.)

Table 8
Measured Sound Pressure Levels in the Community Under Study

Source location	Sound Pressure Level (dB) at Octave Band Center Frequency (Hz)						
	125	250	500	1000	2000	4000	8000
5 ft from cooling tower	89	95	95	93	88	79	72
10 ft from smoke eliminator	92	100	96	89	80	72	66
3 ft from bean tank	84	84	83	89	95	97	96
3 in. from steam ejector	81	78	80	86	88	85	74
3 ft from exhaust fan	105	97	94	91	84	76	80
3 ft from steam vent	96	94	88	83	78	72	70
3 ft from cooling tower	82	87	84	86	84	73	62
6 in. from stack	83	80	75	71	65	63	56
3 ft from blower	98	92	88	86	78	72	73

Source: ref. 1.

tower is quieted or do some of the other sources also have to be quieted? To determine the answer, we will have to treat the other potential noise sources in the same fashion as the cooling tower.

The three bean tanks are each 6 ft tall and have a diameter of 5 ft, and because sound pressure level measurements have been performed very close (3 in.) to the tanks, the total sound power of the three tanks can be determined from

$$PWL = SPL_{3''} + 10 \log S - 10$$

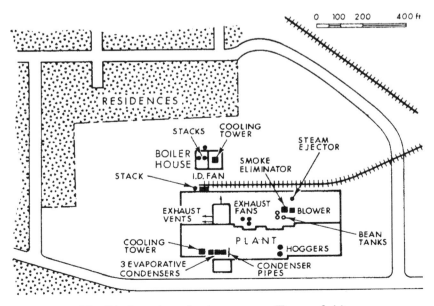

Fig. 24. Location of noise sources. (From ref. 1.)

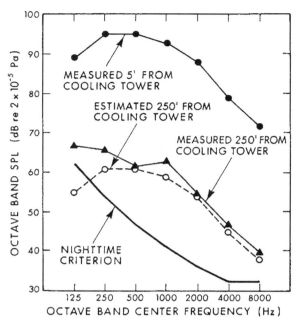

Fig. 25. Noise levels in connection with cooling tower. (From ref. 1.)

$$= SPL_{3''} + 10 \log 3 \times \pi \times 5 \times 6 - 10$$
$$= SPL_{3''} + 10 \log 283 - 10$$
$$= SPL_{3''} + 14\,dB$$

Because the distance to the receiver position is about 800 ft, we use Eq. (31) again to obtain the expected SPL_{800}:

$$SPL_{800} = PWL - 10 \log 4\pi R^2$$
$$= PWL - 10 \log 4\pi 800^2$$
$$= SPL_{3''} + 14\,dB - 10 \log 8 \times 10^6$$
$$= SPL_{3''} + 14\,dB - 69\,dB$$
$$= SPL_{3''} - 55\,dB$$

Figure 26 shows the sound pressure levels associated with the bean tanks and the criterion curve. Because the calculated sound levels protrude above the criterion curve in the 2000–8000-Hz octave-band range, a noise problem will still exist after the cooling tower noise has been controlled.

The same procedure applied to all other noise sources allows us to categorize them as follows:

    Major Noise Sources
        Cooling tower on main roof
        Steam ejector on main roof
        Bean tanks on main roof

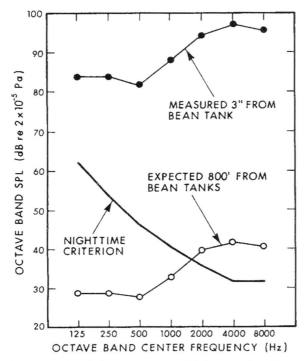

Fig. 26. Noise levels in connection with bean tanks. (From ref. 1.)

Exhaust fan in wall

Secondary Noise Sources
    Stacks
    Induced draft fan
    Cooling tower on boiler roof
    Evaporative condensers
    I-logger mufflers
    Condenser pipes
    Smoke eliminators
    Blower

Tertiary Noise Sources
    Openings in walls
    Small exhaust fans
    Miscellaneous other openings in buildings

For the major noise sources, we must now determine noise control. We decide that for the cooling tower, a barrier is most feasible. How high must the wall or barrier be?

Figure 11 shows the noise reduction that can be expected from a barrier. As shown in Fig. 25, the necessary noise reduction is about 28 dB at 500 Hz (roughly in the center of the frequency range); thus, we find that no barrier can yield such a phenomenal noise reduction. We decide, then, to develop possible alternatives of different height barriers in connection with standard cooling tower mufflers. How high will the barrier have to be under the following circumstances?

Muffler performance: 8 dB at 500 Hz
                      13 dB at 500 Hz
                      18 dB at 500 Hz
                      23 dB at 500 Hz

To determine the barrier performance, it is necessary to select a barrier position between the cooling tower and the receiver. We choose to place the barrier no closer than 4 ft from the side of the cooling tower in order not to disturb the airflow too much, thereby reducing the capacity of the cooling tower. The cooling tower is 8 ft wide.

For the following calculations, it can be assumed that all the noise is radiating out from the center of the cooling tower, located 8 ft from the wall and 8 ft above the roof level. Using English units in connection with Fig. 11, we find that δ has to be at a minimum.

Muffler performance (dB)	Remaining attenuation (dB)	δ	Barrier height (H)
8	20	7.0	20.7
13	15	2.3	14.5
18	10	0.5	10.7
23	5	0.006	8.3

The last column is derived from simple geometrical calculation, based on the equation given in Fig. 11:

$$\delta = a + b - D$$
$$\delta = \sqrt{h^2 + 442^2} + \sqrt{h^2 + 8^2} - 450$$

Because $442^2 \gg h^2$, the equation becomes

$$\delta \cong 442 + \sqrt{h^2 + 8^2} - 450$$
$$\delta \cong \sqrt{h^2 + 8^2} - 8$$

The barrier height H is

$$H = h + 8\,\text{ft}$$

The final solution to this problem—a decision made by management personnel of the plant and based on these alternatives—was to use a muffler with an insertion loss of 13 dB and to build a 15-ft-high barrier that should give a noise reduction of approximately 15 dB.

The development of feasible noise control principles in controlling ejector noise, bean tank noise, and muffler design is beyond the scope of this chapter, but we believe that the example shows some of the complexity involved in determining the offending sources in a community noise situation. In all cases, the extrapolation technique only holds for an ideal situation in which there are no adverse thermal gradients as well as no wind. So rarely does that situation occur that the noise sources should be divided into at least four categories:

Major noise sources—definitely need treatment
Secondary noise sources—need some treatment
Tertiary noise sources—may need treatment
Minor noise sources—do not need any treatment

Noise control steps can then be taken in an orderly manner, with the worst offenders attenuated first, and the least disruptive last.

GLOSSARY

In this chapter we have used terms, the definitions for which are indicated below. Most of these definitions can be found in "Glossary of Terms Frequently Used in Acoustics," issued by the American Institute of Physics as well as refs. 9 and 12.

Absorption coefficient (acoustical absorptivity) (α). The sound-absorption coefficient of a surface exposed to a sound field is the ratio of the sound energy absorbed by the function of both angle of incidence and frequency. Tables of absorption coefficients given in the literature usually list the absorption coefficients at various frequencies, the values being those obtained by averaging overall angles of incidence.

Acoustic, acoustical. The qualifying adjectives acoustic and acoustical mean containing, producing, arising from, actuated by, related to, or associated with sound. Acoustic is used when the term being qualified designates something that has the properties, dimensions, or physical characteristics associated with sound waves; acoustical is used when the term being qualified does not designate explicitly something that has such properties, dimensions, or physical characteristics.

Acoustics. Acoustics is the science of sound, including (1) its production transmission and effects and (2) the qualities that determine the value at a room or other enclosed space with respect to distinct hearing.

Ambient noise. Ambient noise is the all-encompassing noise associated with a given environment, being usually a composite of sounds from many sources near and far.

Angular frequency (ω). The angular frequency of a periodic quantity is its frequency in radians per unit time, usually radians per second. Thus, it is the frequency multiplied by 2π.

Band pressure level. The band pressure level of a sound for a specified frequency band is the effective sound pressure level for the sound energy contained within the band. The width of the band and the reference pressure must be specified. The width of the band may be indicated by the use of a qualifying adjective: for example, octave-band (sound pressure) level, half-octave-band level, third-octave-band level, 50-cps band level. If the sound pressure level is caused by thermal noise, the standard deviation of the band pressure level will not exceed 1 dB if the product of the bandwidth in cycles per second by the integration time in seconds exceeds 20.

Decibel (dB). The decibel is a unit of level which denotes the ratio between two quantities that are proportional to power; the number of decibels corresponding to the ratio of two amounts of power is 10 times the logarithm to the base 10 of this ratio. In many sound fields, the sound pressure ratios are not proportional to the square root of the corresponding power ratios, so that strictly speaking, the term "decibel" should not be used in such cases; however, it is common practice to extend the use of the unit to these cases (*see*, e.g., *Sound pressure level*).

Diffuse sound field (Random-incidence sound field). A diffuse sound field is a sound field such that the sound pressure level is everywhere the same, and all directions of energy flux are equally probable.

Effective sound pressure (p) (root-mean-square sound pressure).   The effective sound
pressure at a point is the root-mean-square value of the instantaneous sound pressures, over a time interval at the point under consideration. In the case of periodic
sound pressure, the interval must be an integral number of periods or an interval long
compared to a period. In the case of nonperiodic sound pressures, the interval should
be long enough to make the value obtained essentially independent of small changes
in the length of the interval. The term "effective sound pressure" is frequently
shortened to "sound pressure."

Free field.   A free sound field is a field in a homogeneous, isotropic medium free from
boundaries. In practice, it is a field in which the effects of the boundaries are negligible over the region of interest. The actual pressure impinging on an object (e.g., a
microphone) placed in an otherwise free sound field will differ from the pressure that
would exist at that point with the object removed, unless the acoustic impedance of
the object matches the acoustic impedance of the medium.

Frequency (f).   The frequency of a function that is periodic in time is the reciprocal of the
period. The unit is the cycle per unit time (e.g., cycles per second [cps] or hertz [Hz]).

Fundamental frequency.   The fundamental frequency of a periodic quantity is equal to the
reciprocal of the shortest period during which the quantity exactly reproduces itself.

Harmonic.   A harmonic is a sinusoidal quantity having a frequency that is an integral
multiple of the fundamental frequency of a periodic quantity to which it is related.

Hertz (Hz).   A unit of frequency; formerly called cycles per second (cps).

Level.   In communication and acoustics, the level of a quantity is the logarithm of the
ratio of that quantity to a reference quantity of the same kind. The base of the logarithm, the reference quantity, and the kind of level must be specified.

Loudness.   Loudness is the intensive attribute of an auditory sensation, in terms of
which sounds may be ordered on a scale extending from soft to loud. Loudness
depends primarily on the sound pressure of the stimulus, but it also depends on the
frequency and waveform of the stimulus.

Loudness level.   The loudness level, in phons, of a sound is numerically equal to the
sound pressure level in decibels, relative to 2×10^{-5} Pa, of a pure tone of frequency
1000 Hz, consisting of a plane-progressive sound wave coming from directly in front
of the observer, which is judged by normal observers to be equivalent in loudness.

Noise.   Unwanted sound.

Noise level.   The acoustical noise level is the sound level.

Normal mode of vibration.   In an undamped multi-degree-of-freedom system undergoing free vibration, a normal mode of vibration is a pattern of motion assumed by
the system in which the motion of every particle is simple harmonic with the same
period and phase. Thus, vibration in a normal mode occurs at a natural frequency of
the system. In general, any composite motion of a system is analyzable into a
summation of normal modes. (The terms *natural mode, characteristic mode*, and
eigen mode are synonymous with *normal mode*.)

Octave-band pressure level.   The octave-band pressure level of a sound is the band
pressure level for a frequency band corresponding to a specified octave. (The location of the octave-band pressure level on a frequency scale is usually specified as the
geometric mean of the upper and lower frequencies of the octave.)

Oscillation. Oscillation is the variation, usually with time, of the magnitude of a quantity with respect to a specified reference when the magnitude is alternately greater and smaller than the reference.

Period. The period of a periodic quantity is the smallest value of the increment of the independent variable for which the function repeats itself.

Phon. The phon is the unit of loudness level.

Power level (PWL). The power level, in decibels, is 10 times the logarithm to the base 10 of the ratio of a given power to a reference power. The form of power (e.g., acoustic) and the reference power must be indicated. The reference power used in this chapter for sound power level is 10^{-12} W.

Pure tone. A pure tone is a sound wave, the sound pressure of which is a simple sinusoidal function of the time.

Rate of decay. The rate of decay is the time rate at which the sound pressure level (or velocity level) is decreasing at a given point and at a given time. The commonly used unit is the decibel per second.

Resonance. The resonance of a system under forced vibration exists when any small increase or decrease in the frequency of excitation causes a decrease in the response of the system.

Resonant frequency (f_r). A resonant frequency is a frequency at which resonance exists.

Reverberation. Reverberation is the sound that persists at a given point after direct reception from the source has stopped.

Reverberation room. A reverberation room is an enclosure in which all the surfaces have been made as sound reflective as possible. Reverberation rooms are used for certain acoustical measurements.

Reverberation time (t). The reverberation time for a given frequency is the time required for the average sound pressure level, originally in a steady state, to decrease 60 dB after the source is stopped. Usually, the pressure level for the upper part of this range is measured and the result extrapolated to cover 60 dB.

Root-mean-square sound pressure. See *Effective sound pressure.*

Sabin (square meter unit of absorption). The Sabin is a measure of the sound absorption of a surface; it is the equivalent of 1 m² (ft²) of perfectly absorptive surface.

Sound. (a) Sound is an alteration in pressure, stress, particle displacement, shear, or so forth in an elastic medium or (b) sound is an auditory sensation evoked by the alterations described in (a). In case of possible confusion, the term "sound wave" or "elastic wave" may be used for concept (a) and the term "sound sensation" for concept (b). Not all sound waves evoke an auditory sensation. The medium in which the sound exists is often indicated by an appropriate adjective (e.g., airborne, structure-borne).

Sound absorption. Sound absorption is the process by which sound energy is diminished in passing through a medium or in striking a surface.

Sound-absorption coefficient. See *Absorption coefficient.*

Sound energy. The sound energy of a given part of a medium is the total energy in this part of the medium minus the energy that would exist in the same part of the medium with no sound waves present.

Sound energy density. The sound energy density at a point in a sound field is the sound energy contained in a given infinitesimal part of the medium divided by the volume of that part of the medium. The commonly used unit is W-s/m^3.

Sound field. A sound field is a region containing sound waves.

Sound level. The sound level, in decibels, is the weighted sound pressure level obtained by use of a sound-level meter whose weighting characteristics are specified in the latest revision of the American Standards Association standard on sound-level meters. The reference pressure is 2×10^{-5} Pa, unless otherwise specified.

Sound-level meter. A sound-level meter is a device used to measure sound pressure level or weighted sound pressure level, constructed in accordance with the standard specifications for sound-level meters set up by the American Standards Association. The sound-level meter consists of a microphone, an amplifier to raise the microphone output to useful levels, a calibrated attenuator to adjust the amplification to values appropriate to the sound levels being measured, and an instrument to indicate the measured sound level; optional weighting networks are included to adjust the overall frequency characteristics of the response and provision is made for an output connection to additional measuring equipment.

Sound power of a source (W). The sound power of a source is the total sound energy radiated by the source per unit of time.

Sound power level (PWL or L_w). The sound power level of a sound source, in decibels, is 10 times the logarithm to the base 10 of the ratio of the sound power radiated by the source to a reference power. The reference power is 10^{-12} W.

Sound pressure level (SPL or L_p). The sound pressure level, in decibels, of a sound is 20 times the logarithm to the base 10 of the ratio of the pressure of this sound to the reference pressure. The reference pressure employed in this chapter is 2×10^{-5} Pa. In many sound fields, the sound pressure ratios are not proportional to the square root of corresponding power ratios and hence cannot be expressed in decibels in the strict sense; however, it is common practice to extend the use of the decibel to these cases.

Sound reduction between rooms. The sound reduction, in decibels, between two rooms is the amount by which the mean-square sound pressure level in the source room exceeds the level in the receiving room. If a common partition separates two rooms, the first of which contains a sound source, the sound reduction between the two rooms is equal to the transmission loss of the partition plus a function of the total absorption in the second room and the area of the common partition.

Spherical wave. A spherical wave is a wave in which the wavefronts are concentric spheres.

Threshold of audibility (threshold of detectability). The threshold of audibility for a specified signal is the minimum effective sound pressure of the signal that is capable of evoking an auditory sensation in a specified fraction of the trials. The characteristics of the signal, the manner in which it is presented to the listener, and the point at which the sound pressure is measured must be specified. The ambient noise reaching the ears is assumed to be negligible, unless otherwise stated.

Transmission coefficient. The sound transmission coefficient of a partition is the fraction of incident sound transmitted through the partition. The angle of incidence and the characteristic of sound observed must be specified (e.g., pressure amplitude at normal incidence).

Transmission loss. Transmission loss is the reduction in the magnitude of some characteristic of a signal between two stated points in a transmission system. The characteristic is often some kind of level, such as power level or voltage level; in acoustics, the characteristic that is commonly measured is sound pressure level. If the levels are expressed in decibels, then the transmission loss is likewise in decibels.

Transmission loss of a partition (TL). The sound transmission loss of a partition, in decibels, is −10 times the logarithm to the base 10 of the transmission coefficient for the partition. It is equal to the number of decibels by which sound incident on a partition is reduced in transmission through it. Thus, a measure of the airborne sound installation of the partition. Unless otherwise specified, it is to be understood that the sound fields on both sides of the partition are diffuse.

Velocity of sound. The speed of sound in air is given by $v = \sqrt{1.4 \times 287 \times T_K}$ where the 1.4 represents the ratio of specific heats for air, 287 is the universal gas constant for air, and T_K is the temperature in Kelvin (°C +273.15).

Wave. A wave is a disturbance propagated in a medium in such a manner that at any point in the medium the quantity serving as a measure of the disturbance is a function of the time, while at any instant the quantity serving as a measure of the disturbance at a point is a function of the position of the point. Any physical quantity that has the same relationship to some independent variable (usually time) that a propagated disturbance has, at a particular instant, with respect to space, may be called a wave.

Wavelength (λ). The wavelength of a periodic wave in an isotropic medium is the perpendicular distance between two wavefronts in which the displacements have a difference in phase of one complete period.

NOMENCLATURE

a	Amplitude (m)
c	Speed of sound (m/s)
d	Depth (m)
f	Frequency (Hz)
g	Gravity acceleration (m/s^2)
l	Length (m)
m	Surface weight (kg/rn^2)
n	Attenuation coefficient
p	Sound pressure (dB)
s	Stiffness (N/m)
t	Time (s)
x	Static deflection (m)
A	Total absorption (m^2-Sabin)
B	Bending stiffness (Nm)
C	Correction factor (dB)
D	Sound energy density (W-s/m^3)
E	Young's modulus (Pa)

IL	Insertion loss (dB)
L_p	Sound pressure level (dB)
M	Total weight (kg)
NR	Noise reduction (dB)
PWL	Sound power level (dB)
S	Area (m^2)
T_c	Celsius temperature (°C)
T_K	Kelvin temperature (K)
T_R	Rankine temperature (°R)
TL	transmission loss (dB)
V	Volume (m$^{3)}$
W	Sound power (W)
α	Absorption coefficient
δ	Path length difference (m)
ε	Transmissibility
η	Loss factor
θ	Geometric angle
λ	Wavelength (m)
ξ	Damping coefficient (kg rad/s)
ρ	Density (kg/rn^3)
σ	Poisson's ratio
τ	Transmission coefficient
ω	Angular frequency (rad/s)
Δ	Noise reduction (dB)
$4m$	Air absorption (1/m or dB/km)

REFERENCES

1. L. K. Wang and N. C. Pereira (eds.), *Handbook of Environmental Engineering, Volume 1, Air and Noise Pollution Control*, Humana, Clifton, NJ 1979.
2. American National Standards Institute, *Preferred Frequencies and Band Numbers for Acoustical Measurements, ANSI S1.6*, American National Standards Institute, New York, 1967.
3. International Electrotechnical Commision, Sound Level Meters; *IEC 60651,* International Electrotechnical Commission, New York, 2001.
4. American National Standards Institute, *Standard for Sound Level Meters, ANSI S1.4*, American National Standards Institute, New York 1983, revised 1987.
5. American National Standards Institute, *Acoustical Levels and Preferred Reference Quantities for Acoustical Levels, ANSI S1.8*, American National Standards Institute, New York, 1969.
6. H. Fletcher and W. A. Munson *J. Acoust. Soc. Am.* **5**, 82–108 (1933–34).
7. L. E. Kinsler, A. R. Frey, A. B. Coppens, et al., *Fundamentals of Acoustics*, 4th Ed., Wiley, New York, 2000.
8. R. W. Young, *J. Acoust. Soc. Am.* **31**, 912–921 (1959).
9. M. J. Crocker, (ed.), *Handbook of Acoustics,*. Wiley–Interscience, New York 1998.
10. L. L. Beranek, *Noise and Vibration Control*, McGraw-Hill, New York, 1971.
11. U. J. Kurze, and G. S. Anderson, *Appl. Acoust.* **4**, 35–53 (1971).
12. C. E. Wilson, *Noise Control*, Harper & Row, New York, 1989.

13. H. Tipton and G. M. Tomlin, *Pollut. Eng.* **8**, 53 (1976).
14. A. A. Hood and D. J. Pines, *J. Acoust. Soc. Am.* **112**(6), 2849–2857 (2002).
15. J. Volante, *Pollut. Eng.* **9**, 36–37 (1977).
16. W. V. Montone, *Pollut. Eng.* **9**, 42–43 (1977).
17. M. E. Delaney and E. N. Bazley, *Appl. Acoust.* **3**, 105–116 (1970).
18. T. F. W. Embleton, J. E. Piercey, and G. A. Daigle, *J. Acoust. Soc. Am.* **74**, 1239–1244 (1983).
19. R. F. Lambert, *J. Acoust. Soc. Am.* **97**, 818–821 (1995).
20. J. A. Zapfe and E. E. Ungar, *J. Acoust. Soc. Am.* **113**(1), 321–326 (2003).
21. K. U. Ingard, *J. Acoust. Soc. Am.* **26**, 151–154 (1954).
22. American National Standards Institute, *Physical Measurement of Sound*; ANSI S1.2, American National Standards Institute, New York, 1962.
23. American Society for Testing and Materials, *Recommended Practice for Laboratory Measurement of Airborne Sound Transmission Loss of Buildings Partitions*; ASTM E90-04 (2004), American Society for Testing and Materials, Philadelphia.
24. J. W. Strutt and Lord Rayleigh, *The Theory of Sound, Vol. II*, Macmillan, London, 1937. England.
25. K. A. Cunefare and S. Shepard, *J. Acoust. Soc. Am.* **93**(5), 2732–2739 (1993).
26. P. A. Nelson and S. J. Elliott, *Active Control of Sound*, Academic, San Diego, CA, 1992.
27. H. Zhu, R. Rajamani and K. A. Stelson, *J. Acoust. Soc. Am.* **113**(2), 852–870 (2003).
28. D. L. Sheadel, *Proceedings of the 92nd Annual Meeting of the Air and Waste Management Association*, 1999, paper 99–574.

Index